Compendium of Theoretical Physics

Armin Wachter Henning Hoeber

Compendium of Theoretical Physics

 Springer

Armin Wachter
Pallas Inc.
Brühl, Germany
awachter@freenet.de

Henning Hoeber
de GeophysiqueCGG house
Compagnie Generale de Geophysique
Feltham, Middlesex TW1401R
United Kingdom
hhoebercgg.com

ISBN 978-1-4419-2069-0 e-ISBN 978-0-387-29198-7

Printed on acid-free paper.

Printed in the United States of America. (SBA)

9 8 7 6 5 4 3 2 1

springeronline.com

Acknowledgements

We'd like to thank the Robert Gordon University, Aberdeen, where a large part of the electrodynamics chapter was written, for its kind support. (Where better than in Aberdeen to write about electrodynamics?) Thanks to Klaus Kämpf for his donation of textbooks; to our colleagues, as well as students, at the Bergische Universität Wuppertal and the Höchstleistungsrechenzentrum Jülich (now John von Neumann Institute) for many discussions and suggestions; to the Pallas GmbH and the Compagnie Générale de Géophysique for their kind support in the final phases of our writing this book.

This translation is based on the 2nd edition of the German book, published in 2004, which has undergone a thorough review in order to minimize remaining errors. For the translation into English, we would like to thank our friends John Mehegan and Andrew Ratcliffe for providing corrections and suggestions to part of the book. The usual disclaimer holds: any errors in this book are ours.

Thanks to Springer for letting us proceed at our own pace in providing the final version of this translation.

Finally, we wish to thank our friends and families, without whom this project would have simply been unthinkable.

Preface

Our book, *Compendium of Theoretical Physics*, contains the topics

- mechanics,

- electrodynamics,

- quantum mechanics, and

- statistical physics and thermodynamics,

which constitute the "canonical curriculum" of theoretical physics, taught at the undergraduate level at most universities. It is oriented mainly toward students of higher levels who are interested in a clearly arranged and coherent presentation of the curriculum, or who are preparing for their degree examinations. Moreover, this book is also suited as an accompanying and complementary textbook for students within the first semesters. For them, it aims to provide a useful guideline to reclassify the content of the various physics courses. Finally, physicists working in industry or research may also benefit from this overview of theoretical physics.

There are, of course, many good textbooks for the above topics (some suggestions can be found in our commented literature list). This book is therefore not to be regarded as a substitute for these books; no student can get by without an extensive study of the curriculum using other, didactically and historically well-edited presentations of theoretical physics. However, it seemed necessary to us to write this book in order to offer a complementary approach to theoretical physics, in which the composition, the structure, and, last but not least, the elegance of physical theories are emphasized and easy to recognize. We hope that this is achieved, among other things, by dispensing with historical-phenomenological explanations.

Throughout, we pursue an axiomatic-deductive approach in that we start the discussion of each theory with its fundamental equations. Subsequently, we derive the various physical relationships and laws in logical (rather than chronological) order. Our aim is to emphasize the connections between the individual theories by consistently using a standardized presentation and notation. Think, for example, of the Hamilton formalism: it constitutes a fundamental concept not only in quantum mechanics but also in statistical physics.

In the first chapter, *Mechanics*, we present the often overemphasized Newtonian approach to mechanics next to the Lagrange and Hamilton formulations. Each of these equivalent representations distinguishes itself by specific advantages: whereas Newton's approach is most easily accessible using the concept of force and by finding the equations of motions, only the Lagrange and Hamilton formalisms provide a suitable platform for a deeper understanding of mechanics and other theoretical concepts. For example, the Lagrange formalism is better suited to seeing the relationship between symmetries and conservation laws. Accordingly, the first three sections of this chapter deal with these three approaches and their connections in equitable fashion. Furthermore, in the section *Relativistic Mechanics*, we introduce the correct Lorentz tensor notation in order to ease the transition to the relativistic theory of electrodynamics, in which the disciplined use of this notation turns out to be very useful and convenient.

The advantage of our deductive method may be particularly apparent in the second chapter, *Electrodynamics*. In contrast to many other textbooks, we start with Maxwell's equations in their most general form. This allows us immediately to see very clearly the structure of this theory. We quickly find the general solutions to Maxwell's equations using the very important concept of gauge invariance. From this, the various laws of electrodynamics follow naturally in a clean and concise manner. For example, the solutions in empty space, or the special cases of electro- and magnetostatics, are easily deduced. Based on our results of relativistic mechanics, we apply the covariant notation to electrodynamics and discuss the Lagrange and Hamilton formalism with respect to the field theoretical character of the theory.

In contrast to the other chapters we begin *Quantum Mechanics* with a mathematical introduction in which some areas of linear algebra are recapitulated using Dirac's notation. In particular, the concepts of operators and eigenvalue problems are discussed. These are of fundamental importance in quantum mechanics. We then present the general structure of quantum theory, where the fundamental concepts are established and discussed in a representation-independent manner. Generally, throughout the whole chapter, we try to avoid overemphasizing a particular representation.

Similarly to mechanics, there are different approaches in statistical mechanics/thermodynamics to describe many-particle systems. First, we have the statistical ansatz that combines (quantum) mechanical laws with a statistical principle. This results in a microscopic description in form of ensemble theories. By contrast, thermodynamics is a purely phenomenological theory based on purely macroscopic experiences. A third approach is given by information theory where a system is considered from the viewpoint of lack of information. In order to highlight the inherent connections of these three concepts, we discuss all of them in our chapter *Statistical Physics and Thermodynamics* and show their mutual equivalence.

Throughout this book important equations and relationships are summarized in boxes containing definitions and theorems. We hope that this facilitates structured learning and makes finding fundamental results easier. Furthermore, we have arranged connected argumentations optically; in principle, the reader should always be able to recognize the end of an argument. At the end of each section we have placed a short summary as well as some applications with solutions. These applications are intended to reaffirm, and sometimes enhance, the understanding of the subject matter. Finally, in the appendix, a short compilation of important and often-used mathematical formulae are given.

It should be obvious that we make no claim of completeness. Instead, the topics of the four chapters are chosen such that, on the one hand, they contain the fundamental ideas and concepts and, on the other hand, cover the areas we have found, quite subjectively, most relevant for examinations and day-to-day work. To complement this book, we make some further literature suggestions in the appendix.

Overall, we hope that we have written a book that works as a broker between textbooks, lecture notes, and formula compilations. It would make us very happy if it helped you to better understand the concepts of theoretical physics.

Cologne and Newcastle
October 2005

Armin Wachter
Henning Hoeber

Table of Contents

List of Applications

Quantum Mechanics

Statistical Physics and Thermodynamics

1. Mechanics

The subject of classical mechanics is the dynamics of material bodies. Up to the beginning of the 20th century, classical mechanics was seen as the fundamental theory of the interaction of material objects. However, the validity of classical mechanics was drastically reduced with the advent of the theory of special relativity in 1905 and with quantum theory in the 1920s; classical mechanics is now to be considered as the limiting case of velocities that are small compared to the speed of light and of energies that are large compared to those on the atomic scale.

One of the most striking characteristics of classical mechanics is the variety of methods available to solve any one given problem. Equivalent formulations of mechanics are

- Newtonian mechanics,

- the d'Alembert principle,

- the Lagrange equations and generalized coordinates,

- Hamilton's principle, and

- Hamilton-Jacobi theory.

In principle, it is possible to solve a given problem with each of these methods. However, we shall see that some of these approaches are much better suited to solve some problems than others. For example, problems with restrained freedom of motion (such as the motion of pearls on a string) are easily tackled in the Lagrangian formulation of mechanics and somewhat cumbersome in Newton's approach. In addition, some of the above methods reveal deeper connections, particularly between symmetries and conservation laws. Finally, some formal structures of mechanics, such as the Poisson bracket and the Hamilton-Jacobi equation, are of particular interest when performing the transition from classical to quantum mechanics.

In the first section of this chapter we lay the foundations of Newtonian mechanics and discuss some immediate implications. Furthermore, we study how Newtonian mechanics must be modified in accelerated coordinate systems. This leads us to the important concepts of inertial systems and Galilei invariance.

The next two sections deal with the Lagrangian and the Hamiltonian formulations of classical mechanics as two examples of alternatives to the Newtonian approach. These formulations lead to great simplifications for systems subject to certain constraints. In the Newtonian formulation these constraints are taken into account by introducing constraining forces into Newton's equations of motion, whereas in Lagrange's and Hamilton's formalisms they can be eliminated through a suitable choice of generalized coordinates, velocities, and momenta. Additionally, these approaches offer a deeper insight into the structure of mechanics and its relation to other physical theories.

In Section 1.4 we discuss the dynamics of rigid bodies. In this context it proves convenient to introduce different coordinate systems in which the kinematic quantities can be formulated in an orderly and transparent fashion.

The important class of central forces is the subject of Section 1.5. Following a discussion of the reduction of two-particle systems to effective one-particle systems, we turn to the equation of motion in central potentials and determine the possible trajectories in $1/r$-potentials. Furthermore, we discuss the scattering of particles. Although this type of problem is more relevant to high-energy physics, and therefore needs to be formulated in a quantum theoretical sense, many of the concepts and structures of scattering theory can already be formulated classically, allowing us to build on these results in later chapters.

This chapter's last section deals with the relativistic generalization of Newtonian mechanics. Our starting point is the experimental observation that the velocity of light is identical in all inertial systems. From this it follows that space and time are not absolute but, contrary to Newton's formulation of mechanics, depend on the reference frame.

1.1 Newtonian Mechanics

In this section we discuss the fundamental terminology and concepts of Newtonian mechanics. A brief recapitulation of some important mathematical results relevant for the description of particle positions is followed by the presentation of Newton's axioms. A number of physical consequences and conservation laws are derived from these axioms, the understanding of which is of great importance for all following sections. Furthermore, we look at the motion of particles in accelerated coordinate systems and show how Newton's equations of motion are modified in these. In this context, we will see that there is a close relation between *Galilei invariance* and the concept of *inertial systems*. Finally, we deal with the dynamical properties of many-particle systems and derive various conservation laws for closed systems of this kind.

1.1.1 Coordinate Systems and Vectors

In order to avoid confusion in what follows, it is useful to begin our discussion of Newtonian mechanics by looking briefly at some mathematical concepts that the reader is undoubtedly already familiar with.

To describe a mechanical system we use, as generally common, the concept of an idealized mathematical mass point (*particle*), whose position in space is represented by a basis-independent vector x in a three-dimensional, real vector space. To quantify this position, one needs a coordinate system, determined by its origin, and three linearly independent basis vectors. The choice of such a basis system is generally suggested by the dynamics of the particular physical system under consideration. If not otherwise stated, we use a Cartesian coordinate system with canonical basis. For this we write

$$K : \{e_1, e_2, e_3\} \, ,$$

with

$$e_i e_j = \delta_{ij} \qquad (\text{orthonormality relation})$$

and

$$\sum_j e_i e_j = 1 \qquad (\text{completeness relation}) \, .$$

The position x of a particle can now be uniquely specified by the projections x_i of x onto the axes of K along the unit vectors e_i:

$$x = \sum_i e_i x_i \, , \quad x_i = x e_i \, . \tag{1.1}$$

The x_i are called the coordinates of x with respect to K. Often, one collects the three coordinates into a coordinate triplet and writes in short-hand notation

$$x = \begin{pmatrix} x_1 \\ x_2 \\ x_3 \end{pmatrix} \, . \tag{1.2}$$

We will follow this convention. However, keep in mind that the basis-independent physical vector x in (1.1) and the basis-dependent triplet x in (1.2) represent two mathematically different objects. The meaning of x, either a physical vector or the projection onto a particular coordinate system, should always be clear from the context.

Temporal differentiation of vectors. Another important aspect in this context is the temporal differentiation of a vector x. We adhere to the general convention, using a single dot for the first derivative, two dots for the second derivative, and so on. If we wish to study the change in time of x relative to two time-dependent coordinate systems, say $K : \{e_1, e_2, e_3\}$ and $K' : \{e_1', e_2', e_3'\}$, whose basis vectors change with time, it is not so much the total temporal derivative

$$\dot{\boldsymbol{x}} = \sum_i (\dot{\boldsymbol{e}}_i x_i + \boldsymbol{e}_i \dot{x}_i) = \sum_i (\dot{\boldsymbol{e}}'_i x'_i + \boldsymbol{e}'_i \dot{x}'_i) \ ,$$

which is of interest, but rather the change in time as seen from K or from K'. For clarity we define the symbols D and D' to denote the following derivatives:

$$D\boldsymbol{x} = \sum_i \boldsymbol{e}_i \dot{x}_i \ , \ D'\boldsymbol{x} = \sum_i \boldsymbol{e}'_i \dot{x}'_i \ . \tag{1.3}$$

Rotating coordinate systems. Consider two coordinate systems K : $\{\boldsymbol{e}_1, \boldsymbol{e}_2, \boldsymbol{e}_3\}$ and K' : $\{\boldsymbol{e}'_1, \boldsymbol{e}'_2, \boldsymbol{e}'_3\}$ with the same origin but rotating relative to one another (i.e., their basis vectors are rotating relative to one another). The transition from K to K' is effected by a proper orthogonal rotation matrix such that

$$\boldsymbol{e}_i = \sum_j \boldsymbol{e}'_j R_{ji} \ , \ \boldsymbol{e}'_i = \sum_j \boldsymbol{e}_j R^{\mathrm{T}}_{ji} \ , \ R_{ij} = \boldsymbol{e}'_i \boldsymbol{e}_j \tag{1.4}$$

and

$$RR^{\mathrm{T}} = R^{\mathrm{T}}R = 1 \ , \ \det R = 1 \ . \tag{1.5}$$

Given these relations, the coordinates x'_i of a vector \boldsymbol{x} with respect to K' can be obtained from the coordinates x_i of the vector with respect to K in the following way:

$$\boldsymbol{x} = \sum_i \boldsymbol{e}_i x_i = \sum_{i,j} \boldsymbol{e}'_j R_{ji} x_i = \sum_j \boldsymbol{e}'_j x'_j \Longrightarrow x'_j = \sum_i R_{ji} x_i \ .$$

The last equation can be put into the matrix notation

$$\boldsymbol{x}' = R\boldsymbol{x} \ ,$$

where again one should keep in mind that this is a relation between the basis-dependent coordinate triplets in K and in K' and not between two physically different vectors.

1.1.2 Newton's Axioms

We now turn to our discussion of the structure of Newtonian mechanics. Before introducing Newton's axioms, it is necessary to introduce three basic mechanical quantities, as well as the concepts of *momentum* and *inertial system*.

Definition: Basic mechanical entities in the SI- (or MKS-)system

All mechanical quantities can be derived from the following three basic entities:

\triangleright

- Length, with the unit meter: m.

- Time, with the unit second: s.

- Mass, with the unit kilogram: kg.

We use the *Système International d'Unités* (SI).

In the Newtonian formulation of classical mechanics, the mechanical state of a system is uniquely determined by the simultaneous specification of all the particles' positions and velocities at any particular time. Given this information, the system's state at any future time can be predicted from *Newton's equations of motion*. These equations follow from *Newton's axioms*, which describe a particle's motion using the concept of momentum.

Definition: Momentum p

A particle's momentum is given by the product of its inertial mass and its momentary velocity vector:

$$p(t) = m(t)\dot{x}(t) \ .$$

Thus, p always points in the same direction as \dot{x}.

The motion of a particle is called *linearly uniform* if its velocity vector is constant.

As we have already stated, the motion of particles is described within given reference frames. It turns out that the equations of motion have different forms in different reference frames. In Newton's formulation of mechanics we always assume the existence of inertial systems in which the equations of motion have their simple well-defined form.

Definition: Inertial System

A frame of reference, with respect to which

- space is *homogeneous* and *isotropic*,

- time is homogeneous,

is called an inertial system.

Homogeneity and isotropy of space imply that no point and no direction in space are particularly favored. Homogeneity of time means, similarly, that there is no moment in time that is particularly favored. From this it follows that reference frames

- that are shifted in space relative to an inertial system,

- that are moving at constant velocity relative to an inertial system,

- that are rotated relative to an inertial system,

- whose time origin is shifted relative to that of an inertial system

are also inertial systems. Thus, the above definition does not single out one particular reference frame, but leads to a whole class of inertial systems that are all connected by the four transformations. They constitute the group of *Galilei transformations*. We shall study this group in more detail in the following subsection and connect it to the fundamental principle of *Galilei invariance*.

We are now prepared to present the fundamental axioms, formulated by Sir Isaac Newton, which form the basis of all of Newtonian mechanics.

Theorem 1.1: Newton's axioms

Inertial law (lex prima)
The momentum of a free particle, i.e., a particle on which no force acts, is conserved in any inertial system (*momentum conservation law*):

$$F = 0 \Longleftrightarrow p(t) = \text{const} .$$

Equation of motion and definition of force (lex secunda)
In an inertial system, the rate of change of momentum of a particle is caused by a force F acting on it, such that

$$F = \frac{dp}{dt} . \tag{1.6}$$

Law of interaction (lex tertia)
The forces F_{ij} and F_{ji} with which two particles i and j act upon one another obey

$$F_{ij} = -F_{ji} .$$

Thus, these forces are equal in strength and opposite in direction (*Actio = Reactio*).

Principle of superposition (lex quarta)
Forces are vectors and are thus superimposed linearly:

$$F = \sum_i F_i .$$

These four axioms imply

- the existence of an absolute time in all inertial systems and

- that the total mass of a closed system is a conserved quantity.

Looking ahead, we mention here that both of these assumptions are dropped in the framework of special relativity (Section 1.6).

To lex prima. The first of Newton's axioms, a special case of the second axiom, allows us to construct an inertial system. To this end, consider the

trajectories of three identical particles on which no forces act, and which are thus moving linearly uniformly away from some common point of origin along the axes of a Cartesian coordinate system. Time can then be fixed by demanding that a force-free particle travel the same distances within the same time intervals.

To lex secunda. Newton's second axiom plays the role of a force definition as well as that of a law. Given the force, (1.6) represents the dynamical equation of motion. For the special but most commonly considered case of constant mass, one obtains the more familiar form of this equation:

$$F = \frac{\mathrm{d}}{\mathrm{d}t}p = m\frac{\mathrm{d}}{\mathrm{d}t}\dot{x} = m\ddot{x} \ , \ m = \text{const} \ . \tag{1.7}$$

For the majority of physical problems the force vector is a function only of the particle's position, its velocity, and time: $F = F(x, \dot{x}, t)$.

The fundamental problem of classical mechanics consists of determining the trajectory $x(t)$ of a particle, given the force vector. In the Newtonian formalism this is achieved by solving the three (generally nonlinear) coupled ordinary differential equations of second order (1.7). Since the fundamental law of Newtonian mechanics is of second order in time, it is necessary to specify exactly two initial conditions in order to uniquely determine the particle's trajectory.

Inertial and heavy (gravitational) mass. According to Newton's second law, a force acting on a particle results in an acceleration proportional to the particle's mass. This mass, thus defined, is called the *inertial mass* of the particle. In addition, there exists a second form of mass. In Subsection 1.5.3 we consider the gravitational force F_G, which a *heavy mass* m experiences due to the presence of another heavy mass M:

$$F_G = -\gamma m M \frac{x}{x^3} \ .$$

It has been experimentally shown that these two kinds of mass are proportional to one another. Thus, it is practical to choose the units such that both masses coincide.[1] This insight, equality of inertial and gravitational mass, is the starting point of Einstein's equivalence principle and the theory of general relativity.

To lex tertia. The third axiom deals with physical processes in which two or more particles interact with one another. Due to this law, a system consisting of many particles can, given certain constraints, be regarded effectively as a single particle.

[1] Proceeding in this manner, one finds the experimental value of Newton's *gravitational constant* as $\gamma = (6.67259 \pm 0.00085) \times 10^{-11}\,\mathrm{m^3 kg^{-1} s^{-2}}$.

1.1.3 Physical Consequences, Conservation Laws

Newton's axioms lead quite directly to a set of important conservation laws, which we now discuss. To start, we define several additional basic mechanical quantities.

Definition: Work W, power P, and kinetic energy T

- The amount of *work* W done by a force \boldsymbol{F} against the inertial mass of a particle along the trajectory $\Gamma_{\boldsymbol{x}} = \{\boldsymbol{x}(t), t \in [t_1 : t_2]\}$ is given by

$$W(\boldsymbol{x}_1, \boldsymbol{x}_2, \Gamma_{\boldsymbol{x}}) = W(t_1, t_2, \Gamma_{\boldsymbol{x}}) = \int\limits_{\boldsymbol{x}_1, \Gamma_{\boldsymbol{x}}}^{\boldsymbol{x}_2} \boldsymbol{F}(\boldsymbol{x}, \dot{\boldsymbol{x}}, t) \mathrm{d}\boldsymbol{x}$$

$$= \int\limits_{t_1}^{t_2} \boldsymbol{F}(\boldsymbol{x}, \dot{\boldsymbol{x}}, t) \dot{\boldsymbol{x}} \mathrm{d}t , \quad [W] = \mathrm{Nm} .$$

- The *power* P is defined as

$$P(t) = \frac{\mathrm{d}W}{\mathrm{d}t} = \boldsymbol{F}(\boldsymbol{x}, \dot{\boldsymbol{x}}, t) \cdot \dot{\boldsymbol{x}}(t) , \quad [P] = \frac{\mathrm{Nm}}{\mathrm{s}} = \mathrm{W} \text{ (Watt)} .$$

- The *kinetic energy* T is given by

$$T(t) = \frac{\boldsymbol{p}^2(t)}{2m} = \frac{m\dot{\boldsymbol{x}}^2(t)}{2} , \quad [T] = \mathrm{kg}\frac{\mathrm{m}^2}{\mathrm{s}^2} = \mathrm{J} \text{ (Joule)} .$$

From this, we immediately find: the total work done by a force against the inertial mass of a particle is equal to the difference of the kinetic energies at the end and the start of the trajectory:

$$W(t) = \int\limits_{t_1}^{t_2} \boldsymbol{F}(\boldsymbol{x}, \dot{\boldsymbol{x}}, t) \dot{\boldsymbol{x}} \mathrm{d}t = \frac{m}{2} \left[\dot{\boldsymbol{x}}(t_2)^2 - \dot{\boldsymbol{x}}(t_1)^2 \right]$$

$$= \frac{1}{2m} \left[\boldsymbol{p}^2(t_2) - \boldsymbol{p}^2(t_1) \right] = T_2 - T_1 .$$

Conservative forces and energy conservation. The differential work done by shifting a particle by $\mathrm{d}\boldsymbol{x}$ is given by

$$\mathrm{d}W = \boldsymbol{F}(\boldsymbol{x}, \dot{\boldsymbol{x}}, t)\mathrm{d}\boldsymbol{x} .$$

In general, $\mathrm{d}W$ is not a total differential, so that

$$\oint \mathrm{d}W \neq 0 .$$

Forces that do not depend explicitly on time nor on the velocity, and for which $\boldsymbol{F}(\boldsymbol{x})\mathrm{d}\boldsymbol{x}$ is a total differential, constitute the important class of *conservative forces*.

Theorem 1.2: Conservative forces

The following statements are equivalent:

- $F(x)$ is a conservative force.

- $F(x)$ has vanishing curl: $\nabla \times F = 0$.

- $F(x)$ is the gradient of a scalar field, the *potential* $V(x)$, such that

$$F(x) = -\nabla V(x) \Longrightarrow V(x) = -\int_{x_0}^{x} dx\, F(x) \ .$$

Here, the potential is only defined up to an arbitrary constant, which we can choose in such a way as to obtain a particularly simple form for V.

- The work is independent of the particle's trajectory; it depends only on its initial and final positions:

$$W(x_1, x_2, \Gamma_x) = W(x_1, x_2) = -\int_{t_1}^{t_2} \nabla V(x)\dot{x}\,dt$$

$$= -[V(x_2) - V(x_1)] \ .$$

- The differential work $dW = F(x)dx$ is a total differential.

From this theorem and the definition of kinetic energy, we obtain the important result:

Theorem 1.3: Energy conservation in conservative systems

The total energy E of a particle moving in a conservative force field is conserved:

$$E = T + V = \text{const} \ .$$

If the energy of a conservative system is known at any particular time, we can use this information as one of the necessary initial conditions needed to define a unique solution of the equations of motion.

One-dimensional motion, conservative forces. One-dimensional motion in a conservative force field can always be reduced to an integral. From

$$E = \frac{1}{2}m\dot{x}^2 + V(x) \Longrightarrow \dot{x} = \pm\sqrt{\frac{2}{m}[E - V(x)]} \ ,$$

we obtain[2]

[2] Note that the particle can only be found in regions for which $E \geq V(x)$. We shall provide a discussion of this relation in the context of central forces, Subsection 1.5.2.

$$t - t_0 = \pm \int_{x_0}^{x} dx' \left(\frac{2}{m}[E - V(x')] \right)^{-\frac{1}{2}} , \tag{1.8}$$

where the sign is determined by the initial condition $\dot{x}(t_0) \gtrless 0$. The linear harmonic oscillator is an example of this case. With $V(x) = kx^2/2$, we find

$$t - t_0 = \pm \sqrt{\frac{m}{k}} \arcsin \left[x \left(\frac{k}{2E} \right)^{\frac{1}{2}} \right]$$

and, with $x_0 = x(t_0) = 0$, $\dot{x}(t_0) > 0$,

$$x = \sqrt{\frac{2E}{k}} \sin\left[\omega(t - t_0) \right] , \quad \omega = \sqrt{\frac{k}{m}} .$$

Angular momentum and torque. In addition to the force F and the momentum p, we define two other fundamental quantities:

Definition: Angular momentum l and torque N

The *angular momentum* is defined as the axial vector

$$l(t) = x(t) \times p(t) , \quad [l] = \mathrm{kg}\frac{\mathrm{m}^2}{\mathrm{s}} .$$

Its temporal derivative defines the *torque*:

$$N(t) = \dot{l}(t) = x(t) \times \dot{p}(t) , \quad [N] = \mathrm{kg}\frac{\mathrm{m}^2}{\mathrm{s}^2} .$$

Thus, angular momentum is conserved if the torque vanishes.

We see that for angular momentum conservation either $F = 0$ or $F \| x$ must be satisfied. In the first case, the coordinate system can always be shifted by a constant vector such that $l = 0$. The second case leads to *central forces*:

Theorem 1.4: Central forces and angular momentum conservation

A central force depends only on the particle's position vector x and acts in the direction of x:

$$F = F(x) = x f(x) , \quad f \text{ arbitrary scalar field} .$$

The angular momentum of a central force is conserved.

Conservative central forces. Central forces are not necessarily conservative. However, in practice it is only the class of conservative central forces we encounter in physical applications. For these, we have

$$x \times F = -x \times \nabla V(x) = 0 .$$

In polar coordinates this equation reads

$$\frac{\partial V}{\partial \varphi} = 0 \; , \quad \frac{\partial V}{\partial \theta} = 0 \; .$$

This gives $V(\boldsymbol{x}) = V(|\boldsymbol{x}|)$.

Theorem 1.5: Rotationally symmetric potentials

For conservative central forces, the potential V exhibits *rotational invariance* (*central symmetry*):

$$V(\boldsymbol{x}) = V(|\boldsymbol{x}|) \; .$$

Conversely, if $V = V(|\boldsymbol{x}|)$, we see from

$$\boldsymbol{F} = -\boldsymbol{\nabla} V = -\frac{\mathrm{d}V}{\mathrm{d}|\boldsymbol{x}|} \frac{\boldsymbol{x}}{|\boldsymbol{x}|}$$

that \boldsymbol{F} is a conservative central force.

Examples of the above are given by the gravitational force and the Coulomb interaction. We consider the equations of motions for this type of force in more detail in Section 1.5.

Homogeneous potentials. Another important class of potentials is defined as follows:

Definition: Homogeneous potential

A *homogeneous potential* is a rotationally symmetric potential of the form

$$V(\boldsymbol{x}) = V(|\boldsymbol{x}|) = \alpha |\boldsymbol{x}|^d \; , \quad d \in \mathbb{R} \; .$$

For a single particle, we have

$$\frac{\mathrm{d}}{\mathrm{d}t}(m\boldsymbol{x}\dot{\boldsymbol{x}}) = m(\dot{\boldsymbol{x}}^2 + \boldsymbol{x}\ddot{\boldsymbol{x}}) = 2T + \boldsymbol{x}\boldsymbol{F} \; . \tag{1.9}$$

Averaging this equation over a time interval τ yields

$$\frac{m}{\tau}\left[\boldsymbol{x}(\tau)\dot{\boldsymbol{x}}(\tau) - \boldsymbol{x}(0)\dot{\boldsymbol{x}}(0)\right] = 2\langle T \rangle + \langle \boldsymbol{x}\boldsymbol{F} \rangle \; , \quad \langle T \rangle = \frac{1}{\tau}\int_0^\tau \mathrm{d}t\, T \; .$$

If the particle's motion is periodic within the interval $[0 : \tau]$, i.e., $\boldsymbol{x}(\tau) = \boldsymbol{x}(0)$, $\dot{\boldsymbol{x}}(\tau) = \dot{\boldsymbol{x}}(0)$, the left-hand side of (1.9) vanishes, and one obtains

$$\langle T \rangle = -\frac{1}{2}\langle \boldsymbol{x}\boldsymbol{F} \rangle \qquad (\textit{virial theorem}) \; . \tag{1.10}$$

Thus, specifically for homogeneous potentials, we find

Theorem 1.6: Virial theorem for homogeneous potentials

For homogeneous potentials, $V(\boldsymbol{x}) = \alpha |\boldsymbol{x}|^d$, $\boldsymbol{F} = -\boldsymbol{\nabla} V(\boldsymbol{x})$, we have

\triangleright

$$\langle \boldsymbol{x} \boldsymbol{F} \rangle = -\frac{\alpha}{\tau} \int_0^\tau dt \boldsymbol{x} \boldsymbol{\nabla} |\boldsymbol{x}|^d = -\frac{\alpha}{\tau} \int_0^\tau dt d |\boldsymbol{x}|^d = -d \langle V \rangle$$

and therefore

$$\langle T \rangle = \frac{d}{2} \langle V \rangle \ .$$

Even if the particle's motion is not periodic, one arrives at (1.10), provided that the position and velocity coordinates are bounded from above, so that the left-hand side of (1.9) vanishes for sufficiently large τ.

Theorem 1.7: Scale transformations for homogeneous potentials

Let $V(\boldsymbol{x}) = \alpha |\boldsymbol{x}|^d$. Then

$$m \frac{d^2 \boldsymbol{x}}{dt^2} = -\boldsymbol{\nabla} V(\boldsymbol{x}) = -\alpha d |\boldsymbol{x}|^{d-2} \boldsymbol{x} \ .$$

Performing a scale transformation of the form

$$\boldsymbol{x} = \lambda \boldsymbol{x}' \ , \ t = \lambda^{(2-d)/2} t' \ ,$$

it follows that

$$\frac{d^2 \boldsymbol{x}}{dt^2} = \lambda^{d-1} \frac{d^2 \boldsymbol{x}'}{dt'^2} \ , \ \boldsymbol{\nabla} V(\boldsymbol{x}) = \lambda^{d-1} \boldsymbol{\nabla}' V(\boldsymbol{x}')$$

$$\Longrightarrow m \frac{d^2 \boldsymbol{x}'}{dt'^2} = -\boldsymbol{\nabla}' V(\boldsymbol{x}') \ .$$

Thus, Newton's equation of motion is form-invariant under such scale transformations.
Suppose that T and T' are specific times (such as oscillation periods) and let R and R' be some specific lengths (e.g., amplitudes) of two forms of motion that are related by the above scale transformation. Then,

$$V(\boldsymbol{x}) = \alpha |\boldsymbol{x}|^d \Longrightarrow \left(\frac{T'}{T} \right)^2 = \left(\frac{R'}{R} \right)^{2-d} \ .$$

In particular, for the gravitational and for the Coulomb force with $d = -1$, we find

$$\left(\frac{T'}{T} \right)^2 = \left(\frac{R'}{R} \right)^3 \ ,$$

so that the squares of the periods of two particles around a force center behave similarly to the cubes of the two distances to the center. This is *Kepler's third law* (see Subsection 1.5.3).

1.1.4 Accelerated Coordinate Systems and Inertial Systems, Galilei Invariance

Newton's equations of motion refer to inertial systems moving linearly uniformly relative to one another. The question we want to tackle in this subsection is: *what is the equation of motion of a particle in a noninertial system K′ that is accelerated relative to the inertial system K* (see Fig. 1.1)?

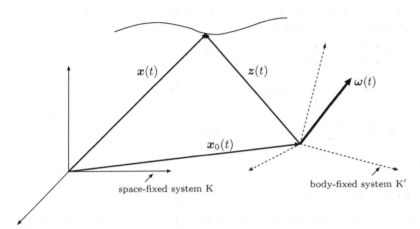

space-fixed system K body-fixed system K′

Fig. 1.1. Definition of space-fixed and body-fixed system.

For the following discussion, and that of rigid bodies in Section 1.4, we introduce the term *space-fixed system* for the inertial system K, whereas the accelerated system K′ is called the *body-fixed system*. In the most general case, the accelerated motion of K′ relative to K consists of a rotation around the origin of K, and a translational motion described by the vector $x_0(t)$. We shall treat these two cases consecutively by looking first at a pure rotation and subsequently adding in the translational motion.

Rotating coordinate systems. As already mentioned in Subsection 1.1.1, two coordinate systems K : $\{e_1, e_2, e_3\}$ and K′ : $\{e'_1, e'_2, e'_3\}$ rotating relative to one another are connected via a time-dependent (3×3)-matrix R satisfying the relations (1.4) and (1.5). Differentiating the orthogonality relation in (1.5) with respect to time, we obtain

$$\dot{R}^{\mathrm{T}}R + R^{\mathrm{T}}\dot{R} = 0 .$$

This implies that $\Omega = R^{\mathrm{T}}\dot{R}$ is skew symmetric:

$$\Omega = R^{\mathrm{T}}\dot{R} = \begin{pmatrix} 0 & \omega_3 & -\omega_2 \\ -\omega_3 & 0 & \omega_1 \\ \omega_2 & -\omega_1 & 0 \end{pmatrix} , \quad \Omega_{jk} = \sum_i \epsilon_{ijk}\omega_i , \ \omega_i \in \mathbb{R}. \qquad (1.11)$$

This relation is invertible, and we find

$$\omega_i = \frac{1}{2} \sum_{j,k} \epsilon_{ijk} \Omega_{jk} \, . \tag{1.12}$$

As we will see in a moment, these are the space-fixed components of a time-dependent vector $\boldsymbol{\omega} = \sum_i \boldsymbol{e}_i \omega_i$, which describes the rotation axis of K$'$ relative to K. For this reason, $\boldsymbol{\omega}$ is termed *momentary rotation axis* or *momentary angular velocity*. Since the bases of K and K$'$ are time-dependent relative to one another, we use the notation introduced in Subsection 1.1.1 for the temporal derivative of a vector \boldsymbol{x} with respect to K and K$'$ [see (1.3)]. For an arbitrary vector \boldsymbol{x}, we calculate

$$\begin{aligned}
D'\boldsymbol{x} &= \sum_i \boldsymbol{e}'_i \dot{x}'_i = \sum_{i,j} \boldsymbol{e}'_i \left(R_{ij} \dot{x}_j + \dot{R}_{ij} x_j \right) \\
&= \sum_j \boldsymbol{e}_j \dot{x}_j + \sum_{k,i,j} \boldsymbol{e}_k R_{ki}^{\mathrm{T}} \dot{R}_{ij} x_j = D\boldsymbol{x} + \sum_{k,j} \boldsymbol{e}_k \Omega_{kj} x_j \\
&= D\boldsymbol{x} - \sum_{k,i,j} \boldsymbol{e}_k \epsilon_{kij} \omega_i x_j = D\boldsymbol{x} - \boldsymbol{\omega} \times \boldsymbol{x} \, ,
\end{aligned}$$

which yields

Theorem 1.8: Theorem of Coriolis

Let $\boldsymbol{\omega}$ denote the momentary angular velocity of a system K$'$ that is rotating relative to a system K, and let $D\boldsymbol{x}$ and $D'\boldsymbol{x}$ be the temporal derivatives of a vector \boldsymbol{x} with respect to K and K$'$. We then have

$$D'\boldsymbol{x} = D\boldsymbol{x} - \boldsymbol{\omega} \times \boldsymbol{x} \, .$$

From this, it follows that the temporal derivative of the angular velocity is independent of the reference frame:

$$D'\boldsymbol{\omega} = D\boldsymbol{\omega} \, .$$

To clarify this theorem, consider the simple case for which $D'\boldsymbol{x} = \boldsymbol{0} \Longrightarrow D\boldsymbol{x} = \boldsymbol{\omega} \times \boldsymbol{x}$. In the reference system K, the vector \boldsymbol{x} changes by $\delta\boldsymbol{x} = \boldsymbol{\omega} \times \boldsymbol{x} \delta t$ in the time δt. This vector is orthogonal to $\boldsymbol{\omega}$ and \boldsymbol{x}. Thus, in K, the change of \boldsymbol{x} is achieved by a (right-handed) rotation of \boldsymbol{x} by the amount $|\boldsymbol{\omega}|\delta t$ around an axis parallel to $\boldsymbol{\omega}$. This is why we call $\boldsymbol{\omega}$ "momentary angular velocity".

Equation of motion in accelerated reference systems. We now consider the most general relative motion of two reference systems K and K$'$ (see Fig. 1.1), where K is again taken to be an inertial system. The vectors $\boldsymbol{x}(t)$ and $\boldsymbol{z}(t)$ describe the trajectories of a particle as seen from K and K$'$, respectively. Furthermore, \boldsymbol{x}_0 shall denote the origin of K$'$ relative to K, so that

$$\boldsymbol{z}(t) = \boldsymbol{x}(t) - \boldsymbol{x}_0(t) \, .$$

Since K is an inertial system, Newton's equation holds:

$$mD^2 \boldsymbol{x} = \boldsymbol{F} \ . \tag{1.13}$$

To find the corresponding equation of motion in the accelerated system K′, we make use of the theorem of Coriolis and calculate as follows:

$$
\begin{aligned}
\frac{\boldsymbol{F}}{m} &= D^2 \boldsymbol{x} = D^2 \boldsymbol{z} + D^2 \boldsymbol{x}_0 \\
&= D\left(D' \boldsymbol{z} + \boldsymbol{\omega} \times \boldsymbol{z}\right) + D^2 \boldsymbol{x}_0 \\
&= D'^2 \boldsymbol{z} + (D' \boldsymbol{\omega}) \times \boldsymbol{z} + 2\boldsymbol{\omega} \times D' \boldsymbol{z} + \boldsymbol{\omega} \times (\boldsymbol{\omega} \times \boldsymbol{z}) + D^2 \boldsymbol{x}_0 \ .
\end{aligned}
$$

From this, we can easily read off the equation of motion, which is the equivalent of (1.13) in the accelerated system K′:

Theorem 1.9: Newton's equation of motion in accelerated systems

Let \boldsymbol{x} and \boldsymbol{z} respectively denote the particle's trajectories in a system K and in a system K′ accelerated relative to K. The equations of motion in these reference systems are

$$mD^2 \boldsymbol{x} = \boldsymbol{F}$$
$$mD'^2 \boldsymbol{z} = \boldsymbol{F} + \boldsymbol{F}_{\mathrm{T}} + \boldsymbol{F}_{\mathrm{z}} + \boldsymbol{F}_{\mathrm{L}} + \boldsymbol{F}_{\mathrm{C}} \ ,$$

with

$$\boldsymbol{F}_{\mathrm{T}} = -mD^2 \boldsymbol{x}_0 \qquad \text{(translational force)}$$

$$\boldsymbol{F}_{\mathrm{F}} = -m\boldsymbol{\omega} \times (\boldsymbol{\omega} \times \boldsymbol{z}) \qquad \text{(centrifugal force)}$$

$$\boldsymbol{F}_{\mathrm{L}} = -m(D'\boldsymbol{\omega}) \times \boldsymbol{z} \qquad \text{(linear force)}$$

$$\boldsymbol{F}_{\mathrm{C}} = -2m\boldsymbol{\omega} \times D'\boldsymbol{z} \qquad \text{(Coriolis force)} \ .$$

Thus, in addition to the original force \boldsymbol{F}, there are four *pseudo forces* in accelerated reference systems. Of these, $\boldsymbol{F}_{\mathrm{T}}$ is due to the translational motion of the coordinate origin \boldsymbol{x}_0, whereas the remaining three stem from the rotational motion.

Galilei invariance of the equations of motion. Using the above result, we can now take a closer look at the concept of inertial systems, which we discussed qualitatively in Subsection 1.1.2. According to Newton's second axiom, the equation of motion is form-invariant in all inertial systems. This is clearly equivalent to the condition

$$D^2 \boldsymbol{x}_0 = \boldsymbol{0} \ , \quad \boldsymbol{\omega} = \boldsymbol{0} \ ,$$

since then all pseudo forces vanish. The condition can be satisfied most generally by letting

$$\boldsymbol{x}_0(t) = \boldsymbol{v}t + \boldsymbol{q} \ , \quad R(t) = R \ , \quad \boldsymbol{v}, \boldsymbol{q}, R = \text{const} \ .$$

In other words: relative to K, the system K$'$ can only be rotated by a fixed amount and move with constant velocity. If, additionally, the force \boldsymbol{F} is not explicitly time-dependent, the time origin of K$'$ may also be shifted relative to that of K. As one can see, demanding form invariance of Newton's equations of motion leads to the four points, listed in Subsection 1.1.2, which characterize inertial systems. Thus, we find the fundamental law:

Theorem 1.10: Inertial systems and Galilei invariance

Inertial systems are connected via coordinate transformations of the form

$$x_i \longrightarrow x_i' = R_{ij}x_j + v_i t + q_i \ , \ t \longrightarrow t' = t + t_0 \ ,$$

with

$$R, \boldsymbol{v}, \boldsymbol{q}, t_0 = \text{const} \ , \ RR^{\mathrm{T}} = 1 \ , \ \det R = 1 \ .$$

These transformations form a proper orthochronous Lie group with 10 parameters. It is called the *group of Galilei transformations*.

Newton's laws are form-invariant in all inertial systems. We say: Newtonian mechanics is *Galilei-invariant*. In inertial systems, force-free particles move linearly uniformly.

Consequently, in mechanics, we cannot distinguish between the state of total stationarity and one of uniform motion.

It is a general principle in physics, not just in classical mechanics, that a symmetry of a system corresponds to some conserved physical quantity. Since the Galilei group is described by 10 parameters, we deduce that in the absence of external forces the motion of a system of particles is characterized by 10 conserved quantities. As we shall see in the following subsection, and in Subsection 1.2.2, these are momentum, angular momentum, energy, and the center of mass movement.

Absolute time. The form of the Galilei group shows that time plays a special role in nonrelativistic mechanics. Time differences are invariant under Galilei transformations. In this sense, time has an absolute character. In contrast, the spatial distance between two points where a particle is located at two different times due to its movement will not be identical in two different inertial systems, since the particle's velocities in the two frames will differ. Therefore, space is not absolute.

Foucault's pendulum. As an example for the use of Newton's equations of motion in accelerated coordinate systems, we study the effect of the earth's rotation on the oscillation plane of a pendulum. Figure 1.2 determines the axes of the inertial system[3] K and that of the relative to K accelerated system K$'$ (earth-fixed system). We consider a *mathematical pendulum* consisting of

[3] Strictly speaking, this system is not an inertial system, in particular due to its accelerated motion around the sun. However, this can be neglected for the argument that follows.

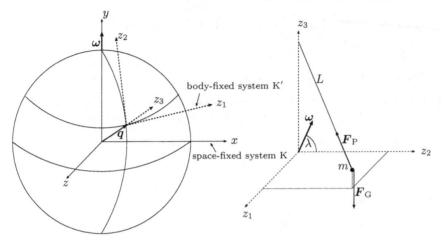

Fig. 1.2. Determination of space-fixed and body-fixed systems on the earth's surface.

a massless rope of length L, fixed at the top, so that it can rotate freely in all directions, and to which a particle of mass m is attached at the other end. The coordinate triplet of the angular velocity $\boldsymbol{\omega}$ in K' reads

$$\boldsymbol{\omega} = \omega \begin{pmatrix} 0 \\ \cos\lambda \\ \sin\lambda \end{pmatrix} \ , \quad \omega \approx \frac{2\pi}{24\,\text{h}} \ .$$

Following Theorem 1.9, we have

$$m\ddot{\boldsymbol{z}} = \boldsymbol{F}_\mathrm{G}(\boldsymbol{z}) + \boldsymbol{F}_\mathrm{P}(\boldsymbol{z})$$
$$+ \boldsymbol{F}_\mathrm{T}(\boldsymbol{q}) + \boldsymbol{F}_\mathrm{F}(\boldsymbol{\omega}, \boldsymbol{z}) + \boldsymbol{F}_\mathrm{L}(\boldsymbol{\omega}, \boldsymbol{z}) + \boldsymbol{F}_\mathrm{C}(\boldsymbol{\omega}, \boldsymbol{z}) \ ,$$

where the coordinate triplets of the gravitational force $\boldsymbol{F}_\mathrm{G}$ and the centripetal force $\boldsymbol{F}_\mathrm{P}$ of the pendulum in K' are given by

$$\boldsymbol{F}_\mathrm{G} = -mg \begin{pmatrix} 0 \\ 0 \\ 1 \end{pmatrix} \ , \quad \boldsymbol{F}_\mathrm{P} = \begin{pmatrix} -\frac{z_1}{L}S \\ -\frac{z_2}{L}S \\ S_3 \end{pmatrix} \ .$$

At this stage, the tensions S and S_3 of the rope are unknown. Given that $\omega^2 \ll 1$ and $\dot{\boldsymbol{\omega}} \approx \boldsymbol{0}$, we can neglect the centrifugal and linear forces $\boldsymbol{F}_\mathrm{F}$ and $\boldsymbol{F}_\mathrm{L}$. The translational force $\boldsymbol{F}_\mathrm{T}$ points in the direction of the z_3-axis and thus diminishes the strength of the gravitational force on the earth's surface, depending on the geographical latitude. $\boldsymbol{F}_\mathrm{T}$ can therefore be absorbed in the gravitational constant g, and we are left with the equation

$$m\ddot{\boldsymbol{z}} = \boldsymbol{F}_\mathrm{G} + \boldsymbol{F}_\mathrm{P}(\boldsymbol{z}) - 2m\boldsymbol{\omega} \times \dot{\boldsymbol{z}}$$

or

$$m\ddot{z}_1 = -\frac{z_1}{L}S - 2m\omega(\dot{z}_3\cos\lambda - \dot{z}_2\sin\lambda)$$

$$m\ddot{z}_2 = -\frac{z_2}{L}S - 2m\omega\dot{z}_1\sin\lambda$$

$$m\ddot{z}_3 = -mg + S_3 + 2m\omega\dot{z}_1\cos\lambda \ .$$

If we consider only small deviations from the vertical z_3-axis, we have $S \approx mg$, $\dot{z}_3 \ll \dot{z}_2$, so that the pendulum's z_1z_2-motion becomes

$$\ddot{z}_1 = -\frac{g}{L}z_1 + 2\omega'\dot{z}_2 \ , \ \omega' = \omega\sin\lambda$$

$$\ddot{z}_2 = -\frac{g}{L}z_2 - 2\omega'\dot{z}_1 \ .$$

The parametrization

$$z_1(t) = u_1(t)\cos\omega't + u_2(t)\sin\omega't$$

$$z_2(t) = u_2(t)\cos\omega't - u_1(t)\sin\omega't$$

leads to

$$\cos\omega't\left(\ddot{u}_1 + \frac{g}{L}u_1\right) + \sin\omega't\left(\ddot{u}_2 + \frac{g}{L}u_2\right) = 0$$

$$\implies \ddot{u}_i + \frac{g}{L}u_i = 0 \ , \ i = 1,2 \ .$$

These are the equations of motion of a simple pendulum, solved by

$$u_i(t) = a_i\cos\Omega t + b_i\sin\Omega t \ , \ \Omega = \sqrt{\frac{g}{L}} \ , \ i = 1,2 \ ,$$

where the four integration constants a_1, b_1, a_2, b_2 are fixed by initial conditions. We see that the pendulum oscillates with frequency $\Omega = \sqrt{g/L}$ around its equilibrium, whereas its oscillation plane rotates with frequency $\omega' = 2\pi/24\,\mathrm{h}\cdot\sin\lambda$. This type of motion results in *rosetta trajectories*, whose exact form depends on the chosen initial conditions. The left-hand picture of Fig. 1.3 shows the trajectory of a pendulum that was released at its maximal amplitude. In the right-hand picture, the pendulum is pushed out of

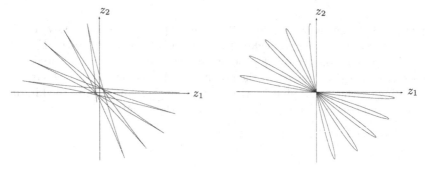

Fig. 1.3. Different rosetta trajectories of Foucault's pendulum.

its equilibrium position. Obviously, at the equator ($\lambda = 0$) the rotation of the earth cannot be demonstrated using a Foucault pendulum. At this location the pendulum moves along the same type of trajectory as in an inertial frame (within the above approximation), which is an ellipse parametrized by $z_i(t) = u_i(t)$, $i = 1, 2$.

1.1.5 N-Particle Systems

We now extend our considerations to systems containing many particles and discuss their kinematic properties. The forces acting on such systems can be classified as follows:

- *Inner forces:* They act exclusively between the particles in the system. To simplify the subsequent discussion, we shall assume that these forces are conservative and central two-particle forces that can be derived from potentials V_{ij}, such that

$$\boldsymbol{F}_{ij} = \boldsymbol{F}_{ij}(\boldsymbol{x}_i - \boldsymbol{x}_j) = -\boldsymbol{\nabla}_i V_{ij}(|\boldsymbol{x}_i - \boldsymbol{x}_j|) \ .$$

- *External forces:* By this, we mean all other forces acting externally on the system. If no external forces are present, the system is called *closed.*

Given these definitions, Newton's equations of motion for an N-particle system are

$$m_i \ddot{\boldsymbol{x}}_i = \sum_{j \neq i} \boldsymbol{F}_{ij} + \boldsymbol{F}_i \ , \ i = 1, \ldots, N \ . \tag{1.14}$$

Before we study the corresponding kinematical quantities, such as momentum, angular momentum, and energy, it is useful to introduce the notion of *center of mass:*

Definition: Center of mass \boldsymbol{x}_C of an N-particle system

$$\boldsymbol{x}_C(t) = \frac{1}{M} \sum_{i=1}^{N} m_i \boldsymbol{x}_i(t) \ , \ M = \sum_{i=1}^{N} m_i \ .$$

In case of a continuous mass distribution $\rho(\boldsymbol{x}, t)$, these equations are replaced by

$$\boldsymbol{x}_C(t) = \frac{1}{M} \int \boldsymbol{x} \rho(\boldsymbol{x}, t) \mathrm{d}^3 x \ , \ M = \int \rho(\boldsymbol{x}, t) \mathrm{d}^3 x = \text{const} \ .$$

Total momentum p and center of mass \boldsymbol{x}_C. If we add the individual equations in (1.14), the contributions of the inner forces are cancelled out due to the principle of "actio=reactio", so that

$$\dot{\boldsymbol{p}} = \sum_i \dot{\boldsymbol{p}}_i = \sum_i m_i \ddot{\boldsymbol{x}}_i = \frac{1}{2} \sum_{i,j} (\boldsymbol{F}_{ij} + \boldsymbol{F}_{ji}) + \sum_i \boldsymbol{F}_i = \sum_i \boldsymbol{F}_i \ .$$

Thus, the total momentum p is determined solely by the external forces. If there are no external forces at all, the total momentum is conserved. In addition, we have

$$M\dot{x}_C = p \, ,$$

so that the center of mass is moving as if all mass were contained in it, and as if all external forces were acting directly on it. In the case of closed systems, this implies that the motion of the center of mass is linearly uniform.

Total angular momentum l and torque N. For the total angular momentum and the torque, we obtain

$$l = \sum_i l_i = \sum_i m_i x_i \times \dot{x}_i$$

$$N = \dot{l} = \sum_i m_i x_i \times \ddot{x}_i = \frac{1}{2} \underbrace{\sum_{i,j} (x_i - x_j) \times F_{ij}}_{= \, 0 \text{ for } F_{ij} \text{ central}} + \sum_i x_i \times F_i \, .$$

Accordingly, the total angular momentum of the system is also a conserved quantity. Decomposing the position vector x_i into the center of mass vector x_C and a vector x_i^C relative to the center of mass, i.e., $x_i = x_C + x_i^C$, we can use $\sum_i m_i x_i^C = 0$ and find the angular momentum to be

$$l = \sum_i m_i x_i \times \dot{x}_i = \sum_i m_i (x_C + x_i^C) \times (\dot{x}_C + \dot{x}_i^C)$$

$$= \sum_i m_i \left[(x_C \times \dot{x}_C) + (x_C \times \dot{x}_i^C) + (x_i^C \times \dot{x}_C) + (x_i^C \times \dot{x}_i^C) \right]$$

$$= l_C + \sum_i x_i^C \times p_i^C \, , \quad l_C = x_C \times p_C \, .$$

So, the total angular momentum is composed of the center of mass's angular momentum with respect to the origin of the coordinate system, and the sum of the particles' angular momenta with respect to the center of mass x_C.

Total energy. Scalar multiplication of (1.14) with \dot{x}_i and summation over all i yield

$$\frac{d}{dt} \left(\frac{1}{2} \sum_i m_i \dot{x}_i^2 \right) = \frac{1}{2} \sum_{i,j} (\dot{x}_i - \dot{x}_j) F_{ij} + \sum_i \dot{x}_i F_i$$

$$\implies \frac{d}{dt} (T + V_{\text{inner}}) = \sum_i \dot{x}_i F_i \, ,$$

where

$$T = \frac{1}{2} \sum_i m_i \dot{x}_i^2 \, , \quad V_{\text{inner}} = \frac{1}{2} \sum_{i,j} V_{ij}(|x_i - x_j|)$$

are the total kinetic and inner potential energy, respectively. Thus, the total change of the inner energy of the system is equal to the power of the external

forces. For closed systems this means that the total energy is conserved. If the external forces are also described by potentials, we have

$$\sum_i \dot{\boldsymbol{x}}_i \boldsymbol{F}_i = -\frac{dV_{\text{ext}}}{dt} \implies \frac{d}{dt}(T + V_{\text{inner}} + V_{\text{ext}}) = 0 \;,$$

so that total energy is again conserved. Using the above vector decomposition $\boldsymbol{x}_i = \boldsymbol{x}_C + \boldsymbol{x}_i^C$, the total kinetic energy can be written as

$$T = \frac{1}{2}\sum_i m_i \dot{\boldsymbol{x}}_i^2 = \frac{1}{2}\sum_i m_i (\dot{\boldsymbol{x}}_C + \dot{\boldsymbol{x}}_i^C)^2 = \frac{M}{2}\dot{\boldsymbol{x}}_C^2 + \frac{1}{2}\sum_i m_i \left(\dot{\boldsymbol{x}}_i^C\right)^2 \;.$$

Just as for angular momentum, this expression consists of the kinetic energy of the center of mass, plus the kinetic energy of the system's movement around the center of mass.

Theorem 1.11: Conserved quantities in N-particle systems

For an N-particle system with no external forces, the following quantities are conserved:

$$\boldsymbol{p} = \text{const} \qquad (\textit{momentum law})$$

$$M\boldsymbol{x}_C - t\boldsymbol{p} = \text{const} \qquad (\textit{center of mass law})$$

$$\boldsymbol{l} = \text{const} \qquad (\textit{angular momentum law})$$

$$E = T + V = \text{const} \qquad (\textit{energy law}) \;.$$

Such a system has 10 conserved quantities, corresponding to the 10 parameters of the Galilei group.

Summary

- All **dynamic variables** of a particle are functions of its position and momentum vectors \boldsymbol{x}, \boldsymbol{p}.

- The temporal evolution of a particle in an **inertial system** is given by **Newton's equation** $\boldsymbol{F} = d\boldsymbol{p}/dt$, where \boldsymbol{F} is the vector of the force acting on the particle.

- Depending on the form of the force, we can formulate conservation laws for specific dynamical quantities.

- Newton's equations of motion are **Galilei-invariant**, i.e., they are valid in **inertial systems** that are related to one another via **Galilei transformations**. In accelerated (noninertial) reference systems, additional

\triangleright

> pseudo forces arise in the equations of motions. They result from the translational and rotational motion of the coordinate system.
>
> • In closed systems, total momentum, total angular momentum, and total energy are conserved.

Applications

1. Rocket problem. Consider a rocket moving linearly uniformly with velocity v relative to an inertial system K in the absence of external force fields. Its total mass consists of the mass of the rocket with empty fuel tank, M_0, and that of the fuel, m_0. At time t_0 the rocket starts ejecting gas in flight direction with an ejection rate $\alpha = dm/dt = \text{const}$, and with constant velocity ω relative to the rocket. When does the rocket come to a complete stop?

Solution. The mass of the rocket at time t is given by

$$M(t) = M_0 + m_0 - \alpha t \ .$$

Thus, its momentum P is

$$P(t) = M(t)\dot{x}(t) \implies \dot{P} = \dot{M}\dot{x} + M\ddot{x} = (M_0 + m_0 - \alpha t)\ddot{x} - \alpha\dot{x} \ .$$

For the fuel momentum p, we find

$$dp(t) = dm(t)[\omega + \dot{x}(t)] \implies \dot{p} = \alpha(\omega + \dot{x}) \ .$$

Given the absence of external forces, it follows that

$$F = \dot{P} + \dot{p} = 0$$

$$\iff (M_0 + m_0 - \alpha t)\ddot{x} = -\alpha\omega$$

$$\iff \int_v^{\dot{x}(t)} d\dot{x} = -\alpha\omega \int_0^t \frac{dt}{M_0 + m_0 - \alpha t}$$

$$\iff \dot{x}(t) - v = \omega \ln\left(1 - \frac{\alpha t}{M_0 + m_0}\right) \ .$$

From this, the time t_1 at which the rocket comes to a complete stop, $\dot{x}(t_1) = 0$, is found to be

$$t_1 = \frac{M_0 + m_0}{\alpha}\left(1 - e^{-v/\omega}\right) \ .$$

Since fuel is limited, the additional constraint $m_0 \geq \alpha t_1$ is also taken into account.

2. Damped harmonic oscillator. Consider a one-dimensional spring, fixed at one end and with a mass m attached at the other end, which is allowed to move along a rail. Additionally, a frictional force proportional to the mass's velocity applies. Find Newton's equation of motion and its solution.

Solution. According to *Hooke's law*, the force of the spring on the mass m is proportional to the displacement of the mass from its equilibrium. Putting the origin of our reference frame into this point of equilibrium, the restoring force F_S due to the spring and the corresponding potential V_S are given by

$$F_S = -\frac{d}{dx}V_S(x) \ , \ V_S(x) = \frac{k}{2}x^2 \ , \ k > 0 \ , \tag{1.15}$$

where k denotes the *spring constant*. For the frictional force, we make the simple ansatz

$$F_F = -c\dot{x} \ , \ c > 0 \ ,$$

with c being the constant *friction coefficient*. The equation of motion for this nonconservative problem reads

$$m\ddot{x} = F_S + F_F \Longrightarrow \ddot{x} + 2\gamma\dot{x} + \omega_0^2 x = 0 \ , \ \gamma = \frac{c}{2m} \ , \ \omega_0^2 = \frac{k}{m} \ . \tag{1.16}$$

To solve this differential equation of the *one-dimensional damped harmonic oscillator*, we let

$$x(t) = e^{i\Omega t}f(t) \ ,$$

which, when inserted into (1.16), leads to the following equation for Ω and the function f:

$$\ddot{f} + \dot{f}(2\gamma + 2i\Omega) + f(\omega_0^2 + 2i\gamma\Omega - \Omega^2) = 0 \ . \tag{1.17}$$

The problem is much simplified if the last bracket vanishes. This can be achieved by setting

$$\Omega_{1,2} = i\gamma \pm \sqrt{\omega_0^2 - \gamma^2} \ .$$

Depending on the value of the square root, three cases now need to be distinguished:

• **Weak damping:** $\omega_0^2 > \gamma^2$. Setting $f(t) = 1$, we find two linearly independent solutions to (1.16):

$$x_1(t) = e^{-\gamma t}e^{i\omega' t} \ , \ x_2(t) = e^{-\gamma t}e^{-i\omega' t} \ , \ \omega' = \sqrt{\omega_0^2 - \gamma^2} \ .$$

The general solution is therefore given by

$$x(t) = e^{-\gamma t}(a\cos\omega' t + b\sin\omega' t) \ ,$$

where a and b are two integration constants to be determined from initial conditions, e.g., $x(0) = x_0$, $\dot{x}(0) = v_0$.

- **Strong damping:** $\omega_0^2 < \gamma^2$. As above, for $f(t) = 1$, one has two solutions,

$$x_1(t) = e^{-(\gamma+\omega')t} \ , \ x_2(t) = e^{-(\gamma-\omega')t} \ , \ \omega' = \sqrt{\gamma^2 - \omega_0^2} \ ,$$

 which can be combined linearly to give the most general solution

$$x(t) = e^{-\gamma t} \left(ae^{-\omega' t} + be^{\omega' t} \right) \ .$$

- **Critical damping:** $\omega_0^2 = \gamma^2$. In this case, setting $f(t) = 1$ yields

$$x_1(t) = e^{-\gamma t} \ .$$

 Since now the first bracket in (1.17) also vanishes, a second solution follows from $f(t) = t$ as

$$x_2(t) = te^{-\gamma t} \ .$$

 In total, we have

$$x(t) = e^{-\gamma t}(a + bt) \ .$$

Figure 1.4 shows different solutions depending of $\omega_0^2 - \gamma^2$, where the initial condition $x(0) = -\dot{x}(0)$ has been chosen throughout. In the absence of friction, $\gamma = 0$, (1.16) reduces to the equation of the *undamped harmonic oscillator*, whose general solution is, according to the above case of weak damping, given by

$$x(t) = a \cos \omega_0 t + b \sin \omega_0 t \ .$$

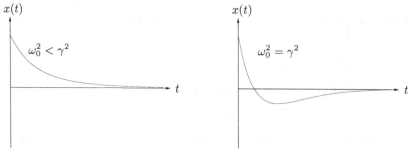

Fig. 1.4. Different types of solutions to the damped harmonic oscillator.

The harmonic oscillator is a kind of showpiece of physics, as small deviations from equilibrium can always be viewed as harmonic oscillations and therefore be described by a potential of the form (1.15).

3. Forced oscillation of an oscillator. Consider the motion of a damped harmonic oscillator whose oscillation is forced by a time-dependent external force $f(t)$:

$$m\ddot{x} + 2\gamma m\dot{x} + m\omega_0^2 x = f(t) .$$

Solution. To find the solution to this problem we use the method of *Green functions*, which we will also find of great use in electrodynamics. According to this, the most general solution for inhomogeneous linear equations of this type can be written in the form

$$x(t) = x_{\text{hom}}(t) + \int G(t, t') f(t') dt' ,$$

where $x_{\text{hom}}(t)$ is the general solution of the corresponding homogeneous problem, and $G(t, t')$ is the Green function that, from the above, must satisfy

$$m\ddot{G}(t, t') + 2\gamma m\dot{G}(t, t') + m\omega_0^2 G(t, t') = \delta(t' - t) . \tag{1.18}$$

Once the Green function has been found, it immediately determines the solution to all inhomogeneous problems. Taking the Fourier decompositions

$$G(t, t') = \frac{1}{\sqrt{2\pi}} \int d\omega \tilde{G}(\omega) e^{i\omega(t-t')}$$

$$\delta(t - t') = \frac{1}{2\pi} \int d\omega e^{i\omega(t-t')}$$

and inserting these terms into (1.18), we find

$$\tilde{G}(\omega) = -\frac{1}{m\sqrt{2\pi}} \frac{1}{\omega^2 - 2i\gamma\omega - \omega_0^2}$$

$$= -\frac{1}{m\sqrt{2\pi}} \frac{1}{(\omega - i\gamma + \omega')(\omega - i\gamma - \omega')} , \quad \omega' = \sqrt{\omega_0^2 - \gamma^2}$$

and therefore

$$G(t, t') = -\frac{1}{2\pi m} \int d\omega \frac{e^{i\omega(t-t')}}{(\omega - i\gamma + \omega')(\omega - i\gamma - \omega')} . \tag{1.19}$$

This integral can best be solved in the complex ω-plane using Cauchy's theorem (see Fig. 1.5), where the following items need to be kept in mind:

- The integrand of (1.19) has two poles of first order for $\omega_0^2 \neq \gamma^2$ and one pole of second order for $\omega_0^2 = \gamma^2$ in the upper half-plane.

- For $t - t' > 0$, the path C along the upper contour must be chosen, so that the contribution of the half-circle is damped exponentially and vanishes in the limit $R \to \infty$. Similarly, for $t - t' < 0$, we need to choose the lower contour of integration along C'.

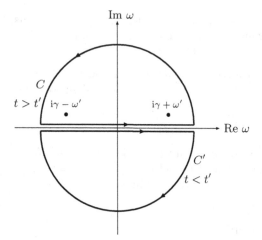

Fig. 1.5. Determination of the integration path within the ω-integration in (1.19).

Thus, for $t - t' < 0$, the integration does not contribute at all. This is in accordance with the *causality principle*, which states that a system, at time t, can only be influenced by the past ($t' < t$). For $t - t' > 0$, we obtain

- **Weak damping:** $\omega_0^2 > \gamma^2$.

$$G(t, t') = \frac{1}{m\omega'} e^{-\gamma(t-t')} \sin[\omega'(t - t')] \ , \ \ \omega' = \sqrt{\omega_0^2 - \gamma^2} \ .$$

- **Strong damping:** $\omega_0^2 < \gamma^2$.

$$G(t, t') = \frac{1}{m\omega'} e^{-\gamma(t-t')} \sinh[\omega'(t - t')] \ , \ \ \omega' = \sqrt{\gamma^2 - \omega_0^2} \ .$$

- **Critical damping:** $\omega_0^2 = \gamma^2$.

$$G(t, t') = \frac{t - t'}{m} e^{-\gamma(t-t')} \ .$$

1.2 Lagrangian Mechanics

In the problems we have encountered so far, our starting point has been the Newtonian equations of motion. Using these, we have deduced the relevant physical information. For this approach to work, it is necessary to know the entire set of forces acting on the system under study. However, in many cases, the determination of these forces can be extremely hard, in particular, if the system's dynamics is restricted by constraints. Consider, for example, a particle moving in the earth's gravitational field along a given trajectory (slide). Here, the restriction of freedom of movement would have to be described by

constraining forces that ensure that the particle moves along the trajectory. Furthermore, in Subsection 1.1.4 we have seen that the validity of Newton's axioms is only given in inertial systems. In accelerated systems the equations of motion tend to be rather more complicated.

In the *Lagrange formulation* of mechanics, no attempt is made at determining the external forces. Instead, their effect is taken into account by reformulating the physical problem in an appropriately chosen geometry (configuration manifold). This results in the *Lagrange equations of motion* formulated in terms of generalized coordinates. Furthermore, the Lagrange equations are chosen in such a way as to be valid in all coordinate systems. This makes them a lot more flexible to use than Newton's equations.

Lagrange's equations can be derived in many different ways. To start, we use the *d'Alembert principle* of virtual displacements. It states that the total sum of work done on the system by the constraining forces vanishes. Additionally, we show how to derive the Lagrange formalism from *Hamilton's principle*, which is an extremal principle of fundamental importance throughout the whole of physics. On our way to this, it will be necessary to take a closer look at some concepts of *variational calculus*.

1.2.1 Constraining Forces, d'Alembert's Principle, and Lagrange Equations

We consider an N-particle system whose motion is constrained such that less than $3N$ degrees of freedom remain. One distinguishes between the following types of constraints:

- *Holonomic constraints:* they can be written as independent equations of the type

$$f_k(\boldsymbol{x}_1,\ldots,\boldsymbol{x}_N,t) = 0 \ , \ k = 1,\ldots,s \ . \tag{1.20}$$

In the case of s holonomic constraints, the $3N$ coordinates of \boldsymbol{x}_i can be reduced to $n = 3N - s$ independent *generalized coordinates* q_j, which implicitly contain the following conditions:

$$\boldsymbol{x}_i = \boldsymbol{x}_i(q_1,\ldots,q_n,t) \ , \ i = 1,\ldots,N \ , \ n = 3N - s \ . \tag{1.21}$$

- *Nonholonomic constraints:* they do not allow a parametrization of type (1.20). If an N-particle system contains s holonomic, as well as r nonholonomic constraints, the q_j are no longer independent. Rather, they are interconnected via nonintegrable constraints in differential form, for example,

$$\sum_j a_{lj}\mathrm{d}q_j + a_{lt}\mathrm{d}t = 0 \ , \ l = 1,\ldots,r \ . \tag{1.22}$$

- *Rheonomic constraints:* they are explicitly time-dependent.

• *Scleronomic constraints:* they are not explicitly time-dependent.

In the following, when considering nonholonomic constraints, we shall restrict ourselves to the special case of (1.22). The following considerations are then valid for holonomic and nonholonomic systems in the above sense.

According to (1.21), the velocity of the ith particle written in terms of the generalized coordinates is

$$\dot{\boldsymbol{x}}_i = \sum_{j=1}^{n} \frac{\partial \boldsymbol{x}_i}{\partial q_j} \dot{q}_j + \frac{\partial \boldsymbol{x}_i}{\partial t} = \dot{\boldsymbol{x}}_i(q_1,\ldots,q_n,\dot{q}_1,\ldots,\dot{q}_n,t) \ , \tag{1.23}$$

where the $\dot{q}_j = \mathrm{d}q_j/\mathrm{d}t$ are called *generalized velocities.*

Virtual displacements and d'Alembert principle. An infinitesimal displacement $\delta \boldsymbol{x}_i$ of the ith particle is called *virtual* if it occurs at fixed time t ($\mathrm{d}t = 0$) and is consistent with the given holonomic and nonholonomic constraints:

$$\delta \boldsymbol{x}_i = \sum_{j=1}^{n} \frac{\partial \boldsymbol{x}_i}{\partial q_j} \delta q_j \ , \quad \sum_{j=1}^{n} a_{lj} \delta q_j = 0 \ , \ l = 1,\ldots,r \ . \tag{1.24}$$

We call such a displacement "virtual", to distinguish it from a real displacement that occurs during a time interval $\mathrm{d}t$, and for which the forces and constraints may change. Let us first consider an N-particle system in equilibrium ($m_i \ddot{\boldsymbol{x}}_i = \boldsymbol{0}$). For the sum of virtual displacements, we then have

$$\sum_{i=1}^{N} \boldsymbol{F}_i \delta \boldsymbol{x}_i = 0 \ , \tag{1.25}$$

where, in the sum, each term vanishes individually. Splitting the forces \boldsymbol{F}_i such that

$$\boldsymbol{F}_i = \boldsymbol{F}_i^{\mathrm{e}} + \boldsymbol{F}_i^{\mathrm{c}} \ ,$$

where $\boldsymbol{F}_i^{\mathrm{e}}$ denote the external forces, and $\boldsymbol{F}_i^{\mathrm{c}}$ the constraining forces that are responsible for upholding the constraints, (1.25) turns into

$$\sum_i \boldsymbol{F}_i^{\mathrm{e}} \delta \boldsymbol{x}_i + \sum_i \boldsymbol{F}_i^{\mathrm{c}} \delta \boldsymbol{x}_i = 0 \ .$$

In many cases, e.g., movement on surfaces, the virtual displacement $\delta \boldsymbol{x}_i$ is perpendicular to the constraining force $\boldsymbol{F}_i^{\mathrm{c}}$ acting on the ith particle, so that no work is performed by any constraining force at all ($\delta \boldsymbol{x}_i \boldsymbol{F}_i^{\mathrm{c}} = 0$). There are, however, examples where the individual displacements do perform work. The *principle of virtual work* states that the sum of all work performed by the constraining forces vanishes.[4]

[4] This is no longer true if frictional forces are present; such systems will not be considered here.

**Theorem 1.12: Principle of virtual work
(equilibrium principle of statics)**

An N-particle system is in equilibrium if the sum of all virtual work performed by the constraining forces vanishes:

$$\sum_{i=1}^{N} \boldsymbol{F}_i^{\text{c}} \delta \boldsymbol{x}_i = 0 \Longrightarrow \sum_{i=1}^{N} \boldsymbol{F}_i^{\text{e}} \delta \boldsymbol{x}_i = 0 .$$

Using a trick introduced by d'Alembert, the above can be transformed into a similar principle that encompasses the general motion of a system; d'Alembert thought of the system with the equations of motion $\boldsymbol{F}_i = \dot{\boldsymbol{p}}_i$ as being a system in equilibrium with forces \boldsymbol{F}_i and compensating opposing forces $-\dot{\boldsymbol{p}}_i$. This reduces the dynamical problem to a static one.

Theorem 1.13: d'Alembert's principle

Due to $(\boldsymbol{F}_i - \dot{\boldsymbol{p}}_i)\delta \boldsymbol{x}_i = (\boldsymbol{F}_i^{\text{e}} + \boldsymbol{F}_i^{\text{c}} - \dot{\boldsymbol{p}}_i)\delta \boldsymbol{x}_i = 0$ and $\sum_{i=1}^{N} \boldsymbol{F}_i^{\text{c}} \delta \boldsymbol{x}_i = 0$, it follows that

$$\sum_{i=1}^{N} (\boldsymbol{F}_i^{\text{e}} - \dot{\boldsymbol{p}}_i) \, \delta \boldsymbol{x}_i = 0 .$$

As one can see, the constraining forces are completely eliminated through this principle. In general, d'Alembert's principle does not follow from Newton's laws. Instead, it must be seen as an additional axiom of classical mechanics. The Lagrangian formalism is based on this principle. Taking into account (1.24), d'Alembert's principle leads to

$$\sum_{j=1}^{n} \delta q_j \sum_{i=1}^{N} m_i \left[\frac{\mathrm{d}}{\mathrm{d}t} \left(\dot{\boldsymbol{x}}_i \frac{\partial \boldsymbol{x}_i}{\partial q_j} \right) - \dot{\boldsymbol{x}}_i \frac{\mathrm{d}}{\mathrm{d}t} \frac{\partial \boldsymbol{x}_i}{\partial q_j} \right] = \sum_{j=1}^{n} Q_j \delta q_j . \qquad (1.26)$$

Herein, Q_j denotes the *generalized force*.

Definition: Generalized force Q_j

$$Q_j = \sum_{i=1}^{N} \boldsymbol{F}_i^{\text{e}} \frac{\partial \boldsymbol{x}_i}{\partial q_j} .$$

Since the q_j do not, in general, have the dimension of length, the Q_j do not generally have the dimension of a force. However, the product $Q_j \delta q_j$ is always work. Using the relation

$$\frac{\partial \boldsymbol{x}_i}{\partial q_j} = \frac{\partial \dot{\boldsymbol{x}}_i}{\partial \dot{q}_j} \, , \quad \frac{\mathrm{d}}{\mathrm{d}t} \left(\dot{\boldsymbol{x}}_i \frac{\partial \boldsymbol{x}_i}{\partial q_j} \right) = \frac{1}{2} \frac{\mathrm{d}}{\mathrm{d}t} \frac{\partial \dot{\boldsymbol{x}}_i^2}{\partial \dot{q}_j} \, , \quad \frac{\mathrm{d}}{\mathrm{d}t} \frac{\partial \boldsymbol{x}_i}{\partial q_j} = \frac{\partial \dot{\boldsymbol{x}}_i}{\partial q_j}$$

following from (1.23), we find from (1.26) that

$$\sum_{j=1}^{n} \delta q_j \left(\frac{\mathrm{d}}{\mathrm{d}t} \frac{\partial T}{\partial \dot{q}_j} - \frac{\partial T}{\partial q_j} - Q_j \right) = 0 \ , \ n = 3N - s \ , \tag{1.27}$$

where

$$T = \sum_{i=1}^{N} \frac{m_i}{2} \dot{x}_i^2 (q_1, \dots, q_n, \dot{q}_1, \dots, \dot{q}_n, t)$$

is the total kinetic energy of the system. Equation (1.27) describes the time dependence of the generalized coordinates q_j. Thanks to d'Alembert's principle, the constraining forces are no longer present; they are hidden in the generalized coordinates.

In the case of purely holonomic constraining forces, all virtual displacements are independent from one another, so that each bracket term of the sum (1.27) can be set to zero. However, we assume that our N-particle system is subject to s holonomic and r nonholonomic constraints of the type (1.22). Of the total of n virtual displacements δq_j, r are then dependent of one another via

$$\sum_{j=1}^{n} a_{lj} \delta q_j = 0 \ , \ l = 1, \dots, r \qquad (\mathrm{d}t = 0) \ ,$$

whereas the remaining $n - r$ displacements are independent. To reduce the number of the displacements δq_j to that of the independent ones, we introduce the *Lagrange multipliers* λ_l , $l = 1, \dots, r$, which can be chosen freely. In general, these are time-dependent functions that may also depend on the generalized coordinates and velocities. Using these multipliers, we rewrite the previous relation as

$$\sum_{j=1}^{n} \delta q_j \sum_{l=1}^{r} \lambda_l a_{lj} = 0 \ . \tag{1.28}$$

The difference of (1.27) and (1.28) yields

$$\sum_{j=1}^{n} \delta q_j \left(\frac{\mathrm{d}}{\mathrm{d}t} \frac{\partial T}{\partial \dot{q}_j} - \frac{\partial T}{\partial q_j} - Q_j - \sum_{l=1}^{r} \lambda_l a_{lj} \right) = 0 \ .$$

For the independent δq_j, the corresponding coefficients (bracket terms) can be set identically zero. Furthermore, we can choose the Lagrange multipliers such that the coefficients of the dependent differentials also vanish. Thus, one obtains

$$\frac{\mathrm{d}}{\mathrm{d}t} \frac{\partial T}{\partial \dot{q}_j} - \frac{\partial T}{\partial q_j} - Q_j - \sum_{l=1}^{r} \lambda_l a_{lj} = 0 \ , \ j = 1, \dots, n \ . \tag{1.29}$$

These are the *Lagrange equations of type I*. If the external forces are all conservative, we find that

$$F_i^{\mathrm{e}} = -\boldsymbol{\nabla}_i V(\boldsymbol{x}_1, \ldots, \boldsymbol{x}_N) \Longrightarrow Q_j = -\frac{\partial V}{\partial q_j} \;,\; \frac{\partial V}{\partial \dot{q}_j} = 0 \;.$$

All in all, we have[5]

Theorem 1.14: Lagrange equations of type I for conservative forces with s holonomic and r nonholonomic constraints

The *Lagrange function* of a conservative N-particle system is given by

$$L = T - V = \sum_{i=1}^{N} \frac{m_i}{2} \dot{\boldsymbol{x}}_i^2 (q_1, \ldots, q_n, \dot{q}_1, \ldots, \dot{q}_n, t) - V(q_1, \ldots, q_n) \;.$$

From this, one obtains the equations of motion (Lagrange equations) in terms of generalized coordinates for the case of s holonomic and r nonholonomic constraints:

$$\frac{\mathrm{d}}{\mathrm{d}t} \frac{\partial L}{\partial \dot{q}_j} - \frac{\partial L}{\partial q_j} - \sum_{l=1}^{r} \lambda_l a_{lj} = 0 \;,\; j = 1, \ldots, 3N - s = n \;, \tag{1.30}$$

$$\sum_{j=1}^{n} a_{lj} \dot{q}_j + a_{lt} = 0 \;,\; l = 1, \ldots, r \;.$$

The Lagrange equations constitute a system of coupled ordinary differential equations of second order for the time dependence of the generalized coordinates. In combination with the r nonholonomic constraints following from (1.22), which are now to be regarded as differential equations, we have a total of $n + r$ equations for the n generalized coordinates and the r Lagrange multipliers. In the case of purely holonomic constraints, Theorem 1.14 reduces to n Lagrange equations for the n generalized coordinates.

Theorem 1.15: Lagrange equations of type II for conservative forces with s holonomic constraints

$$\frac{\mathrm{d}}{\mathrm{d}t} \frac{\partial L}{\partial \dot{q}_j} - \frac{\partial L}{\partial q_j} = 0 \;,\; j = 1, \ldots, 3N - s = n \;.$$

An important corollary, whose proof we postpone to Application 4, is as follows:

Theorem 1.16: Invariance of the Lagrange equations under coordinate transformations

The Lagrange equations are form-invariant under the coordinate transfor- \triangleright

[5] Note that (1.30) is not an equation to determine L (i.e., not a differential equation in L). Rather, it is a functional equation from which we obtain the equations determining the generalized coordinates.

mations $q \to Q(q,t)$. Thus, from the validity of

$$\frac{\mathrm{d}}{\mathrm{d}t}\frac{\partial L}{\partial \dot{q}_j} - \frac{\partial L}{\partial q_j} = 0 \ , \ j = 1,\ldots,3N - s = n \ ,$$

it follows that the Lagrange equations also hold for

$$L'(Q,\dot{Q},t) = L[q(Q,t),\dot{q}(Q,\dot{Q},t),t] \ .$$

Equivalence of Lagrange and Newtonian formalism. In the absence of constraining forces, the Lagrange formalism for an N-particle system must yield Newtons equations of motion. If we choose Cartesian coordinates as the generalized ones, $q = x$, we can see this (in vector notation) as follows:

$$L = T - V = \sum_{i=1}^{N} \frac{m_i}{2}\dot{x}_i^2 - V(x_1,\ldots,x_N,t)$$

$$\Longrightarrow \ \boldsymbol{\nabla}_{x_i} L = -\boldsymbol{\nabla}_{x_i} V = F_i \ , \ \frac{\mathrm{d}}{\mathrm{d}t}\boldsymbol{\nabla}_{\dot{x}_i} L = m_i \ddot{x}_i \ .$$

Therefore, the Lagrange equations are

$$\frac{\mathrm{d}}{\mathrm{d}t}\boldsymbol{\nabla}_{\dot{x}_i} L = \boldsymbol{\nabla}_{x_i} L \Longleftrightarrow m_i \ddot{x}_i = F_i \ , \ i = 1,\ldots,N \ .$$

Interpretation of the Lagrange multipliers. We are free to interpret the constraining forces F_i^{c} as additional external forces F_i^{*} applied in such a way that the motion of the system remains unchanged. The constraints are then eliminated, and we obtain the Lagrange equations

$$\frac{\mathrm{d}}{\mathrm{d}t}\frac{\partial L}{\partial \dot{q}_j} - \frac{\partial L}{\partial q_j} = Q_j^{*} \ .$$

Comparison with Theorem 1.14 yields

$$Q_j^{*} = \sum_{l=1}^{r} \lambda_l a_{lj} \ ,$$

i.e., the Lagrange multipliers determine the generalized constraining forces.

Generalized potential. As one can see from (1.29), Theorems 1.14 and 1.15 are still valid if the generalized forces Q_j can be expressed in terms of a *generalized potential* $V(q_1,\ldots,q_n,\dot{q}_1,\ldots,\dot{q}_n,t)$, such that

$$Q_j = -\frac{\partial V}{\partial q_j} + \frac{\mathrm{d}}{\mathrm{d}t}\frac{\partial V}{\partial \dot{q}_j} \ . \tag{1.31}$$

An example of this case is given by the velocity-dependent *Lorentz force*

$$Q_{\mathrm{L}}(x,\dot{x},t) = F_{\mathrm{L}}^{\mathrm{e}}(x,\dot{x},t) = q\left(E(x,t) + \frac{\dot{x}}{c} \times B(x,t)\right) \ ,$$

which describes the motion of a particle of charge q in an electromagnetic field \boldsymbol{E}, \boldsymbol{B} (c=velocity of light). It is obtained from (1.31), if

$$V(\boldsymbol{x}, \dot{\boldsymbol{x}}, t) = q\phi(\boldsymbol{x}, t) - \frac{q}{c}\boldsymbol{A}(\boldsymbol{x}, t)\dot{\boldsymbol{x}} \ ,$$

where the *scalar* and *vector potentials* ϕ and \boldsymbol{A} are coupled with the fields \boldsymbol{E} and \boldsymbol{B} via (see Subsection 2.2.1)

$$\boldsymbol{B} = \boldsymbol{\nabla} \times \boldsymbol{A} \ , \ \boldsymbol{E} = -\frac{1}{c}\frac{\partial \boldsymbol{A}}{\partial t} - \boldsymbol{\nabla}\phi \ .$$

1.2.2 Conservation Laws

In this subsection we consider transformations of the generalized coordinates, velocities, and of time which leave the Lagrange function invariant, or that yield an equivalent Lagrange function. Such transformations are called *symmetry transformations*. Following a theorem of E. Noether (see Subsection 2.8.3), each symmetry property of a system corresponds to a conservation law. We show this here with the help of the conservation laws of a closed system of N point particles, which we know from Theorem 1.11. The corresponding Lagrange function reads

$$L(\boldsymbol{x}_i, \dot{\boldsymbol{x}}_i, t) = \sum_{i=1}^{N} \frac{1}{2}m_i\dot{\boldsymbol{x}}_i^2 - V \ , \ V = \sum_{i,j} V(|\boldsymbol{x}_i - \boldsymbol{x}_j|) \ .$$

Conservation of energy, homogeneity of time. If the time shift

$$t \to t' = t + \delta t \ , \ \delta \boldsymbol{x}_i = \delta \dot{\boldsymbol{x}}_i = 0$$

is to be a symmetry transformation of the Lagrange function, the following must hold:

$$\delta L = L(\boldsymbol{x}, \dot{\boldsymbol{x}}, t + \delta t) - L(\boldsymbol{x}, \dot{\boldsymbol{x}}, t) = \frac{\partial L}{\partial t}\delta t = 0 \Longrightarrow \frac{\partial L}{\partial t} = 0 \ .$$

On the other hand, we generally have for the Lagrange function (in vector notation)

$$\frac{\mathrm{d}L}{\mathrm{d}t} = \frac{\partial L}{\partial t} + \sum_i \dot{\boldsymbol{x}}_i \boldsymbol{\nabla}_{\boldsymbol{x}_i} L + \sum_i \ddot{\boldsymbol{x}}_i \boldsymbol{\nabla}_{\dot{\boldsymbol{x}}_i} L \ .$$

Taking into account the Lagrange equations

$$\boldsymbol{\nabla}_{\boldsymbol{x}_i} L - \frac{\mathrm{d}}{\mathrm{d}t}\boldsymbol{\nabla}_{\dot{\boldsymbol{x}}_i} L = 0 \ ,$$

it follows that

$$\frac{\mathrm{d}}{\mathrm{d}t}\left(L - \sum_i \dot{\boldsymbol{x}}_i \boldsymbol{\nabla}_{\dot{\boldsymbol{x}}_i} L\right) = \frac{\partial L}{\partial t} \ .$$

In the case at hand, we have $\partial L/\partial t = 0$, so that

$$\sum_i \dot{\boldsymbol{x}}_i \boldsymbol{\nabla}_{\dot{\boldsymbol{x}}_i} L - L = \text{const} .$$

Because

$$\sum_i \dot{\boldsymbol{x}}_i \boldsymbol{\nabla}_{\dot{\boldsymbol{x}}_i} L = \sum_i \dot{\boldsymbol{x}}_i \boldsymbol{\nabla}_{\dot{\boldsymbol{x}}_i} T = \sum_i m_i \dot{\boldsymbol{x}}_i^2 = 2T ,$$

we finally find the conservation of the total energy:

$$T + V = E = \text{const} .$$

Conservation of total momentum, homogeneity of space. Consider next the transformation

$$\boldsymbol{x}_i \longrightarrow \boldsymbol{x}_i' = \boldsymbol{x}_i + \delta \boldsymbol{x} , \quad \delta t = 0 .$$

In the same way as above, one obtains

$$\delta L = L(\boldsymbol{x}_i + \delta \boldsymbol{x}_i, \dot{\boldsymbol{x}}_i, t) - L(\boldsymbol{x}_i, \dot{\boldsymbol{x}}_i, t) = \delta \boldsymbol{x} \sum_i \boldsymbol{\nabla}_{\boldsymbol{x}_i} L = 0 .$$

Since $\delta \boldsymbol{x}$ is an arbitrary displacement, it follows that

$$\sum_i \boldsymbol{\nabla}_{\boldsymbol{x}_i} L = 0 \Longrightarrow \frac{\mathrm{d}}{\mathrm{d}t} \sum_i \boldsymbol{\nabla}_{\dot{\boldsymbol{x}}_i} L = 0 \Longrightarrow \boldsymbol{p} = \sum_i \boldsymbol{p}_i = \text{const} .$$

Invariance of the Lagrange function under displacement of all particles by the same amount $\delta \boldsymbol{x}$ is thus tantamount to conservation of the system's total momentum.

Conservation of total angular momentum, isotropy of space. For an infinitesimal rotation of the type

$$\delta \boldsymbol{x}_i = \delta \boldsymbol{\phi} \times \boldsymbol{x}_i , \quad \delta \dot{\boldsymbol{x}}_i = \delta \boldsymbol{\phi} \times \dot{\boldsymbol{x}}_i , \quad \delta t = 0$$

to be a symmetry transformation, the following must hold:

$$\delta L = \sum_i \delta \boldsymbol{x}_i \boldsymbol{\nabla}_{\boldsymbol{x}_i} L + \sum_i \delta \dot{\boldsymbol{x}}_i \boldsymbol{\nabla}_{\dot{\boldsymbol{x}}_i} L = 0 .$$

This gives the conservation of the total angular momentum:

$$\delta \boldsymbol{\phi} \sum_i (\boldsymbol{x}_i \times \dot{\boldsymbol{p}}_i + \dot{\boldsymbol{x}}_i \times \boldsymbol{p}_i) = \delta \boldsymbol{\phi} \frac{\mathrm{d}}{\mathrm{d}t} \sum_i (\boldsymbol{x}_i \times \boldsymbol{p}_i) = \boldsymbol{0}$$

$$\Longrightarrow \boldsymbol{l} = \sum_i \boldsymbol{l}_i = \text{const} .$$

Center of mass law, invariance under uniform motion. In the case of the transformation

$$\boldsymbol{x}_i \longrightarrow \boldsymbol{x}_i' = \boldsymbol{x}_i + \boldsymbol{v} t , \quad \boldsymbol{v} = \text{const} , \quad \delta t = 0 ,$$

the transformed Lagrange function is given by

$$L' = L + \boldsymbol{v} \sum_i m_i \dot{\boldsymbol{x}}_i + \frac{\boldsymbol{v}^2}{2} \sum_i m_i$$

$$\implies \delta L = \frac{\mathrm{d}}{\mathrm{d}t} \left(M \boldsymbol{x}_\mathrm{C} \boldsymbol{v} \right) + \frac{1}{2} M \boldsymbol{v}^2 . \tag{1.32}$$

Two Lagrange functions differing only by a constant or a time derivative are equivalent and lead to the same equations of motion. On the other hand, we can also write for δL

$$\delta L = \sum_i \left(\delta \boldsymbol{x}_i \boldsymbol{\nabla}_{\boldsymbol{x}_i} L + \delta \dot{\boldsymbol{x}}_i \boldsymbol{\nabla}_{\dot{\boldsymbol{x}}_i} L \right) = \sum_i \left(\boldsymbol{v} t \dot{\boldsymbol{p}}_i + \boldsymbol{v} \boldsymbol{p}_i \right)$$

$$= \frac{\mathrm{d}}{\mathrm{d}t} \left(\boldsymbol{v} t \sum_i \boldsymbol{p}_i \right) . \tag{1.33}$$

Comparing (1.32) and (1.33), we obtain the center of mass law

$$M \boldsymbol{x}_\mathrm{C} - t \sum_i \boldsymbol{p}_i = \mathrm{const} .$$

Overall, we see that the symmetry transformations we have considered here are equivalent to the 10 conserved quantities of Theorem 1.11.

Theorem 1.17: Symmetries and conservation laws

Invariance of the Lagrange function under a symmetry transformation is equivalent to a conserved quantity:

- Invariance under time shifts: conservation of total energy.

- Invariance under spatial translations: conservation of total momentum.

- Invariance under spatial rotations: conservation of total angular momentum.

- Invariance under uniform motion: center of mass law.

1.2.3 Hamilton Principle and Action Functional

With Newtonian mechanics and the Lagrange formalism based on d'Alembert's principle, we have already encountered two approaches to particle mechanics. In this subsection we formulate another principle, the *Hamilton principle*, which is an alternative axiomatic formulation of classical mechanics. Furthermore, we show that the Lagrange equations of type II also follow from this principle.

Action functional. The action functional is defined as the integral over all possible trajectories a given system of particles may take. Hamilton's

principle is a statement about the actual trajectory along which the motion of the system really takes place:

Theorem 1.18: Hamilton's principle for conservative holonomic systems

Consider the Lagrange function $L(q_1, \ldots, q_n, \dot{q}_1, \ldots, \dot{q}_n, t) = T - V$ of a conservative system with holonomic constraints. Its solutions are those for which the corresponding action functional, i.e., the time integral of the Lagrange function, along the actual trajectory is extremal:

$$S[q] = \int_{t_1}^{t_2} dt L[q(t), \dot{q}(t), t] \longrightarrow \text{extremal} .$$

This is equivalent to the condition of vanishing variation δS along the actual trajectory.

Obviously, Hamilton's principle describes an optimization problem. However, we are not faced with the determination of the extrema of a function as in ordinary differential calculus, but, rather, we need to determine functions for which a given functional becomes extremal. This is the subject of *variational calculus*, which we consider briefly before continuing our discussion of Hamiltonian mechanics.

Variational problem with fixed endpoints. An example of variational problems is given by the *brachystochrone*, where the form of a curve between two fixed endpoints is sought, on which a mass particle moving under the influence of gravity alone (no friction) is to reach the endpoint as fast as possible (see Application 5). In problems of this kind, it is often the case that the structure of the functional subject to optimization is given by

$$S[y_1, \ldots, y_n] = \int_{x_1}^{x_2} dx F(y_1, \ldots, y_n, y_1', \ldots, y_n', x) , \qquad (1.34)$$

where F depends only on the functions $y_j(x)$, their first derivatives $y_j'(x)$, and the independent variable x. We consider only the case where the endpoints of the functions y_j remain fixed,

$$[x_1, \boldsymbol{y}(x_1)] , \ [x_2, \boldsymbol{y}(x_2)] \text{ fixed} , \ \boldsymbol{y} = (y_1, \ldots, y_n) ,$$

and seek a Taylor expansion of (1.34). To this end, we vary the functions y_j with the help of auxiliary functions h_j that vanish at the endpoints:

$$\boldsymbol{y}(x) \longrightarrow \boldsymbol{\gamma}(x) = \boldsymbol{y}(x) + \delta \boldsymbol{y}(x) , \ \delta \boldsymbol{y}(x) = \epsilon \boldsymbol{h}(x) , \ \boldsymbol{h}(x_1) = \boldsymbol{h}(x_2) = \boldsymbol{0} .$$

For arbitrarily small $|\epsilon|$, all varied functions $\boldsymbol{\gamma}$ lie in an arbitrarily small neighborhood of \boldsymbol{y}, and we can write

$$S[\boldsymbol{\gamma}] = S[\boldsymbol{y}] + \delta S + \ldots ,$$

with

$$\delta S = \left. \frac{\mathrm{d}S[\boldsymbol{\gamma}]}{\mathrm{d}\epsilon} \right|_{\epsilon=0} \epsilon$$

$$= \epsilon \int_{x_1}^{x_2} \mathrm{d}x \, [\boldsymbol{h}\boldsymbol{\nabla}_{\boldsymbol{\gamma}}F(\boldsymbol{\gamma},\boldsymbol{\gamma}',x) + \boldsymbol{h}'\boldsymbol{\nabla}_{\boldsymbol{\gamma}'}F(\boldsymbol{\gamma},\boldsymbol{\gamma}',x)]_{\epsilon=0}$$

$$= \epsilon \int_{x_1}^{x_2} \mathrm{d}x \, \boldsymbol{h} \left[\boldsymbol{\nabla}_{\boldsymbol{\gamma}}F(\boldsymbol{\gamma},\boldsymbol{\gamma}',x) - \frac{\mathrm{d}}{\mathrm{d}x}\boldsymbol{\nabla}_{\boldsymbol{\gamma}'}F(\boldsymbol{\gamma},\boldsymbol{\gamma}',x) \right]_{\epsilon=0}$$

$$+ \epsilon \left[\boldsymbol{h}\boldsymbol{\nabla}_{\boldsymbol{\gamma}'}F(\boldsymbol{\gamma},\boldsymbol{\gamma}',x)|_{\epsilon=0} \right]_{x_1}^{x_2}$$

$$= \epsilon \int_{x_1}^{x_2} \mathrm{d}x \, \boldsymbol{h} \left[\boldsymbol{\nabla}_{\boldsymbol{y}}F(\boldsymbol{y},\boldsymbol{y}',x) - \frac{\mathrm{d}}{\mathrm{d}x}\boldsymbol{\nabla}_{\boldsymbol{y}'}F(\boldsymbol{y},\boldsymbol{y}',x) \right]$$

$$+ \epsilon \left[\boldsymbol{h}\boldsymbol{\nabla}_{\boldsymbol{y}'}F(\boldsymbol{y},\boldsymbol{y}',x) \right]_{x_1}^{x_2} .$$

Since the auxiliary functions are zero at the endpoints, the last term vanishes. Now suppose \boldsymbol{y} is an extremum of the action functional. Then, $\delta S = 0$, so that the second last term also vanishes. Therefore, we find

Theorem 1.19: Variational formula and Euler-Lagrange equations for fixed endpoints

To a linear approximation, the variation of the action functional with fixed endpoints is given by

$$S[\boldsymbol{y} + \delta \boldsymbol{y}] = S[\boldsymbol{y}] + \delta S + \dots ,$$

with

$$\delta S = \int_{x_1}^{x_2} \mathrm{d}x \delta \boldsymbol{y} \left[\boldsymbol{\nabla}_{\boldsymbol{y}}F(\boldsymbol{y},\boldsymbol{y}',x) - \frac{\mathrm{d}}{\mathrm{d}x}\boldsymbol{\nabla}_{\boldsymbol{y}'}F(\boldsymbol{y},\boldsymbol{y}',x) \right] .$$

From this, the *Euler-Lagrange equations (ELE)* follow as a necessary condition for the action functional to be extremal:

$$\frac{\partial F(\boldsymbol{y},\boldsymbol{y}',x)}{\partial y_j} - \frac{\mathrm{d}}{\mathrm{d}x}\frac{\partial F(\boldsymbol{y},\boldsymbol{y}',x)}{\partial y_j'} = 0 , \; j = 1,\dots,n .$$

Note that this is a necessary but not a sufficient criterion [analogous to the criterion $f'(x) = 0$ in ordinary differential calculus].

In the case of purely holonomic constraints, the ELE are identical to the Lagrange equations of Theorem 1.15 if F is replaced by the Lagrange function L and the functions $y_j(x)$ are interpreted as the generalized coordinates $q_j(t)$. Thus, we have

Theorem 1.20: Hamilton's principle and Lagrange equations of type II

From Hamilton's principle it follows that the actual trajectory of a conservative N-particle system with purely holonomic constraints obeys the Lagrange equations of type II:

$$S = \int_{t_1}^{t_2} dt L \longrightarrow \text{extremal} \implies \frac{d}{dt}\frac{\partial L}{\partial \dot{q}_j} - \frac{\partial L}{\partial q_j} = 0 .$$

Recall that the Lagrange equations are invariant under coordinate transformations. Accordingly, Hamilton's principle is also independent of the coordinate system in which L is expressed. Furthermore, note that Hamilton's formalism does not yield new physics compared to the Lagrange formalism. However, the Hamiltonian formulation of mechanics is of great importance since it represents a generally valid principle of physics, applied not just in mechanics, but particularly also in the formulation of modern field theories.

Equivalence of d'Alembert's and Hamilton's principles. If we consider a holonomic system and its motion in a finite time interval $[t_1 : t_2]$, d'Alembert's principle can be rewritten as

$$\sum_i (\boldsymbol{F}_i^e - m_i \ddot{\boldsymbol{x}}_i)\delta\boldsymbol{x}_i = 0$$

$$\implies \frac{d}{dt}\sum_i m_i \dot{\boldsymbol{x}}_i \delta\boldsymbol{x}_i = \sum_i \boldsymbol{F}_i^e \delta\boldsymbol{x}_i + \delta\sum_i \frac{m_i}{2}\dot{\boldsymbol{x}}_i^2 = \delta W + \delta T ,$$

where δT is the virtual change in kinetic energy. Integration of this relation over $[t_1 : t_2]$ and using $\delta\boldsymbol{x}_i(t_2) = \delta\boldsymbol{x}_i(t_1) = \boldsymbol{0}$, we find

$$\int_{t_1}^{t_2}(\delta W + \delta T)dt = \sum_i m_i\left[\dot{\boldsymbol{x}}_i(t_2)\delta\boldsymbol{x}_i(t_2) - \boldsymbol{x}_i(t_1)\delta\boldsymbol{x}_i(t_1)\right] = 0 .$$

Restricting ourselves to the special case of conservative forces, the virtual work done by the externally applied forces can be written as

$$\delta W = \sum_i \boldsymbol{F}_i^e \delta\boldsymbol{x}_i = -\sum_i \delta\boldsymbol{x}_i \boldsymbol{\nabla}_i V(\boldsymbol{x}_1,\ldots,\boldsymbol{x}_N) = -\delta V ,$$

so that

$$\int_{t_1}^{t_2}(\delta W + \delta T)dt = \int_{t_1}^{t_2}\delta(T - V)dt = \int_{t_1}^{t_2}\delta L dt = 0 .$$

Since the endpoints are held fixed, we can pull out the variation in front of the integral:

$$\delta \int_{t_1}^{t_2} L\mathrm{d}t = \delta S = 0 \ .$$

This shows the equivalence of the d'Alembert and the Hamilton principles for conservative holonomic systems.

Nonconservative systems with holonomic constraints. Using the generalized forces, we can also formulate Hamilton's principle for the case of nonconservative systems. The requirement of stationarity

$$\delta S = \int_{t_1}^{t_2} \mathrm{d}t\delta(T + W) = 0 \ ,$$

with

$$\delta W = \sum_{i=1}^{N} \boldsymbol{F}_i^{\mathrm{e}}\delta\boldsymbol{x}_i = \sum_{j=1}^{n} Q_j\delta q_j$$

and

$$\delta T = \sum_{j=1}^{n} \delta q_j \left(\frac{\partial T}{\partial q_j} - \frac{\mathrm{d}}{\mathrm{d}t}\frac{\partial T}{\partial \dot{q}_j} \right) \ ,$$

then immediately yields (1.27).

Equivalent Lagrange functions, gauge transformations. It is easily seen that with any Lagrange function L, the class of Lagrange functions

$$L' = \alpha L + \frac{\mathrm{d}}{\mathrm{d}t}F(\boldsymbol{q}, t)$$

leads to the same stationary trajectory. This is due to the fact that adding the total temporal differential of a scalar function $F(\boldsymbol{q}, t)$ simply means the addition of a constant term for the new action S':

$$S' = \int_{t_1}^{t_2} \mathrm{d}t \left(\alpha L + \frac{\mathrm{d}}{\mathrm{d}t}F(\boldsymbol{q}, t) \right) = \alpha S + F[\boldsymbol{q}(t_2), t_2] - F[\boldsymbol{q}(t_1), t_1]$$

$$= \alpha S + \mathrm{const} \ .$$

Transformations of this kind, which transform the Lagrange function such that physical results remain unchanged, are called *gauge transformations*. We shall encounter these transformations again in electrodynamics and in quantum mechanics, where they play a fundamental role (just as in theoretical high-energy physics, which are generally formulated as *gauge theories*).

Summary

- **d'Alembert's principle** states that the total sum of work done by the constraining forces vanishes. This postulate can be used to derive the **Lagrange formalism**.

- The **Lagrange equations** are a coupled system of ordinary first-order differential equations. Alternatively to Newton's equation of motion, they describe the dynamics of a mechanical system where the **constraining forces** no longer appear explicitly, but are absorbed in appropriately chosen coordinates. Here, the **generalized coordinates** and **velocities** act as pairs of variables.

- Lagrange's equations are valid in all coordinate systems (form invariance).

- The conservation laws of mechanics follow naturally from the **symmetries** of a mechanical system.

- Lagrange's equations follow from **Hamilton's principle** of an extremal action functional.

Applications

4. Invariance of the Lagrange equations under arbitrary coordinate transformations. Show that Lagrange's equations are invariant under the coordinate transformations

$$q_i \longrightarrow Q_i = Q_i(\boldsymbol{q}, t) .$$

Solution. We have

$$\frac{\mathrm{d}}{\mathrm{d}t} \frac{\partial L}{\partial \dot{q}_i} - \frac{\partial L}{\partial q_i} = 0 , \quad q_i = q_i(\boldsymbol{Q}, t) , \quad \dot{q}_i = \dot{q}_i(\boldsymbol{Q}, \dot{\boldsymbol{Q}}, t) .$$

Because

$$\dot{q}_i = \sum_j \frac{\partial q_i(\boldsymbol{Q}, t)}{\partial Q_j} \dot{Q}_j + \frac{\partial q_i}{\partial t} ,$$

it follows that

$$\frac{\partial q_i}{\partial Q_j} = \frac{\partial \dot{q}_i}{\partial \dot{Q}_j} .$$

Starting from the transformed Lagrange function $L'(\boldsymbol{Q}, \dot{\boldsymbol{Q}}, t)$, we now calculate the Lagrange equations in the new coordinates. To this end, we need

$$\frac{\partial L'}{\partial Q_i} = \sum_j \left(\frac{\partial L}{\partial q_j} \frac{\partial q_j}{\partial Q_i} + \frac{\partial L}{\partial \dot{q}_j} \frac{\partial \dot{q}_j}{\partial Q_i} \right)$$

$$\frac{\partial L'}{\partial \dot{Q}_i} = \sum_j \frac{\partial L}{\partial \dot{q}_j} \frac{\partial \dot{q}_j}{\partial \dot{Q}_i} = \sum_j \frac{\partial L}{\partial \dot{q}_j} \frac{\partial q_j}{\partial Q_i}$$

$$\frac{\mathrm{d}}{\mathrm{d}t} \frac{\partial L'}{\partial \dot{Q}_i} = \sum_j \frac{\mathrm{d}}{\mathrm{d}t}\left(\frac{\partial L}{\partial \dot{q}_j} \right) \frac{\partial q_j}{\partial Q_i} + \sum_j \frac{\partial L}{\partial \dot{q}_j} \frac{\mathrm{d}}{\mathrm{d}t}\left(\frac{\partial q_j}{\partial Q_i} \right)$$

$$\Longrightarrow \frac{\mathrm{d}}{\mathrm{d}t} \frac{\partial L'}{\partial \dot{Q}_i} - \frac{\partial L'}{\partial Q_i} = \sum_j \left(\frac{\mathrm{d}}{\mathrm{d}t} \frac{\partial L}{\partial \dot{q}_j} - \frac{\partial L}{\partial q_j} \right) \frac{\partial q_j}{\partial Q_i}$$

$$+ \sum_j \left(\frac{\mathrm{d}}{\mathrm{d}t} \frac{\partial q_j}{\partial Q_i} - \frac{\partial \dot{q}_j}{\partial Q_i} \right) \frac{\partial L}{\partial \dot{q}_j} .$$

The first sum vanishes since the Lagrange equations hold in the old coordinate system. The second sum is also zero since

$$\frac{\partial \dot{q}_j}{\partial Q_i} = \sum_l \frac{\partial^2 q_j}{\partial Q_i \partial Q_l} \dot{Q}_l + \frac{\partial^2 q_j}{\partial Q_i \partial t} = \frac{\mathrm{d}}{\mathrm{d}t} \frac{\partial q_j}{\partial Q_i} .$$

5. Brachystochrone. This is the standard example of variational calculus. Determine the curve on which a particle moves in the gravitational field between fixed endpoints, such that it reaches the final point as fast as possible. Assume frictionless motion.

Solution. We choose a coordinate system such that the startpoint and endpoint coincide with the coordinates $(0, h)$ and $(a, 0)$. The particle's position vector is

$$\boldsymbol{x}(t) = \begin{pmatrix} x(t) \\ y[x(t)] \end{pmatrix} ,$$

where $y(x)$ describes the trajectory we seek. To find the functional S subject to minimization, we use energy conservation:

$$T = \frac{m}{2} \dot{\boldsymbol{x}}^2 = \frac{m}{2} \dot{x}^2 \left[1 + y'^2(x) \right] , \quad V = mgy(x)$$

$$\Longrightarrow E = T + V = \frac{m}{2} \dot{x}^2 \left[1 + y'^2(x) \right] + mgy(x) = mgh = \text{const}$$

$$\Longrightarrow \frac{\mathrm{d}x}{\mathrm{d}t} = \sqrt{\frac{2g(h - y)}{1 + y'^2}}$$

$$\Longrightarrow \tau = \int_0^{\tau} \mathrm{d}t = S[y] = \int_0^a \mathrm{d}x F(y, y', x) , \quad F(y, y', x) = \frac{1}{\sqrt{2g}} \sqrt{\frac{1 + y'^2}{h - y}} .$$

To find the ELE, we need the following derivatives:

$$\frac{\partial F}{\partial y} = \frac{1}{2\sqrt{2g}} \frac{\sqrt{1+y'^2}}{(h-y)^{3/2}}$$

$$\frac{\partial F}{\partial y'} = \frac{1}{\sqrt{2g}} \frac{y'}{\sqrt{(1+y'^2)(h-y)}}$$

$$\frac{d}{dx}\frac{\partial F}{\partial y'} = \frac{1}{\sqrt{2g}} \frac{y''(h-y) + \frac{y'^2}{2}(1+y'^2)}{[(1+y'^2)(h-y)]^{3/2}} \; .$$

It follows that

$$y''(h-y) = \frac{1}{2}(1+y'^2) \; .$$

This equation no longer depends on x and can be solved, for example, using the substitution $y' = p(y)$, $y'' = (dp/dy)p$. However, we proceed differently and exploit the following identity:

$$\left(\frac{\partial F}{\partial y} - \frac{d}{dx}\frac{\partial F}{\partial y'} \right) = \frac{d}{dx}\left(F - y'\frac{\partial F}{\partial y'} \right) \; ,$$

which holds since F does not depend on x. Along solutions of the ELE, this yields

$$H = y'\frac{\partial F}{\partial y'} - F = \text{const} \; .$$

In the next section we will encounter this relationship again as *conservation of the Hamilton function*. Thus, we have

$$-\frac{1}{\sqrt{2g}} \frac{1}{\sqrt{(1+y'^2)(h-y)}} = c = \text{const}$$

$$\implies \frac{dy}{dx} = -\sqrt{\frac{e-(h-y)}{h-y}} \; , \quad e = \frac{1}{2gc^2}$$

$$\implies x = -\int_h^{y(x)} dy \sqrt{\frac{h-y}{e-(h-y)}} \; .$$

Using the substitution

$$y = h - e\sin^2\psi \implies dy = -2e\sin\psi\cos\psi\, d\psi \; ,$$

it follows that

$$x = 2e\int_0^\psi d\psi \sin^2\psi = e\left(\psi - \frac{1}{2}\sin 2\psi\right) \; .$$

Overall, we obtain a parametrization $[x(\psi), y(\psi)]$ of *cycloids* with one free parameter e, which has to be fixed via the constraint $y(x = a) = 0$. Figure 1.6 shows three possible types of solutions depending on the ratio a/h.

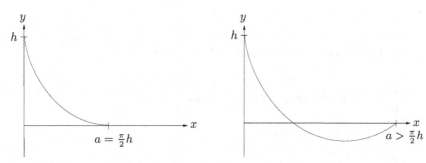

Fig. 1.6. Different types of solutions to the Brachystochrone problem.

6. Mathematical double-pendulum. A planar mathematical double-pendulum with lengths $l_1 = l_2 = l$ and masses $m_1 = m_2 = m$ is moving frictionless in the gravitational field (Fig. 1.7). Determine the oscillation frequencies for small oscillations away from the vertical.

Solution. The position vectors of the masses expressed by the generalized coordinates φ and θ are given by

$$x_1(t) = l \begin{pmatrix} \cos\varphi(t) \\ \sin\varphi(t) \end{pmatrix} \ , \ x_2(t) = l \begin{pmatrix} \cos\varphi(t) + \cos\theta(t) \\ \sin\varphi(t) + \sin\theta(t) \end{pmatrix} \ ,$$

where the purely holonomic constraints

$$x_1^2 - l^2 = 0 \ , \ (x_2 - x_1)^2 - l^2 = 0$$

are already incorporated. For the kinetic and potential energies, we find

$$T = T_1 + T_2 = \frac{m}{2}l^2 \left[2\dot\varphi^2 + \dot\theta^2 + 2\dot\varphi\dot\theta(\cos\varphi\cos\theta + \sin\varphi\sin\theta) \right]$$

$$V = V_1 + V_2 = -mgl(2\cos\varphi + \cos\theta) \ .$$

Accordingly, the Lagrange function reads

$$L = T - V = \frac{m}{2}l^2 \left[2\dot\varphi^2 + \dot\theta^2 + 2\dot\varphi\dot\theta\cos(\varphi - \theta) \right] + mgl(2\cos\varphi + \cos\theta) \ .$$

The Lagrange equations are obtained from the following derivatives:

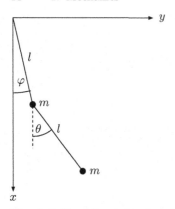

Fig. 1.7. Two-dimensional mathematical double-pendulum.

$$\frac{\partial L}{\partial \varphi} = -ml^2\dot{\varphi}\dot{\theta}\sin(\varphi - \theta) - 2mgl\sin\varphi$$

$$\frac{\partial L}{\partial \dot{\varphi}} = 2ml^2\dot{\varphi} + ml^2\dot{\theta}\cos(\varphi - \theta)$$

$$\frac{\mathrm{d}}{\mathrm{d}t}\frac{\partial L}{\partial \dot{\varphi}} = 2ml^2\ddot{\varphi} + ml^2\ddot{\theta}\cos(\varphi - \theta) - ml^2\dot{\theta}(\dot{\varphi} - \dot{\theta})\sin(\varphi - \theta)$$

$$\frac{\partial L}{\partial \theta} = ml^2\dot{\varphi}\dot{\theta}\sin(\varphi - \theta) - mgl\sin\theta$$

$$\frac{\partial L}{\partial \dot{\theta}} = ml^2\dot{\theta} + ml^2\dot{\varphi}\cos(\varphi - \theta)$$

$$\frac{\mathrm{d}}{\mathrm{d}t}\frac{\partial L}{\partial \dot{\theta}} = ml^2\ddot{\theta} + ml^2\ddot{\varphi}\cos(\varphi - \theta) - ml^2\dot{\varphi}(\dot{\varphi} - \dot{\theta})\sin(\varphi - \theta)$$

$$\Longrightarrow \begin{cases} 2l\ddot{\varphi} + l\ddot{\theta}\cos(\varphi - \theta) + l\dot{\theta}^2\sin(\varphi - \theta) = -2g\sin\varphi \\ l\ddot{\theta} + l\ddot{\varphi}\cos(\varphi - \theta) - l\dot{\varphi}^2\sin(\varphi - \theta) = -g\sin\theta \ . \end{cases}$$

Since we only wish to consider small oscillations such that $\sin\varphi \approx \varphi$, $\cos(\varphi - \theta) \approx 1$, $\dot{\varphi}^2, \dot{\theta}^2 \ll 1$, it follows that

$$2\ddot{\varphi} + \ddot{\theta} = -2\frac{g}{l}\varphi$$

$$\ddot{\theta} + \ddot{\varphi} = -\frac{g}{l}\theta \ .$$

The ansatz

$$\varphi(t) = \alpha \mathrm{e}^{\mathrm{i}\omega t} \ , \quad \theta(t) = \beta \mathrm{e}^{\mathrm{i}\omega t}$$

leads to

$$\left(2\frac{g}{l} - 2\omega^2\right)\alpha - \omega^2\beta = 0$$

$$-\omega^2\alpha + \left(\frac{g}{l} - \omega^2\right)\beta = 0 \ .$$

For this system of linear equations in α and β to have nontrivial solutions, the coefficient determinant must vanish:

$$\begin{vmatrix} 2\frac{g}{l} - 2\omega^2 & -\omega^2 \\ -\omega^2 & \frac{g}{l} - \omega^2 \end{vmatrix} = 0 \Longrightarrow \omega_{1,2}^2 = \frac{g}{l}(2 \pm \sqrt{2}) \ .$$

Inserting this into the above system of equations, one obtains

1. $\omega^2 = \frac{g}{l}(2 + \sqrt{2}) \Longrightarrow \beta = -\sqrt{2}\alpha$

2. $\omega^2 = \frac{g}{l}(2 - \sqrt{2}) \Longrightarrow \beta = \sqrt{2}\alpha \ .$

Thus, in the first case, $\omega = \sqrt{\frac{g}{l}(2 + \sqrt{2})}$, and the pendula oscillate in opposite directions. In the second case, both pendula oscillate with frequency $\omega = \sqrt{\frac{g}{l}(2 - \sqrt{2})}$ in the same direction.

7. Small oscillations and normal modes. In this application, we demonstrate the general mathematical framework corresponding to Application 6 for coupled oscillations given by a Lagrange function of the form

$$L = L(\boldsymbol{q}, \dot{\boldsymbol{q}}) = T(\boldsymbol{q}, \dot{\boldsymbol{q}}) - V(\boldsymbol{q}) \ , \tag{1.35}$$

where the potential depends only on the coordinates. To start, note that the kinetic energy can generally be written as

$$T = c(\boldsymbol{q}, t) + \sum_i b_i(\boldsymbol{q}, t)\dot{q}_i + \frac{1}{2} \sum_{i,j} a_{ij}(\boldsymbol{q}, t)\dot{q}_i\dot{q}_j \ , \quad a_{ij} = a_{ji} \ ,$$

which becomes apparent by squaring (1.23). If the relations between q_i and x_j are time-independent, the kinetic energy is a homogeneous quadratic form:

$$T = \frac{1}{2} \sum_{i,j} a_{ij}(\boldsymbol{q})\dot{q}_i\dot{q}_j \ . \tag{1.36}$$

Using this expression, find a condition for equilibrium, derive the Lagrange equations for small oscillations and determine the corresponding *eigenmodes*.

Solution. From (1.35) and (1.36), one obtains the Lagrange equations

$$\sum_{i,j} \frac{\partial a_{ki}}{\partial q_j} \dot{q}_i\dot{q}_j + \sum_i a_{ki}\ddot{q}_i - \frac{1}{2} \sum_{i,j} \frac{\partial a_{ij}}{\partial q_k}\dot{q}_i\dot{q}_j + \frac{\partial V}{\partial q_k} = 0 \ .$$

For a system in equilibrium ($\dot{q}_i = \ddot{q}_i = 0$), we have the equilibrium conditions

$$\frac{\partial V}{\partial q_i}\bigg|_{\boldsymbol{q}_0} = 0$$

stating that the potential energy must be stationary in q_0.[6] For simplicity, we assume that $q_0 = 0$ and expand the potential as well as the kinetic energy in the following form:

$$V(q) \approx V(0) + \sum_i \underbrace{\left. \frac{\partial V}{\partial q_i} \right|_0}_{=0} q_i + \frac{1}{2} \sum_{i,j} \left. \frac{\partial^2 V}{\partial q_i \partial q_j} \right|_0 q_i q_j$$

$$T(q, \dot{q}) \approx \frac{1}{2} \sum_{i,j} a_{ij}(0) \dot{q}_i \dot{q}_j .$$

Close to its equilibrium the system is therefore described by the Lagrange function

$$L = T - V = \frac{1}{2} \sum_{i,j} A_{ij} \dot{q}_i \dot{q}_j - \frac{1}{2} \sum_{i,j} B_{ij} q_i q_j - V(0) ,$$

with

$$A_{ij} = a_{ij}(0) , \; B_{ij} = \left. \frac{\partial^2 V}{\partial q_i \partial q_j} \right|_0 ,$$

which yields the following Lagrange equations:

$$\sum_i A_{ki} \ddot{q}_i + \sum_i B_{ki} q_i = 0 .$$

Considering only *eigenoscillations*, for which all q_j oscillate with the same frequency ω, the ansatz

$$q_j(t) = Q_j e^{i\omega t}$$

leads to the time-independent *eigenvalue equation*

$$(B - \lambda A)Q = 0 , \; \lambda = \omega^2 , \tag{1.37}$$

which only has nontrivial solutions if the corresponding coefficient determinant vanishes:

$$\det(B - \lambda A) = 0 .$$

This equation determines the possible *eigenfrequencies* ω of the system and, via (1.37), the corresponding *eigenvectors* Q. If we consider the normalized eigenvectors as the columns of the (orthogonal) transformation matrix D, it is possible to show that

$$D^{\mathrm{T}} B D = \begin{pmatrix} \lambda_1 & & 0 \\ & \ddots & \\ 0 & & \lambda_n \end{pmatrix} , \; D^{\mathrm{T}} A D = I .$$

[6] More precisely, the potential must have a local minimum in q_0 for the system to be in stable equilibrium.

This means: D generates a *principal axis transformation* that simultaneously diagonalizes the matrices A and B, and therefore also the Lagrange function

$$L = \frac{1}{2}\dot{q}^{\mathrm{T}}A\dot{q} - \frac{1}{2}q^{\mathrm{T}}Bq - V(\mathbf{0}) \ .$$

1.3 Hamiltonian Mechanics

In this section we consider *Hamilton's formulation* of mechanics. It complements Newton's approach (Section 1.1) and that of Lagrange (Section 1.2) by providing a further concept for describing mechanical systems. Compared to the other two approaches, the Hamilton formulation distinguishes itself in that it reveals a close formal connection between classical mechanics and quantum mechanics. Indeed, quantum mechanics (and also statistical physics) are based mostly on Hamilton's formalism.

We start with a derivation of *Hamilton's equations of motion*, which, in contrast to Newton's and Lagrange's equations, are differential equations of first order. Following this, we discuss conservation laws and introduce the new notation of *Poisson brackets*, through which the formal equivalence of classical mechanics and quantum mechanics becomes particularly apparent. Our subsequent discussion of *canonical transformations* leads to the *Hamilton-Jacobi equation*, which defines a particular canonical transformation, such that all transformed coordinates and momenta are conserved.

1.3.1 Hamilton Equations

The Lagrange function L leads to equations of motion in which the generalized coordinates q_j and their velocities \dot{q}_j are the relevant pairs of variables. In Hamilton's theory the equations of motion are given instead by the generalized coordinates and the associated *generalized momenta*. These are defined as follows:

Definition: Generalized momentum p_j

$$p_j = \frac{\partial L}{\partial \dot{q}_j} \ . \tag{1.38}$$

This definition forms implicit equations for the \dot{q}_j in terms of the q_j, p_j, t:

$$\dot{q}_j = \dot{q}_j(q_1, \ldots, q_n, p_1, \ldots, p_n, t) \ .$$

We consider only the case of all constraints being given by $s = 3N - n$ holonomic conditions. To perform the transformation of variables,

$$L(q_i, \dot{q}_i, t) \longrightarrow H\left(q_i, \frac{\partial L}{\partial \dot{q}_i}, t\right) ,$$

we use the Legendre transformation

$$H = H(q_1, \ldots, q_n, p_1, \ldots, p_n, t) = \sum_{i=1}^{n} p_i \dot{q}_i - L$$

and, respecting (1.38), evaluate the derivatives

$$\frac{\partial H}{\partial p_j} = \dot{q}_j + \sum_{i=1}^{n} p_i \frac{\partial \dot{q}_i}{\partial p_j} - \sum_{i=1}^{n} \frac{\partial L}{\partial \dot{q}_i} \frac{\partial \dot{q}_i}{\partial p_j} = \dot{q}_j$$

$$\frac{\partial H}{\partial q_j} = \sum_{i=1}^{n} p_i \frac{\partial \dot{q}_i}{\partial q_j} - \frac{\partial L}{\partial q_j} - \sum_{i=1}^{n} \frac{\partial L}{\partial \dot{q}_i} \frac{\partial \dot{q}_i}{\partial q_j} = -\frac{\partial L}{\partial q_j} = -\dot{p}_j .$$

This yields

Theorem 1.21: Hamilton equations for s holonomic constraints

The *Hamilton function* of an N-particle system with n generalized coordinates and momenta is given by

$$H = \sum_{i=1}^{n} p_i \dot{q}_i - L(q_1, \ldots, q_n, p_1, \ldots, p_n, t) .$$

From this, we obtain the equations of motion (*Hamilton's equations*) for the generalized coordinates and momenta in the case of s holonomic constraints:

$$\frac{\partial H}{\partial p_j} = \dot{q}_j , \quad \frac{\partial H}{\partial q_j} = -\dot{p}_j , \quad j = 1, \ldots, 3N - s = n .$$

The temporal derivatives of L and H are related as follows:

$$-\frac{\partial L}{\partial t} = \frac{\partial H}{\partial t} .$$

These are the fundamental equations of Hamilton's formulation of mechanics, which are also called *canonical equations*. They form a system of $2n$ ordinary differential equations of first order for the n generalized coordinates and their momenta. For a system of N particles with $n = 3N - s$ degrees of freedom, they describe the motion of the system in an abstract $2n$-dimensional space, called *phase space*, which is spanned by the generalized coordinates and momenta. The following table shows the main results of Lagrangian and Hamiltonian mechanics.

Formulation	Variables	Function	Equations
Lagrange	(q_j, \dot{q}_j)	$L = T - V$	$\dfrac{\mathrm{d}}{\mathrm{d}t}\dfrac{\partial L}{\partial \dot{q}_j} - \dfrac{\partial L}{\partial q_j} = 0$
Hamilton	(q_j, p_j)	$H = \sum_i p_i \dot{q}_i - L$	$\dfrac{\partial H}{\partial p_j} = \dot{q}_j \ , \ \dfrac{\partial H}{\partial q_j} = -\dot{p}_j$

1.3.2 Conservation Laws

From Hamilton's equations one obtains the following conservation laws:

Theorem 1.22: Momentum conservation

If H does not depend explicitly on the generalized coordinate q_j, the corresponding momentum p_j is conserved:

$$\frac{\partial H}{\partial q_j} = 0 \Longrightarrow p_j = \text{const} .$$

Such a coordinate is called *cyclic*.

Differentiating the Hamilton function with respect to time,

$$\frac{\mathrm{d}H}{\mathrm{d}t} = \sum_{j=1}^{n} \left(\frac{\partial H}{\partial q_j} \dot{q}_j + \frac{\partial H}{\partial p_j} \dot{p}_j \right) + \frac{\partial H}{\partial t} \ ,$$

and using the Hamilton equations yields

$$\frac{\mathrm{d}H}{\mathrm{d}t} = \frac{\partial H}{\partial t} \ .$$

Thus, we have

Theorem 1.23: Conservation of the Hamilton function

If H (or L) does not depend explicitly on t, then H (or the corresponding Lagrange equations) is constant along the solutions of Hamilton's equations:

$$\frac{\partial H}{\partial t} = 0 \Longrightarrow H = \text{const} .$$

Energy conservation and interpretation of the Hamilton function.
In the case of scleronomic and holonomic constraints, $\boldsymbol{x}_i = \boldsymbol{x}_i(q_1, \ldots, q_n)$, and conservative external forces, $\partial V / \partial \dot{q}_j = 0$, the conservation of the Hamilton function can be interpreted as energy conservation, since, in this particular case, we have

$$p_j = \frac{\partial L}{\partial \dot{q}_j} = \frac{\partial T}{\partial \dot{q}_j}$$

and

$$T = \sum_{i=1}^{N} \frac{m_i}{2} \dot{\boldsymbol{x}}_i^2 = \sum_{i=1}^{N} \frac{m_i}{2} \sum_{k,l=1}^{n} \frac{\partial \boldsymbol{x}_i}{\partial q_k} \frac{\partial \boldsymbol{x}_i}{\partial q_l} \dot{q}_k \dot{q}_l \Longrightarrow \sum_{j=1}^{n} \frac{\partial T}{\partial \dot{q}_j} \dot{q}_j = 2T \ ,$$

so that

$$H = \sum_{j=1}^{n} \frac{\partial L}{\partial \dot{q}_j} \dot{q}_j - L = \sum_{j=1}^{n} \frac{\partial T}{\partial \dot{q}_j} \dot{q}_j - L = 2T - (T - V) = T + V = E \ .$$

Theorem 1.24: Energy conservation

For holonomic systems, which do not depend explicitly on time, the Hamilton function is identical with the total energy and is a conserved quantity.

$$H(\boldsymbol{q}, \boldsymbol{p}) = T + V = E = \text{const} \ .$$

Complete mechanical information. In Newtonian mechanics we saw that all dynamical variables of a system are functions of the position and momentum vectors. Lagrange theory is formulated in an n-dimensional space of the generalized coordinates (*configuration space*). In the context of Hamiltonian theory we see that the motion of the system can also be described by the *canonical conjugate* variables \boldsymbol{q} and \boldsymbol{p}, i.e., by points in phase space or in the extended phase space, which also contains time t. The state of a mechanical system is then determined by the conjugate variables at any given time, and by an evolution equation in time. Since the state of a system at a later time is uniquely determined by the state at an earlier time, the evolution equation must be a differential equation of first order in time. Hamilton's equations are just of this type. By contrast, Lagrange's equations are second-order differential equations in time, so that one needs two initial conditions, e.g., $q_j(t_0)$ and $\dot{q}_j(t_0)$, per generalized coordinate to uniquely determine the state of the system. Thus, the advantage of the canonical formalism lies in the fact that the system's state can be uniquely determined at any time if the canonical variables are known at a particular time. If we consider the canonical variables as a point π in phase space, the evolution of a mechanical system is described by a suitable function $F[\pi(t), t]$ for which $\dot{\pi}(t) = F[\pi(t), t]$.

1.3.3 Poisson Bracket

Since any quantity F in mechanics can be written as a function of the conjugate variables and of time, we can write an equation of motion as follows:

$$F = F(\boldsymbol{q}, \boldsymbol{p}, t) \Longrightarrow \frac{dF}{dt} = \sum_i \left(\frac{\partial F}{\partial q_i} \dot{q}_i + \frac{\partial F}{\partial p_i} \dot{p}_i \right) + \frac{\partial F}{\partial t}$$

$$= \sum_i \left(\frac{\partial F}{\partial q_i} \frac{\partial H}{\partial p_i} - \frac{\partial F}{\partial p_i} \frac{\partial H}{\partial q_i} \right) + \frac{\partial F}{\partial t} \ . \tag{1.39}$$

This formula can be simplified somewhat with the help of the following definition:

Definition: Poisson bracket

For at least once differentiable functions F and G in the variables x and y, we define the following mapping, called *Poisson bracket*:

$$\{F, G\}_{x,y} = \sum_i \left(\frac{\partial F}{\partial x_i} \frac{\partial G}{\partial y_i} - \frac{\partial F}{\partial y_i} \frac{\partial G}{\partial x_i} \right) . \tag{1.40}$$

It has the following properties:

- Anticommutativity: $\{F, G\} = -\{G, F\}$.

- Linearity and distributivity:

$$\{\alpha F_1 + \beta F_2, G\} = \alpha \{F_1, G\} + \beta \{F_2, G\} .$$

- Identities:

$$\{F, G_1 G_2\} = \{F, G_1\} G_2 + G_1 \{F, G_2\}$$

$$\{F, \{G, J\}\} + \{G, \{J, F\}\} + \{J, \{F, G\}\} = 0 \quad (\textit{Jacobi identity}) .$$

Using this notation, (1.39) can be written as

$$\frac{\mathrm{d}F}{\mathrm{d}t} = \{F, H\}_{q,p} + \frac{\partial F}{\partial t} ,$$

and it follows:

Theorem 1.25: Equation of motion, conserved quantities, and Poisson theorem

The equation of motion for a mechanical quantity $F(q, p, t)$ is

$$\frac{\mathrm{d}F}{\mathrm{d}t} = \{F, H\}_{q,p} + \frac{\partial F}{\partial t} .$$

If F is a conserved quantity, $\mathrm{d}F/\mathrm{d}t = 0$, and not explicitly time-dependent, it follows that

$$\{F, H\}_{q,p} = 0 .$$

The reverse is equally true. Furthermore, if F and G are two such conserved quantities, then $\{F, G\}_{q,p}$ is also conserved. This statement is the *Poisson theorem*. It follows from the Jacobi identity.

Noting that q_i and p_i are independent variables,

$$\frac{\partial q_i}{\partial q_j} = \delta_{ij} , \quad \frac{\partial p_i}{\partial p_j} = \delta_{ij} , \quad \frac{\partial q_i}{\partial p_j} = \frac{\partial p_i}{\partial q_j} = 0 ,$$

we see that (1.40) can also be applied to them, and we obtain the canonical equations in the new notation:

$$\dot{q}_i = \{q_i, H\}_{q,p} \ , \ \dot{p}_i = \{p_i, H\}_{q,p} \ .$$

For the Poisson brackets of momenta and coordinates, one similarly finds

$$\{q_i, p_k\}_{q,p} = \delta_{ik} \ , \ \{q_i, q_k\}_{q,p} = \{p_i, p_k\}_{q,p} = 0 \ . \tag{1.41}$$

Hamilton theory and quantum mechanics. The algebraic properties of the Poisson bracket form the basis of their use and of their transcription to quantum mechanics. In quantum theory, physical quantities are described by linear Hermitean operators, and the Poisson bracket is replaced by the *commutator*, which shares all the algebraic properties of the Poisson bracket:

$$\{F, G\} \longrightarrow -\mathrm{i}\hbar[\boldsymbol{F}, \boldsymbol{G}] = -\mathrm{i}\hbar(\boldsymbol{F}\boldsymbol{G} - \boldsymbol{G}\boldsymbol{F}) \ .$$

One can indeed regard mechanics and quantum mechanics as two different realizations of the same algebraic structure that is defined by the Poisson bracket. An example of this *correspondence principle* is given by *Heisenberg's equation of motion* in quantum mechanics,

$$\mathrm{i}\hbar\frac{\mathrm{d}\boldsymbol{F}}{\mathrm{d}t} = [\boldsymbol{F}, \boldsymbol{H}] + \frac{\partial \boldsymbol{F}}{\partial t} \ ,$$

which is obtained from the equation of motion in Theorem 1.25 by the above replacement.

1.3.4 Canonical Transformations

Having developed the formal structure of the Hamiltonian formalism, we now ask whether it is possible to find transformations under which the canonical equations are invariant. Naturally, we would like to find transformations of this kind that particularly simplify Hamilton's equations. For example, it would be useful to have a transformation that yields a Hamilton function dependent only on the conjugate momenta (see next subsection). In this case, all coordinates would be cyclic, thus making all momenta in this representation-invariant. We define:

Definition: Canonical transformation

A coordinate transformation

$$q_i \longrightarrow Q_i = Q_i(\boldsymbol{q}, \boldsymbol{p}, t) \ , \ p_i \longrightarrow P_i = P_i(\boldsymbol{q}, \boldsymbol{p}, t)$$

is called *canonical* if the form of the transformed canonical equations remains unchanged:

$$H = H(\boldsymbol{q}, \boldsymbol{p}, t): \quad \dot{q}_i = \frac{\partial H}{\partial p_i} \quad , \quad \dot{p}_i = -\frac{\partial H}{\partial q_i}$$

$$H' = H'(\boldsymbol{Q}, \boldsymbol{P}, t): \quad \dot{Q}_i = \frac{\partial H'}{\partial P_i} \quad , \quad \dot{P}_i = -\frac{\partial H'}{\partial Q_i} \ .$$

A necessary condition for a transformation to be canonical is that the old and new Hamilton functions obey the Hamilton principle, i.e.,

$$\delta \int_{t_1}^{t_2} \left(\sum_i p_i \dot{q}_i - H(\boldsymbol{q}, \boldsymbol{p}, t) \right) dt = \delta \int_{t_1}^{t_2} \left(\sum_i P_i \dot{Q}_i - H'(\boldsymbol{Q}, \boldsymbol{P}, t) \right) dt = 0$$

or

$$\delta \int_{t_1}^{t_2} \left[\sum_i \left(p_i \dot{q}_i - P_i \dot{Q}_i \right) + (H' - H) \right] dt = 0 .$$

Disregarding the trivial possibility of a simple multiplication of H and p_i by a constant, the difference between the integrands in the old and new action can only consist of the total time differential dF/dt. Thus, we can write

$$dF = \sum_i (p_i dq_i - P_i dQ_i) + (H' - H)dt . \tag{1.42}$$

The function F is called the generating function (or generator) of the canonical transformation. It is a function of the $4n + 1$ variables \boldsymbol{q}, \boldsymbol{p}, \boldsymbol{Q}, \boldsymbol{P}, and t, of which $2n$ depend upon each other via the above transformations; they can be eliminated in favor of $2n + 1$ independent variables. Thus, F can take on one of the following forms:

$$F_1(\boldsymbol{q}, \boldsymbol{Q}, t) , \ F_2(\boldsymbol{q}, \boldsymbol{P}, t) , \ F_3(\boldsymbol{Q}, \boldsymbol{p}, t) , \ F_4(\boldsymbol{p}, \boldsymbol{P}, t) . \tag{1.43}$$

Obviously, (1.42) is just the total differential of

$$F_1 = F_1(\boldsymbol{q}, \boldsymbol{Q}, t) = \sum_i (p_i q_i - P_i Q_i) + H' - H ,$$

from which it follows that

$$H' = H + \frac{\partial F_1}{\partial t} , \ p_i = \frac{\partial F_1}{\partial q_i} , \ P_i = -\frac{\partial F_1}{\partial Q_i} .$$

With the help of the last two relations, the q_i and p_i can be expressed as functions of the new variables: $q_i = q_i(\boldsymbol{Q}, \boldsymbol{P}, t)$, $p_i = p_i(\boldsymbol{Q}, \boldsymbol{P}, t)$. Inserting these into the right-hand side of the first equation, we obtain the transformed Hamilton function in the new variables.

All other dependencies given in (1.43) are obtained via Legendre transformations of F_1. For F_2, one finds the relations

$$d \left(F_1 + \sum_i Q_i P_i \right) = \sum_i (p_i dq_i + Q_i dP_i) + (H' - H)dt$$

$$\Longrightarrow F_2 = F_2(\boldsymbol{q}, \boldsymbol{P}, t) = F_1 + \sum_i P_i Q_i$$

$$\Longrightarrow H' = H + \frac{\partial F_2}{\partial t} , \ p_i = \frac{\partial F_2}{\partial q_i} , \ Q_i = \frac{\partial F_2}{\partial P_i} , \tag{1.44}$$

and for F_3

$$d\left(F_1 - \sum_i q_i p_i\right) = -\sum_i (q_i dp_i + P_i dQ_i) + (H' - H)dt$$

$$\implies F_3 = F_3(\boldsymbol{Q}, \boldsymbol{p}, t) = F_1 - \sum_i p_i q_i$$

$$\implies H' = H + \frac{\partial F_3}{\partial t} \, , \; q_i = -\frac{\partial F_3}{\partial p_i} \, , \; P_i = -\frac{\partial F_3}{\partial Q_i} \, ,$$

and finally for F_4

$$d\left(F_1 - \sum_i q_i p_i + \sum_i P_i Q_i\right) = -\sum_i (q_i dp_i - Q_i dP_i) + (H' - H)dt$$

$$\implies F_4 = F_4(\boldsymbol{p}, \boldsymbol{P}, t) = F_1 - \sum_i p_i q_i + \sum_i P_i Q_i$$

$$\implies H' = H + \frac{\partial F_4}{\partial t} \, , \; q_i = -\frac{\partial F_4}{\partial p_i} \, , \; Q_i = \frac{\partial F_4}{\partial P_i} \, .$$

We see that there are four different types of canonical transformations, whose generators depend on other $2n + 1$ independent variables. In practice, it is often not so simple to find the right generator that yields a significant simplification of the problem at hand. Sometimes this is only possible for problems where the solution is also easily obtained in a different manner. The actual advantage of the characterization of canonical transformations through generators is given by the insights we can gain into the structure of Hamiltonian mechanics.

Infinitesimal canonical transformations. As a specific example of a canonical transformation, we consider the generating function

$$F_2 = F_2(\boldsymbol{q}, \boldsymbol{P}, \epsilon) = \sum_i q_i P_i + \epsilon f(\boldsymbol{q}, \boldsymbol{P}) + \mathcal{O}(\epsilon^2) \, , \; \epsilon \text{ continuous} \, .$$

It is composed of the identity mapping and a suitable function $f(\boldsymbol{q}, \boldsymbol{P})$ that defines the infinitesimal transformation. According to (1.44), we have

$$Q_i = \frac{\partial F_2}{\partial P_i} = q_i + \epsilon \frac{\partial f}{\partial P_i} + \mathcal{O}(\epsilon^2) \, , \; p_i = \frac{\partial F_2}{\partial q_i} = P_i + \epsilon \frac{\partial f}{\partial q_i} + \mathcal{O}(\epsilon^2) \, .$$

Since $\epsilon \partial f / \partial P_i$ and $\epsilon \partial f / \partial q_i$ are of first order in ϵ, we can replace the variables P_i by their zeroth-order approximations p_i, so that

$$\delta q_i = Q_i - q_i = \epsilon \frac{\partial f(\boldsymbol{q}, \boldsymbol{p})}{\partial p_i} \, , \; \delta p_i = P_i - p_i = -\epsilon \frac{\partial f(\boldsymbol{q}, \boldsymbol{p})}{\partial q_i} \, .$$

This can also be written as

$$\delta q_i = \epsilon \left\{ q_i, f \right\}_{\boldsymbol{q}, \boldsymbol{p}} \, , \; \delta p_i = \epsilon \left\{ p_i, f \right\}_{\boldsymbol{q}, \boldsymbol{p}} \, .$$

Specifically, if we set $\epsilon = dt$ and $f = H$, so that

$$F_2(\boldsymbol{q}, \boldsymbol{P}, \delta t) = \sum_i q_i P_i + H(\boldsymbol{q}, \boldsymbol{p}) dt \ ,$$

we again obtain Hamilton's equations:

$$\delta q_i = dt \{q_i, H\}_{\boldsymbol{q},\boldsymbol{p}} = \dot{q}_i dt = dq_i \ , \ \ \delta p_i = dt \{p_i, H\}_{\boldsymbol{q},\boldsymbol{p}} = \dot{p}_i dt = dp_i \ .$$

These equations signify that the transformation generated by H shifts the coordinates and momenta q_i, p_i at time t to their values at time $t + dt$. Thus, the Hamilton function is the generator of the infinitesimal canonical transformation that corresponds to the actual trajectory in the interval dt.

Invariance of the Poisson bracket under canonical transformations. We are now in a position to see the fundamental importance of the Poisson bracket, to wit, its invariance under canonical transformations. To this end, we first show the invariance of the relations (1.41) between canonical variables. Restricting ourselves to time-independent transformations, we have

$$\dot{P}_i = \sum_j \left(\frac{\partial P_i}{\partial q_j} \dot{q}_j + \frac{\partial P_i}{\partial p_j} \dot{p}_j \right) = \sum_j \left(\frac{\partial P_i}{\partial q_j} \frac{\partial H}{\partial p_j} - \frac{\partial P_i}{\partial p_j} \frac{\partial H}{\partial q_j} \right)$$

$$= \sum_{j,k} \left[\left(\frac{\partial P_i}{\partial q_j} \frac{\partial P_k}{\partial p_j} - \frac{\partial P_i}{\partial p_j} \frac{\partial P_k}{\partial q_j} \right) \frac{\partial H}{\partial P_k} + \left(\frac{\partial P_i}{\partial q_j} \frac{\partial Q_k}{\partial p_j} - \frac{\partial P_i}{\partial p_j} \frac{\partial Q_k}{\partial q_j} \right) \frac{\partial H}{\partial Q_k} \right]$$

$$= \sum_k \frac{\partial H}{\partial P_k} \{P_i, P_k\}_{\boldsymbol{q},\boldsymbol{p}} + \sum_k \frac{\partial H}{\partial Q_k} \{P_i, Q_k\}_{\boldsymbol{q},\boldsymbol{p}} \ .$$

It follows that

$$\{P_i, P_k\}_{\boldsymbol{q},\boldsymbol{p}} = 0 \ , \ \ \{Q_i, P_k\}_{\boldsymbol{q},\boldsymbol{p}} = \delta_{ik} \ .$$

The proof for

$$\{Q_i, Q_k\}_{\boldsymbol{q},\boldsymbol{p}} = 0$$

is carried out analogously. Using the last two relations, we can now write

$$\{F, G\}_{\boldsymbol{Q},\boldsymbol{P}} = \sum_i \left(\frac{\partial F}{\partial Q_i} \frac{\partial G}{\partial P_i} - \frac{\partial F}{\partial P_i} \frac{\partial G}{\partial Q_i} \right)$$

$$= \sum_{i,j,k} \left[\frac{\partial F}{\partial q_j} \frac{\partial G}{\partial q_k} \left(\frac{\partial q_j}{\partial Q_i} \frac{\partial q_k}{\partial P_i} - \frac{\partial q_j}{\partial P_i} \frac{\partial q_k}{\partial Q_i} \right) \right.$$

$$+ \frac{\partial F}{\partial q_j} \frac{\partial G}{\partial p_k} \left(\frac{\partial q_j}{\partial Q_i} \frac{\partial p_k}{\partial P_i} - \frac{\partial q_j}{\partial P_i} \frac{\partial p_k}{\partial Q_i} \right)$$

$$+ \frac{\partial F}{\partial p_j} \frac{\partial G}{\partial q_k} \left(\frac{\partial p_j}{\partial Q_i} \frac{\partial q_k}{\partial P_i} - \frac{\partial p_j}{\partial P_i} \frac{\partial q_k}{\partial Q_i} \right)$$

$$\left. + \frac{\partial F}{\partial p_j} \frac{\partial G}{\partial p_k} \left(\frac{\partial p_j}{\partial Q_i} \frac{\partial p_k}{\partial P_i} - \frac{\partial p_j}{\partial P_i} \frac{\partial p_k}{\partial Q_i} \right) \right]$$

$$= \sum_j \left(\frac{\partial F}{\partial q_j} \frac{\partial G}{\partial p_j} - \frac{\partial F}{\partial p_j} \frac{\partial G}{\partial q_j} \right) = \{F, G\}_{\boldsymbol{q},\boldsymbol{p}} \ ,$$

which shows the invariance of the Poisson bracket under canonical transformations. The proof for time-dependent transformations can be conducted in similar fashion.

Theorem 1.26: Canonical transformations and Poisson bracket

A transformation

$$q_i \longrightarrow Q_i = Q_i(\boldsymbol{q}, \dot{\boldsymbol{q}}, t) \; , \; p_i \longrightarrow P_i = Q_i(\boldsymbol{q}, \dot{\boldsymbol{q}}, t)$$

of the canonical variables $(\boldsymbol{q}, \boldsymbol{p})$ is canonical if, and only if,

$$\{P_i, P_j\}_{\boldsymbol{q},\boldsymbol{p}} = \{Q_i, Q_j\}_{\boldsymbol{q},\boldsymbol{p}} = 0 \; , \; \{P_i, Q_j\}_{\boldsymbol{q},\boldsymbol{p}} = \delta_{ij} \; .$$

The Poisson bracket is invariant under canonical transformations:

$$\{F, G\}_{\boldsymbol{q},\boldsymbol{p}} = \{F, G\}_{\boldsymbol{Q},\boldsymbol{P}} \; .$$

It is therefore generally written without reference to a specific pair of canonical variables.

1.3.5 Hamilton-Jacobi Equation

We now consider a canonical transformation with a generator of type $F_2(\boldsymbol{q}, \boldsymbol{P}, t)$ that is designed to have the property that all of the transformed coordinates and momenta are constant. This is most easily obtained by demanding that the transformed Hamilton function disappears:

$$H' = H + \frac{\partial F_2}{\partial t} = 0 \Longrightarrow \dot{Q}_i = \frac{\partial H'}{\partial P_i} = 0 \; , \; \dot{P}_i = -\frac{\partial H'}{\partial Q_i} = 0 \; . \tag{1.45}$$

Differentiating F_2 with respect to time, and using the fact that [see (1.44)]

$$Q_i = \frac{\partial F_2}{\partial P_i} \; , \; p_i = \frac{\partial F_2}{\partial q_i} \; ,$$

we can write

$$\frac{\mathrm{d}F_2}{\mathrm{d}t} = \sum_i \left(\frac{\partial F_2}{\partial q_i} \dot{q}_i + \frac{\partial F_2}{\partial P_i} \dot{P}_i \right) + \frac{\partial F_2}{\partial t} = \sum_i \left(p_i \dot{q}_i + Q_i \dot{P}_i \right) - H$$

$$= \sum_i p_i \dot{q}_i - H = L \; .$$

It follows that

$$F_2 = \int L(\boldsymbol{q}, \dot{\boldsymbol{q}}, t) \mathrm{d}t + \text{const} \; .$$

Thus, the generator of the canonical transformation that forces the Hamilton function to vanish is, up to a constant, identical to the action functional S along the trajectory. However, this cannot be used to determine the solution, since the coordinates and momenta are just the unknowns of the problem. So, writing S instead of F_2 in (1.45), we obtain the Hamilton-Jacobi equation:

Theorem 1.27: Hamilton-Jacobi equation

The Hamilton-Jacobi equation

$$\frac{\partial S}{\partial t} + H\left(q_i, \frac{\partial S}{\partial q_i}, t\right) = 0 \,,$$

$$S = S(q_i, \beta_i, t) \,, \quad \beta_i = P_i = \text{const} \,, \quad i = 1, \ldots, n$$

is a first-order partial differential equation in the $n+1$ variables \boldsymbol{q} and t. It is equivalent to the system of $2n$ first-order ordinary Hamiltonian differential equations. S is the generator that transforms the Hamilton function H into constant coordinates \boldsymbol{Q} and momenta \boldsymbol{P}. It is called *Hamilton's principal function.*

Once S has been determined using this equation, the coordinates and momenta are obtained from the purely algebraic relations

$$Q_i = \frac{\partial S}{\partial \beta_i} := \alpha_i \,, \quad p_i = \frac{\partial S}{\partial q_i} \implies q_i = q_i(\boldsymbol{\alpha}, \boldsymbol{\beta}, t) \,, \quad p_i = p_i(\boldsymbol{\alpha}, \boldsymbol{\beta}, t) \,.$$

Here, the $2n$ integration constants

$$\boldsymbol{\alpha} = \boldsymbol{Q} = \text{const} \,, \quad \boldsymbol{\beta} = \boldsymbol{P} = \text{const}$$

need to be determined from the initial conditions $\boldsymbol{q}(t_0)$, $\boldsymbol{p}(t_0)$.

Solution via separation of variables. If the Hamilton function is not explicitly time-dependent, we know from Theorem 1.23 that $H = \gamma = \text{const}$. In this case, the ansatz

$$S(\boldsymbol{q}, \boldsymbol{\beta}, t) = S_0(\boldsymbol{q}, \boldsymbol{\beta}) - \gamma(\boldsymbol{\beta})t$$

reduces the Hamilton-Jacobi equation to

$$H\left(q_i, \frac{\partial S_0}{\partial q_i}\right) = \gamma \,,$$

where S_0 is called the *abbreviated action functional* or *Hamilton's characteristic function.* For scleronomic systems, γ equals the total energy. Assuming further that the coordinate q_1 appears only in the combination

$$\phi_1\left(q_1, \frac{\partial S_0}{\partial q_1}\right) \,,$$

so that ϕ_1 is independent of all other coordinates, we have

$$H\left[\phi_1\left(q_1, \frac{\partial S_0}{\partial q_1}\right), q_{i\neq1}, \frac{\partial S_0}{\partial q_{i\neq1}}\right] = \gamma \,,$$

and the separation ansatz

$$S_0(\boldsymbol{q}, \boldsymbol{\beta}, t) = S_1(q_1, \beta_1) + S'(q_{i\neq1}, \alpha_{i\neq1})$$

leads to

$$H\left[\phi_1\left(q_1, \frac{\partial S_1}{\partial q_1}\right), q_{i\neq1}, \frac{\partial S'}{\partial q_{i\neq1}}\right] = \gamma .$$

Assuming that we have found a solution S_0, the previous equation with S_0 inserted must hold for all values of q_1. However, since this coordinate only appears in the function ϕ_1, we can deduce that ϕ_1 is a constant. Thus, from the original partial differential equation with n independent variables, we obtain an ordinary differential equation in q_1,

$$\phi_1\left(q_1, \frac{\partial S_1}{\partial q_1}\right) = \beta_1 ,$$

and a partial differential equation with $n - 1$ independent variables:

$$H\left(\phi_1 = \beta_1, q_{i\neq1}, \frac{\partial S'}{\partial q_{q\neq1}}\right) = \gamma .$$

If this method can be applied successively to all coordinates, the Hamilton-Jacobi equation can be reduced to n ordinary differential equations,

$$\phi_i\left(q_i, \frac{\partial S_i}{\partial q_i}\right) = \alpha_i , \quad H(\phi_i = \beta_i) = \gamma ,$$

and the general solution is the sum of all S_i. Finally, we consider the above method for the case of a cyclic coordinate q_1. The function ϕ_1 is then given by $\partial S_1/\partial q_1$, and it follows that $S_1(q_1, \beta_1) = \beta_1 q_1$, where the constant is equal to the momentum: $p_1 = \beta_1$.

Hamilton-Jacobi equation and quantum mechanics. Given the one-dimensional Hamilton function

$$H(q,p) = \frac{p^2}{2m} + V(q) = E ,$$

we find the Hamilton-Jacobi equation

$$\frac{1}{2m}\left(\frac{\partial S_0}{\partial q}\right)^2 + V(q) = E .$$

In Chapter 3 we shall see that the corresponding quantum mechanical *Hamilton operator* obeys the *Schrödinger equation*

$$-\frac{\hbar^2}{2m}\frac{\partial^2 \psi}{\partial q^2} + V(q)\psi = E\psi ,$$

with the wave function ψ. Inserting the time-independent wave function

$$\psi = e^{iS_0/\hbar} ,$$

we see that

$$\frac{1}{2m}\left(\frac{\partial S_0}{\partial q}\right)^2 - \frac{i\hbar}{2m}\frac{\partial S_0^2}{\partial q^2} + V(q) = E .$$

Obviously, in the limit $\hbar \to 0$, this again results in the above Hamilton-Jacobi equation.

Summary

- From the Lagrange equations, one obtains **Hamilton's equations** via a Legendre transformation. They constitute a system of coupled ordinary first-order differential equations in the **generalized coordinates** and the **generalized momenta.**

- In the Hamiltonian formalism, the dynamics of a mechanical system is uniquely determined by the generalized coordinates and momenta at any one time, as well as by the **Hamilton function.**

- The **Poisson bracket** allows a deeper algebraic understanding of Hamiltonian mechanics. It is invariant under **canonical transformations.**

- The action functional is the generator of the motion.

- The condition for a Hamilton function with purely cyclic coordinates leads to the **Hamilton-Jacobi equation.** It is a partial differential equation of first order.

Applications

8. Independence of the generator from boundary conditions. Show that the variation of the action

$$
\begin{aligned}
S &= \int_{t_1}^{t_2} \mathrm{d}t \left(\sum_i P_i \dot{Q}_i - H'(\boldsymbol{Q}, \boldsymbol{P}, t) + \frac{\mathrm{d}F_1}{\mathrm{d}t} \right) \\
&= \int_{t_1}^{t_2} \mathrm{d}t \left(\sum_i P_i \dot{Q}_i - H'(\boldsymbol{Q}, \boldsymbol{P}, t) \right) + F_1(\boldsymbol{q}, \boldsymbol{Q}, t)\big|_{t_1}^{t_2}
\end{aligned}
$$

determines the new Hamilton function uniquely, independently of the boundary conditions. Show that the transformation is canonical.

Solution. Noting that the startpoint and endpoint $[t_1, \boldsymbol{q}(t_1)]$, $[t_2, \boldsymbol{q}(t_2)]$ are fixed, the action's variation is given by

$$
\delta S = \int_{t_1}^{t_2} \mathrm{d}t \sum_i \left(\dot{Q}_i \delta P_i + P_i \delta \dot{Q}_i - \frac{\partial H'}{\partial Q_i} \delta Q_i - \frac{\partial H'}{\partial P_i} \delta P_i \right) + \sum_i \frac{\partial F_1}{\partial Q_i} \delta Q_i \bigg|_{t_1}^{t_2} .
$$

Partial integration of the $\delta \dot{Q}_i$-terms yields

$$\delta S = \int\limits_{t_1}^{t_2} dt \left[\sum_i \left(\dot{Q}_i - \frac{\partial H'}{\partial P_i} \right) \delta P_i - \sum_i \left(\dot{P}_i + \frac{\partial H'}{\partial Q_i} \right) \delta Q_i \right]$$

$$+ \sum_i \left(P_i + \frac{\partial F_1}{\partial Q_i} \right) \delta Q_i \Bigg|_{t_1}^{t_2} .$$

We see that the last term vanishes independently of δQ_i at its endpoints $t_{1,2}$, since the bracket is already zero. Thus, we obtain Hamilton's equations in the new variables Q_j and P_j, since they (and also their variations) are independent from one another.

9. Poisson bracket of the angular momentum. Determine the Poisson bracket of momentum and angular momentum of a single particle in Cartesian coordinates, as well as the Poisson bracket of the components of the angular momentum.

Solution. According to (1.40), for any function $G(\boldsymbol{q}, \boldsymbol{p}, t)$ of canonically conjugate variables q_j and p_j, we have $\{G, p_j\} = \partial G / \partial q_j$. We use this relation as follows:

$$\{l_x, p_x\} = \frac{\partial}{\partial x}(y p_z - z p_y) = 0$$

$$\{l_x, p_y\} = \frac{\partial}{\partial y}(y p_z - z p_y) = p_z$$

$$\{l_x, p_z\} = \frac{\partial}{\partial z}(y p_z - z p_y) = -p_y .$$

In total, we obtain

$$\{l_i, p_j\} = \epsilon_{ijk} p_k .$$

For the components of the angular momentum, one finds

$$\{l_x, l_x\} = \{l_y, l_y\} = \{l_z, l_z\} = 0$$

$$\begin{aligned} \{l_x, l_y\} &= \{y p_z - z p_y, z p_x - x p_z\} \\ &= \{y p_z, z p_x\} - \{z p_y, z p_x\} - \{y p_z, x p_z\} + \{z p_y, x p_z\} \\ &= \{y p_z, z p_x\} + \{z p_y, x p_z\} = y \{p_z, z\} p_x + x \{z, p_z\} p_y \\ &= -y p_x + x p_y = l_z . \end{aligned}$$

Calculating the remaining Poisson brackets leads to

$$\{l_i, l_j\} = \epsilon_{ijk} l_k .$$

We see that no two components of angular momentum can simultaneously appear as canonical momenta, since their Poisson bracket does not vanish (Theorem 1.26). However, one can easily show that

$$\{\boldsymbol{l}^2, l_i\} = 0 ,$$

so that the absolute value of angular momentum and one of its components can simultaneously be canonical momenta. We will encounter this result in quantum mechanics again.

10. Hamilton-Jacobi equation. Using the Hamilton-Jacobi equation, solve the following systems:

a. Free particle.

b. Harmonic oscillator.

Solution.

To a. In this simple case, from the Hamilton function

$$H = \frac{1}{2m}\left(p_1^2 + p_2^2 + p_3^2\right)$$

and the separation

$$S(q, \beta, t) = S_0(q, \beta) - Et ,$$

we find the Hamilton-Jacobi equation

$$\frac{1}{2m}\left[\left(\frac{\partial S_0}{\partial q_1}\right)^2 + \left(\frac{\partial S_0}{\partial q_2}\right)^2 + \left(\frac{\partial S_0}{\partial q_3}\right)^2\right] = E .$$

All three coordinates are cyclic, so that this equation can be totally separated. The solution is

$$S_0(q, \beta) = \sum_i \beta_i q_i , \quad \beta_i = p_i , \quad \frac{1}{2m}\sum_i \beta_i^2 = E .$$

To b. Consider, to start, the more general one-dimensional Hamilton function

$$H(q, p) = \frac{p^2}{2m} + V(q) .$$

Using the separation

$$S(q, \beta, t) = S_0(q, \beta) - \beta t , \quad \beta = E ,$$

we obtain the Hamilton-Jacobi equation

$$\left(\frac{\partial S_0}{\partial q}\right)^2 + 2mV(q) = 2m\beta ,$$

which is solved by

$$S_0(q, \beta) = \int_{q_0}^{q} dq' \sqrt{2m[\beta - V(q')]} .$$

Furthermore, we have

$$\alpha = \frac{\partial S}{\partial \beta} = \int_{q_0}^{q} dq' \frac{m}{\sqrt{2m[\beta - V(q')]}} - t$$

$$p = \frac{\partial S}{\partial q} = \sqrt{2m[\beta - V(q)]} .$$

For the particular case of the harmonic oscillator, $V(q) = kq^2/2$, the first relation yields the solution

$$q(t) = \sqrt{\frac{2\beta}{k}} \sin\left[\sqrt{\frac{k}{m}}(t - \alpha')\right] ,$$

with integration constants α' and β to be determined by initial conditions. Expanding to three dimensions,

$$H(\boldsymbol{q}, \boldsymbol{p}) = \frac{1}{2m} \sum_i p_i^2 + \frac{1}{2} \sum_i k_i q_i^2 ,$$

and using

$$S(\boldsymbol{q}, \boldsymbol{\beta}, t) = S(\boldsymbol{q}, \boldsymbol{\beta}) - E(\boldsymbol{\beta})t$$

yield the Hamilton-Jacobi equation

$$\sum_i \left(\frac{\partial S_0}{\partial q_i}\right)^2 + \sum_i mk_i q_i^2 = 2mE(\boldsymbol{\beta}) .$$

It is solvable by the ansatz

$$S_0(\boldsymbol{q}, \boldsymbol{\beta}) = \sum_i S_{0,i}(q_i, \beta_i)$$

separating it into three one-dimensional equations

$$\left(\frac{\partial S_{0,i}}{\partial q_i}\right)^2 + mk_i q_i^2 = 2m\beta_i , \quad \sum_i \beta_i = E ,$$

which are the same as we have already solved above.

1.4 Motion of Rigid Bodies

So far, we have limited our discussion to that of individual particles or bodies whose spatial extent can be regarded as negligible. In this section we shall discuss the physical implications for bodies whose spatial structure cannot be disregarded.

The general motion of an N-particle system (body) can be described by the motion of an arbitrary point \boldsymbol{q} (*rotation center*), and the motion of all particles around this point. We distinguish two classes of bodies:

- *Laminar bodies*, which can change their geometry with time, for example, fluids and gases.

- *Rigid bodies*, for which every particle position \boldsymbol{x}_i obeys the condition

$$|\boldsymbol{x}_i(t) - \boldsymbol{x}_j(t)| = \text{const} \quad \forall \ i, j, t$$

(e.g., stones, houses), so that their form does not change with time.

In the following, we restrict ourselves to rigid bodies. After a general discussion of the dynamics of rigid bodies, we turn our attention to that of a purely rotational movement around a fixed point and show how the dynamical quantities for these types of problems can be written in terms of the *inertial tensor* of the body. Furthermore, we use the *Euler angles* to derive the Lagrange equations for the general dynamics of rigid bodies.

1.4.1 General Motion of Rigid Bodies

For the following discussion we need two coordinate systems (see Fig. 1.8), which we have already made use of in Subsection 1.1.4.

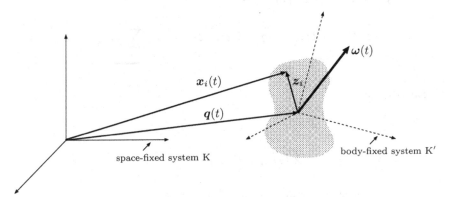

Fig. 1.8. Definition of space-fixed and body-fixed system.

- *Space-fixed system:* K : $\{e_1, e_2, e_3\}$. This system is assumed to be an inertial system, in terms of which the rigid body performs an accelerated movement. The particle coordinates of a body in this system are given by x_i.

- *Body-fixed system:* K' : $\{e_1', e_2', e_3'\}$. This accelerated system with origin in the rotation center q is linked with the rigid body; in this system, the body is at rest. The corresponding time-independent particle vectors shall be denoted by z_i. They are related to x_i via

$$x_i(t) = q(t) + z_i .$$

Just as in Subsection 1.1.4, we use the notation

$$D x = \sum_i e_i \dot{x}_i , \; D' x = \sum_i e_i' \dot{x}_i'$$

in order to avoid confusion. Using $D' z_i = 0$ and Theorem 1.8, we now calculate the kinematic quantities of total momentum, total angular momentum, total torque, and total kinetic energy of the body, where

$$z_C = \frac{1}{M} \sum_i m_i z_i \,, \; M = \sum_i m_i$$

denotes the center of mass of the body in the body-fixed system.

Total momentum p:

$$p = \sum_i m_i \dot{x}_i = M\dot{q} + M\dot{z}_C \,.$$

Total angular momentum l:

$$l = \sum_i m_i x_i \times Dx_i$$

$$= \sum_i m_i (q + z_i) \times (Dq + Dz_i)$$

$$= Mq \times Dq + Mz_C \times Dq + Mq \times (\omega \times z_C) + \underbrace{\sum_i m_i z_i \times (\omega \times z_i)}_{l_{\mathrm{rot}}} \,.$$

Total torque N:

$$N = \sum_i m_i x_i \times D^2 x_i$$

$$= \sum_i m_i (q + z_i) \times \left(D^2 q + D^2 z_i\right)$$

$$= Mq \times D^2 q + Mz_C \times D^2 q + Mq \times D^2 z_C$$
$$+ \underbrace{\sum_i m_i z_i \times \left[(D'\omega) \times z_i + \omega \times (\omega \times z_i)\right]}_{N_{\mathrm{rot}}} \,.$$

Total kinetic energy T:

$$T = \frac{1}{2} \sum_i m_i (Dx_i)^2$$

$$= \frac{1}{2} \sum_i m_i \left[(Dq)^2 + 2(Dq)(Dz_i) + (Dz_i)^2\right]$$

$$= \frac{M}{2}(Dq)^2 + M(Dq)(\omega \times z_C) + \underbrace{\frac{1}{2} \sum_i m_i (\omega \times z_i)^2}_{T_{\mathrm{rot}}} \,. \tag{1.46}$$

The vast majority of problems arising in practice fall into one of two categories. First, the rotation center coincides with the body's center of mass: $z_C = 0$. Second, the rotation axis is fixed (think of a pendulum or a top), so that the origin of the space-fixed system can be shifted to coincide with the rotation center ($q = 0$). In the first case, the quantities p, l, N, and T

can each be decomposed into a purely rotational contribution, and a second one describing the sole rotation around the rotation center (without mixing terms). In the second case, the quantities p, l, N, and T are identical to their rotational parts.

1.4.2 Rotation of Rigid Bodies Around a Point

In the following, we concentrate on the purely rotational motion of the rigid body around a fixed rotation center, and shift the origin of the space-fixed system into this point ($q = 0$). Our aim is to find simpler expressions for l_{rot}, N_{rot}, and T_{rot}. To this end, we introduce the *inertial tensor*, whose name and meaning will become apparent soon.

Definition: Inertial tensor and inertial moments

The body-fixed inertial tensor of a rigid body is defined as the symmetric (3×3)-matrix

$$\Theta_{ab} = \sum_i m_i \left(z_i^2 \delta_{ab} - z_{ia} z_{ib} \right) , \quad \Theta_{ab} = \int d^3 z \rho(z) \left(z^2 \delta_{ab} - z_a z_b \right) ,$$

where the latter equation is to be used in the case of a continuous mass distribution $\rho(z)$. The diagonal elements of the matrix are termed *inertial moments* and the off-diagonals are called *deviation moments*.

Obviously, Θ depends only on the geometry of the rigid body and not on the type of motion. For the rotational part l_{rot} (rotational angular momentum) of the total angular momentum, we can now write

$$
\begin{aligned}
l_{\text{rot}} &= \sum_i m_i z_i \times (\omega \times z_i) \\
&= \sum_i m_i \left[z_i^2 \omega - (z_i \omega) z_i \right] \quad \text{(vector equation)} \\
\Longrightarrow l_{\text{rot}} &= \Theta \omega \quad \text{(coordinate equation in K$'$)} .
\end{aligned}
$$

This leads on to the rotational part N_{rot} (rotational torque) of the total torque:

$$
\begin{aligned}
N_{\text{rot}} &= D l_{\text{rot}} = D' l_{\text{rot}} + \omega \times l_{\text{rot}} \quad \text{(vector equation)} \\
\Longrightarrow N_{\text{rot}} &= \Theta \dot{\omega} + \omega \times l_{\text{rot}} \quad \text{(coordinate equation in K$'$)}.
\end{aligned}
$$

For the rotational part T_{rot} (rotational energy) of the total kinetic energy, we find

$$
\begin{aligned}
T_{\text{rot}} &= \frac{1}{2} \sum_i m_i (\omega \times z_i)^2 \\
&= \frac{1}{2} \sum_i m_i \left[z_i^2 \omega^2 - (z_i \omega)^2 \right] \quad \text{(vector equation)} \\
\Longrightarrow T_{\text{rot}} &= \frac{1}{2} \omega^{\text{T}} \Theta \omega \quad \text{(coordinate equation in K$'$)}.
\end{aligned}
$$

Theorem 1.28: Rotational momentum, torque, and energy

Denoting by ω the momentary angular velocity of a rigid body rotating relative to an inertial system K, and by Θ its inertial tensor in the body-fixed system K$'$, the rotational angular momentum, rotational torque, and rotational energy in coordinate representation with respect to K$'$ are given by

$$l_{\text{rot}} = \Theta\omega \;,\;\; N_{\text{rot}} = \Theta\dot{\omega} + \omega \times l_{\text{rot}} \;,\;\; T_{\text{rot}} = \frac{1}{2}\omega^{\text{T}}\Theta\omega \;.$$

Steiner's theorem. We now derive the relationship between the inertial tensor Θ in the body-fixed system K$'$, with origin in the rotation center, and the inertial tensor Θ^{C} relative to the body-fixed center of mass system KC that is parallel to K$'$ with origin in $\boldsymbol{\Delta} = \frac{1}{M}\sum_i z_i m_i$. For this, we shall assume a continuous mass density in order to get rid of the particle index. The inertial tensor Θ in terms of the KC-vectors $\boldsymbol{Z}_i = \boldsymbol{z}_i - \boldsymbol{\Delta}$, and with

$$\int \mathrm{d}^3 Z \rho(\boldsymbol{Z}) = M \;,\;\; \int \mathrm{d}^3 Z Z \rho(\boldsymbol{Z}) = \mathbf{0}$$

reads

$$\Theta_{ab} = \int \mathrm{d}^3 z \rho(\boldsymbol{z}) \left[z^2 \delta_{ab} - z_a z_b \right]$$

$$= \int \mathrm{d}^3 Z \rho(\boldsymbol{Z}) \left[(\boldsymbol{Z} + \boldsymbol{\Delta})(\boldsymbol{Z} + \boldsymbol{\Delta})\delta_{ab} - (Z_a + \Delta_a)(Z_b + \Delta_b) \right]$$

$$= \int \mathrm{d}^3 Z \rho(\boldsymbol{Z}) \left[\left(\boldsymbol{Z}^2 + \boldsymbol{\Delta}^2 \right) \delta_{ab} - (Z_a Z_b + \Delta_a \Delta_b) \right] \;.$$

From this, we find the following theorem:

Theorem 1.29: Theorem of Steiner

The relation between the inertial tensor in the body-fixed system K$'$, and that in the body-fixed center of mass system KC that is parallel to K$'$ and separated from it by a distance $\boldsymbol{\Delta}$, is

$$\Theta_{ab} = \Theta_{ab}^{\text{C}} + M \left(\boldsymbol{\Delta}^2 \delta_{ab} - \Delta_a \Delta_b \right) \;.$$

The difference between Θ and Θ^{C} is exactly the inertial tensor of a particle with mass M, which is at distance $\boldsymbol{\Delta}$ from the center of mass. So, if we know the inertial tensor of a rigid body relative to its center of mass, we can easily calculate the inertial tensor for any other point via Steiner's law.

Principal inertial moments and principal axes. The form of the inertial tensor Θ clearly depends on the choice of the body-fixed reference system K$'$: $\{e_1', e_2', e_3'\}$. Since the inertial tensor is real and symmetric in any system, there always exists a body-fixed system K$''$: $\{e_1'', e_2'', e_3''\}$, in which the inertial

tensor Ξ is diagonal. The transformation from K′ to such a *principal axis system* K″ (with the same origin) is given by an orthogonal rotation matrix D and the similarity transformation

$$\Xi = D\Theta D^{\mathrm{T}} = \begin{pmatrix} \Xi_1 & 0 & 0 \\ 0 & \Xi_2 & 0 \\ 0 & 0 & \Xi_3 \end{pmatrix} ,$$

where D is determined by the characteristic equation of Θ:

$$\Theta e_i'' = \Xi_i e_i'' , \ 0 \le \Xi_i \in \mathbb{R}, \ i = 1, 2, 3 \Longrightarrow D_{ij} = e_i'' e_j' .$$

The orthonormal eigenvectors of Θ (the basis vectors of K″), i.e., e_i'', are called *principal axes*; the eigenvalues of Θ, i.e., Ξ_i, are called *principal moments*.

1.4.3 Euler Angles and Lagrange Equations

We now come to the Lagrange equations for the general motion of a rigid body. From the above discussion, it is clear that the most general motion is described by six degrees of freedom corresponding to the three coordinates of the rotation center q, as well as the three coordinates of the angular velocity ω.[7]

Our problem, in order to derive the Lagrange function, is to find six independent generalized coordinates that determine the motion of the body uniquely. Three of these are clearly given by the coordinates q_1, q_2, and q_3 of the vector q. If we choose a principal axis system for our body-fixed reference system with origin in the center of mass ($z_{\mathrm{C}} = 0$), we can use (1.46) and Theorem 1.28 to find

$$T = \frac{M}{2} \left(\dot{q}_1^2 + \dot{q}_2^2 + \dot{q}_3^2 \right) + \frac{A}{2} \omega_1'^2 + \frac{B}{2} \omega_2'^2 + \frac{C}{2} \omega_3'^2 , \tag{1.47}$$

where A, B, and C denote the principal moments, and ω_i' the coordinates of ω with respect to K′. Furthermore, we have the *Euler equations*

$$\left. \begin{aligned} N_{\mathrm{rot},1}' &= A\dot{\omega}_1' + (C - B)\omega_2'\omega_3' \\ N_{\mathrm{rot},2}' &= B\dot{\omega}_1' + (A - C)\omega_1'\omega_3' \\ N_{\mathrm{rot},3}' &= C\dot{\omega}_1' + (B - A)\omega_1'\omega_2' . \end{aligned} \right\} \tag{1.48}$$

Our next task is to replace the body-fixed components ω_i' of (1.47) by three generalized coordinates that describe the purely rotational motion of the body. There are several ways of doing this. We use *Euler's method* and express the general rotation by a succession of three independent rotations. If the axes of the space-fixed system K and that of the body-fixed system K′ are

[7] Of course, the rigid body can by subject to more constraints (other than its rigidity), and this can reduce the number of degrees of freedom further.

parallel at some initial time, one can visualize the rotation of K′ relative to K as follows. Initially, a rotation of K by an angle ϕ around its third axis is performed. Subsequently, this interim system is rotated by an angle θ around its second axis. Finally, this system is rotated by an angle ψ around its third axis. In this way, each rotation can be parametrized by three independent quantities.

Definition: Euler angles

Let K and K′ be two orthonormal systems. The (time-dependent) rotation matrix R that turns K into K′ can be written in terms of the Euler angles ϕ, θ, and ψ as follows:

$$R = \begin{pmatrix} \cos\psi & \sin\psi & 0 \\ -\sin\psi & \cos\psi & 0 \\ 0 & 0 & 1 \end{pmatrix} \begin{pmatrix} \cos\theta & 0 & -\sin\theta \\ 0 & 1 & 0 \\ \sin\theta & 0 & \cos\theta \end{pmatrix} \begin{pmatrix} \cos\phi & \sin\phi & 0 \\ -\sin\phi & \cos\phi & 0 \\ 0 & 0 & 1 \end{pmatrix} ,$$

with

$$0 \le \psi \le 2\pi \ , \ 0 \le \theta \le \pi \ , \ 0 \le \phi \le 2\pi \ .$$

The angles are uniquely defined by this rotation if $|R_{33}| \ne 1$.

Note that the order of the individual rotations does matter. Using the relations (1.11) and (1.12), we can now calculate the components of $\boldsymbol{\omega}$ in the space-fixed system K. After some manipulations, we find

$$\omega_1 = -\dot\theta \sin\phi + \dot\psi \sin\theta \cos\phi$$
$$\omega_2 = \dot\theta \cos\phi + \dot\psi \sin\theta \sin\phi$$
$$\omega_3 = \dot\phi + \dot\psi \cos\theta \ .$$

The corresponding components in the body-fixed system K′ are obtained from $\omega_i' = \sum_j R_{ij}\omega_j$ as

$$\omega_1' = \dot\theta \sin\psi - \dot\phi \sin\theta \cos\psi$$
$$\omega_2' = \dot\theta \cos\psi + \dot\phi \sin\theta \sin\psi$$
$$\omega_3' = \dot\psi + \dot\phi \cos\theta \ .$$

Thus, our set of six independent generalized coordinates is $\{q_1, q_2, q_3, \phi, \theta, \psi\}$, and we can write the Lagrange function of the rigid body as follows:

$$\begin{aligned} L(q_1, q_2, q_3, \phi, \theta, \psi) = & \ \frac{M}{2} \left(\dot q_1^2 + \dot q_2^2 + \dot q_3^2 \right) \\ & + \frac{A}{2} \left(\dot\theta \sin\psi - \dot\phi \sin\theta \cos\psi \right)^2 \\ & + \frac{B}{2} \left(\dot\theta \cos\psi + \dot\phi \sin\theta \sin\psi \right)^2 \\ & + \frac{C}{2} \left(\dot\psi + \dot\phi \cos\theta \right)^2 - V \ . \end{aligned}$$

In the case of $V = 0$, the Lagrange equations for the angles are

$$A\dot{\omega}_1' = (B - C)\omega_2'\omega_3' \ , \quad B\dot{\omega}_2' = (C - A)\omega_1'\omega_3' \ , \quad C\dot{\omega}_3' = (A - B)\omega_1'\omega_2' \ ,$$

and if $N_{\text{rot}} = 0$, these equations are identical with the Euler equations (1.48).

Summary

- The motion of a **rigid body** can be split into a translational motion of an arbitrary point and a rotation of the body around this point.

- The dynamical quantities of the rotational motion can be expressed in terms of the **inertial tensor**. It is independent of the dynamics and depends only on the geometry of the body.

- **Steiner's theorem** states that knowledge of the inertial tensor in the center of mass reference system allows for a simple calculation of the inertial tensor for any arbitrary point.

- A purely rotational motion of the rigid body can be parametrized in terms of three independent **Euler angles**.

Applications

11. Physical pendulum. A solid cube of side length a and mass M rotates frictionless around one of its edges under the influence of gravity (Fig. 1.9). Calculate the oscillation frequency of the cube for small deviations from its equilibrium.

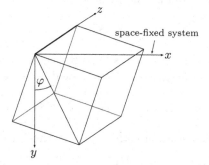

Fig. 1.9. Determination of the space-fixed and body-fixed system, as well as the body-fixed center of mass system of a rotating cube.

Solution. Since the rotation axis (and consequently the rotation center q somewhere along that axis) is fixed, we shift both the space-fixed system and the body-fixed system to one of the cube's corners bordering on the rotation axis. This means that $q = 0$ and $T = T_{rot}$. The body-fixed components of ω and $\mathbf{\Delta}$ are

$$\omega = \dot{\varphi} \begin{pmatrix} 0 \\ 0 \\ 1 \end{pmatrix} \ , \quad \mathbf{\Delta} = \frac{a}{2} \begin{pmatrix} 1 \\ 1 \\ 1 \end{pmatrix} \ .$$

We need the inertial tensor Θ with respect to the body-fixed system. We obtain it via the inertial tensor Θ^C of the body-fixed center of mass system using Steiner's law:

$$\Theta_{11}^C = \rho \int\limits_{-\frac{a}{2}}^{\frac{a}{2}} dZ_1 \int\limits_{-\frac{a}{2}}^{\frac{a}{2}} dZ_2 \int\limits_{-\frac{a}{2}}^{\frac{a}{2}} dZ_3 (Z_2^3 + Z_3^2) = \frac{Ma^2}{6} \ , \ \rho = \frac{M}{a^3}$$

$$= \Theta_{22}^C = \Theta_{33}^C$$

$$\Theta_{ij\neq i}^C = 0 \ .$$

Obviously, the KC-system is a principal axis system. Using Theorem 1.29, we find

$$\Theta = \frac{Ma^2}{12} \begin{pmatrix} 8 & -3 & -3 \\ -3 & 8 & -3 \\ -3 & -3 & 8 \end{pmatrix} \ .$$

The kinetic energy of the system is given by

$$T = T_{rot} = \frac{Ma^2}{3} \dot{\varphi}^2 \ .$$

The potential energy is equal to that of the center of mass,

$$V = -Mg \frac{a}{\sqrt{2}} \cos\varphi \ ,$$

where $a/\sqrt{2}$ denotes the distance between the center of mass and the rotation axis. Energy conservation yields an equation of motion for φ:

$$E = T_{rot} + V = \frac{Ma^2}{3} \dot{\varphi}^2 - Mg \frac{a}{\sqrt{2}} \cos\varphi = \text{const} \ .$$

Differentiation with respect to time gives

$$\frac{2Ma^2}{3} \dot{\varphi}\ddot{\varphi} + \frac{Mga}{\sqrt{2}} \dot{\varphi} \sin\varphi = 0 \Longleftrightarrow \ddot{\varphi} + \frac{3g}{2\sqrt{2}a} \sin\varphi = 0 \ .$$

For small deviations we have $\sin\varphi \approx \varphi$ and therefore

$$\ddot{\varphi} + \frac{g}{L'}\varphi = 0 \ , \ L' = \frac{2\sqrt{2}a}{3} < a \ .$$

This equation of motion is equivalent to that of a *planar mathematical pendulum* consisting of a point mass hanging from a massless rod of reduced length L'. Its general solution is given by

$$\varphi(t) = a \cos \omega t + b \sin \omega t \ , \ \omega = \sqrt{\frac{g}{L'}} \ ,$$

where the integration constants a and b are determined by initial conditions, for example, $\varphi(0) = \varphi_0$, $\dot{\varphi}(0) = \omega_0$.

12. Hollow cylinder rolling down a slope. Consider a massive but hollow cylinder with inner radius a, outer radius b, length h, and mass M rolling down a slope under the influence of gravity (Fig. 1.10). Determine and solve the equation of motion.

Fig. 1.10. Determination of the space-fixed and body-fixed systems of a rolling hollow cylinder.

Solution. For the body-fixed components of $\boldsymbol{\omega}$, we have

$$\boldsymbol{\omega} = \dot{\varphi} \begin{pmatrix} 0 \\ 0 \\ 1 \end{pmatrix} \ .$$

Since $\boldsymbol{\omega}$ only has a z_3-component, we only need to calculate Θ_{33} of the body-fixed inertial tensor in order to determine T_{rot}. Introducing cylindrical coordinates,

$$z_1 = r \cos \phi \ , \ z_2 = r \sin \phi \ , \ z_3 = z \ , \ \mathrm{d}^3 z = r \mathrm{d}r \mathrm{d}\phi \mathrm{d}z \ ,$$

one finds

$$\Theta_{33} = \rho \int_a^b r \mathrm{d}r \int_0^{2\pi} \mathrm{d}\phi \int_0^h \mathrm{d}z r^2 = \frac{\rho}{2}\pi h(b^4 - a^4) \ , \ \rho = \frac{M}{\pi h(b^2 - a^2)}$$

$$= \frac{M}{2}(b^2 + a^2)$$

$$\Longrightarrow T_{\text{rot}} = \frac{M}{4}(b^2 + a^2)\dot{\varphi}^2 \ .$$

Taking into account the rolling constraint

$$\dot{q} = -b\dot{\varphi} \ ,$$

the kinetic and potential energy of the system follows as

$$T = \frac{M}{2}\dot{q}^2 + T_{\text{rot}} = \frac{M}{4}(3b^2 + a^2)\dot{\varphi}^2 \ , \quad V = Mgq\sin\alpha = -Mgb\varphi\sin\alpha \ .$$

Energy conservation yields

$$E = T + V = \frac{M}{4}(3b^2 + a^2)\dot{\varphi}^2 - Mgb\varphi\sin\alpha = \text{const} \ .$$

Differentiation with respect to time results in

$$\frac{M}{2}(3b^2 + a^2)\ddot{\varphi} - Mgb\sin\alpha = 0 \iff \ddot{\varphi} - \frac{2gb\sin\alpha}{3b^2 + a^2} = 0 \ ,$$

which is equivalent to

$$\ddot{q} + g\sin\alpha' = 0 \ , \quad \sin\alpha' = \frac{2b^2}{3b^2 + a^2}\sin\alpha < \sin\alpha \ .$$

This equation is of the same form as that of a point mass rolling down a slope with reduced angle of inclination α'. Its general solution is evaluated by twofold integration over time as

$$q(t) = -\frac{g}{2}\sin\alpha' t^2 + v_0 t + q_0 = -b\varphi(t) \ ,$$

where the integration constants v_0 and q_0 are fixed via initial conditions.

1.5 Central Forces

In this section we discuss the class of central force problems, which are of great importance not just in classical mechanics. In preparation for subsequent sections, we initially deal with two-particle systems and show that their dynamics can, under certain circumstances, be separated into a uniform motion of the center of mass and an effective one-particle motion. We then deduce the *radial equation of motion* in conservative central force fields and determine the particle trajectory in $1/r$-potentials. An example of this type is given by the gravitational potential, whose form is derived using *Kepler's laws*. The last two subsections deal with the scattering of particles in central potentials.

1.5.1 Two-Particle Systems

With a view to following subsections, we first consider the dynamics of two-particle systems in the absence of any external forces. We assume that the particles exert an inner conservative force on one another that depends only on their distance. Newton's equations for the two particles are then given by

$$\left.\begin{aligned} m_1\ddot{\boldsymbol{x}}_1 &= -\boldsymbol{\nabla}_1 V(\boldsymbol{x}_1 - \boldsymbol{x}_2) \\ m_2\ddot{\boldsymbol{x}}_2 &= -\boldsymbol{\nabla}_2 V(\boldsymbol{x}_1 - \boldsymbol{x}_2) = \boldsymbol{\nabla}_1 V(\boldsymbol{x}_1 - \boldsymbol{x}_2) \,. \end{aligned}\right\} \tag{1.49}$$

Introducing the *relative* and *center of mass coordinates*,

$$\boldsymbol{x}_\mathrm{R} = \boldsymbol{x}_1 - \boldsymbol{x}_2 \,, \ \boldsymbol{x}_\mathrm{C} = \frac{m_1\boldsymbol{x}_1 + m_2\boldsymbol{x}_2}{M} \,,$$

as well as the *reduced mass* and the total mass,

$$\mu = \frac{m_1 m_2}{m_1 + m_2} \,, \ M = m_1 + m_2 \,,$$

and subtracting the two equations (1.49) gives the relative motion of the particles:

$$\mu\ddot{\boldsymbol{x}}_\mathrm{R} = \dot{\boldsymbol{p}}_\mathrm{R} = -\boldsymbol{\nabla}_\mathrm{R} V(\boldsymbol{x}_\mathrm{R}) \,, \ \boldsymbol{p}_\mathrm{R} = \mu\dot{\boldsymbol{x}}_\mathrm{R} \,.$$

Bar the substitutions $m \leftrightarrow \mu$, $\boldsymbol{x} \leftrightarrow \boldsymbol{x}_\mathrm{R}$, and $\boldsymbol{p} \leftrightarrow \boldsymbol{p}_\mathrm{R}$, there is formally no difference between the relative motion of the two-particle system and the absolute motion of a single particle. We can therefore interpret the relative motion as an *effective motion* of a single particle of reduced mass μ. Adding the two equations (1.49), we find the motion of the center of mass as

$$M\ddot{\boldsymbol{x}}_\mathrm{C} = \dot{\boldsymbol{p}}_\mathrm{C} = \boldsymbol{0} \implies \dot{\boldsymbol{x}}_\mathrm{C} = \mathrm{const} \,, \ \boldsymbol{p}_\mathrm{C} = M\dot{\boldsymbol{x}}_\mathrm{C} \,.$$

Thus, the center of mass of the two-particle system performs a uniform motion at constant velocity.

Since Newtonian mechanics is Galilei-invariant, we may switch to an inertial system that moves along with the center of mass. In this center of mass system, denoted by a star *, the equations for relative and center of mass motion are given by

$$\mu\ddot{\boldsymbol{x}}_\mathrm{R}^* = -\boldsymbol{\nabla}_\mathrm{R} V(\boldsymbol{x}_\mathrm{R}^*) \,, \ \boldsymbol{x} = \boldsymbol{x}_\mathrm{C} + \boldsymbol{x}^* \,, \ \boldsymbol{x}_\mathrm{C}^* = \boldsymbol{0} \,.$$

If the mass of the second particle is very much larger than that of the first – this is true, for example, in the case of planets orbiting the sun (see the following two subsections) – we find

$$m_2 \gg m_1 \implies \left\{\begin{aligned} M &\to m_2 \\ \mu &\to m_1 \\ \boldsymbol{x}_\mathrm{C}^* &\to \boldsymbol{x}_2^* = \boldsymbol{0} \\ \boldsymbol{x}_\mathrm{R}^* &\to \boldsymbol{x}_1^* \end{aligned}\right\} \implies m_1\ddot{\boldsymbol{x}}_1^* = -\boldsymbol{\nabla}_1 V(\boldsymbol{x}_1^*) \,, \ \boldsymbol{x}_2^* = \boldsymbol{0} \,.$$

In this case, we can view the motion of the two-particle system from the rest system of the heavier particle and apply Newton's equation of motion to the lighter particle.

1.5.2 Conservative Central Problems, $1/r$-Potentials

This subsection deals with the general motion of particles in $1/r$-potentials. Following Theorem 1.4, the angular momentum of a particle in a central force field is a conserved quantity. From this, two statements can be deduced:

1. Since the direction of the angular momentum is conserved, the motion of the particles must lie in a plane. We can therefore choose the coordinate system with its origin in the force center, such that the motion is confined for all times to the xy-plane. In polar coordinates, this means

$$\left.\begin{aligned}
\boldsymbol{x}(t) &= r(t)\begin{pmatrix} \cos\varphi(t) \\ \sin\varphi(t) \\ 0 \end{pmatrix} \\
\dot{\boldsymbol{x}}(t) &= \dot{r}\begin{pmatrix} \cos\varphi \\ \sin\varphi \\ 0 \end{pmatrix} + r\dot{\varphi}\begin{pmatrix} -\sin\varphi \\ \cos\varphi \\ 0 \end{pmatrix} \\
\ddot{\boldsymbol{x}}(t) &= (\ddot{r} - r\dot{\varphi}^2)\begin{pmatrix} \cos\varphi \\ \sin\varphi \\ 0 \end{pmatrix} + (2\dot{r}\dot{\varphi} + r\ddot{\varphi})\begin{pmatrix} -\sin\varphi \\ \cos\varphi \\ 0 \end{pmatrix} \\
\boldsymbol{l}(t) &= m\boldsymbol{x} \times \dot{\boldsymbol{x}} = m\begin{pmatrix} 0 \\ 0 \\ r^2\dot{\varphi} \end{pmatrix} .
\end{aligned}\right\} \tag{1.50}$$

2. Conservation of the absolute value of the angular momentum implies that

$$|\boldsymbol{l}| = mr^2\dot{\varphi} = \text{const} \implies r^2 d\varphi = \text{const} ,$$

which means that the line connecting the force center and the particle covers equal areas in equal times. This is the content of *Kepler's second law*, which we consider in detail in the next subsection.

In addition to the conservation of angular momentum, the total energy of the system is also conserved:

$$E = \frac{m}{2}(\dot{r}^2 + r^2\dot{\varphi}^2) + V(r) = \frac{m}{2}\dot{r}^2 + \frac{l^2}{2mr^2} + V(r) = \text{const} .$$

This relation corresponds to energy conservation of a one-dimensional particle in the *effective potential*

$$V_{\text{eff}}(r) = V(r) + \frac{l^2}{2mr^2} .$$

The last term is equivalent to a (fictitious) centrifugal force that pushes the particle away from the center of motion. This term is therefore called the

centrifugal barrier. Since we are dealing with a one-dimensional problem, we can immediately write down the corresponding equation of motion, analogous to (1.8):

**Theorem 1.30: Radial equation of motion
in a conservative central force field**

The equation of motion of a particle in a conservative central force is given by

$$\dot{r} = \pm\sqrt{\frac{2}{m}[E - V_{\text{eff}}(r)]}$$

$$\Longleftrightarrow dt = \frac{\pm dr}{\sqrt{\frac{2}{m}[E - V_{\text{eff}}(r)]}} \ , \quad V_{\text{eff}}(r) = V(r) + \frac{l^2}{2mr^2} \ . \tag{1.51}$$

The two signs distinguish between the two possible directions of velocity.

In terms of r, the particle can only be found in regions for which $E - V_{\text{eff}}(r) \geq 0$. Consider, for example, the effective potential of Fig. 1.11. The particle can

Fig. 1.11. Effective potential.

either oscillate indefinitely between the points of return r_1 and r_2, or it can be found in the region $r \geq r_3$, eventually escaping from the force field, $r \to \infty$. The sign in the radial equation of motion is determined by the initial conditions, for example, of the type

$$r_1 \leq r \leq r_2 \ , \ \dot{r}(0) = v_0 \gtrless 0 \ ,$$

and remains fixed until the particle reaches the next point of return. If we are more interested in the geometrical form of the particle trajectory, it is more instructive to rewrite (1.51) as

$$d\varphi = \frac{l}{mr^2} dt = \pm\frac{l dr}{r^2\sqrt{2m[E - V_{\text{eff}}(r)]}} \ .$$

Particle trajectories in $1/r$-potentials. Starting from the previous rela-
tion, we will now study all possible trajectories of particles in Coulomb-like
potentials $V(r) = \alpha/r$, i.e.,

$$\varphi - \varphi_0 = \pm \int_{r_0}^{r} \frac{l\, dr'}{r'^2 \sqrt{2m\left(E - \frac{\alpha}{r'}\right) - \frac{l^2}{r'^2}}}. \qquad (1.52)$$

If we set $\varphi_0 = 0$, this standard integral is solved to be

$$\varphi(r) = -\arcsin \frac{l^2/r' + m\alpha}{m\alpha\epsilon}\Big|_{r_0}^{r},$$

with

$$\epsilon = \pm\sqrt{1 + \frac{2El^2}{m\alpha^2}}. \qquad (1.53)$$

Fixing r_0 to $\arcsin \frac{l^2/r_0 + m\alpha}{m\alpha\epsilon} = -\pi/2$ finally yields *conic sections* in polar
coordinates:

$$r(\varphi) = \frac{k}{1 + \epsilon \cos \varphi}, \quad k = -\frac{l^2}{m\alpha}. \qquad (1.54)$$

The sign in (1.53) must be chosen such that $r > 0$ is always satisfied in (1.54).
The following cases can be distinguished:

Attractive case: $\alpha < 0 \Longrightarrow k > 0$.

- $E < 0 \Longrightarrow |\epsilon| < 1 \Longrightarrow r = \frac{k}{1 + |\epsilon| \cos \varphi} > 0$ for $\varphi \in [0 : \pi]$. In the upper
half-plane, φ runs through the interval $[0 : \pi]$. The geometrical form of the
curve is an ellipse (Fig. 1.12a) with large and small axes given by $a = \frac{k}{1-\epsilon^2}$,
$b = \frac{k}{\sqrt{1-\epsilon^2}}$ [see Fig. 1.13 and (1.55) in the next subsection]. r oscillates
between the points of return $r_1 = \frac{k}{1-\epsilon}$ and $r_2 = \frac{k}{1+\epsilon}$. For $\epsilon = 0$, the ellipse
turns into a circle of radius $R = r_1 = r_2 = 1$.

- $E > 0 \Longrightarrow |\epsilon| > 1 \Longrightarrow r = \frac{k}{1 + |\epsilon| \cos \varphi} > 0$ for $\varphi \in [0 : \varphi_{\max}[$,
$\cos \varphi_{\max} = -1/|\epsilon|$. The trajectory is a hyperbola (Fig. 1.12b).

- $E = 0 \Longrightarrow |\epsilon| = 1 \Longrightarrow r = \frac{k}{1 + \cos \varphi} > 0$ for $\varphi \in [0 : \pi[$. The trajectory is a
parabola (Fig. 1.12c).

Repulsive case: $\alpha > 0 \Longrightarrow E > 0$, $k < 0$, $|\epsilon| > 1$
$\Longrightarrow r = \frac{-|k|}{1 - |\epsilon| \cos \varphi} > 0$ for $\varphi \in [0 : \varphi_{\max}[$, $\cos \varphi_{\max} = 1/|\epsilon|$. Again, the
trajectory is a hyperbola (Fig. 1.12d).

Of all possible cases, only the attractive one with $\alpha < 0$ and $E < 0$ leads
to a bounded elliptic motion. In all other cases, the particle approaches the
force center from infinitely far away only to shoot off into the distance again.
In the case of elliptic motion, the quantity ϵ is called *eccentricity*. It is a

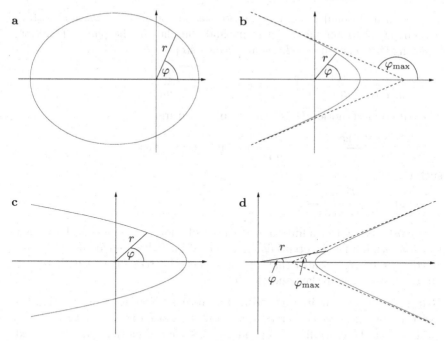

Fig. 1.12. Possible trajectories in a Coulomb potential. **a, b,** and **c** correspond to the attractive case and **d** to the repulsive one.

measure of the deviation of the particle's elliptical trajectory from a pure circle ($\epsilon = 0$).[8] For the half-axes, in terms of physical quantities, we find

$$a = \frac{\alpha}{2E} \,,\ b = \frac{l}{\sqrt{2m|E|}} \,.$$

Thus, the energy is independent of the absolute value of angular momentum and solely determined by the large half-axis a.

Eccentricity and Runge-Lenz vector. We briefly show an alternative derivation of (1.54). For this, consider the vector $l \times p$. Using $\dot{p} = \alpha x / r^3$ and $r = |x|$, we calculate its derivative with respect to time:

$$\frac{d}{dt}(l \times p) = l \times \dot{p} = \frac{\alpha}{r^3} l \times x = -\frac{m\alpha}{r^3} \left[x(x\dot{x}) - \dot{x}r^2 \right] = m\alpha \frac{d}{dt}\left(\frac{x}{r} \right) \,.$$

For the vector of eccentricity, it follows that

$$\epsilon = \frac{l \times p}{m\alpha} - \frac{x}{r} = \text{const} \,.$$

[8] The eccentricity of the moon's motion around the earth and of the earth's motion around the sun is $\epsilon = 0.055$ and $\epsilon = 0.017$, respectively, so that their trajectories are very close to circular.

This is an additional integral of motion called the *Runge-Lenz vector*. It is perpendicular to angular momentum and thus lies in the plane of motion. Squaring this relation, one finds after some steps

$$\epsilon^2 = \frac{2l^2}{m\alpha^2}\left(\frac{p^2}{2m} + \frac{\alpha}{r}\right) + 1 \implies \epsilon = \pm\sqrt{1 + \frac{2El^2}{m\alpha^2}} \ .$$

The equation of motion (1.54) can be obtained from

$$\epsilon\boldsymbol{x} = \frac{(\boldsymbol{l}\times\boldsymbol{p})\boldsymbol{x}}{m\alpha} - r = -\frac{l^2}{m\alpha} - r \quad \text{and} \quad \epsilon\boldsymbol{x} = \epsilon r\cos\varphi \ ,$$

such that

$$r(\varphi) = -\frac{l^2/m\alpha}{1 + \epsilon\cos\varphi} \ .$$

The existence of the additional constant of motion can be used to show that the orbits in $1/r$-potentials for bounded particles are closed. Generally, *Bertrand's theorem* states that the only central forces for which this holds are the inverse square law and Hooke's law.[9]

Runge-Lenz vector in quantum mechanics. Note that the relation between ϵ and the four conserved quantities E and \boldsymbol{l} together with $\epsilon \cdot \boldsymbol{l} = 0$ implies that there are five independent constants of motion. In the context of quantum mechanics (Subsection 3.5.5), we shall find that the spectrum of the hydrogen atom exhibits an "accidental degeneracy": states with the same principal quantum number n and different orbital angular quantum numbers l are degenerate. This indicates that the Hamilton operator \boldsymbol{H} has other symmetries besides rotational invariance and one can construct an operator from the components of the quantum analogue of the Runge-Lenz vector that commutes with \boldsymbol{H}. The effect of this operator is to raise the value of l by one, thus explaining the degeneracy of l at each n.

1.5.3 Kepler's Laws and Gravitational Potential

The gravitational force between two point masses m and M,

$$\boldsymbol{F}_{\mathrm{G}}(\boldsymbol{x}) = -\boldsymbol{\nabla}V_{\mathrm{G}}(|\boldsymbol{x}|) \ , \quad V_{\mathrm{G}}(|\boldsymbol{x}|) = V_{\mathrm{G}}(r) = -\gamma\frac{mM}{r} \ ,$$

is a specific example of attractive $1/r$-potentials of the type we have just discussed. From our discussion so far, we can immediately deduce Kepler's three laws:[10]

1. Every planet of the solar system moves on an elliptical trajectory around the sun, which is one of the focal points of the ellipse.

[9] The interested reader is referred to [6].

[10] Within these laws, the interaction of the planets among each other as well as the motion of the sun are neglected.

2. The connecting vector between the sun and a planet covers equal areas in equal times.

3. The squares of the orbit periods are proportional to the cubes of the larger half-axis of two planets' trajectories.

We shall now follow the reverse path and show that the form of the gravitational force can be deduced completely from Kepler's three laws.

Although the sun and the planets have a spatial structure, it is valid to consider them here as being point particles since the distance between sun and planet is so much larger than their extent. We shall see at the end of the following discussion that this point of view is exact if the sun and the planets are spherical objects.

Choosing an inertial system with origin in the sun (this is allowed since the sun's mass is so much larger than that of the planets), we see from Kepler's first law that the trajectory of a planet lies in a plane, and therefore the direction of its angular momentum relative to the sun is conserved. We can again choose our inertial system such that the relations (1.50) hold. Kepler's second law implies conservation of the absolute value of the angular momentum:

$$l = mr^2\dot{\varphi} = \text{const} .$$

Thus, the force between sun and planets is central,

$$\frac{d(r^2\dot{\varphi})}{dt} = r(2\dot{r}\dot{\varphi} + r\ddot{\varphi}) = 0 ,$$

and we can write

$$\boldsymbol{F}_{\text{G}} = m(\ddot{r} - r\dot{\varphi}^2)\begin{pmatrix} \cos\varphi \\ \sin\varphi \\ 0 \end{pmatrix} .$$

From the geometrical form of the ellipse (see Fig. 1.13), we obtain the desired relation between r and φ. The ellipse is defined as the set of points that have the same distance from the two focal points B and B'. Denoting by a the large, and by b the small half-axis, the ellipse is parametrized by $r + r' = 2a$. After a little algebra, this results in

$$r = \frac{k}{1 + \epsilon\cos\varphi} , \quad k = \frac{b^2}{a} , \quad \epsilon = \frac{\sqrt{a^2 - b^2}}{a} < 1 \qquad (1.55)$$

$$\Longrightarrow \dot{r} = \frac{\epsilon}{k}r^2\dot{\varphi}\sin\varphi = \frac{\epsilon}{k}h\sin\varphi , \quad h = r^2\dot{\varphi}$$

$$\Longrightarrow \ddot{r} = \frac{\epsilon}{k}h\dot{\varphi}\cos\varphi = \frac{\epsilon}{k}\frac{h^2}{r^2}\cos\varphi = \frac{h^2}{kr^2}\left(\frac{k}{r} - 1\right) .$$

For $\boldsymbol{F}_{\text{G}}$, we find

$$\boldsymbol{F}_{\text{G}}(\boldsymbol{x}) = m\left[\frac{h^2}{kr^3}\left(\frac{k}{r} - 1\right) - \frac{h^2}{r^4}\right]\boldsymbol{x} = -m\frac{h^2}{k}\frac{\boldsymbol{x}}{|\boldsymbol{x}|^3}$$

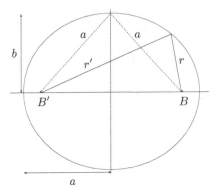

Fig. 1.13. Definition of an ellipse.

and

$$\boldsymbol{F}_G(\boldsymbol{x}) = -\boldsymbol{\nabla} V_G(\boldsymbol{x}) \ , \ V_G(\boldsymbol{x}) = -m\frac{h^2}{k}\frac{1}{|\boldsymbol{x}|} \ .$$

Using Kepler's third law, we can now show that the quantity h^2/k is planet-independent since the area F of the ellipse is given by (T=period of the orbit)

$$F = \pi ab = \int_0^T \mathrm{dt} \frac{\dot{\varphi} r^2}{2} = \frac{Th}{2} \ ,$$

so that

$$\frac{T^2}{a^3} = \frac{4\pi^2 a^2 b^2}{a^3 h^2} = \frac{4\pi^2 b^2}{a h^2} = 4\pi^2 \frac{k}{h^2} = \text{const} \ .$$

Because of the principle "actio=reactio", h^2/k must be proportional to the sun's mass M. In total, we have

Theorem 1.31: Gravitational potential V_G

The gravitational potential between two point masses m and M is obtained from Kepler's laws as

$$V_G(|\boldsymbol{x}|) = V_G(r) = -\gamma\frac{mM}{r} \ , \ \gamma = 6.67 \cdot 10^{-11} \frac{\text{Nm}^2}{\text{kg}^2} \ .$$

γ is a natural constant, called the *gravitational constant*, which must be determined experimentally.

Gravitational potential of a hollow sphere. In the following, we calculate the gravitational potential of a hollow sphere with inner radius a, outer radius b, and homogeneous mass distribution $\rho(\boldsymbol{x},t) = \rho = \frac{3M}{4\pi(b^3-a^3)}$ (Fig. 1.14). To this end, we imagine a probe particle of mass m at distance r to

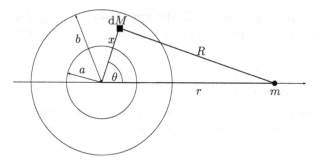

Fig. 1.14. Hollow sphere.

the center of the sphere. In polar coordinates, the potential energy dV due to the mass element dM is given by

$$dV(r) = -\gamma \frac{m\,dM(x,\theta,\varphi)}{R(r,x,\theta)} \ ,$$

with

$$dM(x,\theta,\varphi) = \rho x^2 dx \sin\theta d\theta d\varphi$$

$$R(r,x,\theta) = \sqrt{r^2 + x^2 - 2rx\cos\theta} \Longrightarrow \sin\theta d\theta = \frac{R\,dR}{rx} \ .$$

To find the total potential, we need to integrate over x, θ, and φ:

$$V(r) = -\gamma m\rho \int_a^b dx \int_0^{2\pi} d\varphi \int_0^\pi d\theta \frac{x^2 \sin\theta}{R(r,x,\theta)}$$

$$= -\frac{A}{r} \int_a^b dx\,x \int_{R_{\min}(r,x)}^{R_{\max}(r,x)} dR \ , \quad A = 2\pi\gamma m\rho \ .$$

Now, there are three cases to be distinguished:

1. $r \geq b$: in this case, $R_{\min} = r - x$ and $R_{\max} = r + x$, so that

$$V_1(r) = -2\frac{A}{r} \int_a^b dx\,x^2 = -\frac{2}{3}\frac{A}{r}\left(b^3 - a^3\right) = -\gamma\frac{mM}{r} \ .$$

2. $r \leq a$: $R_{\min} = x - r$, $R_{\max} = x + r$. It follows that

$$V_2(r) = -2A \int_a^b dx\,x = -A\left(b^2 - a^2\right) \ .$$

3. $a \leq r \leq b$: in this case, the potential can be thought of as composed of a
hollow sphere with radii a and r (first case) and a spherical shell with radii
r and b (second case):

$$V_3(r) = -A \left(b^2 - \frac{2}{3}\frac{a^3}{r} - \frac{1}{3}r^2 \right) .$$

From the first case, we see that for $r > b$, a hollow sphere with homogeneous
mass distribution is gravitationally equivalent to a point particle of the same
mass.

Gravitational potential on the earth's surface. For physical problems
in the vicinity of the earth's surface we have

$$r = R + r' , \quad r' \ll R ,$$

where $R = 6.35 \cdot 10^6$ m is the radius of the earth and r' is the radial distance
of the particle to the surface. The gravitational potential of the earth can
therefore be expanded in R, and with the earth's mass $M = 5.98 \cdot 10^{24}$ kg,
we find

$$V_G(r) = -\frac{\gamma m M}{r} = -mgR + mgr' + \dots , \quad g \approx 9.8 \, \frac{\text{N}}{\text{kg}} = 9.8 \, \frac{\text{m}}{\text{s}^2} .$$

Here, g denotes the *gravitational acceleration*. Since the term mgR is just an
irrelevant constant, it can be ignored, giving the gravitational potential with
respect to the earth's surface as

$$V_{GS}(r') \approx mgr' .$$

1.5.4 Elastic One-Particle Scattering by a Fixed Target

The interactions of microscopic objects such as molecules, atoms, nuclei, and
elementary particles are difficult to investigate, since they are not directly
accessible to our senses. Thus, we need methods that enlarge the effects and
make them visible to us. The scattering of particles by one another provides
such an indirect measurement method. Knowing the positions and velocities
of the particles before and after the scattering takes place, one can draw
conclusions with respect to the underlying scattering forces. In practice, this
proceeds by trial and error, in that one calculates the effect of a few physi-
cally plausible types of interaction and then compares theoretical predictions
to experimental results. Although such scattering processes often require a
quantum mechanical treatment (see Section 3.10), in many cases, a purely
classical description provides a very good approximation.

In this subsection we consider the scattering of particles on a fixed tar-
get (*force center*). The subsequent subsection deals with the more general
situation of *two-particle scattering*, where particles (or particle beams) are
fired at one another from opposite directions. We shall always assume that
the interactions can be described by rotationally symmetric potentials that

fall off fast enough, so that both particles can be regarded as quasi-free long before and long after the scattering process.

For one-particle scattering one is generally confronted with the situation shown in Fig. 1.15. A particle is moving linearly in the z-direction with con-

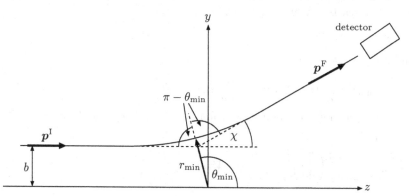

Fig. 1.15. One-particle scattering against a fixed scattering center.

stant initial momentum $p^{\mathrm{I}} = mv_0$ at vertical distance b toward the scattering center. Within a relatively localized area, the particle is deflected, and then continues in a linear motion with final momentum p^{F} until it is registered by a detector situated far away from the scattering zone. For the scattering angle χ, we have

$$\cos\chi = \frac{p^{\mathrm{I}}p^{\mathrm{F}}}{|p^{\mathrm{I}}||p^{\mathrm{F}}|} \, . \tag{1.56}$$

The entire process is determined macroscopically by the initial velocity v_0, the *impact parameter* b, and the interaction potential $V(r)$.

We now put the origin of our coordinate system into the scattering center, and look at the dynamics of the scattering process using polar coordinates

$$\begin{pmatrix} y \\ z \end{pmatrix} = r \begin{pmatrix} \sin\theta \\ \cos\theta \end{pmatrix}.$$

- **Initial conditions:**

$$r(0) = \infty \, , \quad \theta(0) = \pi \, , \quad \dot{r}(0) = -v_0 \, , \quad \dot{\theta}(0) = 0 \, .$$

- **Conservation of total energy:**

$$E = \frac{p^2}{2m} + V(r) = \frac{m}{2}\left(\dot{r}^2 + r^2\dot{\theta}^2\right) + V(r) = \frac{m}{2}v_0^2 = \text{const} \, .$$

In particular, the absolute values of the particle's momentum long before and long after the scattering $[V(r = \infty) = 0]$ are identical: $|p^{\mathrm{I}}| = |p^{\mathrm{F}}|$.

- **Conservation of total angular momentum:**

$$|\boldsymbol{l}| = l = mr^2\dot{\theta} = mbv_0 = \text{const} \implies dt = \frac{mr^2}{l}d\theta .$$

- **Particle trajectory:** the trajectory is symmetric with respect to the r_{\min}-axis. The scattering angle χ is therefore

$$\chi = 2\theta_{\min} - \pi ,$$

and for r_{\min}, we have

$$\dot{r}(r_{\min}) = \frac{dr}{dt}(r_{\min}) = 0 \implies \frac{dr}{d\theta}(r_{\min}) = 0 . \tag{1.57}$$

Taking into account these points, we can now evaluate θ_{\min}, and consequently χ:

$$\frac{m}{2}v_0^2 = \frac{l^2}{2mr^4}\left(\frac{dr}{d\theta}\right)^2 + \frac{l^2}{2mr^2} + V(r)$$

$$\implies \left(\frac{d\theta}{dr}\right)^2 = -\frac{1}{r^2}\left[1 - \frac{r^2}{b^2}\left(1 - \frac{2V(r)}{mv_0^2}\right)\right]^{-1}$$

$$\implies \theta - \theta_0 = \pm \int_{r_0}^{r} \frac{dr'}{r'}\left|1 - \frac{r'^2}{b^2}\left(1 - \frac{2V(r')}{mv_0^2}\right)\right|^{-1/2} .$$

Obviously, the sign of the square root in the integrand is the same as the sign of $d\theta/dr$ or $dr/d\theta$, which, for $\theta_{\min} \le \theta \le \pi$, is positive. Setting $\theta_0 = \pi$, $\theta = \theta_{\min}$, $r_0 = \infty$, $r = r_{\min}$, we find

Theorem 1.32: Scattering angle χ

$$\chi = 2\theta_{\min} - \pi$$

$$\theta_{\min} = \theta_{\min}(b, v_0) = \pi - \int_{r_{\min}}^{\infty} \frac{dr}{r}\left|1 - \frac{r^2}{b^2}\left(1 - \frac{2V(r)}{mv_0^2}\right)\right|^{-1/2} .$$

According to (1.57), the minimal distance $r_{\min} = r_{\min}(b, v_0)$ is determined by the zeros of the integrand.

A scattering experiment will generally not be set up as a single particle hitting a target. Rather, one has a beam of similar particles moving with the same velocity v_0 toward the scattering center. Since this beam shows a small but finite cross section, different particles will generally have different impact parameters and will therefore be deflected at different angles. This circum-

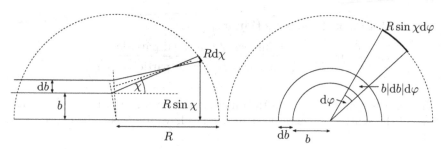

Fig. 1.16. Particle beam of one-particle scattering. *Left:* perpendicular to beam direction. *Right:* in the beam's direction.

stance is clarified in Fig. 1.16. All particles entering the segment $b|db|d\varphi$ will hit the solid angle $R^2 d\Omega = R^2 \sin\chi d\chi d\varphi$, where they are registered by a detector. Thus, the detector actually measures the *cross section*, which is defined as follows:

Definition: Scattering cross section $d\sigma$

$$d\sigma = \frac{(\text{Number of scattered particles toward } d\Omega)/s}{(\text{Number of incoming particles})/s/m^2} = \frac{Ib|db|d\varphi}{I},$$

with $I =$ particle current

$$\Longrightarrow d\sigma = b|db|d\varphi = b(\chi)\left|\frac{db}{d\chi}\right| d\varphi d\chi. \tag{1.58}$$

The *differential cross section* is obtained by normalizing $d\sigma$ to the unit sphere solid angle element $d\Omega$:

Definition: Differential scattering cross section $d\sigma/d\Omega$ for one-particle scattering

$$\frac{d\sigma}{d\Omega} = \frac{1}{\sin\chi} b(\chi)\left|\frac{db}{d\chi}\right|. \tag{1.59}$$

Thinking in terms of the experiment, one obtains the differential cross section by dividing the measured cross section by the detector surface. The differential cross section is therefore independent of the detector's geometry. Note that (1.58) and (1.59) are also valid for potentials that are not rotationally symmetric. However, things are then a little more complicated as the impact parameter b depends also on the azimuth angle φ. Integration of $d\sigma/d\Omega$ over $d\Omega$ yields the *total cross section* σ_{tot}:

Definition: Total cross section σ_{tot}

$$\sigma_{\text{tot}} = \int d\Omega \frac{d\sigma}{d\Omega} = \frac{(\text{Number of scattered particles})/s}{(\text{Number of incoming particles})/s/m^2} \ . \tag{1.60}$$

Like $d\sigma$, the quantity σ_{tot} has the dimension of area. It equals the (fictitious) area of the scattering center that projectiles have to cross to be deflected at all.

1.5.5 Elastic Two-Particle Scattering

We now turn to the more realistic case where a fixed scattering center no longer exists. Instead, particles are shot against one another. As already discussed in Subsection 1.5.1, we can split the two-particle dynamics into a relative motion,

$$\mu\ddot{\boldsymbol{x}}_R = -\boldsymbol{\nabla}V(|\boldsymbol{x}_R|) \ , \ \boldsymbol{x}_R = \boldsymbol{x}_1 - \boldsymbol{x}_2 \ , \ \mu = \frac{m_1 m_2}{M} \ , \ M = m_1 + m_2 \ ,$$

and a center of mass motion,

$$M\dot{\boldsymbol{x}}_C = \text{const} \Longrightarrow E_C = \frac{M}{2}\dot{\boldsymbol{x}}_C^2 = \text{const} \ , \ \boldsymbol{x}_C = \frac{m_1\boldsymbol{x}_1 + m_2\boldsymbol{x}_2}{M} \ .$$

The relative motion corresponds to an effective one-particle problem and can be regarded as the scattering of a single particle with reduced mass μ on a fixed target at $\boldsymbol{x}_R = \boldsymbol{0}$. For this, the same conservation laws hold as discussed in the preceding subsection:

- $E_R = \frac{\mu}{2}\dot{\boldsymbol{x}}_R^2 + V(\boldsymbol{x}_R) = \text{const} \Longrightarrow |\boldsymbol{p}_R^I| = |\boldsymbol{p}_R^F| \ , \ \boldsymbol{p}_R = \mu\dot{\boldsymbol{x}}_R$

- $\boldsymbol{l}_R = \mu\boldsymbol{x}_R \times \dot{\boldsymbol{x}}_R = \text{const} \ .$

Of course, from the conservation of the center of mass and the relative energy, overall energy conservation also follows:

$$E = E_C + E_R = \frac{m_1}{2}\dot{\boldsymbol{x}}_1^2 + \frac{m_2}{2}\dot{\boldsymbol{x}}_2^2 + V(|\boldsymbol{x}_1 - \boldsymbol{x}_2|) = \text{const} \ .$$

The uniform center of mass motion does not influence the dynamics of the scattering process and is simply a consequence of the choice of our space-fixed inertial system from which the scattering is observed (Galilei invariance). We can therefore use the center of mass system, denoted by a star *, in which the center of mass rests at all times:

$$\boldsymbol{x}_i = \boldsymbol{x}_C + \boldsymbol{x}_i^* \ , \ \boldsymbol{x}_C^* = \boldsymbol{0} \ .$$

The center of mass system can be advantageous compared to the laboratory system (where the second particle is at rest long before the scattering takes place), as it allows for a simpler relation between scattering angle χ and impact parameter b. Using $m_1\dot{\boldsymbol{x}}_1^* + m_2\dot{\boldsymbol{x}}_2^* = \boldsymbol{0}$, the center of mass momenta $\boldsymbol{p}_i^* = m_i\dot{\boldsymbol{x}}_i^*$ obey

$$\boldsymbol{p}_1^* = -\boldsymbol{p}_2^* = \frac{m_1 m_2}{m_1 + m_2}\left(\dot{\boldsymbol{x}}_1^* + \frac{m_1}{m_2}\dot{\boldsymbol{x}}_1^*\right) = \frac{m_1 m_2}{m_1 + m_2}(\dot{\boldsymbol{x}}_1^* - \dot{\boldsymbol{x}}_2^*) = \boldsymbol{p}_\mathrm{R}\ .$$

Figure 1.17 shows the process of two-particle scattering in the center of mass reference system. The particles are moving from infinity toward one another

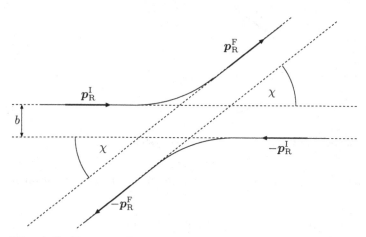

Fig. 1.17. Two-particle scattering in the center of mass system.

with initial momenta $\boldsymbol{p}_1^{\mathrm{I}*} = -\boldsymbol{p}_2^{\mathrm{I}*} = \boldsymbol{p}_\mathrm{R}^{\mathrm{I}}$. After the scattering, the momenta $\boldsymbol{p}_1^{\mathrm{F}*} = -\boldsymbol{p}_2^{\mathrm{F}*} = \boldsymbol{p}_\mathrm{R}^{\mathrm{F}}$ are rotated by the scattering angle χ relative to their initial values [see (1.56)]:

$$\cos\chi = \frac{\boldsymbol{p}_\mathrm{R}^{\mathrm{I}}\boldsymbol{p}_\mathrm{R}^{\mathrm{F}}}{|\boldsymbol{p}_\mathrm{R}^{\mathrm{I}}|^2}\ .$$

In the center of mass system, the relation between the impact parameter b and the scattering angle χ can therefore simply be obtained from (1.59) by replacing $m \to \mu$.

> **Theorem 1.33: Differential cross section**
> **for two-particle scattering in the center of mass system**
>
> $$\frac{\mathrm{d}\sigma}{\mathrm{d}\Omega^*} = \frac{1}{\sin\chi}b(\chi)\left|\frac{\mathrm{d}b}{\mathrm{d}\chi}\right|\ , \quad \text{with } b(\chi) \text{ from Theorem 1.32 and } m \to \mu.$$

Here, the velocity parameter v_0 is the relative velocity of the two particles long before the scattering.

Two-particle scattering in the laboratory system. Using the equations for two-particle scattering in the center of mass frame, we can easily obtain the corresponding relations in the laboratory system (Fig. 1.18), where the second particle is at rest long before the scattering:

$$p_2^I = 0 \Longrightarrow \dot{x}_C = \frac{m_1\dot{x}_1(0)}{m_1 + m_2} = \frac{1}{m_2}p_R^I .$$

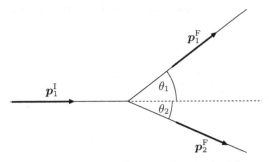

Fig. 1.18. Two-particle scattering in the laboratory system.

The relations between the initial and the final momenta in the two frames are

$$p_1^I = m_1[\dot{x}_C + \dot{x}_1^*(0)] = \frac{m_1}{m_2}p_R^I + p_1^{*I} = \left(\frac{m_1}{m_2} + 1\right)p_R^I$$

$$p_2^I = m_2[\dot{x}_C + \dot{x}_2^*(0)] = p_R^I + p_2^{*I} = 0$$

$$p_1^F = m_1[\dot{x}_C + \dot{x}_1^*(\infty)] = \frac{m_1}{m_2}p_R^I + p_1^{*F} = \frac{m_1}{m_2}p_R^I + p_R^F$$

$$p_2^F = m_2[\dot{x}_C + \dot{x}_2^*(\infty)] = p_R^I + p_2^{*F} = p_R^I - p_R^F .$$

This allows us to express the scattering angles θ_1 and θ_2 in the laboratory frame by the scattering angle χ in the center of mass frame:

$$\cos\theta_1 = \frac{p_1^I p_1^F}{|p_1^I||p_1^F|} = \frac{\left(\frac{m_1}{m_2} + 1\right)p_R^I\left(\frac{m_1}{m_2}p_R^I + p_R^F\right)}{\left(\frac{m_1}{m_2} + 1\right)|p_R^I|\sqrt{\left(\frac{m_1}{m_2}\right)^2 (p_R^I)^2 + (p_R^F)^2 + 2\frac{m_1}{m_2}p_R^I p_R^F}}$$

$$= \frac{\frac{m_1}{m_2} + \cos\chi}{\sqrt{\left(\frac{m_1}{m_2}\right)^2 + 1 + 2\frac{m_1}{m_2}\cos\chi}}$$

$$\cos\theta_2 = \frac{p_1^I p_2^F}{|p_1^I||p_2^F|} = \frac{\left(\frac{m_1}{m_2} + 1\right)p_R^I(p_R^I - p_R^F)}{\left(\frac{m_1}{m_2} + 1\right)|p_R^I||p_R^I - p_R^F|} = \frac{1 - \cos\chi}{\sqrt{2 - 2\cos\chi}} = \sin\frac{\chi}{2}$$

$$\Longrightarrow \quad \theta_2 = \frac{\pi - \chi}{2} .$$

Theorem 1.34: Relation between the differential cross section in the laboratory system and in the center of mass system

$$\frac{d\sigma}{d\Omega_L} = \frac{d\sigma}{d\Omega^*}\frac{d\Omega^*}{d\Omega_L} = \frac{d\sigma}{d\Omega^*}\left(\frac{d\cos\theta_1}{d\cos\chi}\right)^{-1}$$

$$= \frac{d\sigma}{d\Omega^*}\frac{\left[\left(\frac{m_1}{m_2}\right)^2 + 1 + 2\frac{m_1}{m_2}\cos\chi(\theta_1)\right]^{3/2}}{\frac{m_1}{m_2}\cos\chi(\theta_1) + 1},$$

with

$$\cos\theta_1 = \frac{\frac{m_1}{m_2} + \cos\chi}{\sqrt{\left(\frac{m_1}{m_2}\right)^2 + 1 + 2\frac{m_1}{m_2}\cos\chi}}$$

and

$$d\Omega^* = \sin\chi d\chi d\varphi = \text{solid angle element in the center of mass system,}$$
$$d\Omega_L = \sin\theta_1 d\theta_1 d\varphi = \begin{array}{l}\text{solid angle element in the laboratory system,}\\ \text{into which the particles are scattered.}\end{array}$$

(The azimuthal angle dependencies in both frames are identical: $\varphi_L = \varphi$.)

Note that for $m_2 \gg m_1$, the differential cross sections $d\sigma/d\Omega_L$ and $d\sigma/d\Omega^*$ are identical and correspond to the differential cross section $d\sigma/d\Omega$ of one-particle scattering. The transformation from the center of mass to the laboratory system can be graphically visualized as follows:

- $m_1/m_2 < 1$: θ_1 is restricted to the range $[0 : \pi]$ in the laboratory frame. The same is true for the scattering angle χ in the center of mass system (Fig. 1.19). In this range, the mapping $\chi(\theta_1)$ is bijective.

Fig. 1.19. Relationship between two-particle scattering in the center of mass and laboratory system for $m_1 < m_2$.

- $m_1/m_2 > 1$: θ_1 is restricted to the range $[0 : \theta_{\max}]$ for $\chi \in [0 : \pi]$ with $\sin\theta_{\max} = m_2/m_1$. For each θ_1, there are now two possible values of χ

(Fig. 1.20). To uniquely determine χ, we therefore need θ_1 and the final momentum $|\boldsymbol{p}_1^{\mathrm{F}}|$ of the first particle.

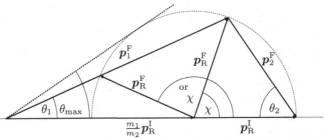

Fig. 1.20. Relationship between two-particle scattering in the center of mass and laboratory system for $m_1 > m_2$.

Summary

- The relevant parameters of **one-particle scattering** are the **differential** and the **total cross section**, as they are most easily determined experimentally. The differential cross section describes the angular distribution of the scattered particles, whereas the total cross section is the area of the scattering center that projectiles must pass through in order to be deflected.

- The mutual scattering of particles (**two-particle scattering**) can be separated into a relative motion and a uniform center of mass motion. The center of mass motion does not influence the dynamics of the scattering, so that the relative motion corresponds to an effective one-particle problem.

- Viewed in the **center of mass system**, the scattering relations of the two-particle system can be obtained from those of the one-particle scattering by substituting $m \longrightarrow \mu$. The corresponding relations in the **laboratory system**, in which the second particle is at rest long before the scattering takes place, follow easily from those of the center of mass system.

Applications

13. Perihelion precession. The gravitational potential of the sun is modified by general relativistic corrections as follows:

$$V(r) = \frac{\alpha}{r} + \frac{a}{r^2} \ , \ \alpha < 0 \ , \ a > 0 \ .$$

Show that this potential results in a rotation of the planets' elliptical orbits.

Solution. The radial equation of motion (1.51) or (1.52) in polar coordinates is (with $\varphi_0 = 0$)

$$\varphi = \pm \int_{r_0}^{r} \frac{l\,dr'}{r'^2 \sqrt{2m\left(E - \frac{\alpha}{r'} - \frac{a}{r'^2}\right) - \frac{l}{r'^2}}}$$

$$= -\frac{1}{\sqrt{d}} \arcsin \frac{\lambda/r + m\alpha}{m\alpha\gamma} + c \ ,$$

where m is the mass of the planet and

$$d = \frac{2ma}{l^2} + 1 \ , \ \gamma = \pm\sqrt{1 + \frac{2E}{m\alpha^2}(2ma + l^2)} \ .$$

Choosing an appropriate integration constant c, we find

$$r = \frac{k}{1 + \gamma \cos(\sqrt{d}\varphi)} \ , \ k = -\frac{\lambda^2}{m\alpha} \ . \tag{1.61}$$

Up to the factor \sqrt{d}, (1.61) is form-identical with the elliptical orbit equation (1.54). However, since $\sqrt{d} > 1$, the minimal distance $r_{\min} = k/(1 + |\gamma|)$ to the sun is not reached after $\Delta\varphi = 2\pi$ as in the pure Coulombic potential, but after $\Delta\varphi = 2\pi/\sqrt{d}$. This results in a rotation of the ellipse called *perihelion precession* (Fig. 1.21).

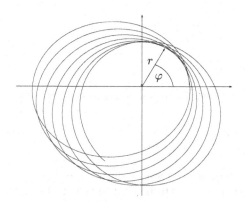

Fig. 1.21. Rotation of a planet's ellipse due to an additional $\frac{1}{r^2}$-term in the gravitational potential.

14. Elastic scattering of spheres. Two hard spheres with masses $m_1 = m_2 = m$ and radii $R_1 = R_2 = R$ are scattered elastically against each other. Calculate the differential cross section in both the center of mass system and the laboratory system. What is the total cross section?

Solution. Consider first the scattering of one of the spheres on a fixed second sphere (Fig. 1.22). Since the scattering is elastic, we can obtain the relation

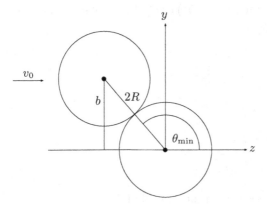

Fig. 1.22. Elastic scattering of a sphere against a fixed sphere.

between the scattering angle χ and the impact parameter b from purely geometrical considerations:

$$\theta_{\min} = \pi - \arcsin \frac{b}{2R}$$

$$\Longrightarrow \chi = 2\theta_{\min} - \pi = 2\arccos \frac{b}{2R}$$

$$\Longrightarrow b = 2R \cos \frac{\chi}{2}$$

$$\Longrightarrow \left| \frac{db}{d\chi} \right| = R \sin \frac{\chi}{2} \ .$$

Using (1.59), the differential cross section of the one-particle scattering is therefore given by

$$\frac{d\sigma}{d\Omega} = \frac{2R^2 \sin \frac{\chi}{2} \cos \frac{\chi}{2}}{\sin \chi} = R^2 = \frac{d\sigma}{d\Omega^*} = \text{const} \ .$$

Since $d\sigma/d\Omega$ is mass-independent, Theorem 1.33 states that this is equal to the differential cross section for two-particle scattering in the center of mass system. The total cross section is obtained using (1.60):

$$\sigma_{\text{tot}} = R^2 \int\limits_{0}^{2\pi} d\varphi \int\limits_{0}^{\pi} d\chi \sin \chi = 4\pi R^2 \ .$$

The differential cross section in the laboratory system is found from Theorem 1.34 as

$$\frac{d\sigma}{d\Omega_L} = R^2 \frac{[2 + 2\cos\chi]^{3/2}}{1 + \cos\chi} = 2R^2\sqrt{2 + 2\cos\chi} = 4R^2\cos\theta_1 \;.$$

Note that for the calculation of the total cross section $\sigma_{tot,L}$, which is of course identical to the total cross section of one-particle scattering, the integration over θ_1 has an upper limit at the first zero of $d\sigma/d\Omega_L$:

$$\sigma_{tot,L} = 8\pi R^2 \int_0^{\frac{\pi}{2}} d\theta_1 \sin\theta_1 \cos\theta_1 = 4\pi R^2 \;.$$

15. Rutherford scattering. Assuming a Coulomb potential $V(r) = \alpha/r$, calculate the differential cross section for electron-proton scattering in the laboratory and in the center of mass reference frame.

Solution. As in the previous application, we start by studying the simpler case of one-particle scattering by a fixed target. We take the proton, with $m_2 = m_p$, to be fixed and scatter the electron ($m_1 = m_e$) on it. The relation between the scattering angle and the impact parameter is provided by Theorem 1.32,

$$\theta_{min} = \frac{\chi + \pi}{2} = \pi - \int_{r_{min}}^{\infty} \frac{dr}{r} \left|1 - \frac{r^2}{b^2}\left(1 - \frac{2\alpha}{m_e v_0^2 r}\right)\right|^{-1/2} ,$$

where r_{min} is the singularity of the integrand. The substitution $u = 1/r \implies dr = -r^2 du$ gives

$$\theta_{min} = \pi - \int_0^{u_{min}} du \left|u^2 + \frac{2\alpha}{m_e b^2 v_0^2}u - \frac{1}{b^2}\right|^{-1/2} . \tag{1.62}$$

Another substitution

$$u^2 + \frac{2\alpha}{m_e b^2 v_0^2}u - \frac{1}{b^2} = -\left[\frac{1}{b^2} + \left(\frac{\alpha}{m_e b^2 v_0^2}\right)^2\right](1 - \omega^2)$$

$$\implies \omega = \left(u + \frac{\alpha}{m_e b^2 v_0^2}\right)\left[\frac{1}{b^2} + \left(\frac{\alpha}{m_e b^2 v_0^2}\right)^2\right]^{-1/2}$$

yields

$$\theta_{min} = \pi + \int_{\omega(0)}^{\omega(u_{min})} d\Omega(1 - \omega^2)^{-1/2} = \pi + \arccos\omega(0) - \arccos\omega(u_{min}) \;.$$

Since u_{min} is a zero of the integrand in (1.62), $\omega(u_{min}) = 1$ and therefore $\arccos\omega(u_{min}) = 0$. This results in

$$\cos\frac{\chi-\pi}{2} = \sin\frac{\chi}{2} = \omega(0) = \frac{\alpha}{m_e b^2 v_0^2}\left[\frac{1}{b^2} + \left(\frac{\alpha}{m_e b^2 v_0^2}\right)^2\right]^{-1/2}$$

$$\Longrightarrow \sin^2\frac{\chi}{2} = \left(1 + \frac{b^2 m_e^2 v_0^4}{\alpha^2}\right)^{-1}$$

$$\Longrightarrow b(\chi) = \frac{|\alpha|}{m_e v_0^2}\cot\frac{\chi}{2}$$

$$\Longrightarrow \left|\frac{db}{d\chi}\right| = \frac{|\alpha|}{2 m_e v_0^2}\frac{1}{\sin^2\frac{\chi}{2}} \ .$$

From this, the differential cross section for the one-particle scattering adds up to

$$\frac{d\sigma}{d\Omega} = \frac{\alpha^2}{4 m_e^2 v_0^4}\frac{1}{\sin^4\frac{\chi}{2}} \ . \tag{1.63}$$

This expression is monotonically decreasing in the energy $m_e v_0^2/2$ of the electrons, as well as in the scattering angle χ, and diverges for $\chi = 0$. This is a consequence of the large reach of the Coulomb potential. Remarkably, there is no difference between the attractive ($\alpha < 0$) and the repulsive case ($\alpha > 0$). The total cross section is

$$\sigma_{\text{tot}} = \frac{\pi\alpha^2}{2 m_e^2 v_0^4}\int_0^\pi \frac{d\chi\sin\chi}{\sin^4\frac{\chi}{2}} = -\frac{\pi\alpha^2}{m_e^2 v_0^4}\frac{1}{\sin^2\frac{\chi}{2}}\bigg|_0^\pi \longrightarrow \infty \ .$$

The divergent behavior is also due to the long reach of the $1/r$-scattering potential. The differential cross section in the center of mass system is found by replacing $m_e \to \mu = m_e m_p/(m_e + m_p)$ to be

$$\frac{d\sigma}{d\Omega^*} = \frac{\alpha^2}{4\mu^2 v_0^4}\frac{1}{\sin^4\frac{\chi}{2}} \ ,$$

with v_0 as the initial relative velocity of electron and proton. Since the ratio of the electron and proton masses is

$$\frac{m_e}{m_p} \approx 5.4\cdot 10^{-4} \ll 1 \ ,$$

we have, to lowest order, $m_e/m_p \approx 0$, $\mu \approx m_e$, so that we can write

$$\frac{d\sigma}{d\Omega} \approx \frac{d\sigma}{d\Omega^*} \approx \frac{d\sigma}{d\Omega_L} \ .$$

1.6 Relativistic Mechanics

At the beginning of the 20th century, physicists believed the entire universe to be filled by a medium called *ether*, which enables the propagation of light. From this ether hypothesis, it follows that the velocity of light in inertial systems moving with different velocity relative to the ether must be different, too. However, even high-precision experiments (e.g., Michelson-Morley experiment) at the time were unable to detect this effect. Instead, all experimental results indicated that light has the same velocity in all inertial systems.

In 1905, Albert Einstein published his *theory of special relativity*, in which he rejected the ether hypothesis totally and proposed axiomatically the constance of the velocity of light. The theory of special relativity – originally formulated within the framework of electrodynamics – provides a self-consistent theory that encompasses Newtonian mechanics as the limiting case of small velocities compared to that of light.

We start this section with a discussion of the basic assumptions and methods of special relativity. In this discussion, *Lorentz transformations* will turn out to be fundamental. We show that the causality principle (i.e., the chronological order of past, present, and future events) remains valid. Furthermore, we derive the adequate definitions of relativistic kinematic *four-quantities*, which ensure a form-invariant formulation of relativistic mechanics. The last part of this section deals with the Lagrange formulation of relativistic mechanics.

1.6.1 Axioms, Minkowski Space, Lorentz Transformations

Special relativity is based on the following axioms:

- **Constance of velocity of light:** the velocity of light in vacuum is the same for all uniformly moving reference frames and is given by $c \approx 3 \cdot 10^8 \, \text{m/s}$.

- **Principle of relativity:** the laws of physics are equally valid in all inertial systems.

The relativity principle incorporates the concepts of *homogeneity of space and time*, according to which there is no specially favored point in space or time, and the *isotropy of space* stating that there is no favored direction in space. In the context of mechanics, the *correspondence principle* can be considered as a further axiom. It claims that, in the limit $v/c \to 0$, the laws of relativistic mechanics turn into the corresponding laws of Newtonian mechanics.

As a consequence of the first axiom, the Newtonian assumption of an absolute time has to be dropped. In this respect, it is advantageous to represent physical events mathematically within a four-dimensional space in which time t (or the product of time and the velocity of light, ct) is regarded as an extra dimension and on an equal footing with the spatial coordinates. This

space is called *Minkowski space*. Before we turn to it, we shall agree upon the following conventions, which allow a more readable formulation of special relativity:

- Two indices, one of which is an upper and the other a lower index, are summed over, so that summation signs can be dropped. Sums with two identical indices, which are both upper or both lower indices, are not defined (*Einstein's summation convention*).

- Vectors with an upper index are called *contravariant*. Vectors with a lower index are called *covariant*. Although there is little need to distinguish between covariant and contravariant indices in the context of relativistic mechanics, we start using this notation here. Its full power will become apparent in the context of the form-invariant formulation of electrodynamics (Section 2.3).

- For matrices, the first index indicates the row and the second one the column.

Definition: Minkowski space

Minkowski space is a four-dimensional linear vector space over the body of real numbers. Its elements x are represented by four-component coordinate vectors or *four-vectors* (in the following also simply: vectors),

$$(x^\mu(t)) = \begin{pmatrix} ct \\ \boldsymbol{x}(t) \end{pmatrix} = \begin{pmatrix} ct \\ x(t) \\ y(t) \\ z(t) \end{pmatrix}.$$

Note the notation: x^μ, $\mu = 0, 1, 2, 3$, stands for the μth contravariant coordinate of x.

Metric of Minkowski space

The scalar product of two four-vectors is defined as

$$(x^\mu) \cdot (y^\nu) = x^\mu g_{\mu\nu} y^\nu = x^\mu y_\mu$$
$$(x_\mu) \cdot (y_\nu) = x_\mu g^{\mu\nu} x_\nu = x_\mu y^\mu$$

$$(g_{\mu\nu}) = (g^{\mu\nu}) = \begin{pmatrix} 1 & 0 & 0 & 0 \\ 0 & -1 & 0 & 0 \\ 0 & 0 & -1 & 0 \\ 0 & 0 & 0 & -1 \end{pmatrix}, \quad g^\mu{}_\nu = g_\nu{}^\mu = \delta^\mu_\nu. \tag{1.64}$$

The following relations result from this:

- $g^{\mu\alpha} g_{\alpha\nu} = \delta^\mu{}_\nu,$

- $x_\mu = g_{\mu\nu}x^\nu$, $x^\mu = g^{\mu\nu}x_\nu$, i.e., co- and contravariant vectors differ just by the sign of the spatial components,

- $x_\mu y^\mu = x^\mu y_\mu$.

$g_{\mu\nu}$ is the (non-Euclidean) *metric tensor* of Minkowski space.

Contrary to the three-dimensional Euclidean case, the norm of a four-vector is not positive definite, and the following cases can occur:

$$x^\mu x_\mu = c^2 t^2 - \boldsymbol{x}^2 \begin{cases} > 0 \\ = 0 \\ < 0 \,. \end{cases}$$

Now, we consider two inertial systems K and K′ moving with constant velocity relative to one another. We assume that, at any one time, a light signal is sent out in all directions. Due to the relativity principle, the propagation of this signal should be perceived in both systems as a uniformly spreading spherical wave. If x_0^μ and $x_0^{\prime\mu}$ denote the space-time origin of the light signal in the systems K and K′, and, accordingly, x^μ and $x^{\prime\mu}$ denote a point of the wave front at a later time, the postulate of constant velocity of light can be expressed as

$$\begin{aligned} (x - x_0)^\mu (x - x_0)_\mu &= c^2(t - t_0)^2 - (\boldsymbol{x} - \boldsymbol{x}_0)^2 \\ &= c^2(t' - t_0')^2 - (\boldsymbol{x}' - \boldsymbol{x}_0')^2 \\ &= (x' - x_0')^\mu (x' - x_0')_\mu = 0 \,. \end{aligned}$$

This means that the four-dimensional distance of any two light vectors is zero in any inertial system. As a consequence of the homogeneity of space-time and the isotropy of space, this statement is generalized, in that the geometrical structure of four-dimensional space-time is given by the invariance of the distance of any two arbitrary four-vectors:

$$(x - y)^\mu (x - y)_\mu = (x' - y')^\mu (x' - y')_\mu \,.$$

Thus, of all the transformations relating K and K′, only those are physically meaningful that respect this equation. These are the *Lorentz transformations*, linear mappings that correspond to translations and rotations in four-dimensional Minkowski space.

Theorem 1.35: Lorentz transformations

Lorentz transformations describe the relativistic transition from one inertial system, K, to another, K′. They are defined by the transformations

$$x^\mu \to x^{\prime\mu} = \Lambda^\mu{}_\nu x^\nu + a^\mu \,, \quad a^\mu = \text{space-time translations}$$

of contravariant vectors. The distance conservation

$$(x - y)^\mu (x - y)_\mu = (x' - y')^\mu (x' - y')_\mu$$

▷

implies that

$$\Lambda^\mu{}_\alpha g_{\mu\nu} \Lambda^\nu{}_\beta = g_{\alpha\beta} \tag{1.65}$$

as well as the transformational behavior of covariant four-vectors,

$$x'_\mu = g_{\mu\nu} x'^\nu = g_{\mu\nu} \left(\Lambda^\nu{}_\alpha x^\alpha + a^\nu \right) = x_\beta [\Lambda^{-1}]^\beta{}_\mu + a_\mu \,,$$

with

$$[\Lambda^{-1}]^\beta{}_\mu = g_{\mu\nu} \Lambda^\nu{}_\alpha g^{\alpha\beta} = \Lambda_\mu{}^\beta \,, \ [\Lambda^{-1}]^\beta{}_\mu \Lambda^\mu{}_\gamma = g_{\mu\nu} \Lambda^\nu{}_\alpha g^{\alpha\beta} \Lambda^\mu{}_\gamma = \delta^\beta_\gamma.$$

In matrix notation, the condition (1.65) reads $\Lambda^{\mathrm{T}} g \Lambda = g$ and corresponds to the condition $R^{\mathrm{T}} R = 1$ for rotational matrices in the three-dimensional Euclidean case. Lorentz transformations with $a^\mu = 0$ constitute the *homogeneous Lorentz group*. Here, not only the distance, but also the scalar product of any two vectors remains unchanged:

$$x^\mu y_\mu = x'^\mu y'_\mu \,.$$

In the general case $(a^\mu) \neq 0$, one obtains the *inhomogeneous Lorentz group* or *Poincaré group*. The homogeneous Lorentz group can be classified further into

homogeneous transformations: $\det(\Lambda) = \pm 1 \,, \ \Lambda^0{}_0 = \pm 1$

 ↪ *orthochronous transformations:* $\det(\Lambda) = \pm 1 \,, \ \Lambda^0{}_0 = +1$

 ↪ *proper transformations:* $\det(\Lambda) = +1 \,, \ \Lambda^0{}_0 = +1.$

Proper transformations are described by three constant (spatial) rotation angles and three constant boost angles. They leave the direction of time, as well as of the three spatial axes unchanged. Together with the space-time translations a^μ they form the 10-parameter *proper orthochronous Poincaré group* (compare to Theorem 1.10).

Examples of proper Lorentz transformations.

- Rotations in space.

$$\Lambda^{(R)} = (\Lambda^{(R)\mu}{}_\nu) = \begin{pmatrix} 1 & 0 & 0 & 0 \\ 0 & & & \\ 0 & & R & \\ 0 & & & \end{pmatrix}$$

- Special boosts:

$$\Lambda^{(1)} = (\Lambda^{(1)\mu}{}_\nu) = \begin{pmatrix} \cosh\alpha & \sinh\alpha & 0 & 0 \\ \sinh\alpha & \cosh\alpha & 0 & 0 \\ 0 & 0 & 1 & 0 \\ 0 & 0 & 0 & 1 \end{pmatrix}$$

$$\Lambda^{(2)} = (\Lambda^{(2)\mu}{}_\nu) = \begin{pmatrix} \cosh\alpha & 0 & \sinh\alpha & 0 \\ 0 & 1 & 0 & 0 \\ \sinh\alpha & 0 & \cosh\alpha & 0 \\ 0 & 0 & 0 & 1 \end{pmatrix}$$

$$\Lambda^{(3)} = (\Lambda^{(3)\mu}{}_\nu) = \begin{pmatrix} \cosh\alpha & 0 & 0 & \sinh\alpha \\ 0 & 1 & 0 & 0 \\ 0 & 0 & 1 & 0 \\ \sinh\alpha & 0 & 0 & \cosh\alpha \end{pmatrix}.$$

The $\Lambda^{(i)}$ describe the transition from one inertial system K to another K', where K' is moving relative to K with constant velocity along the ith spatial axis.

To understand the boost angle α, consider a system K' moving relative to K with constant velocity v (measured in K) along the x-axis. Suppose an event happens at the spatial origin of system K' at time t', which is given in the two systems by the vectors

$$(x^\mu) = \begin{pmatrix} ct \\ vt \\ 0 \\ 0 \end{pmatrix} , \quad (x'^\mu) = \begin{pmatrix} ct' \\ 0 \\ 0 \\ 0 \end{pmatrix}.$$

Using $\Lambda^{(1)}$, we find

$$ct' = ct\cosh\alpha + vt\sinh\alpha \; , \quad 0 = ct\sinh\alpha + vt\cosh\alpha \; , \tag{1.66}$$

from which it follows that

$$\tanh\alpha = -\beta \Longrightarrow \cosh\alpha = \frac{\pm 1}{\sqrt{1-\beta^2}} \; , \quad \sinh\alpha = \frac{\mp\beta}{\sqrt{1-\beta^2}} \; , \quad \beta = \frac{v}{c} \; .$$

To determine the signs we consider the nonrelativistic limiting case $\beta \to 0$, expand $\cosh\alpha$ and $\sinh\alpha$ to lowest order in β,

$$\cosh\alpha \approx \pm 1 \; , \quad \sinh\alpha \approx \mp\beta \; ,$$

and insert these expressions into (1.66):

$$ct' \approx \pm ct \mp \beta vt \; , \quad 0 \approx \mp\beta ct \pm vt = \mp vt \pm vt \; .$$

Respecting the correspondence principle $ct' \xrightarrow{\beta \to 0} ct$, one finally obtains

$$\cosh\alpha = \frac{+1}{\sqrt{1-\beta^2}} \; , \quad \sinh\alpha = \frac{-\beta}{\sqrt{1-\beta^2}} \; .$$

1.6.2 Relativistic Effects

In the following, we discuss some physical consequences arising from the structure of Lorentz transformations. We shall always assume that K' is moving relative to K with constant velocity v in the x-direction.[11]

[11] Note that lower indices do not imply a covariant μ-notation here. Instead, they simply denote different contravariant components.

Time dilation. An observer resting in K emits two light pulses at location x_1 and at times t_1 and t_2. In system K' this corresponds to the times

$$t_1' = t_1 \cosh \alpha + \frac{x_1}{c} \sinh \alpha \ , \ t_2' = t_2 \cosh \alpha + \frac{x_1}{c} \sinh \alpha \ .$$

It follows that

$$\Delta t' = t_2' - t_1' = \frac{\Delta t}{\sqrt{1 - \beta^2}} \geq \Delta t = t_2 - t_1 \ .$$

This means that time intervals in systems moving relative to an observer seem elongated to him.

Length contraction. Consider a pole of length $\Delta x = x_2 - x_1$ aligned along the x-axis in system K. An observer in system K' moving relative to K measures the length of the pole by noting the pole's start and end positions at the same time. Thus, we have

$$x_2' - x_1' = \Delta x' = c\Delta t \sinh \alpha + \Delta x \cosh \alpha \ .$$

Simultaneity of the measurements in the system K' implies

$$c\Delta t' = 0 = c\Delta t \cosh \alpha + \Delta x \sinh \alpha \implies c\Delta t = -\Delta x \tanh \alpha \ ,$$

so that

$$\Delta x' = \Delta x(- \sinh \alpha \tanh \alpha + \cosh \alpha) = \Delta x \sqrt{1 - \beta^2} \leq \Delta x \ .$$

Therefore, an observer perceives the length of objects moving relative to him as being shortened. Since it is irrelevant which system, K or K', is regarded as being in motion, time dilation and length contraction are observed in both systems.

Addition of relativistic velocities. We consider a particle moving with velocity $\boldsymbol{\omega}$ in K and ask for its velocity $\boldsymbol{\omega}'$ in K'. We have

$$\omega_x' = c\frac{dx'^1}{dx'^0} = c\frac{dx^0 \sinh \alpha + dx^1 \cosh \alpha}{dx^0 \cosh \alpha + dx^1 \sinh \alpha} = c\frac{\tanh \alpha + \frac{dx^1}{dx^0}}{1 + \frac{dx^1}{dx^0} \tanh \alpha} = c\frac{-\frac{v}{c} + \frac{1}{c}\omega_x}{1 - \frac{v}{c^2}\omega_x}$$

$$\implies \omega_x' = \frac{\omega_x - v}{1 - \frac{v}{c^2}\omega_x} \ ,$$

$$\omega_{y,z}' = c\frac{dx'^{2,3}}{dx'^0} = c\frac{dx^{2,3}}{dx^0 \cosh \alpha + dx^1 \sinh \alpha} = \frac{c}{\cosh \alpha} \frac{\frac{dx^{2,3}}{dx^0}}{1 + \frac{dx^1}{dx^0} \tanh \alpha}$$

$$= \frac{c}{\cosh \alpha} \frac{\frac{1}{c}\omega_{y,z}}{1 + \frac{\omega_x}{c} \tanh \alpha}$$

$$\implies \omega_{y,z}' = \frac{\omega_{y,z}(1 - \beta^2)}{1 - \frac{v}{c^2}\omega_x}$$

or

$$\omega' = \frac{1}{1 - \frac{v}{c^2}\omega_x} \begin{pmatrix} \omega_x - v \\ \omega_y \sqrt{1 - \beta^2} \\ \omega_z \sqrt{1 - \beta^2} \end{pmatrix} .$$

Obviously, this relation is only physically meaningful for relative velocities for which $|v| \le c$ and $|\omega| \le c$. Furthermore, one easily finds that for any possible combination of $v, |\omega| \le c$, there is no possible velocity in K' larger than the velocity of light. In particular, we have

$$\omega = \begin{pmatrix} c \\ 0 \\ 0 \end{pmatrix} \Longrightarrow \omega' = \begin{pmatrix} c \\ 0 \\ 0 \end{pmatrix} \quad \text{and} \quad v = c \Longrightarrow \omega' = - \begin{pmatrix} c \\ 0 \\ 0 \end{pmatrix} .$$

Thus, the velocity of light is the upper boundary for any type of particle motion.

1.6.3 Causality Principle, Space-, Light- and Time-like Vectors

In Newtonian mechanics, the causality principle states that an event E_2 can only occur after the event E_1 that causes it, so that $t_2 - t_1 \ge 0$. If the causality principle is also to hold in relativistic mechanics, this relation must not be reversed in any inertial system. Consider, therefore, two events

$$(x_1^\mu) = \begin{pmatrix} ct_1 \\ x_1 \\ 0 \\ 0 \end{pmatrix} , \quad (x_2^\mu) = \begin{pmatrix} ct_2 \\ x_2 \\ 0 \\ 0 \end{pmatrix}$$

in K, of which x_2^μ is the reaction of x_1^μ, i.e.,

$$t_2 > t_1 , \ c(t_2 - t_1) \ge |x_2 - x_1| \ge x_2 - x_1 .$$

In system K' moving relative to K with velocity v, one finds ($\cosh \alpha \ge 1$, $|\tanh \alpha| \le 1$)

$$\begin{aligned} c(t_2' - t_1') &= c(t_2 - t_1) \cosh \alpha + (x_2 - x_1) \sinh \alpha \\ &= [c(t_2 - t_1) + (x_2 - x_1) \tanh \alpha] \cosh \alpha \\ &\ge c(t_2 - t_1)(1 + \tanh \alpha) \ge 0 . \end{aligned}$$

We can therefore conclude that the causality principle is still valid in relativistic mechanics. In particular, in any inertial system the chronological sequence past-present-future is maintained. Figure 1.23 shows these three time domains in Minkowski space, where one spatial dimension is suppressed. The past, present, and future of an event E are separated by *light cones*, whose surfaces are generated by vectors, for which

$$\mathrm{d}x^\mu \mathrm{d}x_\mu = 0 \iff |\dot{\boldsymbol{x}}| = c .$$

Vectors of this kind are called *light-like*, as they describe a propagation with the speed of light. Events that can influence E (past), or which can be caused

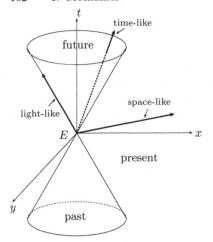

Fig. 1.23. Classification of the three temporal domains past, present, and future.

by E (future) lie on or inside the lower and upper light cones, respectively. The inner area is described by *time-like* vectors with the property

$$\mathrm{d}x^\mu \mathrm{d}x_\mu > 0 \iff |\dot{\boldsymbol{x}}| < c \ .$$

The complementary domain to that past and future of E is the present of E. It is given by the outer part of the light cone described by *space-like* vectors with

$$\mathrm{d}x^\mu \mathrm{d}x_\mu < 0 \iff |\dot{\boldsymbol{x}}| > c \ .$$

This is the area of events that can neither be the cause of, nor the reaction to, E.

1.6.4 Lorentz-Covariant[12] Formulation of Relativistic Mechanics

According to the principle of relativity, there must exist a Lorentz-covariant formulation of relativistic mechanics, so that all physical laws have the same form in all inertial systems, just as in Newtonian mechanics. However, this can only be guaranteed if the corresponding quantities (velocity, momentum, force, etc.) transform as four-vectors. Obviously, this constraint does not apply to the derivative of a four-vector x^μ with respect to its 0th component, as the time differential $\mathrm{d}x^0 = c\mathrm{d}t$ is not a Lorentz scalar, i.e., it is not invariant under arbitrary Lorentz transformations:

[12] In this book, as in many other textbooks, the transformational behavior of relativistic four-quantities is called "Lorentz-covariant", regardless of the subtle difference between "covariant" and "contravariant". The same holds for relativistic equations whose form remains unchanged under Lorentz transformations. Quantities and equations are called "Lorentz-invariant" if their value is unchanged by Lorentz transformations (*Lorentz scalar*).

$$\frac{\mathrm{d}x'^{\mu}}{\mathrm{d}x'^0} \neq \Lambda^{\mu}{}_{\nu} \frac{\mathrm{d}x^{\nu}}{\mathrm{d}x^0} \ .$$

Since, on the other hand, c constitutes the upper velocity limit for any physical motion, x^{μ} is a time-like vector, so that at each time t, there exists an inertial system in which the particle is at rest (*momentary rest system*). This implies that

$$\mathrm{d}x^{\mu}\mathrm{d}x_{\mu} = c^2\mathrm{d}t^2 - \mathrm{d}\boldsymbol{x}^2 = c^2\mathrm{d}t'^2 - \mathrm{d}\boldsymbol{x}'^2 = \mathrm{d}x'^{\mu}\mathrm{d}x'_{\mu} = c^2\mathrm{d}\tau^2 > 0 \ , \quad (1.67)$$

where $\mathrm{d}\tau$ is the time differential of the rest system (*eigentime differential*). Dividing the last equation by $\mathrm{d}\tau^2$ yields

$$\frac{\mathrm{d}x^{\mu}}{\mathrm{d}\tau}\frac{\mathrm{d}x_{\mu}}{\mathrm{d}\tau} = c^2\left(\frac{\mathrm{d}t}{\mathrm{d}\tau}\right)^2 - \left(\frac{\mathrm{d}\boldsymbol{x}}{\mathrm{d}\tau}\right)^2 = \frac{\mathrm{d}x'^{\mu}}{\mathrm{d}\tau}\frac{\mathrm{d}x'_{\mu}}{\mathrm{d}\tau}$$

$$= c^2\left(\frac{\mathrm{d}t'}{\mathrm{d}\tau}\right)^2 - \left(\frac{\mathrm{d}\boldsymbol{x}'}{\mathrm{d}\tau}\right)^2 = c^2 \ .$$

According to this, $\mathrm{d}x^{\mu}/\mathrm{d}\tau$ is a time-like four-vector of length c, thus transforming under arbitrary Lorentz transformations Λ as

$$\frac{\mathrm{d}x'^{\mu}}{\mathrm{d}\tau} = \Lambda^{\mu}{}_{\nu} \frac{\mathrm{d}x^{\nu}}{\mathrm{d}\tau} \ . \tag{1.68}$$

Using this and (1.67), we find

Theorem 1.36: Eigentime differential $d\tau$

The eigentime differential

$$\mathrm{d}\tau = \mathrm{d}t\sqrt{1 - \frac{1}{c^2}\left(\frac{\mathrm{d}\boldsymbol{x}}{\mathrm{d}t}\right)^2} = \mathrm{d}t'\sqrt{1 - \frac{1}{c^2}\left(\frac{\mathrm{d}\boldsymbol{x}'}{\mathrm{d}t'}\right)^2} = \dots$$

is a Lorentz scalar that defines the time scale in the rest system of an object moving relative to inertial systems K, K',

This equation expresses again the time dilation in differential form and generalized to accelerated motion. Due to its transformational behavior (1.68), $\mathrm{d}x^{\mu}/\mathrm{d}\tau$ is called *four-velocity*. It is related to the physical velocity $\dot{\boldsymbol{x}}$ via

Definition: Four-velocity u^{μ}, physical velocity $\dot{\boldsymbol{x}}$

$$(u^{\mu}) = \left(\frac{\mathrm{d}x^{\mu}}{\mathrm{d}\tau}\right) = \frac{\mathrm{d}t}{\mathrm{d}\tau}\left(\frac{\mathrm{d}x^{\mu}}{\mathrm{d}t}\right) = \frac{1}{\sqrt{1 - \frac{\dot{\boldsymbol{x}}^2}{c^2}}}\left(\begin{matrix} c \\ \dot{\boldsymbol{x}} \end{matrix}\right) \ .$$

The derivative of a four-vector with respect to the eigentime τ always yields another four-vector, so that the remaining quantities of Lorentz-covariant relativistic mechanics can be easily constructed. Momentum is defined in analogy to Newtonian mechanics as

Definition: Four-momentum p^μ, physical momentum p

$$(p^\mu) = m_0(u^\mu)$$

$$= \frac{m_0}{\sqrt{1 - \frac{\dot{x}^2}{c^2}}} \left(\frac{dx^\mu}{dt} \right) = \begin{pmatrix} cm \\ p \end{pmatrix} , \quad p = m\dot{x} , \tag{1.69}$$

where m_0 is a Lorentz scalar, i.e., the mass of the particle, measured in its rest system (*rest mass*). This definition implies that the mass m is not constant, but instead depends explicitly on the velocity. It behaves as

$$m = m(t) = \frac{m_0}{\sqrt{1 - \frac{\dot{x}^2}{c^2}}} .$$

Correspondingly, we define force as

Definition: Four-force F^μ, physical force F

$$(F^\mu) = \left(\frac{dp^\mu}{d\tau} \right) = \frac{1}{\sqrt{1 - \frac{\dot{x}^2}{c^2}}} \left(\frac{dp^\mu}{dt} \right)$$

$$= \frac{1}{\sqrt{1 - \frac{\dot{x}^2}{c^2}}} \begin{pmatrix} c\frac{dm}{dt} \\ F \end{pmatrix} , \quad F = \frac{dp}{dt} . \tag{1.70}$$

Equation (1.70) is also the Lorentz-covariant equation of motion of relativistic mechanics. It is easy to verify that, in addition to the relativity principle, (1.69) and (1.70) also obey the correspondence principle: for small velocities, $|\dot{x}| \ll c$, they yield the corresponding equations of Newtonian mechanics.

It is not always possible to provide a complete covariant formulation of a mechanical problem via (1.70), since not all types of forces can be written as four-vectors. An example is the gravitational force. It is a "static far field force" and assumes an infinitely large propagation velocity, thus being in conflict with the first axiom of special relativity. Another example is given by the constraining forces of a rigid body, as they only contain the spatial components of four-vectors. Therefore, the whole area of the dynamics of rigid bodies has no relativistic analogue.

Physical consequences. A particle moving in a conservative force field obeys

$$F = \frac{dp}{dt} = \frac{d}{dt} \left(\frac{m_0 \dot{x}}{\sqrt{1 - \frac{\dot{x}^2}{c^2}}} \right) = -\boldsymbol{\nabla} V(\boldsymbol{x}) .$$

Multiplication with $\dot{\boldsymbol{x}}$ yields

$$\frac{d}{dt}\left(\frac{m_0\dot{\boldsymbol{x}}}{\sqrt{1-\frac{\dot{\boldsymbol{x}}^2}{c^2}}}\right)\dot{\boldsymbol{x}} = \frac{d}{dt}\left(\frac{m_0c^2}{\sqrt{1-\frac{\dot{\boldsymbol{x}}^2}{c^2}}}\right) = -\boldsymbol{\nabla}V(\boldsymbol{x})\dot{\boldsymbol{x}} = -\frac{d}{dt}V(\boldsymbol{x}) .$$

From this, we find

Theorem 1.37: Relativistic energy conservation in conservative force fields

$$mc^2 + V(\boldsymbol{x}) = E = \text{const} .$$

For small velocities, $\dot{\boldsymbol{x}}^2/c^2 \ll 1$, this relation becomes

$$m_0c^2 + \frac{m_0}{2}\dot{\boldsymbol{x}}^2 + \ldots + V(\boldsymbol{x}) = E ,$$

which, bar the irrelevant constant m_0c^2, is equal to classical energy conservation. We therefore define

Definition: Kinetic energy T

$$T = mc^2 - m_0c^2 = \frac{m_0c^2}{\sqrt{1-\frac{\dot{\boldsymbol{x}}^2}{c^2}}} - m_0c^2 .$$

The term m_0c^2 is called the *rest energy* of the particle. Note that the classically valid relation

$$\frac{dT}{dt} = m\dot{\boldsymbol{x}}\ddot{\boldsymbol{x}} = \boldsymbol{F}\dot{\boldsymbol{x}}$$

also holds in relativistic mechanics, since

$$\boldsymbol{F}\dot{\boldsymbol{x}} = m_0\dot{\boldsymbol{x}}\frac{d}{dt}\left(\frac{\dot{\boldsymbol{x}}}{\sqrt{1-\frac{\dot{\boldsymbol{x}}^2}{c^2}}}\right) = \frac{m_0\dot{\boldsymbol{x}}\ddot{\boldsymbol{x}}}{c^2\left(1-\frac{\dot{\boldsymbol{x}}^2}{c^2}\right)^{3/2}} = \frac{d}{dt}(mc^2) = \frac{dT}{dt} .$$

In the absence of external forces, Theorem 1.37 reduces to the famous Einstein equation

$$E = mc^2 = p^0c .$$

It states that energy and mass are equivalent and can be transformed into one another. Combining this with (1.69) yields

Theorem 1.38: Relativistic energy-momentum relation for a free particle

$$p^\mu p_\mu = p^{0^2} - \boldsymbol{p}^2 = m_0^2c^2 \iff E^2 = m_0^2c^4 + \boldsymbol{p}^2c^2 .$$

At this stage, we must point out that the definitions of four-momentum and

four-force are unique if we demand maximal similarity to Newtonian mechanics and take into account the relativity and correspondence principle. However, the definitions of the physical three-quantities p and F do not necessarily follow uniquely from the corresponding four-vectors. Rather, they are a consequence of experimental experience. For example, in particle accelerators, one observes that more and more energy is needed to accelerate a particle close to the velocity of light. This indicates that the mass of the particle grows with velocity, so that the definition (1.69) is more meaningful than $p = m_0 d\boldsymbol{x}/dt$. Another observation is provided by the *mass defect*, stating that the mass of an atom is smaller than the total sum of its constituents. Obviously, this implies that a part of the constituent mass appears as binding energy, which means it is absorbed within the atomic nucleus. With the chosen definition of the physical force (1.70), this mass-energy equivalence is predicted automatically. The fact that, up to the replacement $m_0 \to m$, the relativistic kinematic three-quantities are formally identical to the corresponding Newtonian quantities (simplicity of the theory) has led us, in conjunction with experimental consistency, to the above definitions.

At the end of this subsection we consider the interaction of relativistic particles in the absence of external forces. Energy and momentum conservation can then be combined into a single equation for the four-momentum of the particles:

$$\sum_i p_i^\mu = \sum_j p_j'^\mu \ . \tag{1.71}$$

Here, p_i^μ and $p_j'^\mu$ are the four-momenta of the initial and final states. Due to the energy-mass equivalence, this equation is valid in a very general sense: during the interaction process particles can be created and annihilated. According to Theorem 1.38, it follows that all particles must obey the *mass shell conditions*

$$p_i^\mu p_{\mu,i} = m_i^2 c^2 \ , \ p_j'^\mu p_{\mu,j}' = m_j'^2 c^2 \ , \tag{1.72}$$

where m_i and m_j' are the rest masses of the particles in the initial and final states. Many physical effects, such as the mass defect (Application 18) or the *Compton effect* (Application 19), can be explained as four-momentum conservation.

1.6.5 Lagrange Formulation of Relativistic Mechanics

Following the successful generalization of Newton's equation to a form in accordance with special relativity, we now turn to the Lagrange formulation of relativistic mechanics. The simplest way of finding a relativistic Lagrange function is to use the Hamilton principle, Theorem 1.18, and to search for a Lagrange function L for which the Lagrange equations yield the correct relativistic equations of motion. If we assume that the force can be obtained from a generalized potential according to (1.31),

$$F(\boldsymbol{x}, \dot{\boldsymbol{x}}, t) = -\boldsymbol{\nabla} V_{\boldsymbol{x}}(\boldsymbol{x}, \dot{\boldsymbol{x}}, t) + \frac{\mathrm{d}}{\mathrm{d}t} \boldsymbol{\nabla}_{\dot{\boldsymbol{x}}} V(\boldsymbol{x}, \dot{\boldsymbol{x}}, t) ,$$

then the following ansatz for the Lagrange function of a single particle in this force field leads to the desired result:

$$L = -m_0 c^2 \sqrt{1 - \frac{\dot{\boldsymbol{x}}^2}{c^2}} - V . \tag{1.73}$$

This can be immediately verified by observing that

$$\boldsymbol{\nabla}_{\boldsymbol{x}} L = -\boldsymbol{\nabla}_{\boldsymbol{x}} V , \quad \boldsymbol{\nabla}_{\dot{\boldsymbol{x}}} L = \frac{m_0 \dot{\boldsymbol{x}}}{\sqrt{1 - \frac{\dot{\boldsymbol{x}}^2}{c^2}}} - \boldsymbol{\nabla}_{\dot{\boldsymbol{x}}} V$$

$$\Longrightarrow \frac{\mathrm{d}}{\mathrm{d}t} \left(\frac{m_0 \dot{\boldsymbol{x}}}{\sqrt{1 - \frac{\dot{\boldsymbol{x}}^2}{c^2}}} \right) = -\boldsymbol{\nabla}_{\boldsymbol{x}} V + \frac{\mathrm{d}}{\mathrm{d}t} \boldsymbol{\nabla}_{\dot{\boldsymbol{x}}} V = \boldsymbol{F} .$$

Keep in mind that (1.73) is no longer given by $L = T - V$. The above can easily be generalized to the case of many particles. In addition, we can use any set of generalized coordinates q_i instead of the Cartesian coordinates x_j. As before, the Hamilton function H and the generalized momenta p_i are defined as

$$H = \sum_{i=1}^{n} p_i \dot{q}_i - L , \quad p_i = \frac{\partial L}{\partial \dot{q}_i} ,$$

so that Theorems 1.22 and 1.23 are still valid. Additionally, restricting ourselves to conservative forces ($\partial V / \partial \dot{q}_i = 0$), the Hamilton function is again equal to the total energy of the system. For example, for a single particle we can use Theorem 1.37 to find

$$H = \dot{\boldsymbol{x}} \boldsymbol{\nabla}_{\dot{\boldsymbol{x}}} L - L = \frac{m_0 c^2}{\sqrt{1 - \frac{\dot{\boldsymbol{x}}^2}{c^2}}} + V = E = \mathrm{const} .$$

Lorentz-covariant Lagrange formulation. The above three-dimensional Lagrange method yields the correct relativistic equations of motion within a given inertial system. However, this is not a four-dimensional Lorentz-covariant formulation with the same form in any inertial system. To arrive at such a formalism, we need to start with the Lorentz-invariant Hamilton principle

$$S = \int \mathrm{d}\tau \tilde{L} \longrightarrow \mathrm{extremal} ,$$

from which, for example, the Lagrange equations for a single particle,

$$\frac{\mathrm{d}}{\mathrm{d}\tau} \frac{\partial \tilde{L}}{\partial u^{\mu}} - \frac{\partial \tilde{L}}{\partial x^{\mu}} = 0 , \tag{1.74}$$

can be derived. Here, \tilde{L} is the Lorentz-invariant Lagrange function, $d\tau$ the eigentime differential, x^μ the position four-vector, and u^μ the four-velocity of the particle. In the case of a free particle, \tilde{L} is to be chosen such that (1.74) yields the covariant equation of motion

$$\frac{dp^\mu}{d\tau} = m_0 \frac{du^\mu}{d\tau} = 0 \ .$$

This is true for

$$\tilde{L} = \frac{m_0}{2} u^\alpha u_\alpha = \frac{m_0}{2} u^\alpha g_{\alpha\beta} u^\beta \ ,$$

since

$$\frac{\partial \tilde{L}}{\partial x^\mu} = 0 \ , \quad \frac{\partial \tilde{L}}{\partial u^\mu} = \frac{m_0}{2}(g_{\mu\beta}u^\beta + u^\alpha g_{\alpha\mu}) = m_0 g_{\mu\alpha} u^\alpha = m_0 u_\mu$$

$$\Longrightarrow m_0 \frac{du_\mu}{d\tau} = 0 \Longleftrightarrow m_0 \frac{du^\mu}{d\tau} = 0 \ .$$

In the previous subsection, we have already noted the impossibility of a covariant formulation in the presence of the gravitational force or other "far field forces". However, a force worth mentioning that indeed allows for a covariant formulation is the Lorentz force. It describes the motion of a charged particle in an external electromagnetic field. We will discuss this issue within electrodynamics in Sections 2.3 and 2.8.

Summary

- **Special relativity** is based on two axioms, the **constance of the velocity of light** and the **relativity principle**. In this theory, time represents an additional dimension in Minkowski space, on equal footing to the three spatial dimensions, and not, as in Newtonian mechanics, just an external parameter.

- The transition from one inertial system to another is described by **Lorentz transformations**. They imply various relativistic effects such as **time dilation** and **length contraction**.

- The addition formulae for relativistic velocities show that the velocity of light is the upper bound for any kind of particle motion.

- Using the concept of **four-vectors**, relativistic mechanics can be formulated in a **Lorentz-covariant** fashion, as demanded by the principle of relativity. In this formulation, the **momentary rest system** as well as the **eigentime differential** play an important role.

- For small velocities $|\dot{x}| \ll c$, the laws of relativistic mechanics turn into the corresponding laws of Newtonian mechanics (**correspondence principle**).

▷

- The Lagrange formulation of relativistic mechanics proceeds by using the Hamilton principle and searching for a Lagrange function for which the Lagrange equations yield the correct equations of motion.

Applications

16. Twin paradox. Consider a rocket starting at time $t = 0$ from earth and accelerating uniformly with 20-fold acceleration of gravity to $0.9c$. Subsequently, the rocket continues with constant velocity for a year and then decelerates at the same rate as before to velocity zero. Its return to earth proceeds in similar manner. All values refer to the earth system. Compare the total flight times measured in the earth and the rocket system.

Solution. The rocket's flight contains two acceleration and two deceleration phases, for which ($c = 3 \cdot 10^8$ m/s, $g = 10$ m/s^2)

$$\dot{x}(t) = \pm g't \ , \ g' = 20g = 200 \, \frac{m}{s^2} \ , \ \Delta \dot{x} = \pm 0.9c$$

$$\Longrightarrow \Delta t_1 = 1.35 \cdot 10^6 \, \text{s} = 15.6 \, \text{days} \ ,$$

and two constant phases with

$$\dot{x}(t) = \pm 0.9c \ , \ \Delta t_2 = 365 \, \text{days} \ .$$

The total travel time in the earth system is therefore given by

$$T = 4\Delta t_1 + 2\Delta t_2 = 792.4 \, \text{days} \ .$$

The corresponding times for all phases in the rocket system are obtained using Theorem 1.36:

$$\Delta \tau_1 = \int_0^{\Delta t_1} dt \sqrt{1 - \frac{g'^2}{c^2} t^2} = \frac{1}{2} \left[t\sqrt{1 - \frac{g'^2}{c^2} t^2} + \frac{c}{g'} \arcsin \left(\frac{g'}{c} t \right) \right]_0^{\Delta t_1}$$

$$= 1.13 \cdot 10^6 \, \text{s} = 13.1 \, \text{days}$$

$$\Delta \tau_2 = \int_0^{\Delta t_2} dt \sqrt{1 - 0.81} = 159.1 \, \text{days} \ .$$

The total flight time in the rocket system is therefore

$$\tau = 4\Delta \tau_1 + 2\Delta \tau_2 = 370.6 \, \text{days} \ .$$

According to this, when returning to earth, the astronaut has aged less than half compared to someone who stayed back on earth. However, this statement

implies a paradox, since, due to the relativity principle, we can equally consider the rocket being at rest while the earth moves away and back to it. The astronaut would then have aged faster than an earthling. The solution of this paradox is that only the astronaut has effectively undergone an (absolute) acceleration so that he has not been in an inertial system all the time.

17. Transition to the momentary rest system. Consider a rocket moving at constant acceleration a measured in its rest system K'. Show that its velocity in the system K, at which the rocket starts at $t = 0$ in the x-direction, is given by

$$\dot{x} = \frac{c}{\sqrt{1 + \frac{c^2}{a^2 t^2}}} \ .$$

Solution. We need the momentary Lorentz transformation

$$(\Lambda^\mu{}_\nu) = \begin{pmatrix} \cosh\alpha & \sinh\alpha & 0 & 0 \\ \sinh\alpha & \cosh\alpha & 0 & 0 \\ 0 & 0 & 1 & 0 \\ 0 & 0 & 0 & 1 \end{pmatrix} \ ,$$

which transfers K into K'. For this, we must have

$$u'^\mu = \Lambda^\mu{}_\nu u^\nu \ , \quad (u^\mu) = \frac{dt}{d\tau} \begin{pmatrix} c \\ \dot{x} \\ 0 \\ 0 \end{pmatrix} \ , \quad (u'^\mu) = \begin{pmatrix} c \\ 0 \\ 0 \\ 0 \end{pmatrix}$$

$$\Longrightarrow \cosh\alpha = \frac{dt}{d\tau} \ , \quad \sinh\alpha = -\frac{\dot{x}\,dt}{c\,d\tau} \ .$$

The acceleration b^μ in K is given by

$$(b^\mu) = \frac{d}{d\tau}(u^\mu) = \frac{d^2 t}{d\tau^2} \begin{pmatrix} c \\ \dot{x} \\ 0 \\ 0 \end{pmatrix} + \frac{dt}{d\tau} \begin{pmatrix} 0 \\ \frac{d\dot{x}}{d\tau} \\ 0 \\ 0 \end{pmatrix}$$

$$= \frac{d^2 t}{d\tau^2} \begin{pmatrix} c \\ \dot{x} \\ 0 \\ 0 \end{pmatrix} + \left(\frac{dt}{d\tau}\right)^2 \begin{pmatrix} 0 \\ \ddot{x} \\ 0 \\ 0 \end{pmatrix} \ .$$

From this, the x-component of the rocket's four-acceleration in its rest system follows as

$$b'^1 = \Lambda^1{}_\nu b^\nu = \ddot{x}\left(\frac{dt}{d\tau}\right)^3 = \frac{\ddot{x}}{\left(1 - \frac{\dot{x}^2}{c^2}\right)^{3/2}} \ .$$

For $\dot{x} = c/\sqrt{1 + \frac{c^2}{a^2 t^2}}$, this implies

$$b'^1 = a \ .$$

18. Mass defect. Consider a nucleus of mass M at rest, decaying into two lighter nuclei of rest masses m_1 and m_2. What are the energies of the decay products?

Solution. The four-momentum p^μ of the nucleus before the decay and the four-momenta $p_1'^\mu, p_2'^\mu$ of the two decay products are

$$(p^\mu) = \begin{pmatrix} Mc \\ 0 \end{pmatrix} , \quad (p_1'^\mu) = \begin{pmatrix} p_{10}' \\ \boldsymbol{p}_1' \end{pmatrix} , \quad (p_2'^\mu) = \begin{pmatrix} p_{20}' \\ \boldsymbol{p}_2' \end{pmatrix} ,$$

where the mass shell condition (1.72) for the resting nucleus is already incorporated. Four-momentum conservation (1.71) and the mass shell conditions for the decay products yield a set of four equations for the quantities $p_{10}', p_{20}', \boldsymbol{p}_1', \boldsymbol{p}_2'$:

$$p_{10}' + p_{20}' = Mc$$
$$\boldsymbol{p}_1' + \boldsymbol{p}_2' = 0$$
$$p_{10}'^2 - \boldsymbol{p}_1'^2 = m_1^2 c^2$$
$$p_{20}'^2 - \boldsymbol{p}_2'^2 = m_2^2 c^2 .$$

From this, the energy of the decay products follows as

$$p_{10}'^2 - p_{20}'^2 = (m_1^2 - m_2^2)c^2$$
$$\Longleftrightarrow \quad (p_{10}' + p_{20}')(p_{10}' - p_{20}') = Mc(p_{10}' - p_{20}') = (m_1^2 - m_2^2)c^2$$

$$\Longrightarrow \quad \begin{cases} p_{10}' = \dfrac{E_1}{c} = (M^2 + m_1^2 - m_2^2)\dfrac{c}{2M} \\[2mm] p_{20}' = \dfrac{E_2}{c} = (M^2 - m_1^2 + m_2^2)\dfrac{c}{2M} . \end{cases}$$

Writing the first two conditions of the above four equations as

$$\sqrt{m_1^2 c^2 + \boldsymbol{p}_1'^2} + \sqrt{m_2^2 c^2 + \boldsymbol{p}_1'^2} = Mc ,$$

we find for $|\boldsymbol{p}_1'| > 0$

$$M > m_1 + m_2 ,$$

i.e., the rest mass of the original nucleus is larger than the sum of the rest masses of the end products. The difference $M - (m_1 + m_2)$ is called *mass defect*; it is transformed into kinetic energy of the final nuclei during the decay.

19. Compton effect. Consider a photon (with rest mass 0) hitting an electron at rest. Calculate the photon's momentum after its scattering on the electron in dependence of its original momentum and scattering angle (Fig. 1.24).

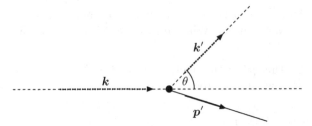

Fig. 1.24. Scattering of a photon on an electron at rest.

Solution. The four-momenta of the photon, k^μ, k'^μ, and that of the electron, p^μ, p'^μ, before and after the scattering are

$$(k^\mu) = \begin{pmatrix} |\boldsymbol{k}| \\ \boldsymbol{k} \end{pmatrix} \;,\; (p^\mu) = \begin{pmatrix} m_e c \\ 0 \end{pmatrix} \;,\; (k'^\mu) = \begin{pmatrix} |\boldsymbol{k}'| \\ \boldsymbol{k}' \end{pmatrix} \;,\; (p'^\mu) = \begin{pmatrix} p'_0 \\ \boldsymbol{p}' \end{pmatrix} \;,$$

where the mass shell conditions are already incorporated in k^μ, p^μ, and k'^μ. Four-momentum conservation and the mass shell condition for the scattered electron yield the equation system

$$|\boldsymbol{k}| + m_e c = |\boldsymbol{k}'| + p'_0$$
$$\boldsymbol{k} = \boldsymbol{k}' + \boldsymbol{p}'$$
$$p_0'^2 - \boldsymbol{p}'^2 = m_e^2 c^2 \;,$$

from which it follows that

$$p_0'^2 = m_e^2 c^2 + \boldsymbol{p}'^2 = m_e^2 c^2 + (\boldsymbol{k} - \boldsymbol{k}')^2$$
$$\Longleftrightarrow \quad (|\boldsymbol{k}| + m_e c - |\boldsymbol{k}'|)^2 = m_e^2 c^2 + (\boldsymbol{k} - \boldsymbol{k}')^2 \;.$$

Using $\boldsymbol{k}\boldsymbol{k}' = |\boldsymbol{k}||\boldsymbol{k}'| \cos\theta$ finally leads to

$$\frac{1}{|\boldsymbol{k}'|} = \frac{1}{|\boldsymbol{k}|} + \frac{1}{m_e c}(1 - \cos\theta) \;.$$

2. Electrodynamics

Electrodynamics is a classical field theory dealing with electromagnetic phenomena. The theory was founded, along with others, on observations going back to the second half of the 18th century, when Coulomb investigated the forces between charged particles. Around 50 years later Faraday studied the interactions between currents and magnetic fields. Today's modern form of electrodynamics is based on the four equations formulated in 1864 by James Clerk Maxwell, from which all electromagnetic effects can be deduced. The connection between the motion of charged particles and the electromagnetic fields is given by the Lorentz force equation.

From a mathematical viewpoint, electrodynamics is an extremely elegant and economical theory. Moreover, contrary to Newtonian mechanics, electrodynamics is a relativistic theory, which explicitly contains the upper limit c for the propagation of information. Accordingly, particles, whose distance vector is space-like, cannot interact with one another. Similar to Newtonian mechanics, electrodynamics makes ample use of the concept of point particles. However, classical electrodynamics is not valid to arbitrarily small distances. Rather, it has to be interpreted as the limiting case of a modern quantum field theory, so-called quantum electrodynamics.

At the start of this chapter, Section 2.1, we present the formal structure of electrodynamics. The fundamental equations of the theory, Maxwell's equations and the Lorentz force, are introduced, interpreted, and their phenomenological basis explained.

Section 2.2 deals with the general solution of Maxwell's equations. We introduce the scalar and vector potentials in such a way that they automatically satisfy the two homogeneous Maxwell equations. With their help, the remaining two inhomogeneous Maxwell equations are written as two inhomogeneous potential equations. Using the gauge freedoms we have for choosing the potentials, these two equations can be decoupled and solved relatively easily. As solutions to the inhomogeneous equations, one obtains the so-called retarded potentials, which explicitly reflect the finite propagation velocity of signals.

Since electrodynamics is a relativistic theory, it can be formulated so that its relativistic form invariance becomes apparent. This is the subject of Section 2.3. We show that all electrodynamic quantities can be suitably combined

to relativistic four-vectors, which transform in a well-defined manner under Lorentz transformations.

Section 2.4 follows up Section 2.2 and shows how to calculate retarded potentials for arbitrarily moving point charges or spatially confined charge and current densities. In particular, we shall find that only accelerated charges emit radiation.

In case of exclusively static (time-independent) charge and current densities, Maxwell's equations decouple into two sets of equations. They constitute the basis of electrostatics and magnetostatics, and these special cases of electrodynamics are discussed in Section 2.5.

Section 2.6 is dedicated to electrodynamics in matter. In principle, Maxwell's equations are valid in vacuum as well as in matter of any kind. However, in the latter case, given the large number of charged particles in a medium (and their variation on atomic length scale), it is sensible to reformulate Maxwell's equations in terms of macroscopic fields. To this end, we introduce two additional, matter-dependent fields, which are related to the macroscopic fields via phenomenological material equations.

In Section 2.7, we discuss the propagation of electromagnetic waves in conducting and nonconducting media. We investigate, in particular, the reflection and refraction of electromagnetic waves on the boundary of two different media. An interesting effect is the dissemination of wave packets in dispersive media due to the varying propagational velocities of the wave packet's Fourier components.

In the last section of this chapter, Section 2.8, we introduce the Lagrangian formulation of electrodynamics. Its importance does not lie in its practical ability to find solutions to given problems (as in mechanics). Rather, it allows for a deeper understanding of fundamental symmetry principles, in particular that of gauge symmetry and its implications. The formulation presented here is to be found in all (quantum) field theories of modern physics.

2.1 Formalism of Electrodynamics

In this section the general formalism of electrodynamics is introduced. We begin our discussion with *Maxwell's equations* and the *Lorentz force*, which establishes a connection between electrodynamics and mechanics. Furthermore, we study the *continuity equation* stating that charge is conserved in any closed system. Subsequently, the Maxwell equations are interpreted with respect to their mathematical meaning and phenomenological implications. This section concludes with a derivation of *energy* and *momentum conservation in electrodynamics*.

2.1.1 Maxwell's Equations and Lorentz Force

Electromagnetic phenomena are described by two fundamental vector fields:

- the *electric field vector* $\boldsymbol{E}(\boldsymbol{x}, t)$ and

- the *magnetic induction field vector* $\boldsymbol{B}(\boldsymbol{x}, t)$.[1]

These fields are caused by

- the *electric charge density* $\rho(\boldsymbol{x}, t)$ and

- the *electric current density vector* $\boldsymbol{j}(\boldsymbol{x}, t)$.

The fields \boldsymbol{E} and \boldsymbol{B} are coupled to the quantities ρ and \boldsymbol{j} via a system of first-order partial differential equations, which we take (axiomatically) as the starting point of the theory.

Theorem 2.1: Maxwell equations

$$\boldsymbol{\nabla} \boldsymbol{E}(\boldsymbol{x}, t) = 4\pi\rho(\boldsymbol{x}, t) \tag{I}$$

$$\boldsymbol{\nabla} \times \boldsymbol{E}(\boldsymbol{x}, t) + \frac{1}{c}\frac{\partial \boldsymbol{B}(\boldsymbol{x}, t)}{\partial t} = 0 \tag{II}$$

$$\boldsymbol{\nabla} \boldsymbol{B}(\boldsymbol{x}, t) = 0 \tag{III}$$

$$\boldsymbol{\nabla} \times \boldsymbol{B}(\boldsymbol{x}, t) - \frac{1}{c}\frac{\partial \boldsymbol{E}(\boldsymbol{x}, t)}{\partial t} = \frac{4\pi}{c}\boldsymbol{j}(\boldsymbol{x}, t) \ . \tag{IV}$$

In this chapter, we use exclusively the *Gaussian unit system*; it will be discussed further at the end of this section.

The theory of electrodynamics is completed by noting the force exerted on a charged particle moving in an electromagnetic field:

Theorem 2.2: Lorentz force

The electromagnetic force on a particle with charge q moving with velocity $\dot{\boldsymbol{x}}$ through the fields \boldsymbol{E} and \boldsymbol{B} is

$$\boldsymbol{F}_{\mathrm{L}}(\boldsymbol{x}, t) = q\left(\boldsymbol{E}(\boldsymbol{x}, t) + \frac{\dot{\boldsymbol{x}}(t)}{c} \times \boldsymbol{B}(\boldsymbol{x}, t)\right) \ . \tag{2.1}$$

The first term in this equation describes the force originating from a pure electric field and is always pointed in the direction of the field \boldsymbol{E}. The force due to the magnetic field, given by the second term, is perpendicular to the field \boldsymbol{B} and to the velocity $\dot{\boldsymbol{x}}$ of the charge. Thus, magnetic forces do not perform any work on the particle.

Interpretation of the Lorentz force. Most problems in electrodynamics fall into one of two groups. First, given the charge and current densities,

[1] In the following, we use the terms "magnetic induction field" and "magnetic field" synonymously, although the latter is actually reserved for the (macroscopic) field \boldsymbol{H}. See footnote 15 on page 190.

we may wish to determine the fields. Alternatively, given the fields, we may wish to study their effect on a test charge interacting with them. At first it may seem that these two problems are completely uncoupled. However, the fields in (2.1) are given by the superposition of all fields present. Thus, if we consider a test charge, we also need to consider the effect of the fields created by this charge itself, in addition to the fields already present before the test charge was introduced. In a totally self-consistent formalism the effect of this feedback system would have to be included. However, it is possible to show that these effects are generally small enough to be discarded: letting λ denote the characteristic length scale of a given problem, feedback effects can be neglected if

$$\lambda \gtrsim \frac{e^2}{m_e c^2} = 2.8 \cdot 10^{-15} \text{ m} ,$$

where e is the charge and m_e the mass of the electron. Therefore, only on a very small length scale do these effects become significant.[2] In the following we will always neglect them.

Continuity equation. Another important equation in electrodynamics of fundamental importance is the *continuity equation*. It reflects the experimental fact that any change of charge in a closed system of volume V is accompanied by a flow of charge through the surface F of the volume (*charge conservation*):

$$\frac{\mathrm{d}}{\mathrm{d}t} \int_V \mathrm{d}V \rho(\boldsymbol{x}, t) = - \oint_F \mathrm{d}\boldsymbol{F} \boldsymbol{j}(\boldsymbol{x}, t) .$$

By combining the divergence of (IV) with (I) we see that the continuity equation (in differential form) is indeed respected by Maxwell's equations.

Theorem 2.3: Continuity equation

Maxwell's equations are in agreement with the fundamental law of charge conservation

$$\frac{\partial \rho(\boldsymbol{x}, t)}{\partial t} + \boldsymbol{\nabla} \boldsymbol{j}(\boldsymbol{x}, t) = 0 .$$

2.1.2 Interpretation of Maxwell's Equations

The four Maxwell equations reflect experimental results, which are usually formulated in the form of the following laws:

[2] See the discussion of the self-energy in Subsection 2.5.1.

(I) Gauss's law.

$$\nabla E(\boldsymbol{x},t) = 4\pi\rho(\boldsymbol{x},t) \Longleftrightarrow \oint_F E(\boldsymbol{x},t)\mathrm{d}\boldsymbol{F} = 4\pi Q(t) \ ,$$

with

$$Q(t) = \int_V \rho(\boldsymbol{x},t)\mathrm{d}V \ .$$

This states that the total electric flux through the surface F enclosing a volume V is proportional to the total charge Q contained in V.

(II) Faraday's induction law.

$$\nabla \times E(\boldsymbol{x},t) + \frac{1}{c}\frac{\partial B(\boldsymbol{x},t)}{\partial t} = 0$$

$$\Longleftrightarrow V(t) = \oint_C E(\boldsymbol{x},t)\mathrm{d}\boldsymbol{l} = -\frac{1}{c}\int_F \mathrm{d}\boldsymbol{F}\frac{\partial B(\boldsymbol{x},t)}{\partial t} \ . \tag{2.2}$$

This states that a temporally varying magnetic field produces an electric field that circulates around the direction of the magnetic change. These *induced currents* are directed in such a manner as to weaken their own cause (*Lenz's law*). In other words: the temporal change of the magnetic field through a constant surface F leads to an *electromotive force (voltage)* V, which is determined by the contour integral of the induced electric field along the boundary C of F. Imagine a conducting loop enclosing the surface F (Fig. 2.1). A change of the magnetic field through F induces an electric field, which leads to a movement of the free charges within the conductor, i.e., to an electric current. One obtains the resulting voltage by cutting off the conductor and measuring the quantity $V = \int_1^2 \mathrm{d}\boldsymbol{l}E$ at the cutoff points 1 and 2. Note that Faraday's law is in actual fact more general, covering also the case of

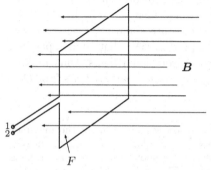

Fig. 2.1. Conducting loop within a magnetic field.

$$V(t) = \oint_{C'} \boldsymbol{E}'(\boldsymbol{x}',t)\mathrm{d}\boldsymbol{l}' = -\frac{1}{c}\frac{\mathrm{d}\Phi_{\mathrm{m}}(t)}{\mathrm{d}t} \; ,$$

where

$$\Phi_{\mathrm{m}}(t) = \int_{F(t)} \mathrm{d}\boldsymbol{F}(t)\boldsymbol{B}(\boldsymbol{x},t)$$

is the *magnetic flux* through the (not necessarily constant) surface F.[3] Here, primed quantities refer to the rest system of the conductor. We shall consider this case in connection with the Lorentz-covariant formulation of electrodynamics in Application 25.

(III) Absence of magnetic monopoles.

$$\boldsymbol{\nabla}\boldsymbol{B}(\boldsymbol{x},t) = 0 \Longleftrightarrow \oint_{F} \boldsymbol{B}(\boldsymbol{x},t)\mathrm{d}\boldsymbol{F} = 0 \; .$$

This states that the total magnetic flux through a surface F enclosing a volume vanishes. In other words: there are no *magnetic monopoles*, i.e., no sources or sinks for magnetic fields. In contrast to electric fields, magnetic field lines are always closed curves.[4]

(IV) Maxwell's displacement current and Ampère's law.

$$\boldsymbol{\nabla} \times \boldsymbol{B}(\boldsymbol{x},t) - \frac{1}{c}\frac{\partial \boldsymbol{E}(\boldsymbol{x},t)}{\partial t} = \frac{4\pi}{c}\boldsymbol{j}(\boldsymbol{x},t) \; .$$

This equation contains two parts. The first, *Maxwell's displacement current*

$$-\frac{1}{c}\frac{\partial \boldsymbol{E}(\boldsymbol{x},t)}{\partial t} \; , \tag{2.3}$$

was introduced by Maxwell as an additional term in (IV) when he realized that without it the complete set of equations would be in contradiction to the continuity equation. Furthermore, this term is necessary to explain electromagnetic radiation phenomena in the vacuum. This can be seen by considering the case where neither sources nor currents are present ($\rho = 0$, $\boldsymbol{j} = \boldsymbol{0}$); given Maxwell's equations without the displacement current in (IV), one quickly finds that both \boldsymbol{E} and \boldsymbol{B} are free of sources and rotations and thus vanish. It is only through the addition of (2.3) in (IV) that time-dependent electromagnetic fields become nonvanishing rotational fields, thus making their propagation in the vacuum possible. The second part, *Ampère's* or *Oersted's law*, reads

[3] Imagine, for example, that the conducting loop is moved out of the magnetic field or that it is rotating around an axis perpendicular to the magnetic field.

[4] Note, in this context, that Maxwell's equations are not all independent. Taking, for example, the divergence of (II) leads to $\boldsymbol{\nabla}\boldsymbol{B}(\boldsymbol{x},t) = f(\boldsymbol{x})$, with a scalar function f, that is determined experimentally to be 0.

$$\boldsymbol{\nabla} \times \boldsymbol{B}(\boldsymbol{x},t) = \frac{4\pi}{c}\boldsymbol{j}(\boldsymbol{x},t) \iff \oint_C \boldsymbol{B}(\boldsymbol{x},t)\mathrm{d}\boldsymbol{l} = \frac{4\pi}{c}I(t) \,, \tag{2.4}$$

with

$$I(t) = \int_F \boldsymbol{j}(\boldsymbol{x},t)\mathrm{d}\boldsymbol{F} \quad (\textit{electric current}).$$

This law is a generalization of Ampère's law of magnetostatics and can be obtained in the *quasi-static approximation*, where the displacement current is neglected. According to this law, a current induces a magnetic field whose closed field lines circulate around the current (Fig. 2.2).

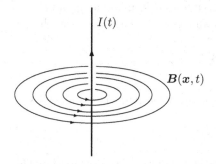

Fig. 2.2. Magnetic field induced by a current.

Uniqueness of the solutions. According to Helmholtz's integration law, any vector field \boldsymbol{V} defined over a onefold joint domain with a piecewise smooth boundary can be decomposed into a divergence-free and a curl-free part. Such a decomposition is unique, once the boundaries of both terms have been fixed. In particular, it is unique at infinity up to an additive constant as long as \boldsymbol{V} falls off asymptotically at least as fast as $1/r$. In this case, we have

$$\boldsymbol{V}(\boldsymbol{x}) = \frac{1}{4\pi}\left[\underbrace{\boldsymbol{\nabla} \times \int \mathrm{d}^3x' \frac{\boldsymbol{\nabla}' \times \boldsymbol{V}(\boldsymbol{x}')}{|\boldsymbol{x} - \boldsymbol{x}'|}}_{\text{divergence-free}} - \underbrace{\boldsymbol{\nabla} \int \mathrm{d}^3x' \frac{\boldsymbol{\nabla}'\boldsymbol{V}(\boldsymbol{x}')}{|\boldsymbol{x} - \boldsymbol{x}'|}}_{\text{curl-free}}\right] . \tag{2.5}$$

In other words: given the boundary conditions (or, equivalently, a fast enough asymptotic fall-off), as well as the sources and rotations of a vector field, this field is uniquely determined. Since we do not expect to find fields created from charges and currents at very large distances, it seems reasonable to assume that \boldsymbol{E} and \boldsymbol{B} fall off as $1/r^2$ toward infinity and thus are uniquely determined by Maxwell's equations.

Superposition principle. One of the most important characteristics of the Maxwell equations is their linearity. Just as for normal differential equations, linearity implies the principle of superposition, so that solutions of Maxwell's equations may be combined linearly to form new solutions. In particular, the most general solution is obtained from the general solution of the homogeneous equations plus one specific solution to the inhomogeneous equations. We will come back to this point in Section 2.2.

2.1.3 Energy and Momentum Conservation

It is intuitively clear that electromagnetic fields contain energy. In this subsection we determine the energy and momentum densities of electromagnetic fields by considering a system of point-like charges q_i situated at \boldsymbol{x}_i, and electromagnetic fields \boldsymbol{E} and \boldsymbol{B}. Using the δ-function, we can write the charge and current densities as[5]

$$\rho(\boldsymbol{x}, t) = \sum_i q_i \delta(\boldsymbol{x} - \boldsymbol{x}_i) \ , \ \ \boldsymbol{j}(\boldsymbol{x}, t) = \sum_i q_i \dot{\boldsymbol{x}}_i \delta(\boldsymbol{x} - \boldsymbol{x}_i) \ .$$

Energy conservation. Following (2.1) the force on particle i is given by

$$\boldsymbol{F}_\mathrm{L}(\boldsymbol{x}_i, t) = q_i \left(\boldsymbol{E}(\boldsymbol{x}_i, t) + \frac{\dot{\boldsymbol{x}}_i}{c} \times \boldsymbol{B}(\boldsymbol{x}_i, t) \right) \ .$$

The electric part of $\boldsymbol{F}_\mathrm{L}$ is exerting work on the particle, so that the mechanical energy E_mech is changed:

$$\frac{\mathrm{d}E_\mathrm{mech}}{\mathrm{d}t} = \sum_i \boldsymbol{F}_\mathrm{L}(\boldsymbol{x}_i, t) \dot{\boldsymbol{x}}_i = \sum_i q_i \dot{\boldsymbol{x}}_i \boldsymbol{E}(\boldsymbol{x}_i, t) = \int \mathrm{d}^3 x \boldsymbol{j}(\boldsymbol{x}, t) \boldsymbol{E}(\boldsymbol{x}, t) \ .$$

Using (IV), this can be written as

$$\frac{\mathrm{d}E_\mathrm{mech}}{\mathrm{d}t} = \frac{c}{4\pi} \int \mathrm{d}^3 x \boldsymbol{E} \boldsymbol{\nabla} \times \boldsymbol{B} - \frac{1}{4\pi} \int \mathrm{d}^3 x \boldsymbol{E} \frac{\partial \boldsymbol{E}}{\partial t} \ . \tag{2.6}$$

This energy balance equation implicitly contains the energy of the electromagnetic field E_em. To see this we subtract from (2.6) a total of 0 in form of (II) multiplied by $c\boldsymbol{B}/4\pi$. This yields

$$\frac{\mathrm{d}E_\mathrm{mech}}{\mathrm{d}t} = \frac{c}{4\pi} \int \mathrm{d}^3 x (\boldsymbol{E} \boldsymbol{\nabla} \times \boldsymbol{B} - \boldsymbol{B} \boldsymbol{\nabla} \times \boldsymbol{E}) - \int \mathrm{d}^3 x \frac{\partial}{\partial t} \left(\frac{\boldsymbol{E}^2 + \boldsymbol{B}^2}{8\pi} \right)$$

$$= -\frac{c}{4\pi} \oint_F (\boldsymbol{E} \times \boldsymbol{B}) \mathrm{d}\boldsymbol{F} - \frac{\partial}{\partial t} \int \mathrm{d}^3 x \left(\frac{\boldsymbol{E}^2 + \boldsymbol{B}^2}{8\pi} \right) \ , \tag{2.7}$$

where we have used the identity

[5] There are some fundamental problems with the electrodynamic concept of "point charges" and the use of the δ-function to describe them; they are discussed in Subsection 2.5.1.

$$\nabla(p \times q) = q\nabla \times p - p\nabla \times q$$

and Stokes's law. In order to interpret the individual terms in (2.7) we consider first the case of an infinitely large volume assuming that the fields fall off toward infinity faster than $1/r$. In this case (2.7) simplifies to

$$\frac{dE_{\text{mech}}}{dt} = -\frac{\partial}{\partial t}\int d^3x \left(\frac{E^2 + B^2}{8\pi}\right).$$

This suggests identifying

$$\frac{E^2 + B^2}{8\pi}$$

with the energy density ϵ_{em} of the electromagnetic field. Now, using Gauss's law, we find from (2.7) for finite volumes

$$\frac{dE_{\text{mech}}}{dt} = -\int d^3x \left(\nabla S + \frac{\partial \epsilon_{\text{em}}}{\partial t}\right), \tag{2.8}$$

where

$$S = \frac{c}{4\pi}(E \times B)$$

defines the *Poynting vector*, which we identify with the energy density of the electromagnetic field. Since (2.8) is valid for arbitrary volumes, we obtain

Theorem 2.4: Energy law of electrodynamics (Poynting theorem)

In a system consisting of charges and electromagnetic fields, the following energy balance equation holds:

$$\frac{\partial \epsilon_{\text{mech}}}{\partial t} + \frac{\partial \epsilon_{\text{em}}}{\partial t} = -\nabla S,$$

with

$$\frac{\partial \epsilon_{\text{mech}}(x,t)}{\partial t} = j(x,t)E(x,t) \qquad \text{Temporal derivative of the mechanical energy density}$$

$$\epsilon_{\text{em}} = \frac{E^2(x,t) + B^2(x,t)}{8\pi} \qquad \text{Electromagnetic energy density}$$

$$S(x,t) = \frac{c}{4\pi}E(x,t) \times B(x,t) \qquad \text{Poynting vector, energy current density.}$$

Poynting's theorem is a kind of a continuity equation for the energy of the system. It states that the temporal change of the total energy (mechanical plus electromagnetic) in a volume V is equal to the energy flux through the surface boundary F of V:

$$\frac{d}{dt}(E_{\text{mech}} + E_{\text{em}}) = -\oint_F S dF.$$

If the volume V is chosen sufficiently large, so that all charges and fields lie inside F, then the right-hand side vanishes and we obtain the energy conservation law for closed systems:

$$E_{\text{mech}} + E_{\text{em}} = \text{const} . \tag{2.9}$$

Momentum conservation. The derivation of the law of momentum conservation proceeds along the same lines as that of energy conservation. Here, our starting point is the temporal change of the mechanical momentum:

$$\frac{\mathrm{d}\boldsymbol{P}_{\text{mech}}}{\mathrm{d}t} = \sum_i \frac{\mathrm{d}\boldsymbol{P}_{\text{mech},i}}{\mathrm{d}t} = \sum_i \boldsymbol{F}_{\text{L},i}(\boldsymbol{x}_i, t)$$

$$= \int \mathrm{d}^3 x \left(\rho(\boldsymbol{x}, t)\boldsymbol{E}(\boldsymbol{x}, t) + \frac{\boldsymbol{j}(\boldsymbol{x}, t)}{c} \times \boldsymbol{B}(\boldsymbol{x}, t) \right) .$$

Expressing ρ and \boldsymbol{j} in terms of \boldsymbol{E} and \boldsymbol{B} and then symmetrizing the equation by adding [compare to (II) and (III)]

$$\frac{1}{4\pi} \left(\boldsymbol{\nabla} \times \boldsymbol{E} + \frac{1}{c}\frac{\partial \boldsymbol{B}}{\partial t} \right) \times \boldsymbol{E} + \frac{1}{4\pi}\boldsymbol{B}(\boldsymbol{\nabla}\boldsymbol{B}) = \boldsymbol{0} ,$$

we find

$$\frac{\mathrm{d}\boldsymbol{P}_{\text{mech}}}{\mathrm{d}t} = -\frac{1}{4\pi c}\frac{\partial}{\partial t} \int \mathrm{d}^3 x (\boldsymbol{E} \times \boldsymbol{B}) + \frac{1}{4\pi} \int \mathrm{d}^3 x [\boldsymbol{E}(\boldsymbol{\nabla}\boldsymbol{E}) + \boldsymbol{B}(\boldsymbol{\nabla}\boldsymbol{B})$$
$$- \boldsymbol{E} \times (\boldsymbol{\nabla} \times \boldsymbol{E}) - \boldsymbol{B} \times (\boldsymbol{\nabla} \times \boldsymbol{B})] .$$

The integrand of the first term,

$$\frac{1}{4\pi c}(\boldsymbol{E} \times \boldsymbol{B}) = \frac{\boldsymbol{S}}{c^2} ,$$

is identified with the momentum density $\boldsymbol{g}_{\text{em}}$ of the electromagnetic field. The components of the integrand of the second term can be written one by one as the divergence of a vector field, so that we finally obtain

Theorem 2.5: Momentum law of electrodynamics

In a system consisting of charged particles and electromagnetic fields, we have the momentum balance equation

$$\left[\frac{\partial \boldsymbol{g}_{\text{mech}}}{\partial t} \right]_i + \left[\frac{\partial \boldsymbol{g}_{\text{em}}}{\partial t} \right]_i = \boldsymbol{\nabla} T_i , \tag{2.10}$$

with

$$\frac{\partial \boldsymbol{g}_{\text{mech}}(\boldsymbol{x}, t)}{\partial t} = \rho(\boldsymbol{x}, t)\boldsymbol{E}(\boldsymbol{x}, t) + \frac{\boldsymbol{j}(\boldsymbol{x}, t)}{c} \times \boldsymbol{B}(\boldsymbol{x}, t)$$

Temp. derivative of the mechanical momentum density

\triangleright

2.1 Formalism of Electrodynamics 123

$$g_{\text{em}}(\boldsymbol{x},t) = \frac{\boldsymbol{S}(\boldsymbol{x},t)}{c^2}$$

Electromagnetic
momentum density

$$\boldsymbol{T}_i = (T_{i1}, T_{i2}, T_{i3})$$

$$T_{ik} = \frac{1}{4\pi}\left[E_i E_k + B_i B_k - \frac{\delta_{ik}}{2}\left(\boldsymbol{E}^2 + \boldsymbol{B}^2\right)\right]$$

Maxwell's
stress tensor.

Integrating (2.10) over a sufficiently large volume, which contains all particles and fields, we find the momentum conservation law for closed systems

$$\boldsymbol{P}_{\text{mech}} + \boldsymbol{P}_{\text{em}} = \text{const}.$$

This is analogous to the energy law (2.9). In the following we shall often be considering \boldsymbol{E} and \boldsymbol{B} fields oscillating in time. In this case, it is sensible to consider the averages of ϵ_{em} and \boldsymbol{S} over a single oscillatory period since this eliminates the oscillating terms.

Definition: Time-averaged electromagnetic energy density $\overline{\epsilon_{\text{em}}}$ and energy current density $\overline{\boldsymbol{S}}$

For oscillating fields of the form

$$\boldsymbol{E}(\boldsymbol{x},t) = \text{Re}\left[\boldsymbol{E}(\boldsymbol{x})\mathrm{e}^{-\mathrm{i}\omega t}\right] \ , \ \ \boldsymbol{B}(\boldsymbol{x},t) = \text{Re}\left[\boldsymbol{B}(\boldsymbol{x})\mathrm{e}^{-\mathrm{i}\omega t}\right] \ ,$$

it is useful to consider the time-averaged quantities $\overline{\epsilon_{\text{em}}}$ and $\overline{\boldsymbol{S}}$ defined as

$$\overline{\epsilon_{\text{em}}} = \frac{1}{T}\int\limits_{t}^{t+T} \mathrm{d}t \epsilon_{\text{em}} = \frac{|\boldsymbol{E}(\boldsymbol{x})|^2 + |\boldsymbol{B}(\boldsymbol{x})|^2}{16\pi} \ , \ T = \frac{2\pi}{\omega} \tag{2.11}$$

$$\overline{\boldsymbol{S}} = \frac{1}{T}\int\limits_{t}^{t+T} \mathrm{d}t \boldsymbol{S} = \frac{c}{8\pi}\text{Re}[\boldsymbol{E}(\boldsymbol{x}) \times \boldsymbol{B}^*(\boldsymbol{x})] \ . \tag{2.12}$$

2.1.4 Physical Units

Maxwell's equations (I) to (IV) describe the functional relationships between the charge and current densities ρ and \boldsymbol{j} as well as the fields \boldsymbol{E} and \boldsymbol{B}. However, the constants of proportionality appear a little arbitrary and depend on the chosen unit system. Before making such a choice, Maxwell's equations could be written as

$$\left.\begin{array}{l} \boldsymbol{\nabla E}(\boldsymbol{x},t) = 4\pi k_1 \rho(\boldsymbol{x},t) \\[2mm] \boldsymbol{\nabla} \times \boldsymbol{E}(\boldsymbol{x},t) + k_2 \dfrac{\partial \boldsymbol{B}(\boldsymbol{x},t)}{\partial t} = 0 \\[2mm] \boldsymbol{\nabla B}(\boldsymbol{x},t) = 0 \\[2mm] \boldsymbol{\nabla} \times \boldsymbol{B}(\boldsymbol{x},t) - k_3 \dfrac{\partial \boldsymbol{E}(\boldsymbol{x},t)}{\partial t} = 4\pi k_4 \boldsymbol{j}(\boldsymbol{x},t) \ , \end{array}\right\} \tag{2.13}$$

with four constants k_1, \ldots, k_4. Assuming that the continuity equation holds in any unit system and is always given by

$$\frac{\partial \rho}{\partial t} + \boldsymbol{\nabla j} = 0 \ , \tag{2.14}$$

the combination of the first and the last equation of (2.13) with (2.14) yields the condition

$$k_1 k_3 = k_4 \ .$$

Another constraint follows from the experimental fact that the propagational velocity of electromagnetic waves in vacuum is equal to the velocity of light. The corresponding *wave equations* are obtained by the combination of the two curl equations of (2.13) as

$$\boldsymbol{\nabla}^2 \begin{pmatrix} \boldsymbol{E} \\ \boldsymbol{B} \end{pmatrix} - k_2 k_3 \frac{\partial^2}{\partial t^2} \begin{pmatrix} \boldsymbol{E} \\ \boldsymbol{B} \end{pmatrix} = 0$$

and thus yields the constraint

$$k_2 k_3 = \frac{1}{c^2} \ .$$

All in all, we find that only two of our four constants of proportionality are in fact independent. Their choice uniquely defines a unit system. The most common choices are the MKSA-system (meter, kilogram, second, Ampère) and the Gauss system:

System	k_1	k_2	k_3	k_4
MKSA	$\dfrac{1}{4\pi\epsilon_0}$	1	$\epsilon_0 \mu_0$	$\dfrac{\mu_0}{4\pi}$
Gauss	1	$\dfrac{1}{c}$	$\dfrac{1}{c}$	$\dfrac{1}{c}$

The quantities $\epsilon_0 = 8.854 \cdot 10^{-12} \, \mathrm{A}^2 \mathrm{s}^4 \mathrm{m}^{-3} \mathrm{kg}^{-1}$ and $\mu_0 = 1/\epsilon_0 c^2$ are termed *dielectric constant of the vacuum* and *permeability constant of the vacuum*, respectively.

The MKSA-system is simply the MKS-system of mechanics, supplemented by the fundamental quantity "current". Its unit "A (Ampère)" is

defined via the force that two current-carrying conductors exert upon one another at a certain distance. Since the electric current in a conductor is equal to the charge $\mathrm{d}q$ that flows through its cross section per unit time $\mathrm{d}t$,

$$I(t) = \frac{\mathrm{d}q}{\mathrm{d}t} \ ,$$

the composed unit of charge, "C (Coulomb)" is

$$1\,\mathrm{C} = 1\,\mathrm{As} \ .$$

In the Gauss system, the three fundamental units of centimeter, gram, and second are supplemented by the charge unit "ESU" (electrostatic unit). Again, this is defined via the force that two static charges of one ESU exert upon one another over a certain distance. Here the unit of current is composed of the units ESU and second.[6] In macroscopic experimental physics, the MKSA-system is mostly used due to practical reasons, whereas in atomic, nuclear, and many textbooks of theoretical physics, the Gauss system is preferred. The main advantage of the latter, relative to the MKSA-system (or others), lies in the fact that it exhibits the relativistic structure of electrodynamics most clearly through factors of v/c. As we shall see in later sections, E and B fields are transformed into one another when changing the reference frame. In the Gauss system this is reflected quite naturally since E and B have identical units. Due to these considerations, we shall adhere to the Gauss system throughout the whole chapter.

Summary

- Electromagnetic phenomena are due to two types of sources, the **charge density** ρ and the **current density vector** j, which cause the **electromagnetic fields** E and B. These quantities are coupled via a system of first-order partial differential equations called **Maxwell's equations**.

- Maxwell's equations and the **Lorentz force** define the theory of classical electrodynamics.

- Maxwell's equations are consistent with charge conservation in a closed system, as expressed by the **continuity equation**.

- Between the **mechanical** and **electromagnetic energy** and the **momentum densities** there exist **balance equations**, which can be formulated as continuity equations. The sum of mechanical and electromagnetic energy, as well as that of mechanical and electromagnetic momentum, are conserved inside a closed system of fields and charges.

\triangleright

[6] The conversion between different unit systems is discussed in detail in [12].

> • In this chapter we always use the **Gauss unit system** as it exhibits the relativistic structure of electrodynamics most clearly.

Applications

20. Magnetic monopoles. Suppose that, in addition to the electric charge and current densities $\rho = \rho_e$ and $\boldsymbol{j} = \boldsymbol{j}_e$, there were magnetic counterparts ρ_m and \boldsymbol{j}_m, so that Maxwell's equations were to take on the symmetric form

$$\boldsymbol{\nabla} \boldsymbol{E} = 4\pi \rho_e$$

$$\boldsymbol{\nabla} \times \boldsymbol{E} + \frac{1}{c} \frac{\partial \boldsymbol{B}}{\partial t} = -\frac{4\pi}{c} \boldsymbol{j}_m$$

$$\boldsymbol{\nabla} \boldsymbol{B} = 4\pi \rho_m$$

$$\boldsymbol{\nabla} \times \boldsymbol{B} - \frac{1}{c} \frac{\partial \boldsymbol{E}}{\partial t} = \frac{4\pi}{c} \boldsymbol{j}_e \ .$$

Given the existence of magnetic monopoles and by only assuming that the ratio of electric and magnetic charges of all particles is the same, show that Maxwell's equations (I) to (IV) would still be valid.

Tip: study the above equations under the duality transformation

$$\begin{pmatrix} \boldsymbol{E}' \\ \boldsymbol{B}' \\ \rho'_e \\ \rho'_m \\ \boldsymbol{j}'_e \\ \boldsymbol{j}'_m \end{pmatrix} = \begin{pmatrix} \cos\alpha & -\sin\alpha & 0 & 0 & 0 & 0 \\ \sin\alpha & \cos\alpha & 0 & 0 & 0 & 0 \\ 0 & 0 & \cos\alpha & -\sin\alpha & 0 & 0 \\ 0 & 0 & \sin\alpha & \cos\alpha & 0 & 0 \\ 0 & 0 & 0 & 0 & \cos\alpha & -\sin\alpha \\ 0 & 0 & 0 & 0 & \sin\alpha & \cos\alpha \end{pmatrix} \begin{pmatrix} \boldsymbol{E} \\ \boldsymbol{B} \\ \rho_e \\ \rho_m \\ \boldsymbol{j}_e \\ \boldsymbol{j}_m \end{pmatrix} \ .$$

Solution. It is straightforward to verify that the symmetrized Maxwell equations are invariant under this duality transformation. Since the electromagnetic charge ratio is taken to be constant for all particles, one may choose the angle α in such a way that

$$\rho'_m = \rho_e \left(\sin\alpha + \frac{\rho_m}{\rho_e} \cos\alpha \right) = 0$$

and

$$|\boldsymbol{j}'_m| = |\boldsymbol{j}_e| \left(\sin\alpha + \frac{|\boldsymbol{j}_m|}{|\boldsymbol{j}_e|} \cos\alpha \right) = |\boldsymbol{j}_e| \left(\sin\alpha + \frac{\rho_m}{\rho_e} \cos\alpha \right) = 0 \ .$$

For this particular choice of α, the symmetrized Maxwell equations reduce to the previous equations (I) to (IV). In other words, if magnetic monopoles existed, such that the electromagnetic charge ratio of all particles was identical, one could choose the magnetic charge to vanish identically. Thus, the interesting question in connection with magnetic monopoles is whether there exist

particles with differing electromagnetic charge ratios. If this were the case, one would have to allow a magnetic charge and to discuss the above symmetrized set of equations. In addition, a magnetic charge of this type would explain, as first pointed out by Dirac, the quantization of electric charge.[7]

21. Conducting loop with plate condenser. Consider a plate condenser connected via a conducting loop in a homogeneous magnetic field, as shown in Fig. 2.3. The magnetic field vector points into the plane of the paper, and its absolute value is growing with time. Which of the two condenser plates is charged positively?

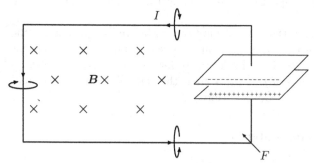

Fig. 2.3. Conducting loop with plate condenser in an external magnetic field. The arrows indicate the direction of the induced current and the induced magnetic field, given that the external field points into the plane of the paper and is growing with time.

Solution. Due to Faraday's induction law, a current is induced in the conductor around which a magnetic field circulates (Ampère's law) and directed in such a way as to weaken its origin, i.e., the growth of the external magnetic field (Lenz's law). Thus, the current circulates in mathematically positive direction (counterclockwise) and the top condenser plate is charged negatively.[8] Mathematically, we can see this as follows: looking at the voltage in the mathematically positive direction with the normal vector of the plane F pointing out of the plane of the paper, we find

$$BF = -|B||F| \implies V = -\frac{1}{c}\frac{\partial B}{\partial t}F = +\frac{1}{c}\frac{\partial |B|}{\partial t}|F| > 0 \ .$$

Therefore, a positive voltage is induced; the current flows in the mathematically positive direction.

[7] Practically all particles have a charge, which is an integer multiple of the elementary charge e of the electron.

[8] Note: the direction of the current is opposite to the flow of the negatively charged free electrons (*technical current direction*).

2.2 Solutions of Maxwell's Equations in the Form of Potentials

Maxwell's equations are a system of four coupled first-order differential equations for the fields \boldsymbol{E} and \boldsymbol{B}. By introducing a *vector potential* \boldsymbol{A} and a *scalar potential* ϕ, these equations can be reduced to two differential equations of the second order. Often it is easier to evaluate these potential fields first and then derive the fields \boldsymbol{E} and \boldsymbol{B} from these. However, it turns out that the potentials are not uniquely defined; whole classes of potentials related to one another by *gauge transformations* yield the same electromagnetic fields. Such transformations play an important role in modern formulations of quantum field theories.

This section deals with the general solution of Maxwell's equations in the form of potentials. After their introduction, we discuss gauge transformations and associated *gauge conditions*. Finally, we derive solutions to Maxwell's equations in terms of *retarded potentials*, first, the general homogeneous and, second, a specific inhomogeneous solution.

2.2.1 Scalar and Vector Potential

The potentials \boldsymbol{A} and ϕ are defined as follows:

Definition: Vector potential \boldsymbol{A} and scalar potential ϕ

$$\left. \begin{aligned} \boldsymbol{B}(\boldsymbol{x},t) &= \boldsymbol{\nabla} \times \boldsymbol{A}(\boldsymbol{x},t) \\ \boldsymbol{E}(\boldsymbol{x},t) + \frac{1}{c}\frac{\partial \boldsymbol{A}(\boldsymbol{x},t)}{\partial t} &= -\boldsymbol{\nabla}\phi(\boldsymbol{x},t) \ . \end{aligned} \right\} \tag{2.15}$$

It is easy to verify by insertion that through these definitions, the homogeneous Maxwell equations (II) and (III) are satisfied automatically. For the two remaining inhomogeneous equations (I) and (IV), we find the *potential equations*

$$\boldsymbol{\nabla}^2\phi(\boldsymbol{x},t) + \frac{1}{c}\frac{\partial}{\partial t}\boldsymbol{\nabla}\boldsymbol{A}(\boldsymbol{x},t) = -4\pi\rho(\boldsymbol{x},t) \tag{2.16}$$

$$\boldsymbol{\nabla}^2\boldsymbol{A}(\boldsymbol{x},t) - \frac{1}{c^2}\frac{\partial^2 \boldsymbol{A}(\boldsymbol{x},t)}{\partial t^2} - \boldsymbol{\nabla}\left(\boldsymbol{\nabla}\boldsymbol{A}(\boldsymbol{x},t) + \frac{1}{c}\frac{\partial\phi(\boldsymbol{x},t)}{\partial t}\right)$$
$$= -\frac{4\pi}{c}\boldsymbol{j}(\boldsymbol{x},t) \ . \tag{2.17}$$

The problem of finding the six components of the fields \boldsymbol{E} and \boldsymbol{B} has been transformed into finding the four components of \boldsymbol{A} and ϕ. Nevertheless, thus far, (2.16) and (2.17) do not seem to yield a simplification since they are still coupled equations in the potential fields. However, the potentials exhibit certain *gauge freedoms*, which we can use to adjust the vector potential in

such a way that (2.16) and (2.17) decouple. This is demonstrated in the following subsection.

2.2.2 Gauge Transformations

Gauge transformations in electrodynamics are defined as follows:

Definition: Gauge transformations

Transformations of the form

$$A(x,t) \longrightarrow A'(x,t) = A(x,t) + \nabla\chi(x,t)$$

$$\phi(x,t) \longrightarrow \phi'(x,t) = \phi(x,t) - \frac{1}{c}\frac{\partial\chi(x,t)}{\partial t},$$

with an arbitrary scalar function $\chi(x,t)$, are called *gauge transformations*. They leave the electromagnetic fields E and B invariant.

Notice that all field equations, and therefore all physical predictions, are invariant under gauge transformations. As the potentials are not directly observable, they are often described as unphysical. However, this is only true on a classical level. In the context of quantum theory, we shall encounter situations in which the vector potential itself plays an important physical role (quantization of the magnetic flux, Bohm-Aharanov effect, Subsection 3.6.2).

The choice of the gauge transformation, i.e., the *gauge function* χ, depends on the problem at hand. In the following we consider two of the most commonly used gauges, which simplify significantly the inhomogeneous potential equations (2.16) and (2.17).

Coulomb gauge. The Coulomb gauge is defined by the following condition:

Definition: Coulomb gauge (transverse gauge)

$$\nabla A(x,t) = 0 .$$

In this gauge the potential equations (2.16) and (2.17) become

$$\nabla^2\phi(x,t) = -4\pi\rho(x,t) \qquad \text{(\textit{Poisson equation})} \tag{2.18}$$

$$\nabla^2 A(x,t) - \frac{1}{c^2}\frac{\partial^2 A(x,t)}{\partial t^2} = -\frac{4\pi}{c}j(x,t) + \frac{1}{c}\nabla\frac{\partial\phi(x,t)}{\partial t} . \tag{2.19}$$

The solution of the Poisson equation (2.18) is given by the *instantaneous Coulomb potential*

$$\phi(x,t) = \int \frac{\rho(x',t)}{|x - x'|}d^3x' . \tag{2.20}$$

Note that both sides of this equation contain the same temporal argument, which implies that a charge at location x' has an instantaneous effect (i.e.,

no time delay) on the potential ϕ at location x. In other words, the Coulomb gauge is not relativistically invariant.

It is not immediately clear that the Coulomb gauge leads to a decoupling of (2.16) and (2.17). In order to see this, we use Helmholtz's integration law (2.5) to write the current density vector j as

$$j(x,t) = j_T + j_L ,$$

where

$$j_T(x,t) = \frac{1}{4\pi}\nabla \times \int d^3x' \frac{\nabla' \times j(x',t)}{|x - x'|} \tag{2.21}$$

$$j_L(x,t) = -\frac{1}{4\pi}\nabla \int d^3x' \frac{\nabla' j(x',t)}{|x - x'|} \tag{2.22}$$

denotes the transverse and the longitudinal components of j, respectively. Combining (2.20) with the continuity equation,

$$\nabla \frac{\partial \phi(x,t)}{\partial t} = -\nabla \int \frac{\nabla' j(x',t)}{|x - x'|} d^3x' ,$$

and comparing this with (2.22), it follows that

$$\nabla \frac{\partial \phi}{\partial t} = 4\pi j_L .$$

The right-hand side of (2.19) is therefore proportional to the transverse current density (2.21), so that (2.19) can be rewritten as the inhomogeneous wave equation

$$\nabla^2 A - \frac{1}{c^2}\frac{\partial^2 A}{\partial t^2} = -\frac{4\pi}{c} j_T .$$

This explains why the Coulomb gauge is also called the "transverse gauge".

Lorentz gauge. The second class of gauge transformations we consider is defined by

Definition: Lorentz gauge

$$\nabla A(x,t) = -\frac{1}{c}\frac{\partial \phi(x,t)}{\partial t} .$$

This transformation leads to the symmetric and decoupled inhomogeneous wave equations in A and ϕ,

$$\left(\nabla^2 - \frac{1}{c^2}\frac{\partial^2}{\partial t^2}\right)\phi(x,t) = -4\pi\rho(x,t)$$

$$\left(\nabla^2 - \frac{1}{c^2}\frac{\partial^2}{\partial t^2}\right)A(x,t) = -\frac{4\pi}{c}j(x,t) .$$

The advantage of the Lorentz gauge compared to the Coulomb gauge is its relativistic invariance. We will show this explicitly in Subsection 2.3.2.

Existence of the gauges. To show that both types of gauge transformations always exist, we now determine the corresponding gauge functions χ.

- **Coulomb gauge:** we wish to find a function χ such that

$$\nabla A' = \nabla A + \nabla^2 \chi = 0 \implies \nabla^2 \chi = -\nabla A .$$

This, again, is the Poisson equation, which is solved by

$$\chi(\boldsymbol{x}, t) = \frac{1}{4\pi} \int \frac{\nabla' A(\boldsymbol{x}', t)}{|\boldsymbol{x} - \boldsymbol{x}'|} d^3 x' .$$

- **Lorentz gauge:** here the function χ must be chosen such that

$$\nabla A' + \frac{1}{c}\frac{\partial \phi'}{\partial t} = \nabla A + \nabla^2 \chi + \frac{1}{c}\frac{\partial \phi}{\partial t} - \frac{1}{c^2}\frac{\partial^2 \chi}{\partial t^2} = 0$$

$$\implies \left(\nabla^2 - \frac{1}{c^2}\frac{\partial^2}{\partial t^2} \right) \chi = -\left(\nabla A + \frac{1}{c}\frac{\partial \phi}{\partial t} \right) .$$

The solution χ of this inhomogeneous wave equation is not unique since an arbitrary solution Λ of the corresponding homogeneous equation

$$\left(\nabla^2 - \frac{1}{c^2}\frac{\partial^2}{\partial t^2} \right) \Lambda = 0$$

can be added to χ. This equation defines a particular type of *restricted gauge transformations*. It can always be used to achieve $\phi(\boldsymbol{x}, t) = 0$.

Theorem 2.6: Maxwell's equations using potential functions

Using a vector potential A and a scalar potential ϕ, Maxwell's equations can be written as two coupled second-order differential equations (potential equations). The potentials have certain gauge freedoms, which can be used to decouple these two equations. In Coulomb gauge

$$\nabla A(\boldsymbol{x}, t) = 0 ,$$

the potential equations read

$$\nabla^2 \phi(\boldsymbol{x}, t) = -4\pi \rho(\boldsymbol{x}, t)$$

$$\nabla^2 A(\boldsymbol{x}, t) - \frac{1}{c^2}\frac{\partial^2 A(\boldsymbol{x}, t)}{\partial t^2} = -\frac{4\pi}{c} j_{\mathrm{T}}(\boldsymbol{x}, t) ,$$

with

$$j_{\mathrm{T}}(\boldsymbol{x}, t) = \frac{1}{4\pi} \nabla \times \int d^3 x' \frac{\nabla' \times j(\boldsymbol{x}', t)}{|\boldsymbol{x} - \boldsymbol{x}'|} .$$

▷

The corresponding relations in Lorentz gauge

$$\nabla A(x,t) = -\frac{1}{c}\frac{\partial \phi(x,t)}{\partial t} \tag{2.23}$$

are given by

$$\left.\begin{array}{l}\left(\nabla^2 - \dfrac{1}{c^2}\dfrac{\partial^2}{\partial t^2}\right)\phi(x,t) = -4\pi\rho(x,t) \\[2mm] \left(\nabla^2 - \dfrac{1}{c^2}\dfrac{\partial^2}{\partial t^2}\right)A(x,t) = -\dfrac{4\pi}{c}j(x,t) \ .\end{array}\right\} \tag{2.24}$$

General solution of Maxwell's equations. Using gauge transformations we were able to significantly simplify the problem of solving Maxwell's equations. The wave equations (2.24) for the scalar potential and the components of the vector potential are of the same structure,

$$\left(\nabla^2 - \frac{1}{c^2}\frac{\partial^2}{\partial t^2}\right)g(x,t) = -4\pi f(x,t) \ , \tag{2.25}$$

where f is a known source or current function. This is a linear equation, with its solution given by the superposition of the general homogeneous and a particular inhomogeneous solution, such that the overall solution must satisfy the Lorentz gauge.

In the following two subsections we derive the general solution of the homogeneous and a particular solution of the inhomogeneous wave equation (2.24) in the Lorentz gauge (2.23).

2.2.3 General Solution of the Homogeneous Wave Equations

In the homogeneous case, (2.25) reduces to

$$\left(\nabla^2 - \frac{1}{c^2}\frac{\partial^2}{\partial t^2}\right)g_{\text{hom}}(x,t) = 0 \ . \tag{2.26}$$

To find its solution we decompose the function g_{hom} into its complex Fourier components,

$$g_{\text{hom}}(x,t) = \frac{1}{\sqrt{2\pi}^4}\int d^3k \int d\omega e^{i(kx-\omega t)}\tilde{g}(k,\omega) \ , \tag{2.27}$$

and insert this expression into (2.26):

$$\left(k^2 - \frac{\omega^2}{c^2}\right)\tilde{g}(k,\omega) = 0 \ .$$

According to this, \tilde{g} must vanish everywhere, except at $\omega = \pm c|k|$. Therefore, we can write

$$\tilde{g}(k,\omega) = \tilde{g}_1(k)\delta(\omega - c|k|) + \tilde{g}_2(k)\delta(\omega + c|k|) \ ,$$

where \tilde{g}_1, \tilde{g}_2 are arbitrarily selectable complex coefficient functions. Now, (2.27) becomes

$$g_{\text{hom}}(\boldsymbol{x}, t) = \frac{1}{(2\pi)^2} \int \mathrm{d}^3 k \left(\tilde{g}_1(\boldsymbol{k}) \mathrm{e}^{\mathrm{i}[\boldsymbol{kx} - \omega(\boldsymbol{k})t]} + \tilde{g}_2(\boldsymbol{k}) \mathrm{e}^{\mathrm{i}[\boldsymbol{kx} + \omega(\boldsymbol{k})t]} \right) , \quad (2.28)$$

with

$$\omega(\boldsymbol{k}) = c|\boldsymbol{k}| .$$

This is the general homogeneous solution of (2.26). As the potentials \boldsymbol{A} and ϕ are real, we only need to consider the real part of (2.28), so that[9]

Theorem 2.7: Solutions of the homogeneous wave equations

The most general form of the solutions of the homogeneous wave equations (2.24) is given by

$$\phi_{\text{hom}}(\boldsymbol{x}, t) = \mathrm{Re} \int \mathrm{d}^3 k \tilde{\phi}(\boldsymbol{k}) \mathrm{e}^{\mathrm{i}[\boldsymbol{kx} - \omega(\boldsymbol{k})t]} , \quad \omega(\boldsymbol{k}) = c|\boldsymbol{k}|$$

$$A_{\text{hom},i}(\boldsymbol{x}, t) = \mathrm{Re} \int \mathrm{d}^3 k \tilde{A}_i(\boldsymbol{k}) \mathrm{e}^{\mathrm{i}[\boldsymbol{kx} - \omega(\boldsymbol{k})t]} , \quad i = 1, 2, 3 .$$

The complex Fourier coefficients $\tilde{\phi}, \tilde{A}_i$ are determined by initial conditions, for example, $\phi(\boldsymbol{x}, 0) = \phi_0(\boldsymbol{x})$, $\dot{\phi}(\boldsymbol{x}, 0) = \psi_0(\boldsymbol{x})$, $\boldsymbol{A}(\boldsymbol{x}, 0) = \boldsymbol{A}_0(\boldsymbol{x})$, $\dot{\boldsymbol{A}}(\boldsymbol{x}, 0) = \boldsymbol{B}_0(\boldsymbol{x})$, as well as by the Lorentz condition (2.23).

The solutions ϕ_{hom} and $\boldsymbol{A}_{\text{hom}}$ represent waves. These are investigated in detail in Section 2.7.

2.2.4 Specific Solution of the Inhomogeneous Wave Equation, Retarded Potentials

The solution of the inhomogeneous equation (2.25) can be written most generally as

$$g(\boldsymbol{x}, t) = \int \mathrm{d}^3 x' \int \mathrm{d}t' G(\boldsymbol{x}, t, \boldsymbol{x}', t') f(\boldsymbol{x}', t') \quad (2.29)$$

if G, the Green function of our problem, satisfies

$$\left(\boldsymbol{\nabla}^2 - \frac{1}{c^2} \frac{\partial^2}{\partial t^2} \right) G(\boldsymbol{x}, t, \boldsymbol{x}', t') = -4\pi \delta(\boldsymbol{x} - \boldsymbol{x}') \delta(t - t') . \quad (2.30)$$

To determine G, we use the Fourier-transformed equivalent of (2.30). With

$$G(\boldsymbol{x}, t, \boldsymbol{x}', t') = \int \mathrm{d}^3 k \int \mathrm{d}\omega \tilde{G}(\boldsymbol{k}, \omega) \mathrm{e}^{\mathrm{i}\boldsymbol{k}(\boldsymbol{x} - \boldsymbol{x}')} \mathrm{e}^{\mathrm{i}\omega(t - t')} \quad (2.31)$$

and

[9] Because $\mathrm{Re}(a_1 + \mathrm{i}a_2)\mathrm{e}^{\pm \mathrm{i}\omega t} = a_1 \cos \omega t \mp a_2 \sin \omega t$, it is sufficient to restrict oneself to the solution with $\mathrm{e}^{-\mathrm{i}\omega t}$.

$$\delta(\boldsymbol{x} - \boldsymbol{x}')\delta(t - t') = \frac{1}{(2\pi)^4} \int d^3k \int d\omega e^{i\boldsymbol{k}(\boldsymbol{x}-\boldsymbol{x}')} e^{i\omega(t-t')} \ ,$$

one finds

$$\left(-\boldsymbol{k}^2 + \frac{\omega^2}{c^2}\right) \tilde{G}(\boldsymbol{k}, \omega) = -\frac{1}{4\pi^3}$$

$$\Longrightarrow \tilde{G}(\boldsymbol{k}, \omega) = \frac{1}{4\pi^3} \frac{1}{\boldsymbol{k}^2 - \frac{\omega^2}{c^2}} = \frac{1}{8\pi^3} \frac{c}{k} \left(\frac{1}{\omega + ck} - \frac{1}{\omega - ck}\right) \ , \quad k = |\boldsymbol{k}| \ .$$

For (2.31), this implies that

$$G(\boldsymbol{x}, t, \boldsymbol{x}', t') = \frac{c}{(2\pi)^3} \int d^3k \frac{e^{i\boldsymbol{k}\Delta\boldsymbol{x}}}{k}$$

$$\times \int d\omega e^{i\omega\Delta t} \left(\frac{1}{\omega + ck} - \frac{1}{\omega - ck}\right) \ , \tag{2.32}$$

with

$$\Delta\boldsymbol{x} = \boldsymbol{x} - \boldsymbol{x}' \ , \quad \Delta t = t - t' \ .$$

Obviously, the integrand in (2.32) has two poles of first order at $\omega = \mp ck$. To determine the ω-integral, we make use of Cauchy's theorem and proceed in similar fashion as for the forced harmonic oscillator in Application 3. This means we choose a closed semicircle with radius R in the complex ω-plane (see Fig. 1.5). As in Application 3, we have to remember that for $\Delta t > 0$ ($\Delta t < 0$) the upper path C (lower path C') must be chosen. Additionally, due to the principle of causality, we must demand that the integration in (2.29) does not contribute for $\Delta t < 0$, i.e.,

$$G(\boldsymbol{x}, t, \boldsymbol{x}', t') = 0 \ \forall \ t < t' \ .$$

This condition can be mathematically realized by shifting the poles into the upper half-plane. This means performing the replacements $ck \longrightarrow ck \mp i\epsilon$ with $0 < \epsilon \ll 1$ in (2.32), since for $\Delta t < 0$ the integration in the lower half-plane is then identical to zero. For $\Delta t > 0$ and in the limit $\epsilon \to 0$, it follows that

$$G(\boldsymbol{x}, t, \boldsymbol{x}', t') = \frac{c}{2\pi^2} \int d^3k \frac{e^{i\boldsymbol{k}\Delta\boldsymbol{x}}}{k} \sin(ck\Delta t)$$

$$= \frac{c}{\pi} \int_0^\infty dk k \sin(ck\Delta t) \int_{-1}^1 d\cos\theta e^{ik|\Delta\boldsymbol{x}|\cos\theta}$$

$$= \frac{2c}{\pi|\Delta\boldsymbol{x}|} \int_0^\infty dk \sin(ck\Delta t) \sin(k|\Delta\boldsymbol{x}|) \ . \tag{2.33}$$

Since the integrand in (2.33) is even, we can extend the integration range to $[-\infty : \infty]$. Substituting $\kappa = ck$, we find

$$G(\boldsymbol{x},t,\boldsymbol{x}',t') = \frac{1}{2\pi|\Delta\boldsymbol{x}|} \int\limits_{-\infty}^{\infty} d\kappa \left[e^{i\kappa\left(\Delta t - \frac{|\Delta\boldsymbol{x}|}{c}\right)} - e^{i\kappa\left(\Delta t + \frac{|\Delta\boldsymbol{x}|}{c}\right)} \right]$$

$$= \frac{1}{|\Delta\boldsymbol{x}|} \left[\delta\left(\Delta t - \frac{|\Delta\boldsymbol{x}|}{c}\right) - \delta\left(\Delta t + \frac{|\Delta\boldsymbol{x}|}{c}\right) \right] ,$$

where only the first term contributes since, for $\Delta t > 0$, the argument of the second δ-function never vanishes. The explicit form of the Green function is therefore given by

$$G(\boldsymbol{x},t,\boldsymbol{x}',t') = \frac{\delta\left(t' - t + \frac{|\boldsymbol{x}-\boldsymbol{x}'|}{c}\right)}{|\boldsymbol{x}-\boldsymbol{x}'|} . \tag{2.34}$$

This function is also called the *retarded Green function* since it honors the causality requirement that an effect observed at time t and location x is due to a disturbance at location \boldsymbol{x}' at an earlier time $t' = t - |\boldsymbol{x} - \boldsymbol{x}'|/c$. Substituting (2.34) in (2.29), one finally obtains the retarded solution of (2.25) in the absence of boundary conditions:

$$g(\boldsymbol{x},t) = \int d^3x' \int dt' \frac{f(\boldsymbol{x}',t')}{|\boldsymbol{x}-\boldsymbol{x}'|} \delta\left(t' - t + \frac{|\boldsymbol{x}-\boldsymbol{x}'|}{c}\right) .$$

Therefore, we have

Theorem 2.8: Solution of the inhomogeneous wave equations in terms of retarded potentials

Specific solutions of the inhomogeneous potential equations (2.24) are given by the retarded potentials

$$\phi_{\text{ret}}(\boldsymbol{x},t) = \int d^3x' \int dt' \frac{\rho(\boldsymbol{x}',t')}{|\boldsymbol{x}-\boldsymbol{x}'|} \delta(t' - t_{\text{ret}})$$

$$= \int d^3x' \frac{\rho(\boldsymbol{x}',t_{\text{ret}})}{|\boldsymbol{x}-\boldsymbol{x}'|} \tag{2.35}$$

$$\boldsymbol{A}_{\text{ret}}(\boldsymbol{x},t) = \frac{1}{c} \int d^3x' \int dt' \frac{\boldsymbol{j}(\boldsymbol{x}',t')}{|\boldsymbol{x}-\boldsymbol{x}'|} \delta(t' - t_{\text{ret}})$$

$$= \frac{1}{c} \int d^3x' \frac{\boldsymbol{j}(\boldsymbol{x}',t_{\text{ret}})}{|\boldsymbol{x}-\boldsymbol{x}'|} , \tag{2.36}$$

with

$$t_{\text{ret}} = t - \frac{|\boldsymbol{x}-\boldsymbol{x}'|}{c} \quad (\textit{retarded time}) .$$

These solutions are in agreement with the theory of special relativity since they account for the fact that a change in the charge and current density requires the time $|\boldsymbol{x} - \boldsymbol{x}'|/c$ to propagate from the source of disturbance \boldsymbol{x}' to the measurement point \boldsymbol{x}.

As will be shown in Application 22, ϕ_{ret} and $\boldsymbol{A}_{\text{ret}}$ automatically satisfy the Lorentz condition (2.23). Note also that boundary conditions are incorporated by adding suitably chosen homogeneous solutions ϕ_{hom} and $\boldsymbol{A}_{\text{hom}}$, which must also satisfy the Lorentz condition.

If in (2.35) and (2.36) the retarded time t_{ret} is replaced by $t_{\text{av}} = t + |\boldsymbol{x} - \boldsymbol{x}'|/c$, one obtains the *advanced potentials* ϕ_{av} and $\boldsymbol{A}_{\text{av}}$. They are also specific solutions of the inhomogeneous wave equations (2.24). However, these solutions on their own are not in agreement with our causality requirement. The difference between retarded and advanced solutions, $(\phi_{\text{ret}}, \boldsymbol{A}_{\text{ret}}) - (\phi_{\text{av}}, \boldsymbol{A}_{\text{av}})$, is a solution of the homogeneous wave equation and, therefore, is already contained in $(\phi_{\text{hom}}, \boldsymbol{A}_{\text{hom}})$.

Summary

- By introducing a **vector potential** and a **scalar potential**, Maxwell's equations can be transformed into two coupled partial differential equations of second order (**potential equations**).

- These potentials are not unique. They possess certain **gauge freedoms**, which can be used to completely decouple the potential equations. The most commonly used gauges are the **Coulomb gauge** and the **Lorentz gauge**. The Coulomb gauge is not Lorentz-invariant.

- In the Lorentz gauge, one obtains two **inhomogeneous wave equations**. Their most general solution is composed of the homogeneous solution given by a superposition of plane monochromatic waves and a specific inhomogeneous solution given in terms of **retarded potentials**, which satisfy the principle of causality.

- The Maxwell and potential equations are **gauge-invariant**.

Applications

22. Retarded potentials and Lorentz condition. Show that the retarded potentials (2.35) and (2.36) satisfy the Lorentz condition (2.23).

Solution. We need to show the validity of the relation

$$\boldsymbol{\nabla} \boldsymbol{A}_{\text{ret}}(\boldsymbol{x}, t) = -\frac{1}{c}\frac{\partial}{\partial t}\phi_{\text{ret}}(\boldsymbol{x}, t)$$

$$\Longleftrightarrow \int \mathrm{d}^3 x' \,\hat{\boldsymbol{\nabla}}\left(\frac{\boldsymbol{j}(\boldsymbol{x}', t_{\text{ret}})}{|\boldsymbol{x} - \boldsymbol{x}'|}\right) = -\int \mathrm{d}^3 x' \,\frac{\partial}{\partial t_{\text{ret}}}\left(\frac{\rho(\boldsymbol{x}', t_{\text{ret}})}{|\boldsymbol{x} - \boldsymbol{x}'|}\right), \quad (2.37)$$

with

$$t_{\mathrm{ret}} = t - \frac{|\boldsymbol{x} - \boldsymbol{x}'|}{c} \ .$$

Herein, $\hat{\boldsymbol{\nabla}}$ denotes the total vectorial derivative with respect to \boldsymbol{x}. For the derivation of the above relation, we need the continuity equation

$$\boldsymbol{\nabla}' \boldsymbol{j}(\boldsymbol{x}', t_{\mathrm{ret}}) = -\frac{\partial \rho(\boldsymbol{x}', t_{\mathrm{ret}})}{\partial t_{\mathrm{ret}}}$$

as well as the identities

$$\boldsymbol{\nabla} t_{\mathrm{ret}} = \frac{1}{c} \boldsymbol{\nabla} |\boldsymbol{x} - \boldsymbol{x}'| = -\frac{1}{c} \boldsymbol{\nabla}' |\boldsymbol{x} - \boldsymbol{x}'| = -\boldsymbol{\nabla}' t_{\mathrm{ret}}$$

and

$$\begin{aligned}
\hat{\boldsymbol{\nabla}} \boldsymbol{j}(\boldsymbol{x}', t_{\mathrm{ret}}) &= \frac{\partial \boldsymbol{j}(\boldsymbol{x}', t_{\mathrm{ret}})}{\partial t_{\mathrm{ret}}} \boldsymbol{\nabla} t_{\mathrm{ret}} = -\frac{\partial \boldsymbol{j}(\boldsymbol{x}', t_{\mathrm{ret}})}{\partial t_{\mathrm{ret}}} \boldsymbol{\nabla}' t_{\mathrm{ret}} \\
&= -\hat{\boldsymbol{\nabla}}' \boldsymbol{j}(\boldsymbol{x}', t_{\mathrm{ret}}) + \boldsymbol{\nabla}' \boldsymbol{j}(\boldsymbol{x}', t_{\mathrm{ret}}) \\
&= -\hat{\boldsymbol{\nabla}}' \boldsymbol{j}(\boldsymbol{x}', t_{\mathrm{ret}}) - \frac{\partial \rho(\boldsymbol{x}', t_{\mathrm{ret}})}{\partial t_{\mathrm{ret}}} \ .
\end{aligned}$$

With these, the integrand on the left-hand side of (2.37) can be transformed as follows:

$$\begin{aligned}
\hat{\boldsymbol{\nabla}} \left(\frac{\boldsymbol{j}(\boldsymbol{x}', t_{\mathrm{ret}})}{|\boldsymbol{x} - \boldsymbol{x}'|} \right) &= \frac{|\boldsymbol{x} - \boldsymbol{x}'| \hat{\boldsymbol{\nabla}} \boldsymbol{j}(\boldsymbol{x}', t_{\mathrm{ret}}) - \boldsymbol{j}(\boldsymbol{x}', t_{\mathrm{ret}}) \boldsymbol{\nabla} |\boldsymbol{x} - \boldsymbol{x}'|}{|\boldsymbol{x} - \boldsymbol{x}'|^2} \\
&= \frac{-|\boldsymbol{x} - \boldsymbol{x}'| \hat{\boldsymbol{\nabla}}' \boldsymbol{j}(\boldsymbol{x}', t_{\mathrm{ret}}) + \boldsymbol{j}(\boldsymbol{x}', t_{\mathrm{ret}}) \boldsymbol{\nabla}' |\boldsymbol{x} - \boldsymbol{x}'|}{|\boldsymbol{x} - \boldsymbol{x}'|^2} \\
&\quad - \frac{1}{|\boldsymbol{x} - \boldsymbol{x}'|} \frac{\partial \rho(\boldsymbol{x}', t)}{\partial t_{\mathrm{ret}}} \\
&= -\hat{\boldsymbol{\nabla}}' \left(\frac{\boldsymbol{j}(\boldsymbol{x}', t_{\mathrm{ret}})}{|\boldsymbol{x} - \boldsymbol{x}'|} \right) - \frac{\partial}{\partial t_{\mathrm{ret}}} \left(\frac{\rho(\boldsymbol{x}', t_{\mathrm{ret}})}{|\boldsymbol{x} - \boldsymbol{x}'|} \right) \ . \quad (2.38)
\end{aligned}$$

Inserting (2.38) in (2.37) and using the fact that, due to Gauss's theorem, the divergence term vanishes, the above assertion follows.

23. Vector potential of a closed current. Consider a thin wire bent into two semicircles with radii a and b as shown in Fig. 2.4. Suppose a temporally varying current $I(t)$ flows through the closed loop. Calculate the retarded vector potential at the origin $\boldsymbol{x} = \boldsymbol{0}$.

Solution. For a current flowing through a conductor we have

$$\boldsymbol{j} \mathrm{d}^3 x = \boldsymbol{j} |\mathrm{d}\boldsymbol{l}| \mathrm{d}F = I \mathrm{d}\boldsymbol{l} = I \boldsymbol{t}(s) \mathrm{d}s \ , \quad \boldsymbol{t}(s) = \frac{\mathrm{d}\boldsymbol{l}}{\mathrm{d}s} \ ,$$

where $\mathrm{d}\boldsymbol{l}$ is a line element, $\mathrm{d}F$ the cross-sectional area, and $\boldsymbol{t}(s)$ the tangential vector of the conductor, parametrized by s. The retarded vector potentials of the individual segments of the conducting wire can then be written as

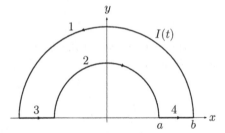

Fig. 2.4. Current-carrying loop bent into two semicircles.

$$A_{\text{ret},i}(\mathbf{0},t) = \frac{1}{c}\int\limits_{s_1}^{s_2} ds \frac{I(t_{\text{ret}})t_i(s)}{|\mathbf{0}-l_i(s)|} \;,\; t_i(s) = \frac{dl_i(s)}{ds} \;,\; t_{\text{ret}} = t - \frac{|l_i(s)|}{c} \;,$$

with

$$l_1(s) = b\begin{pmatrix} \cos s \\ \sin s \\ 0 \end{pmatrix} \;,\; s_1 = 0, s_2 = \pi$$

$$l_2(s) = a\begin{pmatrix} \cos s \\ \sin s \\ 0 \end{pmatrix} \;,\; s_1 = \pi, s_2 = 0$$

$$l_3(s) = \begin{pmatrix} s \\ 0 \\ 0 \end{pmatrix} \;,\; s_1 = -b, s_2 = -a$$

$$l_4(s) = \begin{pmatrix} s \\ 0 \\ 0 \end{pmatrix} \;,\; s_1 = a, s_2 = b \;.$$

This gives

$$A_{\text{ret},1}(\mathbf{0},t) = \frac{2I\left(t-\frac{b}{c}\right)}{c}\begin{pmatrix} 1 \\ 0 \\ 0 \end{pmatrix}$$

$$A_{\text{ret},2}(\mathbf{0},t) = -\frac{2I\left(t-\frac{a}{c}\right)}{c}\begin{pmatrix} 1 \\ 0 \\ 0 \end{pmatrix}$$

$$A_{\text{ret},3}(\mathbf{0},t) = -\frac{1}{c}\begin{pmatrix} 1 \\ 0 \\ 0 \end{pmatrix}\int\limits_{-b}^{-a} ds \frac{I\left(t+\frac{s}{c}\right)}{s}$$

$$A_{\text{ret},4}(\mathbf{0},t) = \frac{1}{c}\begin{pmatrix} 1 \\ 0 \\ 0 \end{pmatrix}\int\limits_{a}^{b} ds \frac{I\left(t-\frac{s}{c}\right)}{s} \;.$$

In total, we have

$$A_{\text{ret}}(\mathbf{0}, t) = \sum_{i=1}^{4} A_{\text{ret},i}(\mathbf{0}, t)$$

$$= \frac{2}{c} \begin{pmatrix} 1 \\ 0 \\ 0 \end{pmatrix} \left[I\left(t - \frac{b}{c}\right) - I\left(t - \frac{a}{c}\right) + \int_{a}^{b} ds \frac{I\left(t - \frac{s}{c}\right)}{s} \right].$$

2.3 Lorentz-Covariant[10] Formulation of Electrodynamics

Contrary to Newtonian mechanics, it has been experimentally verified that electrodynamics satisfies the principles of special relativity. For example, this is apparent by the fact that the electric charge of a particle, as opposed to its mass, is independent of its velocity. Accordingly, the form of Maxwell's equations (and the Lorentz force) does not change under Lorentz transformations. This can be shown explicitly by writing Maxwell's equations in a manifestly Lorentz-covariant form, which is the subject of this section.

Following a brief mathematical interlude on the transformational properties of Lorentz tensors, we use the *electromagnetic field strength tensor* to bring Maxwell's equations into a manifestly covariant notation and discuss the properties of electromagnetic fields under Lorentz transformations. Furthermore, we show that the Lorentz force yields the correct relativistic description of particle motion. Finally, we use the covariantly generalized form of Maxwell's stress tensor to rewrite the energy and momentum conservation discussed previously in Subsection 2.1.3.

2.3.1 Lorentz Tensors

In connection with our discussion of relativistic mechanics, we have already touched upon some elements of special relativity (Subsection 1.6.1). We now extend this formalism by putting Lorentz transformations onto a broader mathematical foundation. This will help us rewrite electrodynamics in a covariant notation.

Definition: Contravariant tensor of rank n

A contravariant tensor of rank n is the totality $T^{\alpha_1 \cdots \alpha_n}$ of all n-fold indexed quantities (all indices are upper indices), which, under Lorentz transformations $\Lambda^{\alpha}{}_{\beta}$, behave as

$$T'^{\alpha_1 \cdots \alpha_n} = \Lambda^{\alpha_1}{}_{\beta_1} \cdots \Lambda^{\alpha_n}{}_{\beta_n} T^{\beta_1 \cdots \beta_n}.$$

Note that the matrix $\Lambda^{\alpha}{}_{\beta}$ is not a tensor as it is not defined in an inertial system. Rather, it describes the transition between two inertial systems.

[10] See footnote 12 on page 102.

With the help of the metric tensor $g_{\alpha\alpha'}$ [see (1.64)], covariant quantities (with lower indices) can be constructed from contravariant tensors via

$$T_{\alpha_1\ldots\alpha_n} = g_{\alpha_1\beta_1} \cdots g_{\alpha_n\beta_n} T^{\beta_1\ldots\beta_n} \, .$$

Using Theorem 1.35, the covariant tensors are seen to transform as

$$
\begin{aligned}
T'_{\alpha_1\ldots\alpha_n} &= g_{\alpha_1\beta_1} \cdots g_{\alpha_n\beta_n} T'^{\beta_1\ldots\beta_n} \\
&= g_{\alpha_1\beta_1} \cdots g_{\alpha_n\beta_n} \Lambda^{\beta_1}{}_{\gamma_1} \cdots \Lambda^{\beta_n}{}_{\gamma_n} T^{\gamma_1\ldots\gamma_n} \\
&= g_{\alpha_1\beta_1} \cdots g_{\alpha_n\beta_n} \Lambda^{\beta_1}{}_{\gamma_1} \cdots \Lambda^{\beta_n}{}_{\gamma_n} g^{\gamma_1\epsilon_1} \cdots g^{\gamma_n\epsilon_n} T_{\epsilon_1\ldots\epsilon_n} \\
&= T_{\epsilon_1\ldots\epsilon_n} [\Lambda^{-1}]^{\epsilon_1}{}_{\alpha_1} [\Lambda^{-1}]^{\epsilon_n}{}_{\alpha_n} \, ,
\end{aligned}
$$

with

$$[\Lambda^{-1}]^{\epsilon}{}_{\alpha} = g_{\alpha\beta} \Lambda^{\beta}{}_{\gamma} g^{\gamma\epsilon} \, .$$

In addition to co- and contravariant tensors, we also have mixed tensors, whose indices are both upper and lower. Their transformational behavior can be deduced from those of co- and contravariant tensors. For example, a mixed tensor of rank 2 (one co- and one contravariant index) transforms as

$$
\begin{aligned}
T'^{\alpha}{}_{\beta} = T'^{\alpha\gamma} g_{\gamma\beta} = \Lambda^{\alpha}{}_{\mu} \Lambda^{\gamma}{}_{\nu} T^{\mu\nu} g_{\gamma\beta} &= \Lambda^{\alpha}{}_{\mu} \Lambda^{\gamma}{}_{\nu} T^{\mu}{}_{\epsilon} g^{\epsilon\nu} g_{\gamma\beta} \\
&= \Lambda^{\alpha}{}_{\mu} T^{\mu}{}_{\epsilon} [\Lambda^{-1}]^{\epsilon}{}_{\beta} \, .
\end{aligned}
$$

Given two tensors A and B, the following algebraic operations are defined:

- Addition: $aA^{\alpha_1\ldots\alpha_n} + bB^{\beta_1\ldots\beta_n}$ is a contravariant tensor of rank n, provided that a and b are Lorentz scalars.

- Multiplication: $A^{\alpha_1\ldots\alpha_n} B^{\beta_1\ldots\beta_m}$ is a contravariant tensor of rank $n+m$.

- Contraction: $A^{\alpha_1\ldots\alpha_n}_{\beta_1\ldots\beta_m} B^{\beta_1\ldots\beta_m\gamma_1\ldots\gamma_r}$ is a contravariant tensor of rank $n+r$. Compared to multiplication, the number of degrees is reduced by the number of indices that are being summed over.

Tensor fields and differential operators. The above can be extended in a straightforward manner to tensor fields, which are functions of the space-time four-vector x^μ.

Definition: Contravariant tensor field of rank n

A contravariant tensor field of rank n is the totality $T^{\alpha_1\ldots\alpha_n}(x^\mu)$ of all n-fold indexed functions of x^μ (all indices are upper indices), which, under Lorentz transformations $\Lambda^{\alpha}{}_{\beta}$, behave as

$$T'^{\alpha_1\ldots\alpha_n}(x'^\mu) = \Lambda^{\alpha_1}{}_{\beta_1} \cdots \Lambda^{\alpha_n}{}_{\beta_n} T^{\beta_1\ldots\beta_n}([\Lambda^{-1}]^{\mu}{}_{\nu} x'^{\nu}) \, .$$

Note that the argument is transformed as well.

Tensor fields[11] can be differentiated with respect to their arguments. Using $x^\nu = [\Lambda^{-1}]^{\nu}{}_{\mu} x'^{\mu}$, we find for the partial derivative $\partial/\partial x^\alpha$

[11] In the following, we will not distinguish linguistically between "tensor" and "tensor field".

$$\frac{\partial x^\nu}{\partial x'^\mu} = [\Lambda^{-1}]^\nu{}_\mu \implies \frac{\partial}{\partial x'^\mu} = \frac{\partial}{\partial x^\nu}\frac{\partial x^\nu}{\partial x'^\mu} = \frac{\partial}{\partial x^\nu}[\Lambda^{-1}]^\nu{}_\mu \ .$$

Therefore,

$$\frac{\partial}{\partial x^\mu} = \partial_\mu$$

transforms as a covariant and

$$\frac{\partial}{\partial x_\mu} = \partial^\mu$$

as a contravariant four-vector. Furthermore, the *d'Alembert operator*

$$\Box = \partial^\mu\partial_\mu = \frac{1}{c^2}\frac{\partial^2}{\partial t^2} - \boldsymbol{\nabla}^2$$

is a Lorentz scalar.

2.3.2 Lorentz-Covariant Formulation of Maxwell's Equations

Since the continuity equation

$$\frac{\partial \rho(\boldsymbol{x},t)}{\partial t} + \boldsymbol{\nabla}\boldsymbol{j}(\boldsymbol{x},t) = 0$$

follows from Maxwell's equations, which we know are Lorentz-covariant, this equation itself must be invariant under Lorentz transformations. This implies that the charge density $\rho(\boldsymbol{x},t)$ and the current density vector $\boldsymbol{j}(\boldsymbol{x},t)$ form the four-vector

$$(j^\mu(x)) = \begin{pmatrix} c\rho(\boldsymbol{x},t) \\ \boldsymbol{j}(\boldsymbol{x},t) \end{pmatrix} \ , \quad x = (x^\mu) \ ,$$

such that the continuity equation can be written in the manifestly invariant form

$$\partial_\mu j^\mu(x) = 0 \ .$$

Since the charge density is the temporal component of a four-vector, we see that $\mathrm{d}q = \mathrm{d}^3x\rho = \mathrm{d}^3x j^0/c$ and, therefore, the charge q of a particle is a Lorentz scalar (unlike its mass).

Consider now the inhomogeneous Maxwell equations (I) and (IV). Their right-hand sides form a four-vector j^μ, whereas the left-hand sides contain partial derivatives. The simplest manifestly covariant ansatz for these equations is therefore given by

$$\partial_\mu F^{\mu\nu} = 4\pi j^\mu \ , \ (F^{\mu\nu}) = \begin{pmatrix} 0 & -E_x & -E_y & -E_z \\ E_x & 0 & -B_z & B_y \\ E_y & B_z & 0 & -B_x \\ E_z & -B_y & B_x & 0 \end{pmatrix} \ , \tag{2.39}$$

where the first Maxwell equation is obtained from $\nu = 0$ and the last from $\nu = i$. As (I) and (IV) yield a total of 6 equations, $F^{\mu\nu}$ must be an antisymmetric contravariant tensor of rank 2. To see this we return to the potential equations (2.24),

$$\Box\phi(\boldsymbol{x},t) = 4\pi\rho(\boldsymbol{x},t) \;, \quad \Box\boldsymbol{A}(\boldsymbol{x},t) = \frac{4\pi}{c}\boldsymbol{j}(\boldsymbol{x},t) \;, \tag{2.40}$$

in Lorentz gauge (2.23),

$$\boldsymbol{\nabla}\boldsymbol{A}(\boldsymbol{x},t) = -\frac{1}{c}\frac{\partial\phi(\boldsymbol{x},t)}{\partial t} \;. \tag{2.41}$$

Since the right-hand sides of (2.40) again form the four-vector j^μ and the d'Alembert operator \Box is a Lorentz scalar, it follows that the potentials ϕ and \boldsymbol{A} must also combine to give the four-vector potential

$$(A^\mu(x)) = \begin{pmatrix} \phi(\boldsymbol{x},t) \\ \boldsymbol{A}(\boldsymbol{x},t) \end{pmatrix} \;,$$

so that (2.40) and (2.41) can be brought into the manifestly covariant form

$$\Box A^\mu(x) = \frac{4\pi}{c}j^\mu(x) \;, \quad \partial_\mu A^\mu(x) = 0 \;.$$

By expressing the definitions (2.15) for the potentials in terms of the contravariant quantities ∂^μ and A^ν, a comparison with (2.39) shows that

$$F^{\mu\nu} = \partial^\mu A^\nu - \partial^\nu A^\mu \;.$$

This means that $F^{\mu\nu}$ is indeed a contravariant tensor of rank 2. The remaining homogeneous Maxwell equations (II) and (III) can also be expressed covariantly by introducing the *pseudo tensor* of rank 2,

$$G^{\mu\nu} = \frac{1}{2}\epsilon^{\mu\nu\alpha\beta}F_{\alpha\beta} \;,$$

which is the dual of $F^{\mu\nu}$. Equations (II) and (III) can then be combined to

$$\partial_\mu G^{\mu\nu} = 0 \;,$$

where

$$\epsilon^{\mu\nu\alpha\beta} = \begin{cases} +1 & \text{for } (\mu\nu\alpha\beta) \text{ an even permutation of (0123)} \\ -1 & \text{for } (\mu\nu\alpha\beta) \text{ an odd permutation of (0123)} \\ 0 & \text{otherwise} \end{cases}$$

denotes the *Levi-Civita symbol*.

Theorem 2.9: Maxwell's equations in Lorentz-covariant form

Since electrodynamics is in agreement with special relativity, the pairs ρ and \boldsymbol{j} as well as ϕ and \boldsymbol{A} can be combined to give the four-vectors

$$(j^\mu) = \begin{pmatrix} c\rho \\ \boldsymbol{j} \end{pmatrix} \;, \quad (A^\mu) = \begin{pmatrix} \phi \\ \boldsymbol{A} \end{pmatrix} \;.$$

\triangleright

The Lorentz-covariant Maxwell equations read

$$\partial_\mu F^{\mu\nu}(x) = \frac{4\pi}{c} j^\nu \begin{cases} \nu = 0 : \ \boldsymbol{\nabla E} = 4\pi\rho \\ \nu = i : \ \boldsymbol{\nabla} \times \boldsymbol{B} - \frac{1}{c}\frac{\partial \boldsymbol{E}}{\partial t} = \frac{4\pi}{c}\boldsymbol{j} \end{cases}$$

$$\partial_\mu G^{\mu\nu}(x) = 0 \begin{cases} \nu = 0 : \ \boldsymbol{\nabla B} = 0 \\ \nu = i : \ \boldsymbol{\nabla} \times \boldsymbol{E} + \frac{1}{c}\frac{\partial \boldsymbol{B}}{\partial t} = 0 \ , \end{cases}$$

with

$$(F^{\mu\nu}) = \begin{pmatrix} 0 & -E_x & -E_y & -E_z \\ E_x & 0 & -B_z & B_y \\ E_y & B_z & 0 & -B_x \\ E_z & -B_y & B_x & 0 \end{pmatrix} \ , \ F^{\mu\nu} = \partial^\mu A^\nu - \partial^\nu A^\mu$$

and

$$G^{\mu\nu} = \frac{1}{2}\epsilon^{\mu\nu\alpha\beta} F_{\alpha\beta} \ .$$

The totally antisymmetric tensor $F^{\mu\nu}$ is called the *field strength tensor* and the totally antisymmetric pseudo tensor $G^{\mu\nu}$ is called the *dual field strength tensor*.

The homogeneous Maxwell equations can also be written as

$$\partial_\alpha F^{\beta\gamma} + \partial_\gamma F^{\alpha\beta} + \partial_\beta F^{\gamma\alpha} = 0 \ .$$

This equation is also form-invariant because each of the terms transforms like a mixed tensor of rank 3 (with one co- and two contravariant indices).

2.3.3 Transformational Behavior of Electromagnetic Fields

The transformational properties of \boldsymbol{E} and \boldsymbol{B} are determined by the transformational behavior $F'^{\mu\nu} = \Lambda^\mu{}_\alpha \Lambda^\nu{}_\beta F^{\alpha\beta}$ of the second-rank field strength tensor $F^{\mu\nu}$. In the case of a general Lorentz transformation from an inertial system K to an inertial system K' moving with velocity \boldsymbol{v} (measured in K), the fields \boldsymbol{E} and \boldsymbol{B} transform as

$$\boldsymbol{E}' = \gamma\left(\boldsymbol{E} + \frac{\boldsymbol{v}}{c} \times \boldsymbol{B}\right) - \frac{\gamma^2}{\gamma+1}\frac{\boldsymbol{v}(\boldsymbol{vE})}{c^2} \ , \ \gamma = \frac{1}{\sqrt{1 - \frac{v^2}{c^2}}}$$

$$\boldsymbol{B}' = \gamma\left(\boldsymbol{B} - \frac{\boldsymbol{v}}{c} \times \boldsymbol{E}\right) - \frac{\gamma^2}{\gamma+1}\frac{\boldsymbol{v}(\boldsymbol{vB})}{c^2}$$

or

$$\left. \begin{array}{ll} \boldsymbol{E}'_\parallel = \boldsymbol{E}_\parallel & , \ \boldsymbol{B}'_\parallel = \boldsymbol{B}_\parallel \\ \boldsymbol{E}'_\perp = \gamma\left(\boldsymbol{E}_\perp + \frac{\boldsymbol{v}}{c} \times \boldsymbol{B}\right) & , \ \boldsymbol{B}'_\perp = \gamma\left(\boldsymbol{B}_\perp - \frac{\boldsymbol{v}}{c} \times \boldsymbol{E}\right) \ , \end{array} \right\} \quad (2.42)$$

where \parallel and \perp denote the field components parallel and perpendicular to v, respectively. From this we see that the fields E and B do not transform independently. A pure E or B field in a system K is transformed into a combination of electric and magnetic fields in another system K'. Therefore, since the distinction between E and B fields depends on the chosen inertial system, it makes more sense to speak of the electromagnetic field $F^{\mu\nu}$, rather than of E and B independently.

Lorentz invariants. As explained in Subsection 2.3.1, each tensor of rank 2 can be contracted with itself (in several ways) to form a set of Lorentz scalars. Some of them are

$$T_{\mu\nu}T^{\nu\mu} \ , \ T_{\mu\nu}T^{\nu}{}_{\gamma}T^{\gamma\mu} \ , \ T_{\mu\nu}T^{\nu\gamma}T_{\gamma\epsilon}T^{\epsilon\mu} \ , \ \cdots \ .$$

In case of the electromagnetic field strength tensor $F^{\mu\nu}$, the cubic invariant vanishes. The other invariants yield

$$
\begin{aligned}
F_{\mu\nu}F^{\nu\mu} &= 2\left(E^2 - B^2\right) = I_1 = \text{invariant} \\
F_{\mu\nu}F^{\nu\gamma}F_{\gamma\epsilon}F^{\epsilon\mu} &= 2\left(E^2 - B^2\right)^2 + 4(EB)^2 = \text{invariant} \\
&\implies (EB) = \text{invariant} \ .
\end{aligned}
$$

For the E and B fields, this implies:

- If $E \perp B$ holds in any one inertial system, then this relation holds in all inertial systems.
 If $I_1 > 0$, there exists an inertial system in which $B = 0$.
 If $I_1 < 0$, there exists an inertial system in which $E = 0$.

- The first item is even true in opposite direction: if $E = 0$ or $B = 0$ holds in any one inertial system, then $E \perp B$ holds in all inertial systems.

- If $|E| = |B|$ holds in any one inertial system, then it is true in all inertial systems.

2.3.4 Lorentz Force and Covariance

Having shown the compatibility of Maxwell's equations with special relativity by using a Lorentz-covariant formulation, we still have to perform this exercise for the Lorentz force F_L and the corresponding equation of motion. Here, we can only initially assume that it is valid in the nonrelativistic limit $|\dot{x}|/c \to 0$, i.e.,

$$\frac{\mathrm{d}}{\mathrm{d}t}m_0\dot{x} = F_L = q\left(E + \frac{\dot{x}}{c} \times B\right) \ , \tag{2.43}$$

where q denotes the charge, \dot{x} the velocity, and m_0 the rest mass of the particle. From Subsection 1.6.4, we know that the covariant equation of motion of relativistic mechanics is given by

$$\frac{dp^\mu}{d\tau} = \frac{1}{\sqrt{1 - \frac{\dot{x}^2}{c^2}}} \frac{dp^\mu}{dt} = F^\mu \,,$$

with

$$(p^\mu) = \begin{pmatrix} cm \\ m\dot{x} \end{pmatrix} \,, \quad (F^\mu) = \frac{1}{\sqrt{1 - \frac{\dot{x}^2}{c^2}}} \begin{pmatrix} c\frac{dm}{dt} \\ F \end{pmatrix} \,.$$

In the above,

$$d\tau = dt\sqrt{1 - \frac{\dot{x}^2}{c^2}} \,, \quad m = \frac{m_0}{\sqrt{1 - \frac{\dot{x}^2}{c^2}}}$$

are the eigentime differential and the relativistic mass of the particle, respectively. Since the expression (2.43) is linear in the fields E and B as well as in the particle velocity \dot{x}, its simplest possible covariant ansatz reads

$$\frac{dp^\mu}{d\tau} = \frac{q}{c} F^{\mu\nu} \frac{dx_\nu}{d\tau} \,, \quad \frac{dx^\mu}{d\tau} = \text{four-velocity} \,. \tag{2.44}$$

Note that both sides are contravariant four-vectors, since q and c are Lorentz scalars. Expressing this equation in terms of the three-vectors E, B, and \dot{x}, we find

$$\mu = 0 : \quad \frac{d}{dt} \frac{m_0 c^2}{\sqrt{1 - \frac{\dot{x}^2}{c^2}}} = qE\dot{x}$$

$$\mu = i : \quad \frac{d}{dt} \frac{m_0 \dot{x}}{\sqrt{1 - \frac{\dot{x}^2}{c^2}}} = q\left(E + \frac{\dot{x}}{c} \times B \right) \,.$$

Obviously, the spatial μ-components give the right nonrelativistic relationship (2.43). We therefore conclude that the ansatz (2.44) is the correct relativistic generalization of (2.43). Additionally, if m_0 is replaced by m in the equation of motion, the Lorentz force F_L also holds for relativistic velocities. The equation for the $\mu = 0$ component is a conservation law: the temporal change of the particle's energy is equal to the power transferred from the electromagnetic field to the particle.

Theorem 2.10: Covariant equation of motion, Lorentz force

In covariant notation, the relativistic generalization of Newton's equation of motion for a particle in an electromagnetic field is

$$\frac{dp^\mu}{d\tau} = \frac{q}{c} F^{\mu\nu} \frac{dx_\nu}{d\tau} \,.$$

▷

The spatial components of this equation are

$$\frac{\mathrm{d}}{\mathrm{d}t} \frac{m_0 \dot{\boldsymbol{x}}}{\sqrt{1 - \frac{\dot{x}^2}{c^2}}} = \boldsymbol{F}_{\mathrm{L}} = q \left(\boldsymbol{E} + \frac{\dot{\boldsymbol{x}}}{c} \times \boldsymbol{B} \right) ,$$

i.e., $\boldsymbol{F}_{\mathrm{L}}$ yields the correct particle motion to all orders of $|\dot{\boldsymbol{x}}|/c$.

2.3.5 Energy and Momentum Conservation

In the derivation of the momentum law of electrodynamics (Theorem 2.5), we introduced the Maxwell stress tensor

$$T_{ik} = \frac{1}{4\pi} \left[E_i E_k + B_i B_k - \frac{\delta_{ik}}{2} \left(\boldsymbol{E}^2 + \boldsymbol{B}^2 \right) \right] . \tag{2.45}$$

Its relativistic version is given by the second-rank tensor

$$T^{\mu\nu} = \frac{1}{4\pi} \left(g^{\mu\alpha} F_{\alpha\beta} F^{\beta\nu} + \frac{1}{4} g^{\mu\nu} F_{\alpha\beta} F^{\alpha\beta} \right) , \tag{2.46}$$

whose spatial components are the negative of the above three-stress tensor. Using the electromagnetic energy and momentum densities ϵ_{em} and $\boldsymbol{g}_{\mathrm{em}}$, it can be written as the matrix

$$(T^{\mu\nu}) = \begin{pmatrix} \dfrac{\boldsymbol{E}^2 + \boldsymbol{B}^2}{8\pi} & \left[\dfrac{\boldsymbol{E} \times \boldsymbol{B}}{4\pi} \right]^{\mathrm{T}} \\ \dfrac{\boldsymbol{E} \times \boldsymbol{B}}{4\pi} & -(T_{ik}) \end{pmatrix} = \begin{pmatrix} \epsilon_{\mathrm{em}} & [c\boldsymbol{g}_{\mathrm{em}}]^{\mathrm{T}} \\ c\boldsymbol{g}_{\mathrm{em}} & -(T_{ik}) \end{pmatrix} ,$$

with T_{ik} from (2.45). Now, with the help of some tensor algebra, one can show that the differential energy and momentum conservation laws of Theorems 2.4 and 2.5 are incorporated in the following Lorentz-covariant relation:

$$\partial_\mu T^{\mu\nu} = -\frac{1}{c} F^{\nu\rho} j_\rho \begin{cases} \nu = 0 : \text{energy conservation, Theorem 2.4} \\ \nu = i : \text{momentum conservation, Theorem 2.5.} \end{cases}$$

For a system free of charges, this implies

$$0 = \partial_\mu T^{\mu\nu} = \frac{1}{c} \partial_t T^{0\nu} + \partial_i T^{i\nu} .$$

By integrating this equation over a sufficiently large volume, the second term on the right-hand side vanishes and we obtain

$$p_{\mathrm{em}}^\nu = \frac{1}{c} \int \mathrm{d}^3 x\, T^{0\nu} = \text{const} .$$

Thus, the four-momentum $(p_{\mathrm{em}}^\mu) = \begin{pmatrix} E_{\mathrm{em}}/c \\ \boldsymbol{P}_{\mathrm{em}} \end{pmatrix}$ contains the energy and the momentum of the electromagnetic field and is a conserved quantity for a closed system containing no charges.

Summary

- Electrodynamics is a **relativistic field theory**. Using the **electromagnetic field strength tensor** and the **dual field strength tensor**, Maxwell's equations can be written in manifestly Lorentz-covariant form.

- This is because charge and current density as well as scalar and vector potential can both be combined to four-vectors.

- Lorentz covariance of the theory implies that E and B fields are transformed into one another when changing the inertial system. In other words, the distinction between E and B depends on the chosen reference system.

- The Lorentz force yields the correct description for the movement of charged particles in electromagnetic fields to all orders of $|\dot{x}|/c$.

- Using the **Maxwell stress tensor**, the conservation laws for energy and momentum can be merged into a single Lorentz-covariant equation.

Applications

24. Uniformly moving charge. Calculate the electromagnetic field of a charge q moving in an inertial system K at constant velocity v in direction of x.

Solution. A simple and elegant way of solving this problem is to move to the rest system K$'$ of the charge. In this reference frame, the fields E' and B', as well as the potentials ϕ' and A', can be written down immediately:

$$\left.\begin{array}{ll} \text{(I):} & \boldsymbol{\nabla'} E'(x',t') = 4\pi q\delta(x') \\[2ex] \text{(IV):} & \boldsymbol{\nabla'} \times B'(x',t') - \dfrac{1}{c}\dfrac{\partial E(x',t')}{\partial t'} = 0 \end{array}\right\} \Longrightarrow \left\{\begin{array}{l} E'(x',t') = q\dfrac{x'}{|x'|^3} \\[2ex] B'(x',t') = 0 \end{array}\right.$$

$$\Longrightarrow (A'^\mu(x')) = \begin{pmatrix} \phi'(x',t') \\ A'_x(x',t') \\ A'_y(x',t') \\ A'_z(x',t') \end{pmatrix} = \begin{pmatrix} \frac{q}{|x'|} \\ 0 \\ 0 \\ 0 \end{pmatrix}.$$

To return to the original system K, we use the Lorentz transformation (see Subsection 1.6.1)

$$(\Lambda^{(1)\mu}{}_\nu)^{-1} = \begin{pmatrix} \gamma & \frac{\gamma v}{c} & 0 & 0 \\ \frac{\gamma v}{c} & \gamma & 0 & 0 \\ 0 & 0 & 1 & 0 \\ 0 & 0 & 0 & 1 \end{pmatrix}, \quad \gamma = \frac{1}{\sqrt{1 - \frac{v^2}{c^2}}}.$$

For the potentials in K, it follows that

$$(A^\mu(x')) = \begin{pmatrix} \phi(x',t') \\ A_x(x',t') \\ A_y(x',t') \\ A_z(x',t') \end{pmatrix} = \gamma \begin{pmatrix} \frac{q}{|x'|} \\ \frac{vq}{c|x'|} \\ 0 \\ 0 \end{pmatrix} ,$$

with

$$\begin{pmatrix} ct' \\ x' \\ y' \\ z' \end{pmatrix} = \begin{pmatrix} \gamma & -\frac{\gamma v}{c} & 0 & 0 \\ -\frac{\gamma v}{c} & \gamma & 0 & 0 \\ 0 & 0 & 1 & 0 \\ 0 & 0 & 0 & 1 \end{pmatrix} \begin{pmatrix} ct \\ x \\ y \\ z \end{pmatrix} = \begin{pmatrix} \gamma(ct - \frac{vx}{c}) \\ \gamma(x - vt) \\ y \\ z \end{pmatrix}$$

$$\Longrightarrow \phi(x,t) = \frac{\gamma q}{\sqrt{\gamma^2(x - vt)^2 + y^2 + z^2}} , \quad A(x,t) = \frac{v}{c}\phi(x,t)\begin{pmatrix} 1 \\ 0 \\ 0 \end{pmatrix} .$$

Finally, for the E and B fields in K, we obtain

$$B(x,t) = \nabla \times A(x,t) = \frac{v\gamma q}{c[\gamma^2(x - vt)^2 + y^2 + z^2]^{3/2}}\begin{pmatrix} 0 \\ z \\ -y \end{pmatrix}$$

$$E(x,t) = -\nabla\phi(x,t) - \frac{1}{c}\frac{\partial A(x,t)}{\partial t}$$

$$= \frac{\gamma q}{[\gamma^2(x - vt)^2 + y^2 + z^2]^{3/2}}\begin{pmatrix} x - vt \\ y \\ z \end{pmatrix} .$$

25. Generalized Faraday's induction law. In Subsection 2.1.2, we discussed Faraday's induction law

$$\oint_C E(x,t)dl = -\frac{1}{c}\int_F dF\frac{\partial B(x,t)}{\partial t} ,$$

where C is a closed contour, for example, that of a conducting loop, and F is its surface area. Using this law and combining it with (2.42), show that Lorentz covariance of electrodynamics implies the *generalized Faraday's induction law*

$$V(t) = \oint_{C'} E'(x',t)dl' = -\frac{1}{c}\frac{d}{dt}\int_{F(t)} dF B(x,t) = -\frac{1}{c}\frac{d\Phi_m(t)}{dt} ,$$

where primed quantities refer to the rest system of the conducting loop.

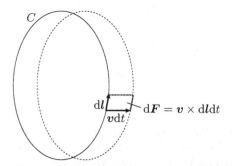

Fig. 2.5. Conducting loop moving with constant velocity \boldsymbol{v}.

Solution. In the unprimed observational reference frame we have

$$\frac{\mathrm{d}\Phi_{\mathrm{m}}}{\mathrm{d}t} = \int_{F(t)} \mathrm{d}\boldsymbol{F}\,\frac{\partial \boldsymbol{B}}{\partial t} + \int \frac{\mathrm{d}\boldsymbol{F}}{\mathrm{d}t}\boldsymbol{B} = -c \oint_{C(t)} \boldsymbol{E}\mathrm{d}\mathrm{l} + \int \frac{\mathrm{d}\boldsymbol{F}}{\mathrm{d}t}\boldsymbol{B} \,.$$

Assuming that the conductor moves at constant velocity \boldsymbol{v} (see Fig. 2.5), we have $\mathrm{d}\boldsymbol{F} = \boldsymbol{v} \times \mathrm{d}\boldsymbol{l}\mathrm{d}t$. This implies that

$$\frac{\mathrm{d}\Phi_{\mathrm{m}}}{\mathrm{d}t} = -c \oint_{C(t)} \mathrm{d}\boldsymbol{l}\left(\boldsymbol{E} + \frac{\boldsymbol{v}}{c} \times \boldsymbol{B}\right)$$

$$= -c \oint_{C(t)} (\mathrm{d}\boldsymbol{l}_{\parallel} + \mathrm{d}\boldsymbol{l}_{\perp})\left(\boldsymbol{E}_{\parallel} + \boldsymbol{E}_{\perp} + \frac{\boldsymbol{v}}{c} \times \boldsymbol{B}\right) \,, \qquad (2.47)$$

where \parallel and \perp respectively indicate the components parallel and perpendicular to the velocity vector \boldsymbol{v}. Now, by using (2.42) and

$$\mathrm{d}l_{\parallel} = \frac{1}{\gamma}\mathrm{d}l'_{\parallel} \,, \;\; \mathrm{d}l_{\perp} = \mathrm{d}l'_{\perp} \,, \;\; \gamma = \frac{1}{\sqrt{1 - \frac{v^2}{c^2}}} \,,$$

the right-hand side of (2.47) can be expressed in terms of the primed quantities:

$$\frac{\mathrm{d}\Phi_{\mathrm{m}}}{\mathrm{d}t} = -c \oint_{C'} \left(\frac{1}{\gamma}\mathrm{d}l'_{\parallel} + \mathrm{d}l'_{\perp}\right)\left(\boldsymbol{E}_{\parallel} + \boldsymbol{E}_{\perp} + \frac{\boldsymbol{v}}{c} \times \boldsymbol{B}\right)$$

$$= -c \oint_{C'} (\mathrm{d}l'_{\parallel} + \mathrm{d}l'_{\perp})\left[\frac{1}{\gamma}\boldsymbol{E}_{\parallel} + \frac{1}{\gamma}\gamma\left(\boldsymbol{E}_{\perp} + \frac{\boldsymbol{v}}{c} \times \boldsymbol{B}\right)\right]$$

$$= -\frac{c}{\gamma} \oint_{C'} \mathrm{d}\boldsymbol{l}'(\boldsymbol{E}'_{\parallel} + \boldsymbol{E}'_{\perp}) \,.$$

This finally results in

$$\frac{\mathrm{d}\Phi_{\mathrm{m}}}{\mathrm{d}t} = \frac{1}{\gamma}\frac{\mathrm{d}\Phi_{\mathrm{m}}}{\mathrm{d}t'} = -\frac{c}{\gamma} \oint_{C'} \mathrm{d}\boldsymbol{l}'\,\boldsymbol{E}'(\boldsymbol{x}',t')$$

$$\Longleftrightarrow \quad \frac{\mathrm{d}\Phi_{\mathrm{m}}(t)}{\mathrm{d}t} = -c \oint_{C'} \mathrm{d}\boldsymbol{l}' \, \boldsymbol{E}'(\boldsymbol{x}',t) \ .$$

2.4 Radiation Theory

This section deals with electromagnetic radiation phenomena induced by temporally varying charges and currents. Generally, for arbitrary distributions, the calculation of the corresponding electromagnetic fields via retarded potentials is only possible using a numerical approach, since each point \boldsymbol{x}' is associated with a retarded time t_{ret}, which, moreover, depends explicitly on \boldsymbol{x}'. Therefore, we consider here only the case of accelerated point charges and charge distributions with a small extent, for which an analytical determination of the retarded potentials is possible.

We start by calculating the retarded potentials for arbitrarily moving point charges (*Liénard-Wiechert potentials*) and then derive the corresponding electromagnetic fields \boldsymbol{E} and \boldsymbol{B}. As it turns out, each of them can be written as a sum of two terms, one of which corresponds to a uniformly moving charge and the other to an accelerated motion. We also calculate the energy and power that are radiated by the accelerational parts of these fields. Finally, we consider the radiation fields of small charge and current density distributions in the *dipole approximation*.

2.4.1 Liénard-Wiechert Potentials

A point charge q moving along the trajectory $\boldsymbol{x}_0(t)$ with velocity $\dot{\boldsymbol{x}}_0(t)$ possesses the charge and current density distributions

$$\rho(\boldsymbol{x},t) = q\delta[\boldsymbol{x} - \boldsymbol{x}_0(t)] \ , \ \boldsymbol{j}(\boldsymbol{x},t) = q\dot{\boldsymbol{x}}_0\delta[\boldsymbol{x} - \boldsymbol{x}_0(t)] \ .$$

Inserting these quantities into the retarded potentials (2.35) and carrying out the spatial integration gives

$$\phi(\boldsymbol{x},t) = q \int \mathrm{d}t' \frac{1}{|\boldsymbol{x} - \boldsymbol{x}_0(t')|} \delta\left(t' - t + \frac{|\boldsymbol{x} - \boldsymbol{x}_0(t')|}{c}\right)$$

$$\boldsymbol{A}(\boldsymbol{x},t) = \frac{q}{c} \int \mathrm{d}t' \frac{\dot{\boldsymbol{x}}_0(t')}{|\boldsymbol{x} - \boldsymbol{x}_0(t')|} \delta\left(t' - t + \frac{|\boldsymbol{x} - \boldsymbol{x}_0(t')|}{c}\right) \ .$$

For the temporal integration one has to take into account that the argument of the δ-function is a function of the integration variable t'. In this case the following identity holds:

$$\int \mathrm{d}t' g(t')\delta[f(t')] = \sum_k \frac{g(t_k)}{\left|\frac{\mathrm{d}f}{\mathrm{d}t'}\right|_{t_k}} \ , \ t_k = \text{zeros of } f.$$

Using this, we find

Theorem 2.11: Liénard-Wiechert potentials for an arbitrarily moving point charge

The retarded potentials of a point charge q moving along an arbitrary trajectory $\boldsymbol{x}_0(t)$ with velocity $\dot{\boldsymbol{x}}_0(t)$ are given by the *Liénard-Wiechert potentials*

$$\phi(\boldsymbol{x},t) = \frac{q}{R(t_{\mathrm{ret}}) - \frac{1}{c}\boldsymbol{R}(t_{\mathrm{ret}})\dot{\boldsymbol{x}}_0(t_{\mathrm{ret}})}$$

$$\boldsymbol{A}(\boldsymbol{x},t) = \frac{q}{c}\frac{\dot{\boldsymbol{x}}_0(t_{\mathrm{ret}})}{R(t_{\mathrm{ret}}) - \frac{1}{c}\boldsymbol{R}(t_{\mathrm{ret}})\dot{\boldsymbol{x}}_0(t_{\mathrm{ret}})} \; ,$$

$$\boldsymbol{R}(t_{\mathrm{ret}}) = \boldsymbol{x} - \boldsymbol{x}_0(t_{\mathrm{ret}}) \; , \quad R(t_{\mathrm{ret}}) = |\boldsymbol{R}(t_{\mathrm{ret}})| \; . \tag{2.48}$$

Here, $\boldsymbol{R}(t_{\mathrm{ret}})$ denotes the distance vector between the observation point \boldsymbol{x} and the particle's location \boldsymbol{x}_0 at the retarded time

$$t_{\mathrm{ret}} = t - \frac{|\boldsymbol{x} - \boldsymbol{x}_0(t_{\mathrm{ret}})|}{c} = t - \frac{R(t_{\mathrm{ret}})}{c} \; . \tag{2.49}$$

Before we turn to the calculation of the \boldsymbol{E} and \boldsymbol{B} fields, we show how this result can be obtained in a more elegant way by using the Lorentz-covariant formalism. We start by first noting that

$$(R^\mu) = \begin{pmatrix} R \\ \boldsymbol{R} \end{pmatrix} = \begin{pmatrix} c(t - t_{\mathrm{ret}}) \\ |\boldsymbol{x} - \boldsymbol{x}_0(t_{\mathrm{ret}})| \end{pmatrix}$$

is a four-vector since, according to (2.48) and (2.49), we have in any inertial system

$$R^\mu R_\mu = c^2(t_{\mathrm{ret}} - t)^2 - \boldsymbol{R}^2(t_{\mathrm{ret}}) = 0 \; .$$

The retarded potentials can be easily specified in the (primed) momentary rest system of the particle:

$$\phi'(\boldsymbol{x}',t') = \frac{q}{|\boldsymbol{x}' - \boldsymbol{x}_0'(t'_{\mathrm{ret}})|} = \frac{q}{R'(t'_{\mathrm{ret}})} \; , \quad \boldsymbol{A}'(\boldsymbol{x}',t') = \boldsymbol{0} \; ,$$

or

$$A'^\mu = q\frac{u'^\mu}{R'_\nu u'^\nu} \; , \quad (u'^\mu) = \begin{pmatrix} c \\ 0 \\ 0 \\ 0 \end{pmatrix} \; , \tag{2.50}$$

where u'^μ is the four-velocity in the rest system. Both sides of (2.50) transform as a contravariant four-vector, so that this equation is form-invariant and thus valid in any inertial system. Transforming back to the original (unprimed) inertial system we find

$$A^\mu = q\frac{u^\mu}{R_\nu u^\nu} \; , \quad (u^\mu) = \frac{1}{\sqrt{1 - \frac{\dot{\boldsymbol{x}}_0^2}{c^2}}}\begin{pmatrix} c \\ \dot{\boldsymbol{x}}_0 \end{pmatrix} \; ,$$

which yields the same potentials as in Theorem 2.11.

Determination of the electromagnetic fields. To calculate

$$E = -\nabla\phi - \frac{1}{c}\frac{\partial A}{\partial t} \ , \quad B = \nabla \times A \tag{2.51}$$

from Theorem 2.11, we must know $\partial t_{\text{ret}}/\partial t$ and ∇t_{ret}, since the Liénard-Wiechert potentials are given in terms of the retarded time t_{ret}, which additionally depends on x. To obtain $\partial t_{\text{ret}}/\partial t$ and ∇t_{ret}, we calculate as follows:

$$\frac{\partial R^2}{\partial t_{\text{ret}}} = 2R\frac{\partial R}{\partial t_{\text{ret}}} = 2R\frac{\partial R}{\partial t_{\text{ret}}} = -2R\dot{x}_0$$

$$\Longrightarrow \quad \frac{\partial R}{\partial t} = \frac{\partial R}{\partial t_{\text{ret}}}\frac{\partial t_{\text{ret}}}{\partial t} = -\frac{R\dot{x}_0}{R}\frac{\partial t_{\text{ret}}}{\partial t} \tag{2.52}$$

$$R(t_{\text{ret}}) = c(t - t_{\text{ret}}) \Longrightarrow \frac{\partial R}{\partial t} = c\left(1 - \frac{\partial t_{\text{ret}}}{\partial t}\right) . \tag{2.53}$$

Combining (2.52) and (2.53), we obtain

$$\frac{\partial t_{\text{ret}}}{\partial t} = \frac{1}{1 - n\beta} \ , \quad \text{with} \quad n = \frac{R}{R} \ , \ \beta = \frac{\dot{x}_0}{c} \ . \tag{2.54}$$

Differentiating R with respect to its components yields on the one hand

$$\nabla R = \nabla c(t - t_{\text{ret}}) = -c\nabla t_{\text{ret}} \ ,$$

and on the other

$$\nabla R = \nabla|x - x_0(t_{\text{ret}})| = \frac{R}{R} + \frac{\partial R}{\partial t_{\text{ret}}}\nabla t_{\text{ret}} = n - n\dot{x}_0\nabla t_{\text{ret}} \ ,$$

so that

$$\nabla t_{\text{ret}} = \frac{1}{c}\frac{n}{n\beta - 1} \ . \tag{2.55}$$

Using (2.54) and (2.55) (taken at time t_{ret}), we can now perform the calculations in (2.51). After a few steps, we arrive at

Theorem 2.12: E and B fields of an arbitrarily moving point charge

The electromagnetic fields of an arbitrarily moving point charge q with velocity \dot{x}_0 are given by

$$E(x,t) = q\left[\frac{(n - \beta)(1 - \beta^2)}{R^2(1 - n\beta)^3}\right]_{t_{\text{ret}}} + \frac{q}{c}\left[\frac{n \times [(n - \beta) \times \dot{\beta}]}{R(1 - n\beta)^3}\right]_{t_{\text{ret}}}$$

$$= \qquad E_0(x,t) \qquad + \qquad E_a(x,t)$$

$$B(x,t) = \qquad n \times E_0(x,t) \qquad + \qquad n \times E_a(x,t)$$

$$= \qquad B_0(x,t) \qquad + \qquad B_a(x,t) \ ,$$

\triangleright

with

$$n = \frac{R}{R} \; , \; \beta = \frac{\dot{x}_0}{c} \; .$$

The magnetic field is perpendicular to the electric field as well as to the vector connecting the observation point and the retarded particle position.

In this theorem the E and B fields are both given as the sum of two terms. The first terms are independent of the particle acceleration $\dot{\beta}$ and behave as $1/R^2$ for large distances, whereas the second terms fall off as $1/R$ and vanish for $\dot{\beta} = 0$. Therefore, E_0 and B_0 are the fields of a uniformly moving charge. We can verify this explicitly by considering a point charge q moving along the x-axis with constant velocity v. Putting the observation point at the origin, we find

$$x_0(t) = \begin{pmatrix} vt \\ 0 \\ 0 \end{pmatrix} \; , \; x = 0 \; , \; R(t_{\text{ret}}) = vt_{\text{ret}} \; , \; n = \begin{pmatrix} -1 \\ 0 \\ 0 \end{pmatrix} \; , \; \beta = \begin{pmatrix} \frac{v}{c} \\ 0 \\ 0 \end{pmatrix}$$

$$\Longrightarrow E_0(0, t) = -\frac{q}{\gamma^2 v^2 t_{\text{ret}}^2 \left(1 + \frac{v}{c}\right)^2} \begin{pmatrix} 1 \\ 0 \\ 0 \end{pmatrix} \; , \; \gamma = \frac{1}{\sqrt{1 - \beta^2}} \; .$$

The retarded time is calculated as

$$t_{\text{ret}} = t - \frac{|x - x_0(t_{\text{ret}})|}{c} = t - \frac{vt_{\text{ret}}}{c} \Longrightarrow t_{\text{ret}} = \frac{t}{1 + \frac{v}{c}} \; ,$$

so that

$$E_0(0, t) = -\frac{q}{\gamma^2 v^2 t^2} \begin{pmatrix} 1 \\ 0 \\ 0 \end{pmatrix} \; , \; B_0(0, t) = n \times E(0, t) = 0 \; .$$

These are the same as the fields we found in Application 24 for $x = 0$. There we calculated the E and B fields of a uniformly moving point charge in x-direction using the Lorentz-covariant formalism.

2.4.2 Radiation Energy

In this subsection, we calculate the power P radiated by a moving point charge, where all quantities are taken at the retarded time t_{ret}. Consider an area element $R^2 d\Omega$ through which we have the energy flux or power $dP = Sn R^2 d\Omega$:

$$\frac{dP}{d\Omega} = \frac{dE}{dt d\Omega} = R^2 Sn = \frac{cR^2}{4\pi} n(E \times B) \; .$$

Inserting this equation into the fields from Theorem 2.12 and analyzing the asymptotic R behavior of the individual terms, we realize that only the com-

bination $E_a \times B_a$ yields a nonvanishing contribution. In other words, only accelerated charges emit radiation. Therefore, we have

$$\frac{dP}{d\Omega} = \frac{cR^2}{4\pi} n(E_a \times B_a) = \frac{q^2}{4\pi c} \frac{\left(n \times [(n - \beta) \times \dot{\beta}] \right)^2}{(1 - n\beta)^6} .$$

From the charge's point of view, it is useful to consider the energy emitted in the retarded time interval dt_{ret}. Taking into account (2.54), this is given by

$$\frac{dP'}{d\Omega} = \frac{dP}{d\Omega} \frac{\partial t}{\partial t_{\text{ret}}} = \frac{q^2}{4\pi c} \frac{\left(n \times [(n - \beta) \times \dot{\beta}] \right)^2}{(1 - n\beta)^5} . \tag{2.56}$$

To simplify this equation, we first consider the two limiting cases $\beta \ll 1$ and $\beta \parallel \dot{\beta}$ and also set

$$\cos \theta = \frac{n\beta}{\beta} , \quad \cos \theta' = \frac{n\dot{\beta}}{\dot{\beta}} .$$

First limit: $\beta \ll 1$. In this case, (2.56) reduces to

$$\frac{dP'}{d\Omega} = \frac{dP}{d\Omega} = \frac{q^2}{4\pi c} \dot{\beta}^2 \sin^2 \theta' . \tag{2.57}$$

This term is independent of the direction of the particle's velocity. The angular component of the radiated power is depicted in Fig. 2.6. Performing the angular integration in (2.57), we find the *Larmor formula*

$$P = \frac{2q^2}{3c} \dot{\beta}^2 . \tag{2.58}$$

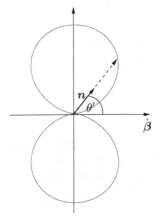

Fig. 2.6. Radiated power $dP/d\Omega$ of an accelerated point charge in the limit $\beta \ll 1$. θ' denotes the angle between the particle acceleration $\dot{\beta}$ and the direction of radiation, n.

Second limit: $\boldsymbol{\beta} \parallel \dot{\boldsymbol{\beta}}$. Now, $\theta = \theta'$, and (2.56) becomes

$$\frac{\mathrm{d}P}{\mathrm{d}\Omega} = \frac{q^2}{4\pi c} \dot{\beta}^2 \frac{\sin^2 \theta}{(1 - \beta \cos \theta)^5} \; .$$

The direction of maximal radiation emission is determined as

$$\frac{\mathrm{d}}{\mathrm{d}\cos\theta} \frac{\mathrm{d}P}{\mathrm{d}\Omega} = 0 \Longrightarrow \cos\theta_{\mathrm{max}} = \frac{1}{3\beta} \left(\sqrt{15\beta^2 + 1} - 1 \right) \; .$$

This means that the radiation cone is inclined more and more forward as the particle's velocity increases (Fig. 2.7).

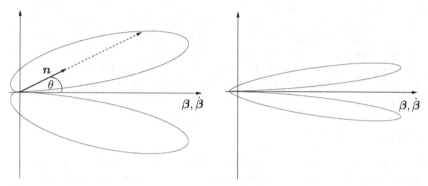

Fig. 2.7. Radiated power $\mathrm{d}P/\mathrm{d}\Omega$ of an accelerated point charge in the limit $\boldsymbol{\beta} \parallel \dot{\boldsymbol{\beta}}$ for $\beta = 0.5$ (*left*) and $\beta = 0.81$ (*right*). For a scaled comparison, the right plot must be magnified by a factor of 100.

Lorentz-invariant generalization. Using the results of the first limiting case and applying the Lorentz-covariant formalism to it, we can easily calculate the radiated power of a point charge moving with arbitrary velocity. Since the energy E as well as the time t are 0-components of four-vectors, the expression $P = \mathrm{d}E/\mathrm{d}t$ is a Lorentz scalar. To generalize the first limiting case to arbitrary $\boldsymbol{\beta}$, it is therefore sufficient to find a Lorentz scalar that turns into (2.58) for $\beta \ll 1$. Consequently, we write (2.58) in terms of the nonrelativistic momentum $\boldsymbol{p}_{\mathrm{nr}}$ and generalize the result in invariant manner by replacing $\boldsymbol{p}_{\mathrm{nr}}$ and $\mathrm{d}t$ with the four-momentum $(p^\mu) = \frac{cm_0}{\sqrt{1-\beta^2}} \begin{pmatrix} 1 \\ \beta \end{pmatrix}$ and the eigentime $\mathrm{d}\tau = \mathrm{d}t\sqrt{1 - \beta^2}$:

$$P = \frac{2q^2}{3m_0^2 c^3} \frac{\mathrm{d}\boldsymbol{p}_{\mathrm{nr}}}{\mathrm{d}t} \frac{\mathrm{d}\boldsymbol{p}_{\mathrm{nr}}}{\mathrm{d}t} \longrightarrow -\frac{2q^2}{3m_0^2 c^3} \frac{\mathrm{d}p^\mu}{\mathrm{d}\tau} \frac{\mathrm{d}p_\mu}{\mathrm{d}\tau} \; .$$

The right-hand side of this equation is clearly a Lorentz scalar and can be seen to converge to (2.58) for $\beta \ll 1$. Therefore, for arbitrary particle velocities we have

Theorem 2.13: Radiation power of an arbitrarily moving point charge

The radiation power of a point charge q moving with arbitrary velocity $\dot{\boldsymbol{x}}_0$ is given by

$$P = \frac{2q^2}{3c}\gamma^6\left[\dot{\boldsymbol{\beta}}^2 - (\boldsymbol{\beta}\times\dot{\boldsymbol{\beta}})^2\right] \ , \ \boldsymbol{\beta} = \frac{\dot{\boldsymbol{x}}_0}{c} \ , \ \gamma = \frac{1}{\sqrt{1-\beta^2}} \ .$$

This equation implies that only accelerated charges emit radiation.

The radiation power P of an accelerated point charge reduces its kinetic energy. Therefore, one also speaks of *radiation losses*. These losses occur, for example, in linear or circular particle accelerators. Röntgen radiation is another example of radiation losses. The latter is created by applying a large voltage across the anode and cathode of an electron tube. When running through this potential difference, the electrons are accelerated and hit the anode with high energy. During their collision the electrons are very quickly decelerated ($|\dot{\boldsymbol{\beta}}|$ is large), thus emitting electromagnetic energy ("Bremsstrahlung").

2.4.3 Dipole Radiation

We now wish to calculate the radiation field of a system with temporally varying charge and current densities confined to an area with radius R_0:

$$\rho(\boldsymbol{x},t), \boldsymbol{j}(\boldsymbol{x},t) = \begin{cases} \text{arbitrary for } |\boldsymbol{x}| \leq R_0 \\ \qquad 0 \qquad \text{for } |\boldsymbol{x}| > R_0 \ . \end{cases}$$

Due to the linearity of Maxwell's equations, it will suffice to consider a single temporal Fourier component of these distributions that is varying sinusoidally in time,

$$\rho(\boldsymbol{x},t) = \rho(\boldsymbol{x})\mathrm{e}^{-\mathrm{i}\omega t} \ , \ \boldsymbol{j}(\boldsymbol{x},t) = \boldsymbol{j}(\boldsymbol{x})\mathrm{e}^{-\mathrm{i}\omega t} \ ,$$

where physical quantities are obtained by taking the real parts. The corresponding retarded vector potential is given by

$$\boldsymbol{A}_{\mathrm{ret}}(\boldsymbol{x},t) = \frac{1}{c}\int \mathrm{d}^3x \frac{\boldsymbol{j}(\boldsymbol{x}',t_{\mathrm{ret}})}{|\boldsymbol{x}-\boldsymbol{x}'|} = \boldsymbol{A}_{\mathrm{ret}}(\boldsymbol{x})\mathrm{e}^{-\mathrm{i}\omega t} \ ,$$

with

$$\boldsymbol{A}_{\mathrm{ret}}(\boldsymbol{x}) = \frac{1}{c}\int \mathrm{d}^3x' \frac{\boldsymbol{j}(\boldsymbol{x}')\mathrm{e}^{\mathrm{i}k|\boldsymbol{x}-\boldsymbol{x}'|}}{|\boldsymbol{x}-\boldsymbol{x}'|} \ , \ k = \frac{\omega}{c} \ . \tag{2.59}$$

Due to (IV) and (2.15), the \boldsymbol{E} and \boldsymbol{B} fields in the outer area $|\boldsymbol{x}| > R_0$ are given by

$$\left.\begin{array}{l} \boldsymbol{B}(\boldsymbol{x},t) = \boldsymbol{B}(\boldsymbol{x})\mathrm{e}^{-\mathrm{i}\omega t} \ , \ \boldsymbol{B}(\boldsymbol{x}) = \boldsymbol{\nabla}\times\boldsymbol{A}_{\mathrm{ret}}(\boldsymbol{x}) \\[2mm] \boldsymbol{E}(\boldsymbol{x},t) = \boldsymbol{E}(\boldsymbol{x})\mathrm{e}^{-\mathrm{i}\omega t} \ , \ \boldsymbol{E}(\boldsymbol{x}) = \frac{\mathrm{i}}{k}\boldsymbol{\nabla}\times\boldsymbol{B}(\boldsymbol{x}) \ . \end{array}\right\} \tag{2.60}$$

To evaluate (2.59) further, we assume that $R_0 \ll |x|$ and $R_0 \ll \lambda$, where $\lambda = 2\pi/k$ is the wavelength of the radiation field described by (2.60). With this assumption, we can expand the term $\frac{e^{ik|x-x'|}}{|x-x'|}$ in (2.59) as follows:

$$|x - x'| = |x| \left(1 + \frac{x'^2}{x^2} - 2\frac{xx'}{x^2}\right)^{1/2} \approx |x| \left(1 - 2\frac{xx'}{x^2}\right)^{1/2}$$

$$\approx |x| \left(1 - \frac{xx'}{x^2}\right)$$

$$\approx |x| \left(1 + \frac{xx'}{x^2}\right)^{-1} \qquad (2.61)$$

$$\implies \frac{e^{ik|x-x'|}}{|x - x'|} \approx \frac{e^{ik|x|}e^{-ikxx'/|x|}}{|x|} \left(1 + \frac{xx'}{x^2}\right) \approx \frac{e^{ikr}}{r}e^{-iknx'} , \qquad (2.62)$$

with

$$r = |x| , \quad n = \frac{x}{|x|} .$$

Due to $2\pi/k \gg R_0 \iff k \ll 2\pi/R_0$, we can use the *long wavelength* or *dipole approximation*

$$e^{-iknx'} \approx 1 - iknx' + \dots \approx 1 . \qquad (2.63)$$

Within this approximation, (2.59) becomes

$$A_{\mathrm{ret}}(x) = \frac{1}{c}\frac{e^{ikr}}{r} \int d^3x' j(x') .$$

Using the continuity equation

$$\nabla j(x, t) = -\frac{\partial \rho(x, t)}{\partial t} \implies \nabla j(x) = i\omega\rho(x) , \qquad (2.64)$$

and the identity $(j\nabla')x' = j$, we can rewrite the integrand of this equation as

$$\int d^3x' j(x') = \int d^3x' [j(x')\nabla']x' = -\int d^3x' x'[\nabla' j(x')]$$

$$= -i\omega \int d^3x' x'\rho(x') .$$

Therefore, we find

$$A_{\mathrm{ret}}(x) = -ikp\frac{e^{ikr}}{r} , \qquad (2.65)$$

where

$$p = \int d^3x' x'\rho(x') , \quad p(t) = \int d^3x' x'\rho(x')e^{-i\omega t}$$

is the *electric dipole moment* of the charge distribution. We see that (2.65) describes an outgoing spherical wave whose wave number $k = \omega/c$ is determined by the frequency of the charge distribution. Inserting (2.65) into (2.60), one finally obtains

Theorem 2.14: E and B fields of temporally oscillating charge and current densities

Consider a current and a charge density distribution confined to the area $r = |\boldsymbol{x}| < R_0$:

$$\rho(\boldsymbol{x},t) = \mathrm{Re}\left[\rho(\boldsymbol{x})e^{-i\omega t}\right] \; , \quad \boldsymbol{j}(\boldsymbol{x},t) = \mathrm{Re}\left[\boldsymbol{j}(\boldsymbol{x})e^{-i\omega t}\right] \; .$$

For $R_0 \ll r$ and $R_0 \ll 2\pi c/\omega$, the corresponding electromagnetic fields, in dipole approximation, are given by

$$\boldsymbol{B}(\boldsymbol{x},t) = \mathrm{Re}\left[\boldsymbol{B}(\boldsymbol{x})e^{-i\omega t}\right] \; , \quad \boldsymbol{E}(\boldsymbol{x},t) = \mathrm{Re}\left[\boldsymbol{E}(\boldsymbol{x})e^{-i\omega t}\right] \; ,$$

with

$$\left.\begin{array}{l} \boldsymbol{B}(\boldsymbol{x}) = \dfrac{e^{ikr}}{r}k^2\left(1 - \dfrac{1}{ikr}\right)(\boldsymbol{n} \times \boldsymbol{p}) \\[4mm] \boldsymbol{E}(\boldsymbol{x}) = \dfrac{e^{ikr}}{r}\left\{k^2[(\boldsymbol{n} \times \boldsymbol{p}) \times \boldsymbol{n}] + \dfrac{1}{r}\left(\dfrac{1}{r} - ik\right)[3\boldsymbol{n}(\boldsymbol{n}\boldsymbol{p}) - \boldsymbol{p}]\right\} \; . \end{array}\right\} \quad (2.66)$$

Here, $\boldsymbol{n} = \boldsymbol{x}/|\boldsymbol{x}|$, and

$$\boldsymbol{p} = \int d^3x'\, \boldsymbol{x}'\rho(\boldsymbol{x}') \; , \quad \boldsymbol{p}(t) = \boldsymbol{p}e^{-i\omega t} \qquad (2.67)$$

is the electric dipole moment of the charge distribution.

The electric field E has a longitudinal and a transverse component relative to the propagation direction \boldsymbol{n}, whereas the magnetic field B is transversely polarized to \boldsymbol{n}. It is instructive to consider two special cases:

- **Far or radiation zone:** $R_0 \ll \lambda \ll r$. In this case, (2.66) simplifies to

$$\boldsymbol{B}(\boldsymbol{x}) = \frac{e^{ikr}}{r}k^2(\boldsymbol{n} \times \boldsymbol{p}) \; , \quad \boldsymbol{E}(\boldsymbol{x}) = \frac{e^{ikr}}{r}k^2(\boldsymbol{n} \times \boldsymbol{p}) \times \boldsymbol{n} = \boldsymbol{B}(\boldsymbol{x},t) \times \boldsymbol{n} \; .$$

E and B together with \boldsymbol{n} form an orthogonal trihedral (in this order). The time-averaged energy current density is calculated as

$$\overline{\boldsymbol{S}} = \frac{c}{8\pi}\mathrm{Re}(\boldsymbol{E} \times \boldsymbol{B}^*) = \frac{ck^4}{8\pi r^2}[(\boldsymbol{n} \times \boldsymbol{p})(\boldsymbol{n} \times \boldsymbol{p}^*)]\boldsymbol{n} \; ,$$

and points in the direction of \boldsymbol{n}, i.e., away from the oscillating charge density and radially outward. For the time-averaged radiation power, it follows that

$$\frac{\overline{dP}}{d\Omega} = r^2\boldsymbol{n}\overline{\boldsymbol{S}} = \frac{ck^4}{8\pi}(\boldsymbol{n} \times \boldsymbol{p})(\boldsymbol{n} \times \boldsymbol{p}^*) \; . \qquad (2.68)$$

The $1/r$-behavior of E and B is a characteristic of radiation fields. It implies that the radiation power becomes independent of r in the limit $r \to \infty$, i.e., that radiation is effectively emitted.

- **Near zone:** $R_0 \ll r \ll \lambda$. Here we have

$$kr \ll 1 \Longrightarrow \frac{1}{krr^m} \gg \frac{1}{r^m} \ , \ \mathrm{e}^{\mathrm{i}kr} \approx 1 \ ,$$

and (2.66) reduces to

$$B(x) = \frac{\mathrm{i}k}{r^2}(n \times p) \ , \ E(x) = \frac{3n(np) - p}{r^3} \ .$$

If we ignore the term $\mathrm{e}^{-\mathrm{i}\omega t}$, the electric field is identical to that of an electrostatic dipole (see Subsection 2.5.2). The magnetic field is smaller than the electric one by a factor $kr \ll 1$, so that the near fields are predominantly electric.

Taking into account the next-to-leading-order term $-\mathrm{i}knx'$ in (2.63), the fields (2.66) are supplemented by magnetic dipole and electric quadrupole fields. In general, the various types of radiation fields are denoted by E1 (electric dipole field), E2 (electric quadrupole field), M1 (magnetic dipole field), and so on.

Summary

- **Liénard-Wiechert potentials** describe the retarded potentials of an arbitrarily moving point charge. The corresponding E and B fields are each composed of two terms. The first terms are the fields of a uniformly moving charge, while the second ones are proportional to the particle acceleration.

- Only the accelerational parts contribute to the **radiation power** of an arbitrarily moving charge, which means that only accelerated charges emit radiation.

- An interesting special case is given by spatially confined and temporally oscillating charge and current density distributions. In this case the corresponding electromagnetic fields can easily be calculated within the dipole approximation.

Applications

26. Linear dipole antenna. Consider a linear dipole antenna of length L oriented along the z-axis from $z = -L/2$ to $z = L/2$ through which flows an electric current (Fig. 2.8). The current is known to be I_0 at the antenna's

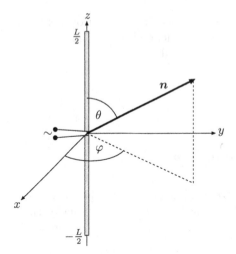

Fig. 2.8. Dipole antenna. The alternating current means both wires are charged alternately positive and negative.

center and to fall off linearly toward both ends:

$$I(z)e^{-i\omega t} = I_0\left(1 - \frac{2|z|}{L}\right)e^{-i\omega t} \ , \quad L \ll \lambda = \frac{2\pi c}{\omega} \ .$$

What is the time-averaged radiation power the antenna emits in the far zone?

Solution. Using the continuity equation (2.64), we find the linear charge density $\rho'(z)$ (i.e., the charge per unit length) to be

$$\rho'(z) = -\frac{i}{\omega}\frac{\mathrm{d}I(z)}{\mathrm{d}z} = \pm\frac{2iI_0}{\omega L} \ ,$$

where the upper (lower) sign holds for positive (negative) z. The dipole moment (2.67) is parallel to the z-axis and is given by

$$p = \int_{-L/2}^{L/2} \mathrm{d}z z\rho'(z) = \frac{iI_0 L}{2\omega} \ .$$

The time-averaged angular distribution (2.68) of the radiation power in the far zone is found using $n = \begin{pmatrix} \cos\varphi\sin\theta \\ \sin\varphi\sin\theta \\ \cos\theta \end{pmatrix}$:

$$\frac{\overline{\mathrm{d}P}}{\mathrm{d}\Omega} = \frac{I_0^2(\omega L)^2}{32\pi c^3}\sin^2\theta \ .$$

Integrating this equation over the angle yields the total radiation power

$$\overline{P} = \frac{I_0^2(\omega L)^2}{12c^3} \ .$$

27. Circular motion of a point charge. What is the far zone radiation power of a point charge q moving with constant angular velocity ω on a circular orbit with radius r_0 ($r_0 \ll \lambda = 2\pi c/\omega$) in the xy-plane?

Solution. The charge density is given by

$$\rho(\boldsymbol{x},t) = q\delta[\boldsymbol{x} - \boldsymbol{x}_0(t)] \, , \quad \boldsymbol{x}_0(t) = r_0 \begin{pmatrix} \cos \omega t \\ \sin \omega t \\ 0 \end{pmatrix} \, .$$

This yields the dipole moment

$$\boldsymbol{p}(t) = q \int \mathrm{d}^3 x' \boldsymbol{x}' \delta[\boldsymbol{x}' - \boldsymbol{x}_0(t)] = q\boldsymbol{x}_0(t) = \mathrm{Re}\left[\boldsymbol{p} e^{-i\omega t} \right] \, , \quad \boldsymbol{p} = qr_0 \begin{pmatrix} 1 \\ i \\ 0 \end{pmatrix} \, ,$$

which already has the correct time dependency as assumed in (2.67). By inserting \boldsymbol{p} in (2.68) and using $\boldsymbol{n} = \begin{pmatrix} \cos\varphi\sin\theta \\ \sin\varphi\sin\theta \\ \cos\theta \end{pmatrix}$, we finally find

$$\frac{\overline{\mathrm{d}P}}{\mathrm{d}\Omega} = \frac{\omega^4 q^2 r_0^2}{8\pi c^3}\left(1 + \cos^2\theta\right) \Longrightarrow \overline{P} = \frac{\omega^4 q^2 r_0^2}{3c^3} \, .$$

2.5 Time-Independent Electrodynamics

If we are dealing with static charge and current densities, then the four Maxwell equations decompose into two decoupled systems of equations. These form the basis of *electrostatics* and *magnetostatics*. In many textbooks of electrodynamics these static equations are discussed and phenomenologically motivated prior to the general equations. Following our axiomatic-deductive approach, we choose the opposite strategy and consider the static case as a special case of (I) to (IV) from the beginning. This allows us to derive many of the static physical laws from the results of previous sections by simply eliminating all time dependencies.

After deriving the electrostatic and magnetostatic relationships, we discuss the *multipole expansion* of static potentials for large distances. Furthermore, we consider boundary problems associated with electric conductors within electrostatics using Green function calculus as well as other calculation methods. At the end of this section, we turn to a standard example of magnetostatic field distribution problems.

2.5.1 Electrostatics and Magnetostatics

If charge and current densities are constant in time, then all time dependencies disappear from Maxwell's equations and we are left with the static case of electrodynamics:

**Definition: Static case of electrodynamics
(electrostatics, magnetostatics)**

In the static case of electrodynamics, Maxwell's equations (I) to (IV) decouple into two differential equation systems:

$$\left.\begin{array}{l} \nabla E(x) = 4\pi\rho(x) \\[2mm] \nabla \times E(x) = 0 \end{array}\right\} \text{electrostatics} \tag{2.69}$$

$$\left.\begin{array}{l} \nabla B(x) = 0 \\[2mm] \nabla \times B(x) = \dfrac{4\pi}{c} j(x) \end{array}\right\} \text{magnetostatics .} \tag{2.70}$$

The first system constitutes the basis of electrostatics and the second one that of magnetostatics. Since the charge density is time-independent, the continuity equation simplifies to

$$\nabla j(x) = 0 .$$

Thus, there is no overall charge flux.

Since the static case is a special case of electrodynamics, many of the implications following from (2.69) and (2.70) can be derived from the general electrodynamical relationships of previous sections. We shall do this by going consecutively through these sections in ascending order. From Section 2.2 all definitions and theorems can be carried over by simply eliminating all time dependencies.

**Definition: Static scalar potential ϕ
and static vector potential A**

In the static case the scalar and vector potentials ϕ and A are defined implicitly by

$$E(x) = -\nabla\phi(x) , \; B(x) = \nabla \times A(x) .$$

E are B invariant under the gauge transformations

$$A(x) \longrightarrow A'(x) = A(x) + \nabla\chi(x)$$
$$\phi(x) \longrightarrow \phi'(x) = \phi(x) + \text{const} .$$

In the static case the Coulomb and Lorentz gauges are identical and, therefore, lead to the same potential equations with solutions provided by Theorems 2.7 and 2.8:

Theorem 2.15: Static Maxwell equations in form of potentials

In the static case the Coulomb and Lorentz gauge $\nabla A(x) = 0$ leads to the potential equations

$$\nabla^2 \phi(x) = -4\pi \rho(x) \qquad \text{(static Poisson equation)}$$

$$\nabla^2 A(x) = -\frac{4\pi}{c} j(x) \,.$$

Their general solution is

$$\phi_{\text{hom}}(x) = \text{Re} \int d^3k \tilde{\phi}(k) e^{ikx} \,, \quad A_{\text{hom},i}(x) = \text{Re} \int d^3k \tilde{A}_i(k) e^{ikx} \,.$$

An inhomogeneous solution of the potential equations is given by

$$\phi(x) = \int d^3x' \frac{\rho(x')}{|x - x'|} \,, \quad A(x) = \frac{1}{c} \int d^3x' \frac{j(x')}{|x - x'|} \,. \tag{2.71}$$

Due to the absence of time, it does not make sense to denote the inhomogeneous solution by "retarded potential". Combining the last equation with $B = \nabla \times A$ yields

Theorem 2.16: Biot Savart's law

Given the current density, the magnetic induction field is calculated as

$$B(x) = \frac{1}{c} \int d^3x' j(x') \times \frac{x - x'}{|x - x'|^3} \,.$$

It is clear that the Lorentz-covariant formulation of electrodynamics (Section 2.3) cannot by transferred to the static case, since time and space are no longer on equal footing. Therefore, the static case can only refer to one particular inertial system. In Section 2.4, only the first two Theorems 2.11 and 2.12 have static analogues, which we summarize as follows:

Theorem 2.17: Potentials and fields of a static point charge q at rest at x_0 (Coulomb law)

$$\phi(x) = \frac{q}{|x - x_0|} \,, \quad E(x) = q \frac{x - x_0}{|x - x_0|^3} \,, \quad A(x) = 0 \,, \quad B(x) = 0 \,.$$

As expected, the potentials of this theorem are a special case of (2.71) for $\rho(x) = q\delta(x - x_0)$ and $j(x) = 0$. The subsequent Theorems 2.13 and 2.14 in Section 2.4 all deal with moving charges and, therefore, do not contribute to the static case.

Electrostatic field energy and self-energy problem. Now, we consider a charge distribution $\rho(x)$ at rest. According to Theorem 2.4, the correspond-

ing electromagnetic (or in this case: electrostatic) field energy reads

$$E_{\text{em}} = \frac{1}{8\pi} \int d^3x \, \boldsymbol{E}^2(\boldsymbol{x})$$

$$= \frac{1}{8\pi} \int d^3x [\boldsymbol{\nabla}\phi(\boldsymbol{x})] \boldsymbol{\nabla}\phi(\boldsymbol{x}) = \frac{1}{2} \int d^3x \rho(\boldsymbol{x})\phi(\boldsymbol{x}) ,$$

where ϕ is the electrostatic potential generated by the charge density itself. Inserting ϕ from Theorem 2.15 into this equation yields

$$E_{\text{em}} = \frac{1}{2} \int d^3x \int d^3x' \frac{\rho(\boldsymbol{x})\rho(\boldsymbol{x}')}{|\boldsymbol{x} - \boldsymbol{x}'|} . \tag{2.72}$$

Note that this equation is only sensible for a continuous charge distribution. For a set of discrete point charges q_i at locations \boldsymbol{x}_i,

$$\rho(\boldsymbol{x}) = \sum_i q_i \delta(\boldsymbol{x} - \boldsymbol{x}_i) ,$$

(2.72) turns into

$$E_{\text{em}} = \frac{1}{2} \sum_{i,j} \int d^3x \int d^3x' \frac{q_i q_j \delta(\boldsymbol{x} - \boldsymbol{x}_i)\delta(\boldsymbol{x}' - \boldsymbol{x}_j)}{|\boldsymbol{x} - \boldsymbol{x}'|}$$

$$= \frac{1}{2} \sum_{i,j} \frac{q_i q_j}{|\boldsymbol{x}_i - \boldsymbol{x}_j|} . \tag{2.73}$$

In this sum terms with $i = j$ diverge. They correspond to the *self-energy* of the charges q_i due to their own fields at their positions \boldsymbol{x}_i. The reason for this unphysical behavior lies in the concept of point charges and in the use of δ-functions. Electrodynamics – as a classical field theory – is not valid down to arbitrarily small distances. With respect to a broad estimation of the range of validity of electrodynamics, it is reasonable to assume that the self-energy of a particle is comparable to its rest energy. Assuming further that the particle, say an electron, has a finite radius R_0, then its self-energy is of magnitude e^2/R_0 and we can estimate R_0 via

$$\frac{e^2}{R_0} \approx m_e c^2 \implies R_0 \approx \frac{e^2}{m_e c^2} = 2.8 \cdot 10^{-15} \text{ m} ,$$

where e is the charge and m_e the rest mass of the electron. This is called the *classical electron radius*. However, from quantum mechanics, we know that quantum effects are significant on a much larger length scale of magnitude $\hbar^2/(m_e e^2) = 0.5 \cdot 10^{-10}$ m (*Bohr's radius*). Thus, we have to accept that the self-energy problem cannot be solved within electrodynamics. The adequate framework is given by *quantum electrodynamics*, a quantum field theory, which combines classical electrodynamics and relativistic quantum mechanics.

By subtracting the unphysical self-energy contributions from (2.73), we obtain the potential energy of a discrete charge distribution composed of the interaction energies between different charges:

$$E_{\text{em}} = \frac{1}{2} \sum_{i \neq j} \frac{q_i q_j}{|\boldsymbol{x}_i - \boldsymbol{x}_j|} \; .$$

2.5.2 Multipole Expansion of Static Potentials and Fields

In the following, we consider static charge and current density distributions that are spatially confined to $|\boldsymbol{x}| \leq R_0$. Given this constraint, it is possible to evaluate in an approximate manner the corresponding potentials and fields at large distances $|\boldsymbol{x}| \gg R_0$ using the technique of *multipole expansion*. According to Theorem 2.15, the scalar and vector potentials have the same $1/|\boldsymbol{x} - \boldsymbol{x}'|$-dependency, which we expand around $\boldsymbol{x}' = \boldsymbol{0}$ as follows:

$$\frac{1}{|\boldsymbol{x} - \boldsymbol{x}'|} = \frac{1}{|\boldsymbol{x}|} + \sum_i x_i' \frac{\partial}{\partial x_i'} \frac{1}{|\boldsymbol{x} - \boldsymbol{x}'|} \bigg|_{\boldsymbol{x}'=0}$$

$$+ \frac{1}{2} \sum_{i,j} x_i' x_j' \frac{\partial}{\partial x_i'} \frac{\partial}{\partial x_j'} \frac{1}{|\boldsymbol{x} - \boldsymbol{x}'|} \bigg|_{\boldsymbol{x}'=0} + \dots \; .$$

Using

$$\frac{\partial}{\partial x_i'} \frac{1}{|\boldsymbol{x} - \boldsymbol{x}'|} \bigg|_{\boldsymbol{x}'=0} = - \frac{\partial}{\partial x_i} \frac{1}{|\boldsymbol{x} - \boldsymbol{x}'|} \bigg|_{\boldsymbol{x}'=0} = - \frac{\partial}{\partial x_i} \frac{1}{|\boldsymbol{x}|} \; ,$$

it follows that

$$\frac{1}{|\boldsymbol{x} - \boldsymbol{x}'|} = \frac{1}{|\boldsymbol{x}|} + \sum_i \frac{x_i' x_i}{|\boldsymbol{x}|^3} + \frac{1}{2} \sum_{i,j} x_i' x_j' \left(\frac{3 x_i x_j}{|\boldsymbol{x}|^5} - \frac{\delta_{ij}}{|\boldsymbol{x}|^3} \right) + \dots \; . \quad (2.74)$$

Electrostatic multipole expansion. Inserting this expansion into (2.71), we obtain for ϕ in the first three leading orders:

- *Electric monopole moment* (charge):

$$\phi_0(\boldsymbol{x}) = \frac{1}{|\boldsymbol{x}|} \int \mathrm{d}^3 x' \rho(\boldsymbol{x}') = \frac{Q}{|\boldsymbol{x}|} \; , \; Q = \int \mathrm{d}^3 x' \rho(\boldsymbol{x}') \; .$$

 Therefore, when viewed from a large distance, a static charge distribution behaves as a point charge.

- *Electric dipole moment:*

$$\phi_1(\boldsymbol{x}) = \frac{\boldsymbol{x}}{|\boldsymbol{x}|^3} \int \mathrm{d}^3 x' \boldsymbol{x}' \rho(\boldsymbol{x}') = \frac{\boldsymbol{x} \boldsymbol{p}}{|\boldsymbol{x}|^3} \; , \; \boldsymbol{p} = \int \mathrm{d}^3 x' \boldsymbol{x}' \rho(\boldsymbol{x}') \; ,$$

 where \boldsymbol{p} denotes the electric dipole moment of the charge distribution.

- *Electric quadrupole moment:*

$$\phi_2(\boldsymbol{x}) = \sum_{i,j} \frac{3 x_i x_j - |\boldsymbol{x}|^2 \delta_{ij}}{2 |\boldsymbol{x}|^5} \int \mathrm{d}^3 x' x_i' x_j' \rho(\boldsymbol{x}') \; .$$

This expression can be simplified by subtracting a total of 0 written as

$$\sum_{i,j} \frac{3x_i x_j - |x|^2 \delta_{ij}}{6|x|^5} \int d^3x' |x'|^2 \delta_{ij} \rho(x') \ .$$

We then have

$$\phi_2(x) = \sum_{i,j} \frac{3x_i x_j - |x|^2 \delta_{ij}}{6|x|^5} \int d^3x' (3x_i' x_j' - |x'|^2 \delta_{ij}) \rho(x')$$

$$= \sum_{i,j} \frac{3x_i x_j - |x|^2 \delta_{ij}}{6|x|^5} Q_{ij} \ , \tag{2.75}$$

where

$$Q_{ij} = \int d^3x' (3x_i' x_j' - |x'|^2 \delta_{ij}) \rho(x')$$

defines the electric quadrupole moment. Since Q_{ij} has vanishing trace, the second term in (2.75) is zero and we finally arrive at

$$\phi_2(x) = \sum_{i,j} \frac{x_i x_j}{2|x|^5} Q_{ij} \ .$$

Theorem 2.18: Multipole expansion of the scalar potential

For large distances $|x| \gg R_0$, the scalar potential of a static charge distribution $\rho(x)$ spatially confined to $|x| \le R_0$ can be expanded as

$$\phi(x) = \frac{Q}{|x|} + \frac{xp}{|x|^3} + \frac{1}{2} \sum_{i,j} \frac{x_i x_j}{|x|^5} Q_{ij} + \dots \ , \tag{2.76}$$

with

$$Q = \int d^3x' \rho(x') \qquad \text{(electric monopole moment, charge)}$$

$$p = \int d^3x' x' \rho(x') \qquad \text{(electric dipole moment)}$$

$$Q_{ij} = \int d^3x' (3x_i' x_j' - |x'|^2 \delta_{ij}) \rho(x') \qquad \text{(electric quadrupole moment)} \ .$$

The electric fields corresponding to the monopole and dipole moments are

$$E_{\text{Mo}}(x) = Q \frac{x}{|x|^3} \ , \quad E_{\text{Di}}(x) = \frac{3x(xp) - p|x|^2}{|x|^5} \ .$$

Under orthogonal transformations, the matrix Q_{ij} behaves as a tensor of rank 2 and has 9 components. Only five of these are independent due to their symmetry properties $Q_{ij} = Q_{ji}$, $Q_{11} + Q_{22} + Q_{33} = 0$. Transforming Q_{ij} to its principal axis system further reduces the number of independent components to two.

The next leading-order term of the expansion (2.76) contains the *octupole moment*. It consists of a third-rank tensor with 27 components, of which

also just a few are independent. If higher moments need to be calculated, the multipole expansion in spherical coordinates provides a much easier approach. This is discussed in Application 28.

Magnetostatic multipole expansion. Taking into account the first two terms in (2.74), we find from (2.71) for A:

- *Magnetic monopole moment:*

$$A_0(x) = \frac{1}{c|x|} \int d^3x' j(x') .$$

Since $\nabla j = 0$, any scalar- or vector-like function $f(x')$ satisfies

$$0 = \int d^3x' f(x') \nabla' j(x') = - \int d^3x' [\nabla' f(x')] j(x') . \tag{2.77}$$

Setting $f(x') = x'$, it follows that $\int d^3x' j = 0 \implies A_0(x) = 0$. This means the current density distribution j has no monopole component.

- *Magnetic dipole moment:*

$$A_1(x) = \frac{1}{c|x|^3} \int d^3x' (xx') j(x') .$$

Using the integral relation (2.77), this expression can be transformed as follows:

$$f(x') = x'_k x'_l \implies \int d^3x' (x'_l j_k + x'_k j_l) = 0$$

$$\implies \int d^3x' x'_l j_k = \frac{1}{2} \int d^3x' (x'_l j_k - x'_k j_l)$$

$$\implies \sum_l \int d^3x' x_l x'_l j_k = \frac{1}{2} \sum_l \int d^3x' (x_l x'_l j_k - x_l x'_k j_l)$$

$$\implies \int d^3x' (xx') j = \frac{1}{2} \int d^3x' [(xx') j - x'(xj)]$$

$$= -\frac{1}{2} x \times \int d^3x' x' \times j \tag{2.78}$$

$$\implies A_1(x) = \frac{\mu \times x}{|x|^3} , \quad \mu = \frac{1}{2c} \int d^3x' x' \times j ,$$

where μ denotes the magnetic dipole moment.

Theorem 2.19: Multipole expansion of the vector potential

For large distances $|x| \gg R_0$, the vector potential of a static current density distribution $j(x)$ spatially confined to $|x| \leq R_0$ can be expanded as

$$A(x) = \frac{\mu \times x}{|x|^3} + \cdots ,$$

\triangleright

with

$$\boldsymbol{\mu} = \frac{1}{2c} \int d^3x' \boldsymbol{x}' \times \boldsymbol{j}(\boldsymbol{x}') \quad \text{(magnetic dipole moment)} .$$

Contrary to the scalar potential, the vector potential has no monopole component. The corresponding magnetic dipole field is

$$\boldsymbol{B}_{\mathrm{Di}}(\boldsymbol{x}) = \frac{3\boldsymbol{x}(\boldsymbol{x}\boldsymbol{\mu}) - \boldsymbol{\mu}|\boldsymbol{x}|^2}{|\boldsymbol{x}|^5} .$$

It has the same structure as the electric dipole field.

If the charge or current density distribution has a certain spatial symmetry, some moments in the multipole expansion of ϕ and \boldsymbol{A} vanish. For example, if there is an equal number of negative and positive charges, then the electric monopole must vanish. Also, if for every dipole, there exists an equal-valued dipole in the opposite direction, then the electric dipole moment vanishes, and so on.

Magnetic dipole in an external magnetic field. We now calculate the force and torque on a magnetic dipole in an external magnetic field \boldsymbol{B}. We again assume that the current density distribution \boldsymbol{j} is spatially confined around $\boldsymbol{x} = \boldsymbol{0}$ and, additionally, we assume that the magnetic field is only weakly varying in space. Expanding the magnetic field around the origin,

$$\boldsymbol{B}(\boldsymbol{x}') = \boldsymbol{B}(\boldsymbol{0}) + (\boldsymbol{x}'\boldsymbol{\nabla})\boldsymbol{B}(\boldsymbol{x})|_{\boldsymbol{x}=0} + \cdots ,$$

we can write for the force \boldsymbol{F} acting on the current density distribution:

$$\begin{aligned} \boldsymbol{F} &= \frac{1}{c} \int d^3x' \boldsymbol{j}(\boldsymbol{x}') \times \boldsymbol{B}(\boldsymbol{x}') \\ &= \frac{1}{c} \underbrace{\int d^3x' \boldsymbol{j}(\boldsymbol{x}') \times \boldsymbol{B}(\boldsymbol{0})}_{=0} + \frac{1}{c} \int d^3x' \boldsymbol{j}(\boldsymbol{x}') \times (\boldsymbol{x}'\boldsymbol{\nabla})\boldsymbol{B}(\boldsymbol{x})|_{\boldsymbol{x}=0} \\ &= \frac{1}{c} \int d^3x' \boldsymbol{j}(\boldsymbol{x}') \times (\boldsymbol{x}'\boldsymbol{\nabla})\boldsymbol{B}(\boldsymbol{x})|_{\boldsymbol{x}=0} \\ &= \frac{1}{c} \left[\int d^3x' (\boldsymbol{\nabla}\boldsymbol{x}')\boldsymbol{j}(\boldsymbol{x}') \right] \times \boldsymbol{B}(\boldsymbol{x})|_{\boldsymbol{x}=0} . \end{aligned}$$

This equation can be evaluated further with the help of (2.78), where \boldsymbol{x} is replaced by the $\boldsymbol{\nabla}$ operator, which also represents a constant vector with respect to the \boldsymbol{x}'-integration. It then follows that

$$\begin{aligned} \boldsymbol{F} &= -\frac{1}{2c} \left[\boldsymbol{\nabla} \times \int d^3x' \boldsymbol{x}' \times \boldsymbol{j}(\boldsymbol{x}') \right] \times \boldsymbol{B}(\boldsymbol{x})|_{\boldsymbol{x}=0} \\ &= -(\boldsymbol{\nabla} \times \boldsymbol{\mu}) \times \boldsymbol{B}(\boldsymbol{x})|_{\boldsymbol{x}=0} = \boldsymbol{\nabla}[\boldsymbol{\mu}\boldsymbol{B}(\boldsymbol{x})]|_{\boldsymbol{x}=0} - \boldsymbol{\mu} \underbrace{[\boldsymbol{\nabla}\boldsymbol{B}(\boldsymbol{x})]|_{\boldsymbol{x}=0}}_{=0} \\ &= \boldsymbol{\nabla}[\boldsymbol{\mu}\boldsymbol{B}(\boldsymbol{x})]|_{\boldsymbol{x}=0} . \end{aligned}$$

To lowest order, the torque N on μ is given by

$$
\begin{aligned}
N &= \frac{1}{c} \int d^3 x' x' \times [j(x') \times B(0)] \\
&= \frac{1}{c} \int d^3 x' \{[x' B(0)] j(x') - [x' j(x')] B(0)\} \\
&= \frac{1}{c} \int d^3 x' [x' B(0)] j(x') - \frac{B(0)}{c} \int d^3 x' x' j(x') .
\end{aligned}
\tag{2.79}
$$

Using

$$
\nabla(x'^2 j) = 2x' j + x'^2 \nabla j = 2x' j
$$

and Gauss's law, we see that the second term in (2.79) vanishes. A second application of (2.78) finally yields

$$
N = \frac{1}{c} \int d^3 x' [x' B(0)] j(x') = -B(0) \times \int d^3 x' \frac{x' \times j(x')}{2c} = \mu \times B(0).
$$

Theorem 2.20: Magnetic dipole in an external magnetic field

Consider a magnetic dipole μ of a static charge current distribution confined around $x = 0$ within a spatially weakly varying magnetic field B. The force and torque on μ are given by

$$
F = \nabla[\mu B(x)]|_{x=0} , \quad N = \mu \times B(0) .
$$

The energy of the magnetic dipole is

$$
W = -\int dx\, F = -\mu B(0) .
$$

2.5.3 Boundary Problems in Electrostatics I

In electrostatics one is often confronted with the following problem: given a volume V bounded by a surface F and with an internal static charge distribution ρ, what are the scalar potential ϕ and the corresponding electric field E inside V? If there is no confining area (or if the volume is infinitely large), we can immediately write down the solution from the static Poisson equation:

$$
\phi(x) = \int d^3 x' \frac{\rho(x')}{|x - x'|} .
\tag{2.80}
$$

However, the presence of a confining area implies certain boundary conditions, which have to be taken into account. In this case, to ensure compliance with the boundary conditions, a homogeneous solution must be added to the inhomogeneous one (2.80).

In the following, we deal with the solution of such boundary problems where the boundary is an electric conductor. By this we mean materials (normally metals) in which freely moving electrons are present. Those charges generally lead to time-dependent fields. However, after a time, a closed system will reach an equilibrium state, which we are interested in. Inside electric conductors this means $E = 0$. Otherwise, there will be forces acting on the charges that result in charge displacements, which are contradictory to the static case.[12]

We start our discussion with the following question: what continuity conditions must an electric field E obey at conducting surfaces? To answer this, we consider a conducting body with a boundary surface, in which we put a volume element ΔV or an area element ΔF with height h (Fig. 2.9). Applying

Fig. 2.9. Integration domain at a surface of an electric conductor.

Gauss's law on the divergence equation and Stokes's law on the curl equation of (2.69), then, in the limit $h \to 0$, only those area elements δF and those line elements δl contribute, which are parallel to the boundary surface. Therefore, we obtain

$$(E_2 - E_1)n\delta F = 4\pi\delta q , \quad (E_2 - E_1)t\delta l = 0 .$$

Due to $E_1 = 0$, it follows

Theorem 2.21: Continuity conditions at conductors

At the boundary surface F of a conductor, the electrostatic field E and the corresponding scalar potential ϕ obey the following continuity conditions:

• The tangential components of E are zero:

$$E(x)t = 0 \Longleftrightarrow \phi(x) = \text{const} , \quad x \in F .$$

\triangleright

[12] Strictly speaking, this statement is only true for electric fields averaged over many atomic length units. However, for our purposes, this point can be neglected. It is discussed in the next section.

- The normal component of E is proportional to the surface charge density σ (charge/unit area) of the conductor:

$$E(x)n = 4\pi\sigma(x) \iff \frac{\partial\phi}{\partial n}(x) = -4\pi\sigma(x) \,, \quad x \in F \,.$$

Here, n is the normal vector pointing into the vacuum, and t represents a tangential vector of the boundary surface. $\partial/\partial n$ is an abbreviation for $n\nabla$.

If the conductor's surface charge density is given, the charge distribution is known everywhere. This is because ρ is supposed to be known inside V and, inside the conductor, we have $\rho = \nabla E/4\pi = 0$ due to $E = 0$. In this case, the solution can again be determined with the help of (2.80).

Dirichlet and Neumann boundary conditions. It is often the case that only the surface potential is given, instead of the surface charge density. It is then advantageous to rewrite the static Poisson equation in integral form, in which the surface potential and its normal derivative appear explicitly. To this end, we use the second Green identity (A.2). Inserting $\psi(x') = 1/|x-x'|$ into this equation and interpreting ϕ as the electrostatic potential, then, using $\nabla'^2|x-x'|^{-1} = -4\pi\delta(x-x')$ and $\nabla'^2\phi(x') = -4\pi\rho(x')$, we find

$$\int_V \mathrm{d}^3x' \left[-4\pi\phi(x')\delta(x-x') + \frac{4\pi}{|x-x'|}\rho(x') \right]$$

$$= \oint_F \mathrm{d}F' \left[\phi(x')\frac{\partial}{\partial n'}\frac{1}{|x-x'|} - \frac{1}{|x-x'|}\frac{\partial\phi(x')}{\partial n'} \right] \,.$$

If the observation point x lies inside V, it follows further that

$$\phi(x) = \int_V \mathrm{d}^3x' \frac{\rho(x')}{|x-x'|}$$

$$+ \frac{1}{4\pi}\oint_F \mathrm{d}F' \left[\frac{1}{|x-x'|}\frac{\partial\phi(x')}{\partial n'} - \phi(x')\frac{\partial}{\partial n'}\frac{1}{|x-x'|} \right] \,; \qquad (2.81)$$

otherwise $\phi(x) = 0$. As we mentioned at the start of this subsection, in the limit $V \to \infty$ this equation turns into the known expression (2.80). If, on the other hand, we have $\rho(x) = 0$ inside the whole volume V, then $\phi(x)$ is solely determined by its values and derivatives on the surface F. Now, given (2.81), one distinguishes between two types of boundary conditions:

- *Dirichlet boundary condition:*[13] $\phi(x)|_{x\in F} = \phi_0(x)$.

- *Neumann boundary condition:* $\left.\dfrac{\partial\phi(x)}{\partial n}\right|_{x\in F} = -4\pi\sigma(x)$.

[13] Note that this boundary condition is a generalization of the first condition in Theorem 2.21, since here ϕ does not (unlike for metals) need to be constant on the surface.

Each of these conditions leads to unique solutions. This can be seen as follows: assume there are two solutions ϕ_1 and ϕ_2 of the Poisson equation $\nabla^2\phi = -4\pi\rho$, both of which obey a Dirichlet or Neumann boundary condition. Then, for $u = \phi_1 - \phi_2$, we have

$$\nabla^2 u(\boldsymbol{x})\big|_{\boldsymbol{x}\in V} = 0 \;, \quad u(\boldsymbol{x})\big|_{\boldsymbol{x}\in F} = 0 \;, \quad \frac{\partial u(\boldsymbol{x})}{\partial n}\bigg|_{\boldsymbol{x}\in F} = 0 \;. \tag{2.82}$$

Due to the first Green identity (A.1), we further have with $\psi = \phi = u$

$$\int_V \mathrm{d}^3 x \left[u\nabla^2 u + (\nabla u)^2 \right] = \oint_F \mathrm{d}F u \frac{\partial u}{\partial n} \implies \int_V \mathrm{d}^3 x (\nabla u)^2 = 0$$

$$\implies \nabla u = \boldsymbol{0} \;,$$

i.e., u is constant inside V. Combining this with (2.82), it follows that for the Dirichlet boundary condition: $\phi_1 = \phi_2$ and for the Neumann condition: $\phi_1 = \phi_2 + \text{const}$. An example for the uniqueness of these solutions is given by the *Faraday cage*. It consists of an arbitrarily formed closed metal surface, inside which no charges are present, i.e.,

$$\nabla^2\phi(\boldsymbol{x})\big|_{\boldsymbol{x}\in V} = 0 \;, \quad \phi(\boldsymbol{x})\big|_{\boldsymbol{x}\in F} = \text{const} \;.$$

Obviously, the expression $\phi(\boldsymbol{x})\big|_{\boldsymbol{x}\in V} = \text{const}$ is a solution and, due to uniqueness, it is already the desired one.

Formal solution of electrostatic boundary problems via Green functions. In general, there exists no solution if both a Dirichlet and a Neumann boundary condition are simultaneously present, as they both have unique solutions generally differing from one another. In this respect, the integral form (2.81) is disadvantageous, as it contains both types of boundary conditions. Therefore, we shall rewrite this equation such that one of these boundary conditions is eliminated. In the derivation of (2.81), we have set $\psi(\boldsymbol{x}') = 1/|\boldsymbol{x}-\boldsymbol{x}'|$ in the second Green identity, since it is a solution of $\nabla'^2\psi(\boldsymbol{x}') = -4\pi\delta(\boldsymbol{x}-\boldsymbol{x}')$. However, this ψ is just a special Green function G solving the equation

$$\nabla'^2 G(\boldsymbol{x},\boldsymbol{x}') = -4\pi\delta(\boldsymbol{x}-\boldsymbol{x}') \;.$$

The general solution is given by

$$G(\boldsymbol{x},\boldsymbol{x}') = \frac{1}{|\boldsymbol{x}-\boldsymbol{x}'|} + g(\boldsymbol{x},\boldsymbol{x}') \;,$$

where g fulfills the *Laplace equation*

$$\nabla'^2 g(\boldsymbol{x},\boldsymbol{x}') = 0$$

inside V. Now, repeating our derivation starting from the second Green identity with $\psi = G$, we obtain

$$\phi(\boldsymbol{x}) = \int\limits_V \mathrm{d}^3 x' \rho(\boldsymbol{x}') G(\boldsymbol{x}, \boldsymbol{x}')$$

$$+\frac{1}{4\pi} \oint\limits_F \mathrm{d}F' \left[G(\boldsymbol{x}, \boldsymbol{x}') \frac{\partial \phi(\boldsymbol{x}')}{\partial n'} - \phi(\boldsymbol{x}') \frac{\partial G(\boldsymbol{x}, \boldsymbol{x}')}{\partial n'} \right] \ .$$

With the help of the function g, we are now free to choose G on the surface, such that one of the two boundary conditions vanishes:

- **Dirichlet boundary condition:** here, we set

$$G(\boldsymbol{x}, \boldsymbol{x}')|_{\boldsymbol{x}' \in F} = 0 \ .$$

Then,

$$\phi(\boldsymbol{x}) = \int\limits_V \mathrm{d}^3 x' \rho(\boldsymbol{x}') G(\boldsymbol{x}, \boldsymbol{x}') - \frac{1}{4\pi} \oint\limits_F \mathrm{d}F' \phi(\boldsymbol{x}') \frac{\partial G(\boldsymbol{x}, \boldsymbol{x}')}{\partial n'}$$

is a solution of the problem, provided that we have a solution of Laplace's equation with the boundary condition

$$g(\boldsymbol{x}, \boldsymbol{x}')|_{\boldsymbol{x}' \in F} = -\frac{1}{|\boldsymbol{x} - \boldsymbol{x}'|} \ .$$

- **Neumann boundary condition:** similar to the Dirichlet problem, it is tempting to set $\partial G / \partial n'|_{\boldsymbol{x}' \in F} = 0$. However, this approach is incompatible with

$$\boldsymbol{\nabla}'^2 G(\boldsymbol{x}, \boldsymbol{x}') = -4\pi\delta(\boldsymbol{x} - \boldsymbol{x}') \implies \oint\limits_F \mathrm{d}F' \frac{\partial G}{\partial n'} = -4\pi \ .$$

Therefore, the correct ansatz is

$$\frac{\partial G}{\partial n'}\bigg|_{\boldsymbol{x}' \in F} = -\frac{4\pi}{F} \ .$$

Provided that we are able to find a solution to Laplace's equation with the boundary condition

$$\frac{\partial g(\boldsymbol{x}, \boldsymbol{x}')}{\partial n'}\bigg|_{\boldsymbol{x}' \in F} = -\frac{4\pi}{F} - \frac{\partial}{\partial n'} \frac{1}{|\boldsymbol{x} - \boldsymbol{x}'|} \ ,$$

the expression

$$\phi(\boldsymbol{x}) = \int\limits_V \mathrm{d}^3 x' \rho(\boldsymbol{x}') G(\boldsymbol{x}, \boldsymbol{x}') + \frac{1}{4\pi} \oint\limits_F \mathrm{d}F' G(\boldsymbol{x}, \boldsymbol{x}') \frac{\partial \phi(\boldsymbol{x}')}{\partial n'} + \langle \phi \rangle_F$$

is a solution of the Neumann problem, where $\langle \phi \rangle_F = \frac{1}{F} \oint_F \mathrm{d}F' \phi(\boldsymbol{x}')$ denotes the average value of ϕ on the surface F.

Theorem 2.22: Boundary problems in electrostatics

Consider a volume V confined by a surface F. Inside V, the charge density ρ is known. Furthermore, let

$$G(\boldsymbol{x}, \boldsymbol{x}') = \frac{1}{|\boldsymbol{x} - \boldsymbol{x}'|} + g(\boldsymbol{x}, \boldsymbol{x}') \ .$$

One usually distinguishes between two types of boundary problems:

- **Dirichlet problem:** the electrostatic potential ϕ is given on F: $\phi(\boldsymbol{x})|_{\boldsymbol{x} \in F} = \phi_0(\boldsymbol{x})$. In this case, the solution inside V is given by

$$\phi(\boldsymbol{x})|_{\boldsymbol{x} \in V} = \int\limits_V \mathrm{d}^3 x' \rho(\boldsymbol{x}') G(\boldsymbol{x}, \boldsymbol{x}') - \frac{1}{4\pi} \oint\limits_F \mathrm{d}F' \phi(\boldsymbol{x}') \frac{\partial G(\boldsymbol{x}, \boldsymbol{x}')}{\partial n'} \ ,$$

with

$$\boldsymbol{\nabla}'^2 g(\boldsymbol{x}, \boldsymbol{x}')\big|_{\boldsymbol{x}' \in V} = 0 \ , \quad g(\boldsymbol{x}, \boldsymbol{x}')\big|_{\boldsymbol{x}' \in F} = -\frac{1}{|\boldsymbol{x} - \boldsymbol{x}'|} \ . \tag{2.83}$$

- **Neumann problem:** the normal derivative of ϕ is given on F: $\frac{\partial \phi(\boldsymbol{x})}{\partial n}\big|_{\boldsymbol{x} \in F} = -4\pi\sigma(\boldsymbol{x})$. The corresponding solution inside V reads

$$\phi(\boldsymbol{x}) = \int\limits_V \mathrm{d}^3 x' \rho(\boldsymbol{x}') G(\boldsymbol{x}, \boldsymbol{x}') + \frac{1}{4\pi} \oint\limits_F \mathrm{d}F' G(\boldsymbol{x}, \boldsymbol{x}') \frac{\partial \phi(\boldsymbol{x}')}{\partial n'} + \langle \phi \rangle_F \ ,$$

with

$$\boldsymbol{\nabla}'^2 g(\boldsymbol{x}, \boldsymbol{x}')\big|_{\boldsymbol{x}' \in V} = 0 \ , \quad \frac{\partial g(\boldsymbol{x}, \boldsymbol{x}')}{\partial n'}\bigg|_{\boldsymbol{x}' \in F} = -\frac{4\pi}{F} - \frac{\partial}{\partial n'} \frac{1}{|\boldsymbol{x} - \boldsymbol{x}'|} \ . \tag{2.84}$$

On the one hand, using Green function calculus leads to a simplification, since boundary conditions no longer depend on special Dirichlet or Neumann boundary values. On the other hand, it often turns out to be very difficult to find a function g (and therefore G) with the correct boundary behavior (2.83) or (2.84).

The function $g(\boldsymbol{x}, \boldsymbol{x}')$ solves the Laplace equation inside the volume V. It therefore represents the potential of a charge distribution lying outside V. This external charge distribution is conditioned such that the Green function can take on the values $G = 0$ or $\partial G / \partial n' = -4\pi/F$ on the boundary surface F. For many boundary problems with relatively simple geometry, we can therefore proceed by determining a charge distribution outside V, which, in conjunction with the charges inside V, yields a potential fulfilling the boundary conditions on F. This is the *method of mirror charges*, which, along with other methods, is considered in the next subsection.

2.5.4 Boundary Problems in Electrostatics II

In this subsection we study two standard examples of electrostatic boundary problems that are discussed in many textbooks on electrodynamics. The first example illustrates the use of the method of mirror charges. The second one deals with the solution of Laplace's equation in spherical coordinates.

Point charge in front of a grounded metal sphere. Consider a grounded metal sphere with radius R in the field of a point charge situated at distance $a > R$ to the sphere's center (Fig. 2.10). We look for the electrostatic po-

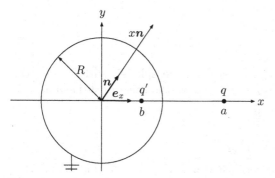

Fig. 2.10. Grounded metal sphere in front of a point charge.

tential outside the sphere (i.e., inside V) as well as for the influenced charge density σ on the sphere's surface. Due to the grounding, the sphere's potential is equal to that of earth ($\phi_0 = 0$). Inside V, we have to solve the Poisson equation

$$\nabla^2 \phi(\boldsymbol{x}) = -4\pi\delta(\boldsymbol{x} - a\boldsymbol{e}_x) \;,$$

with the Dirichlet boundary condition

$$\phi(\boldsymbol{x})|_{|\boldsymbol{x}|=R} = 0 \;. \tag{2.85}$$

We will use the method of mirror charges and try to place a suitable charge distribution outside V (i.e., inside the sphere), such that the boundary condition (2.85) is fulfilled. Due to symmetry reasons, we make an ansatz in which the mirror charge is a point charge q' located at distance $b < R$ on the x-axis, so that

$$\phi(\boldsymbol{x}) = \frac{q}{|x\boldsymbol{n} - a\boldsymbol{e}_x|} + \frac{q'}{|x\boldsymbol{n} - b\boldsymbol{e}_x|} \;, \quad x\boldsymbol{n} = \boldsymbol{x} \;.$$

On the sphere's surface, it follows that

$$\phi(\boldsymbol{x})|_{|\boldsymbol{x}|=R} = \frac{q}{R\left|\boldsymbol{n} - \frac{a}{R}\boldsymbol{e}_x\right|} + \frac{q'}{b\left|\frac{R}{b}\boldsymbol{n} - \boldsymbol{e}_x\right|} \;.$$

The Dirichlet boundary condition is fulfilled if we set

$$b = \frac{R^2}{a} \ , \ q' = -\frac{Rq}{a} \ .$$

Therefore, the electrostatic potential inside and outside the sphere is given by

$$\phi(x) = \begin{cases} \dfrac{q}{|xn - ae_x|} - \dfrac{Rq}{a\left|xn - \frac{R^2}{a}e_x\right|} & \text{for } x \in V \\[2ex] 0 & \text{for } x \notin V \ . \end{cases} \tag{2.86}$$

Using

$$\left.\frac{\partial \phi}{\partial x}\right|_{x=R} = -4\pi\sigma(x) \ ,$$

we obtain the influenced charge density on the sphere's surface (i.e., the amount of charge transferred from the earth to the sphere's surface) as

$$\sigma(\gamma) = -\frac{q}{4\pi a R} \frac{\left(1 - \frac{R^2}{a^2}\right)}{\left[1 - \frac{2R}{a}\cos\gamma + \frac{R^2}{a^2}\right]^{3/2}} \ ,$$

where γ denotes the angle between n and e_x. As expected, this distribution has a maximum at $\gamma = 0$, i.e., in the direction of the point charge q. Integration of σ over the whole surface of the sphere yields the mirror charge q', in accordance with Gauss's law.

Based on these results, we can extend our problem by asking for the electrostatic potential of an isolated metal sphere with radius R and charge Q in the field of a point charge q situated at distance $a > R$ to the sphere's center. With this in mind, we initially consider the grounded case again, for which we already know the distribution of the influenced charge q'. Now, we remove the grounding and place the rest charge $Q - q'$ onto the isolated sphere. Since the sphere has already reached an equilibrium state with respect to the forces between q' and q, the rest charge will be distributed uniformly over the sphere's surface. Inside V, the charge $Q - q'$ will therefore act in such a way as to be concentrated at the sphere's center.[14] Thus, inside V, an extra term has to be added to the potential of (2.86), resulting in

$$\phi(x)|_{x \in V} = \frac{q}{|xn - ae_x|} - \frac{Rq}{a\left|xn - \frac{R^2}{a}e_x\right|} + \frac{Q + \frac{Rq}{a}}{|x|} \ .$$

[14] This can be shown in complete analogy to the calculation of the gravitational potential of a hollow sphere in Subsection 1.5.3 with the masses replaced by charges and the gravitational force by the Coulomb force.

Metal sphere in a homogeneous electric field, Laplace equation in spherical coordinates. Before tackling the problem, we turn briefly to the solution of Laplace's equation in spherical coordinates. This is useful whenever dealing with spherically symmetric problems, where no charges are present inside the volume V. In spherical coordinates, the Laplace equation is

$$\frac{1}{r}\frac{\partial^2}{\partial r^2}(r\phi) + \frac{1}{r^2 \sin\theta}\frac{\partial}{\partial\theta}\left(\sin\theta\frac{\partial\phi}{\partial\theta}\right) + \frac{1}{r^2 \sin^2\theta}\frac{\partial^2\phi}{\partial\varphi^2} = 0 .$$

Using the ansatz

$$\phi(\boldsymbol{x}) = \frac{U(r)}{r}P(\theta)Q(\varphi) ,$$

it follows that

$$PQ\frac{d^2U}{dr^2} + \frac{UQ}{r^2 \sin\theta}\frac{d}{d\theta}\left(\sin\theta\frac{dP}{d\theta}\right) + \frac{UP}{r^2 \sin^2\theta}\frac{d^2Q}{d\varphi^2} = 0 .$$

Multiplying this equation by $r^2 \sin^2\theta/(UPQ)$ yields

$$r^2 \sin^2\theta\left[\frac{1}{U}\frac{d^2U}{dr^2} + \frac{1}{r^2 \sin\theta P}\frac{d}{d\theta}\left(\sin\theta\frac{dP}{d\theta}\right)\right] + \frac{1}{Q}\frac{d^2Q}{d\varphi^2} = 0 . \tag{2.87}$$

The last term has only a φ-dependency and is therefore constant:

$$\frac{1}{Q}\frac{d^2Q}{d\varphi^2} = -m^2 = \text{const} .$$

Similar considerations lead to two separate equations in U and P:

$$\left.\begin{array}{l} \dfrac{d^2U}{dr^2} - \dfrac{l(l+1)}{r^2}U = 0 , \quad l = \text{const} \\[12pt] \dfrac{1}{\sin\theta}\dfrac{d}{d\theta}\left(\sin\theta\dfrac{dP}{d\theta}\right) + \left[l(l+1) - \dfrac{m^2}{\sin^2\theta}\right]P = 0 . \end{array}\right\} \tag{2.88}$$

Without proof, we assert that, for physically meaningful solutions, the constants l and m can only take on the integer values

$$l = 0, 1, 2, \dots , \quad m = -l, -l+1, \dots, l-1, l .$$

The solutions of (2.88) are given by

$$U(r) = Ar^{l+1} + Br^{-l} , \quad Q(\varphi) = e^{\pm im\varphi} .$$

Substituting $x = \cos\theta$ in (2.87) leads to the *Legendre differential equation* (see Section A.6)

$$\frac{d}{dx}\left((1-x^2)\frac{dP}{dx}\right) + \left(l(l+1) - \frac{m^2}{1-x^2}\right)P = 0 ,$$

which is solved by the *Legendre functions* $P_{l,m}(x)$. Overall, we obtain

Theorem 2.23: Solution of Laplace's equation in spherical coordinates

The solution of Laplace's equation $\nabla^2\phi(\boldsymbol{x}) = 0$ in spherical coordinates is given by

$$\phi(\boldsymbol{x}) = \sum_{l=0}^{\infty} \sum_{m=-l}^{l} \left[A_{lm}r^l + B_{lm}r^{-l-1} \right] e^{im\varphi} P_{l,m}(\cos\theta) .$$

If the problem has azimuthal symmetry (no φ-dependency), then $m = 0$ and

$$\phi(\boldsymbol{x}) = \sum_{l=0}^{\infty} \left[A_l r^l + B_l r^{-l-1} \right] P_l(\cos\theta) . \tag{2.89}$$

$P_l = P_{l,0}$ denote the *Legendre polynomials*.

After these initial considerations, we now turn to the following problem: what is the potential outside a grounded metal sphere that is placed in a homogeneous electric field $\boldsymbol{E}_0 = E_0\boldsymbol{e}_z$? If we put the origin of our coordinate system at the sphere's center this problem becomes azimuthally symmetric, so that (2.89) can be applied. The coefficients A_l and B_l are determined by the following boundary conditions:

- At $z \to \infty$, the electric field is equal to the original homogeneous field:

$$\phi(z \to \infty) = -E_0 r \cos\theta \implies A_1 = -E_0 , \ A_{l\neq 1} = 0 .$$

- The potential vanishes on the sphere's surface:

$$\phi(R,\theta) = 0 \implies B_1 = E_0 R^3 , \ B_{l\neq 1} = 0 .$$

Therefore, we have

$$\phi(r,\theta) = \begin{cases} E_0 \cos\theta \left(-r + \dfrac{R^3}{r^2} \right) & \text{for } r > R \\ 0 & \text{for } r < R . \end{cases}$$

The corresponding electric field for $x = 0$ is displayed in Fig. 2.11. For the influenced surface charge density, we obtain

$$\sigma(\theta) = -\frac{1}{4\pi} \frac{\partial\phi}{\partial r}\bigg|_{r=R} = \frac{3}{4\pi} E_0 \cos\theta .$$

The integral of σ over the whole surface of the sphere vanishes, which means that no charge is influenced overall. Hence, there is no need to distinguish between a grounded and an isolated sphere in this type of problem.

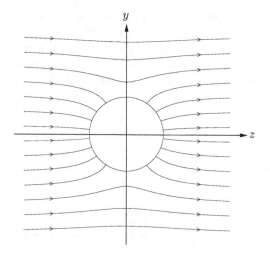

Fig. 2.11. Electric field resulting from a metal sphere placed in a field that was originally homogeneous in the z-direction.

2.5.5 Field Distributions in Magnetostatics

In magnetostatics, we first consider the case of boundary problems similar to those of electrostatics. To this end, we assume that no currents are present inside the volume V. Therefore, we have $\nabla \times \boldsymbol{B} = \boldsymbol{0}$, and \boldsymbol{B} can be written as the gradient of a scalar field:

$$\boldsymbol{B}(\boldsymbol{x}) = \nabla\psi(\boldsymbol{x}) \ , \ \boldsymbol{j}(\boldsymbol{x}) = \boldsymbol{0} \ , \ \boldsymbol{x} \in V \ .$$

From this and $\nabla \boldsymbol{B} = 0$, we again find that the Laplace equation holds inside V,

$$\nabla^2\psi(\boldsymbol{x}) = 0 \ .$$

Given certain Dirichlet or Neumann boundary conditions, the solutions to this equation can, in principle, be constructed in the same way as discussed in the previous two subsections.

Within electrostatics, we have assumed that the confining area of V is a conductor, so that, inside it, the electrostatic field vanishes (if we ignore the electric fields stemming from the bounded charges within the conductor crystal and compensating each other on a macroscopic level). In magnetostatics, the analogous situation would be a boundary surface, inside which the magnetostatic field vanishes. However, this scenario is unrealistic, since the bounded charges are associated with electric and magnetic dipoles, resulting in a nonvanishing electric current on a macroscopic level. In turn, this current produces a nonvanishing \boldsymbol{B} field inside the confining material. Therefore, the determination of boundary conditions in magnetostatics requires a discussion of *polarization* and *magnetization* of matter, which is the subject of the next section.

Magnetic field of a current-carrying coil. We now turn our attention to a typical magnetostatic field distribution problem, the determination of the magnetic field of an (infinitely) long current-carrying coil. To prepare, we consider a circular thin wire with radius R, through which a constant electric current I flows (Fig. 2.12, left). We ask for the induced magnetic field \boldsymbol{B} far

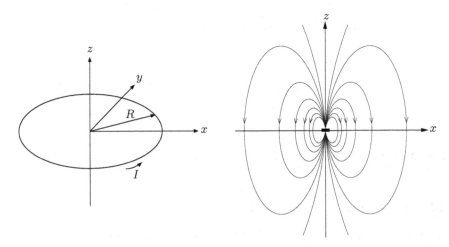

Fig. 2.12. Circular current-carrying conducting loop (*left*) and its magnetic dipole field (*right*).

away from the conductor, as well as at an arbitrary point on the z-axis. To answer the first question, we can utilize Theorem 2.19 since the conducting loop behaves like a magnetic dipole for large distances $|\boldsymbol{x}| \gg R$. Using

$$\mathrm{d}^3 x' \boldsymbol{j}(\boldsymbol{x}') = I t(s)\mathrm{d}s \; , \; t(s) = \frac{\mathrm{d}\boldsymbol{l}(s)}{\mathrm{d}s} = R \begin{pmatrix} -\sin s \\ \cos s \\ 0 \end{pmatrix} \; , \; \boldsymbol{l}(s) = R \begin{pmatrix} \cos s \\ \sin s \\ 0 \end{pmatrix} ,$$

we evaluate this dipole to be

$$\boldsymbol{\mu} = \frac{I}{2c} \int_0^{2\pi} \mathrm{d}s\, \boldsymbol{l}(s) \times t(s) = \frac{\pi I R^2}{c} \begin{pmatrix} 0 \\ 0 \\ 1 \end{pmatrix} .$$

The magnetic dipole field is

$$\boldsymbol{B}(\boldsymbol{x}) = \frac{I\pi R^2}{c} \frac{3z\boldsymbol{x} - \boldsymbol{e}_z|\boldsymbol{x}|^2}{|\boldsymbol{x}|^5} \; , \; |\boldsymbol{x}| \gg R \; .$$

Its shape is depicted on the right-hand side of Fig. 2.12. The second question can be easily solved using Biot-Savart's law, Theorem 2.16:

$$\boldsymbol{B}(0,0,z) = \frac{I}{c}\int_{0}^{2\pi} \mathrm{d}s \boldsymbol{t}(s) \times \frac{z\boldsymbol{e}_z - \boldsymbol{l}(s)}{|z\boldsymbol{e}_z - \boldsymbol{l}(s)|^3} = \frac{2\pi I R^2}{c\sqrt{R^2 + z^2}^3}\begin{pmatrix} 0 \\ 0 \\ 1 \end{pmatrix}. \qquad (2.90)$$

Now we consider a current-carrying coil of length L. This is wound around the z-axis and for simplicity is considered to be composed of N circular conductors placed on top of each other (Fig. 2.13, left). Looking at the right-

Fig. 2.13. *Left:* idealized coil composed of circular conducting segments placed on top of each other. *Right:* determination of the integration path for the evaluation of the magnetic field strength inside the coil.

hand side of Fig. 2.12, it is intuitively clear that inside the coil and in the limit $L \to \infty$, a magnetic field emerges in the z-direction, whereas the field outside the coil vanishes. Moreover, we assert (without proof) that the inner field is homogeneous. Its value at the origin can be obtained by inserting the z-coordinates of the individual conducting loops into (2.90) and summing up these contributions. If the conducting loops lie close together, the sum can be replaced by an integral,

$$\sum \longrightarrow n\int \mathrm{d}z \ , \ n = \frac{N}{L} \ , \ N = \text{winding number,}$$

so that

$$\boldsymbol{B}(0) = \frac{2\pi I R^2 n}{c}\begin{pmatrix} 0 \\ 0 \\ 1 \end{pmatrix}\int_{-L/2}^{L/2}\frac{\mathrm{d}z}{\sqrt{R^2 + z^2}^3} = \frac{2\pi I L n}{c\sqrt{R^2 + \frac{L^2}{4}}}\begin{pmatrix} 0 \\ 0 \\ 1 \end{pmatrix}$$

$$\overset{L\to\infty}{=} \frac{4\pi I n}{c}\begin{pmatrix} 0 \\ 0 \\ 1 \end{pmatrix} = \boldsymbol{B}(\sqrt{x^2 + y^2} < R, z) \ . \qquad (2.91)$$

Note that we also arrive at this result with the help of Ampère's law (2.4). If we choose the integration path as indicated on the right-hand side of Fig. 2.13, it encompasses N_l windings along the partial length l, i.e., the current

I flows N_l times through the integration domain. Neglecting the external magnetic field, we find

$$\oint \mathbf{B}dl = |\mathbf{B}|l = \frac{4\pi I N_l}{c} \implies |\mathbf{B}| = \frac{4\pi I N_l}{cl} = \frac{4\pi I n}{c} .$$

Summary

- For the case of time-independent charge and current density distributions, Maxwell's equations decompose into two decoupled differential equation systems. They are the basic equations of **electrostatics** and **magnetostatics**.

- The relationships of previous sections can be carried over to the static case by eliminating all time dependencies. The Coulomb and Lorentz gauges are identical and lead to the **static Poisson equation** in the scalar and vector potentials.

- If the static charge and current density distributions are spatially confined, then, for large distances, the corresponding scalar and vector potentials can be expanded in powers of $1/|\mathbf{x}|$ (**multipole expansion**). Contrary to the scalar potential, the vector potential has no monopole part, in accordance with (III).

- Within electrostatics, one usually distinguishes between two types of boundary problems, called the **Dirichlet** and **Neumann boundary conditions**. They correspond to different continuity conditions of the E field at (conducting) boundary surfaces. Each of these boundary conditions leads to a unique solution that can be formally determined using Green function calculus. From this calculus, one obtains the **method of mirror charges**, which is often of great practical use.

- Due to the permanent polarization and magnetization of matter, magnetostatic boundary problems can be discussed only with a detailed knowledge of the confining material.

- Inside a long current-carrying coil, the induced magnetic field is proportional to the current strength, as well as to the coil's winding density.

Applications

28. Multipole expansion in spherical representation. What is the multipole expansion of the electrostatic potential ϕ in spherical representation, and how are the individual spherical moments related to the Cartesian ones?

Solution. Using spherical coordinates

$$\boldsymbol{x} = r \begin{pmatrix} \cos\varphi\sin\theta \\ \sin\varphi\sin\theta \\ \cos\theta \end{pmatrix} , \quad \boldsymbol{x}' = r' \begin{pmatrix} \cos\varphi'\sin\theta' \\ \sin\varphi'\sin\theta' \\ \cos\theta' \end{pmatrix} ,$$

we have according to (A.15)

$$\frac{1}{|\boldsymbol{x}-\boldsymbol{x}'|} = \sum_{l=0}^{\infty} \sum_{m=-l}^{l} \frac{4\pi}{2l+1} \frac{r'^{l}}{r^{l+1}} Y^*_{l,m}(\theta',\varphi') Y_{l,m}(\theta,\varphi) ,$$

which, in conjunction with (2.71), yields the following representation of the scalar potential:

$$\phi(\boldsymbol{x}) = \sum_{l=0}^{\infty} \sum_{-l}^{l} \frac{4\pi}{2l+1} \frac{q_{l,m}}{r^{l+1}} Y_{l,m}(\theta,\varphi) .$$

The expansion coefficients

$$q_{l,m} = \int d^3x' r'^{l} \rho(\boldsymbol{x}') Y^*_{l,m}(\theta',\varphi')$$

are the electric multipole moments in spherical representation. In the case of a real charge density, the moments with $m < 0$ are related, using (A.13), via $q_{l,-m} = (-1)^m q^*_{l,m}$ to the corresponding moments with $m > 0$. Expressing the first few moments in Cartesian coordinates,

$$q_{0,0} = \frac{1}{4\pi} \int d^3x' \rho(\boldsymbol{x}') = \frac{1}{4\pi} Q$$

$$q_{1,1} = -\sqrt{\frac{3}{8\pi}} \int d^3x' (x'-iy')\rho(\boldsymbol{x}') = -\sqrt{\frac{3}{8\pi}} (p_x - i p_y)$$

$$q_{1,0} = \sqrt{\frac{3}{4\pi}} \int d^3x' z' \rho(\boldsymbol{x}') = \sqrt{\frac{3}{4\pi}} p_z ,$$

we see the relationship between spherical and Cartesian moments: the spherical $l=0$-moment corresponds to the Cartesian monopole moment, the spherical $l=1$-moments contain the Cartesian dipole moments, and so on.

29. Capacity of a plate condenser. Consider a condenser consisting of two conducting plates of area F placed at distance d parallel to each other (Fig. 2.14). One of the plates has charge Q and the other $-Q$. If the electric field outside the condenser can be neglected, what is the potential difference (voltage) between the plates?

Solution. At both condenser plates, we have the boundary conditions

$$\boldsymbol{e}_x \boldsymbol{E}(x=0) = \boldsymbol{e}_x \boldsymbol{E}(x=d) = \frac{4\pi Q}{F} \implies \frac{\partial\phi}{\partial x}\Big|_{x=0} = \frac{\partial\phi}{\partial x}\Big|_{x=d} = -\frac{4\pi Q}{F} .$$

Obviously, the ansatz

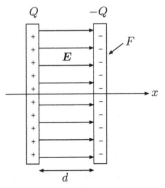

Fig. 2.14. Charged plate condenser.

$$\phi(x) = -\frac{4\pi Q}{F}x \implies E(x) = \frac{4\pi Q}{F}\begin{pmatrix} 1 \\ 0 \\ 0 \end{pmatrix}$$

obeys these conditions (and $\nabla^2\phi = 0$ inside the condenser). From this, the voltage follows as

$$V = \frac{4\pi Q}{F}\int\limits_0^d dx = -[\phi(x=d) - \phi(x=0)] = \frac{Q}{C}, \quad C = \frac{F}{4\pi d}, \tag{2.92}$$

where C denotes the *capacity* of the condenser. This quantity depends only on the condenser's geometry and describes its capacity of charge at a given voltage. Even if the condenser's charges vary in time, (2.92) remains valid under certain circumstances, namely if the induction term $\frac{1}{c}\frac{\partial B}{\partial t}$ in (II) can be neglected (quasi-static approximation). In this case, (I) and (II) are identical in form to the electrostatic equations (2.69).

30. Self-inductivity of a coil. Consider a cylindrically winded coil with winding number N and length L, through which a temporally varying current $I(t)$ flows. Using Faraday's induction law, calculate the induced voltage at the ends of the coil in the quasi-static approximation.

Solution. If the displacement current $\frac{1}{c}\frac{\partial E}{\partial t}$ can be neglected in (IV) (quasi-static approximation), then (III) and (IV) are form-identical to the magnetostatic equations (2.70). In this approximation, due to (2.91), we have for the magnetic field inside the coil

$$B(t) = \frac{4\pi I(t)N}{cL}\begin{pmatrix} 0 \\ 0 \\ 1 \end{pmatrix}.$$

The voltage at the ends of the coil equals the N-fold of the voltage, which is produced inside one single conducting loop by the magnetic field. Therefore, in conjunction with (2.2), it follows that

$$V(t) = -\frac{N}{c} \int\limits_{F} \mathrm{d}\boldsymbol{F} \frac{\partial B(t)}{\partial t} = -S \frac{\mathrm{d}I}{\mathrm{d}t} \; , \; S = \frac{4\pi^2 R^2 N^2}{c^2 L} \; .$$

The quantity S is called *self-inductivity* and depends only on the coil's geometry.

2.6 Electrodynamics in Matter

Maxwell's equations and the Lorentz force law are fundamental laws of nature with a large range of validity. In particular, they are also valid in matter. In this case, the fundamental microscopic fields \boldsymbol{E} and \boldsymbol{B} are induced by freely moving charges, as well as by the intrinsic dipole moments of bound particles (such as electrons, protons, and neutrons). However, due to the large number of particles in matter ($\approx 10^{23}$), Maxwell's equations (I) to (IV) are practically unsolvable. If one is interested in electromagnetic phenomena in matter on a macroscopic length scale, it is sensible to assume that a detailed knowledge of the charge and current density distributions, as well as of the fields \boldsymbol{E} and \boldsymbol{B}, is not necessary. Of interest are only the macroscopic quantities averaged over many atomic length units.

This section deals with the formulation of macroscopic electrodynamics in matter. We show how to describe macroscopic phenomena by using the *macroscopic Maxwell equations*, where the fundamental microscopic distributions and fields are replaced by suitably chosen spatially averaged expressions. These equations describe the properties of matter in terms of *polarization* and *magnetization*, and we will see that, in many cases, a linear relation can be found empirically between these quantities and the macroscopic fields. At the end of this section, we discuss the continuity conditions for macroscopic fields at boundaries.

2.6.1 Macroscopic Maxwell Equations

To derive a macroscopic formulation of electrodynamics, we first need a suitable spatial averaging procedure for the microscopic quantities ρ, \boldsymbol{j}, \boldsymbol{E}, and \boldsymbol{B}. A meaningful ansatz is given by

Definition: Macroscopic average $\langle \cdots \rangle$

The macroscopic average $\langle G(\boldsymbol{x}, t) \rangle$ of an arbitrary field $G(\boldsymbol{x}, t)$ is given by its convolution with a suitable function $f(\boldsymbol{x})$,

$$\langle G(\boldsymbol{x}, t) \rangle = \int \mathrm{d}^3 x' G(\boldsymbol{x}', t) f(\boldsymbol{x} - \boldsymbol{x}') \; ,$$

where f possesses the following properties:

\triangleright

- $f(x)$ is localized at $x = 0$.

- If a denotes the microscopic length scale (such as atomic distance) and b is the macroscopic length scale (for example, 1 cm), the width Δ of f satisfies $a \ll \Delta \ll b$.

This definition guarantees that, on the one hand, atomic fluctuations are averaged out and, on the other hand, the actually interesting spatial dependencies on the macroscopic level remain unaffected. Furthermore, the averaging procedure commutes with partial derivatives. For the temporal derivative this is obvious. For the spatial ones it follows from

$$
\begin{aligned}
\frac{\partial \langle G(x,t) \rangle}{\partial x} &= \int d^3 x' G(x',t) \frac{\partial f(x - x')}{\partial x} \\
&= -\int d^3 x' G(x',t) \frac{\partial f(x - x')}{\partial x'} \\
&= \int d^3 x' \frac{\partial G(x',t)}{\partial x'} f(x - x') \\
&= \left\langle \frac{\partial G(x,t)}{\partial x} \right\rangle .
\end{aligned}
$$

By applying this averaging procedure to the homogeneous Maxwell equations (II) and (III), we obtain the macroscopic equations

$$
\nabla \times \langle E(x,t) \rangle + \frac{1}{c} \frac{\partial \langle B(x,t) \rangle}{\partial t} = 0 , \quad \nabla \langle B(x,t) \rangle = 0 .
$$

The corresponding averaging of the inhomogeneous Maxwell equations (I) and (IV) requires some extra considerations: as noted in this section's introduction, the charges in matter are made up of two parts. One part results from the particles bound within atoms and molecules, the latter being electrically neutral overall. The second part is due to charges, which can move almost freely within the crystal lattice. For this reason, we can split both the charge and current density into a bound neutral (nt) and a free (fr) part:

$$
\rho(x,t) = \rho_{\text{nt}}(x,t) + \rho_{\text{fr}}(x,t) , \quad \rho_{\text{nt}}(x,t) = \sum_i \rho_i [x - x_i(t), t] \tag{2.93}
$$

$$
j(x,t) = j_{\text{nt}}(x,t) + j_{\text{fr}}(x,t) , \quad j_{\text{nt}}(x,t) = \sum_i j_i [x - x_i(t), t] . \tag{2.94}
$$

Herein, the component ρ_i of ρ_{nt} denotes the charge density of a neutral unit at location x_i, for example, the charge density of the nucleus and the electron cloud of an overall neutral atom. For this, we have

$$
\int d^3 x' \rho_i(x',t) = 0 .
$$

Since bound as well as free charges are conserved quantities, we have two continuity equations:

$$\frac{\partial \rho_{\mathrm{fr}}(\boldsymbol{x},t)}{\partial t} + \boldsymbol{\nabla} \boldsymbol{j}_{\mathrm{fr}}(\boldsymbol{x},t) = 0 \; , \quad \frac{\partial \rho_{\mathrm{nt}}(\boldsymbol{x},t)}{\partial t} + \boldsymbol{\nabla} \boldsymbol{j}_{\mathrm{nt}}(\boldsymbol{x},t) = 0 \; . \qquad (2.95)$$

Expanding the convolutional function f and keeping terms only up to first order give

$$f(\boldsymbol{x} - \boldsymbol{x}') \approx f(\boldsymbol{x}) + \boldsymbol{x}' \boldsymbol{\nabla}' \, f(\boldsymbol{x} - \boldsymbol{x}')|_{\boldsymbol{x}'=0} = f(\boldsymbol{x}) - \boldsymbol{x}' \boldsymbol{\nabla} f(\boldsymbol{x}) \; ,$$

while the averaging of ρ_{nt} yields

$$
\begin{aligned}
\langle \rho_{\mathrm{nt}}(\boldsymbol{x},t) \rangle &= \sum_i \int \mathrm{d}^3 x' \rho_i(\boldsymbol{x}' - \boldsymbol{x}_i) f(\boldsymbol{x} - \boldsymbol{x}') \\
&= \sum_i \int \mathrm{d}^3 x'' \rho_i(\boldsymbol{x}'') f(\boldsymbol{x} - \boldsymbol{x}_i - \boldsymbol{x}'') \\
&= \sum_i f(\boldsymbol{x} - \boldsymbol{x}_i) \underbrace{\int \mathrm{d}^3 x'' \rho_i(\boldsymbol{x}'')}_{0} \\
&\quad - \sum_i \boldsymbol{\nabla} f(\boldsymbol{x} - \boldsymbol{x}_i) \int \mathrm{d}^3 x'' \boldsymbol{x}'' \rho_i(\boldsymbol{x}'') \\
&= - \sum_i \boldsymbol{p}_i \boldsymbol{\nabla} f(\boldsymbol{x} - \boldsymbol{x}_i) \; , \quad \boldsymbol{p}_i = \int \mathrm{d}^3 x'' \boldsymbol{x}'' \rho_i(\boldsymbol{x}'') \; ,
\end{aligned}
$$

where \boldsymbol{p}_i is the electric dipole moment of the neutral charge density ρ_i. Further manipulation leads to

$$
\begin{aligned}
\langle \rho_{\mathrm{nt}}(\boldsymbol{x},t) \rangle &= - \sum_i \boldsymbol{p}_i \boldsymbol{\nabla} f(\boldsymbol{x} - \boldsymbol{x}_i) \\
&= - \boldsymbol{\nabla} \int \mathrm{d}^3 x'' \sum_i \boldsymbol{p}_i \delta(\boldsymbol{x}'' - \boldsymbol{x}_i) f(\boldsymbol{x} - \boldsymbol{x}'') \\
&= - \boldsymbol{\nabla} \langle \boldsymbol{P}(\boldsymbol{x},t) \rangle \; . \qquad (2.96)
\end{aligned}
$$

Here,

$$\langle \boldsymbol{P}(\boldsymbol{x},t) \rangle = \left\langle \sum_i \boldsymbol{p}_i \delta(\boldsymbol{x} - \boldsymbol{x}_i) \right\rangle = \frac{\text{mean electric dipole moment}}{\text{volume}}$$

denotes the *mean polarization*, i.e., the mean density of the electric dipole moment of the neutral charge density. We are now in a position to apply the averaging procedure to the first Maxwell equation (I). Using (2.93) and (2.96), we find

$$\boldsymbol{\nabla} \langle \boldsymbol{E}(\boldsymbol{x},t) \rangle = 4\pi \langle \rho_{\mathrm{fr}}(\boldsymbol{x},t) \rangle - 4\pi \boldsymbol{\nabla} \langle \boldsymbol{P}(\boldsymbol{x},t) \rangle \; .$$

In order to apply the averaging to the remaining Maxwell equation (IV), a calculation for $\langle \boldsymbol{j}_{\mathrm{nt}} \rangle$ is carried out, similar to that for $\langle \rho_{\mathrm{nt}} \rangle$. We again consider the linear approximation of the convolutional function f, so that

$$\langle j_{\mathrm{nt}}(\boldsymbol{x},t)\rangle = \sum_i \int \mathrm{d}^3 x' \, j_i(\boldsymbol{x}' - \boldsymbol{x}_i) f(\boldsymbol{x} - \boldsymbol{x}')$$

$$= \sum_i \int \mathrm{d}^3 x'' \, j_i(\boldsymbol{x}'') f(\boldsymbol{x} - \boldsymbol{x}_i - \boldsymbol{x}'')$$

$$= \sum_i f(\boldsymbol{x} - \boldsymbol{x}_i) \int \mathrm{d}^3 x'' \, j_i(\boldsymbol{x}'')$$

$$- \sum_i \int \mathrm{d}^3 x'' \, j_i(\boldsymbol{x}'')(\boldsymbol{x}'' \boldsymbol{\nabla}) f(\boldsymbol{x} - \boldsymbol{x}_i) \; . \tag{2.97}$$

Using $\boldsymbol{j}_i = (\boldsymbol{j}_i \boldsymbol{\nabla}'')\boldsymbol{x}''$ and (2.95), we can rewrite the first term as

$$\sum_i f(\boldsymbol{x} - \boldsymbol{x}_i) \int \mathrm{d}^3 x'' \, j_i(\boldsymbol{x}'') = \sum_i f(\boldsymbol{x} - \boldsymbol{x}_i) \int \mathrm{d}^3 x'' \, (\boldsymbol{j}_i \boldsymbol{\nabla}'')\boldsymbol{x}''$$

$$= -\sum_i f(\boldsymbol{x} - \boldsymbol{x}_i) \int \mathrm{d}^3 x'' \, (\boldsymbol{\nabla}'' \boldsymbol{j}_i)\boldsymbol{x}''$$

$$= \sum_i f(\boldsymbol{x} - \boldsymbol{x}_i) \int \mathrm{d}^3 x'' \, \frac{\partial \rho_i}{\partial t} \boldsymbol{x}''$$

$$= \frac{\partial \langle \boldsymbol{P}(\boldsymbol{x},t)\rangle}{\partial t} \; .$$

To simplify the second term in (2.97), we initially concentrate on a single term of the sum and a single component $(\boldsymbol{j}_i)_k = j_k$. Ignoring possible quadrupole moments, we find

$$\int \mathrm{d}^3 x'' \, j_k(\boldsymbol{x}'' \boldsymbol{\nabla}) = \int \mathrm{d}^3 x'' \, j_k \sum_l x_l'' \partial_l$$

$$= \sum_{l,n} \int \mathrm{d}^3 x'' \, x_l'' \partial_n''(x_k'' j_n) \partial_l - \underbrace{\sum_l \int \mathrm{d}^3 x'' \, x_l'' x_k''(\boldsymbol{\nabla}'' \boldsymbol{j})\, \partial_l}_{\text{quadrupole} \approx 0}$$

$$= -\sum_l \int \mathrm{d}^3 x'' \, x_k'' j_l \partial_l \; .$$

If we use the integral relation

$$\int \mathrm{d}^3 x'' \, g(\boldsymbol{x}'') \boldsymbol{\nabla}'' \boldsymbol{j}(\boldsymbol{x}'') = -\int \mathrm{d}^3 x'' \, [\boldsymbol{\nabla}'' g(\boldsymbol{x}'')] \boldsymbol{j}(\boldsymbol{x}'') \; ,$$

it follows further that

$$g(\boldsymbol{x}'') = x_k'' x_l'' \implies \int \mathrm{d}^3 x'' \, (x_l'' j_k + x_k'' j_l) = \underbrace{-\int \mathrm{d}^3 x'' \, x_k'' x_l'' \boldsymbol{\nabla}'' \boldsymbol{j}}_{\text{quadrupole} \approx 0}$$

$$\implies \int d^3x'' j_k(x''\nabla) = \frac{1}{2}\sum_l \int d^3x''(x_l'' j_k \partial_l - x_k'' j_l \partial_l)$$

$$= \frac{1}{2}\left(\int d^3x''[x'' \times j(x'')] \times \nabla\right)_k .$$

Consequently, the second term in (2.97) can finally be written as

$$-\sum_i \int d^3x'' j_i(x'')(x''\nabla)f(x - x_i) = -c\sum_i \mu_i \times \nabla f(x - x_i) ,$$

where

$$\mu_i = \frac{1}{2c}\int d^3x'' x'' \times j_i(x'')$$

denotes the magnetic dipole moment of the neutral current j_i. Overall, we find the following expression for $\langle j_{\mathrm{nt}}\rangle$:

$$\langle j_{\mathrm{nt}}(x,t)\rangle = \frac{\partial \langle P(x,t)\rangle}{\partial t} + c\nabla \times \langle M(x,t)\rangle . \tag{2.98}$$

The term

$$\langle M(x,t)\rangle = \left\langle \sum_i \mu_i \delta(x - x_i)\right\rangle = \frac{\text{mean magnetic dipole moment}}{\text{volume}}$$

is the *mean magnetization*, i.e., the mean density of the magnetic dipole moments of the neutral currents. By using (2.94) and (2.98), the averaging procedure of Maxwell's equation (IV) yields

$$\nabla \langle B(x,t)\rangle - \frac{1}{c}\frac{\partial \langle E(x,t)\rangle}{\partial t} = \frac{4\pi}{c}\langle j_{\mathrm{fr}}(x,t)\rangle + \frac{4\pi}{c}\frac{\partial \langle P(x,t)\rangle}{\partial t}$$
$$+ 4\pi\nabla \times \langle M(x,t)\rangle .$$

We summarize:

Theorem 2.24: Macroscopic Maxwell equations

The macroscopic Maxwell equations in matter are

$$\nabla \langle D(x,t)\rangle = 4\pi \langle \rho_{\mathrm{fr}}(x,t)\rangle \tag{I'}$$

$$\nabla \times \langle E(x,t)\rangle + \frac{1}{c}\frac{\partial \langle B(x,t)\rangle}{\partial t} = 0 \tag{II'}$$

$$\nabla \langle B(x,t)\rangle = 0 \tag{III'}$$

$$\nabla \times \langle H(x,t)\rangle - \frac{1}{c}\frac{\partial \langle D(x,t)\rangle}{\partial t} = \frac{4\pi}{c}\langle j_{\mathrm{fr}}(x,t)\rangle , \tag{IV'}$$

where

▷

$$\langle D \rangle = \langle E \rangle + 4\pi \langle P \rangle \ , \ \langle P \rangle \begin{cases} = \dfrac{\text{mean electric dipole moment}}{\text{volume}} \\ = \text{mean polarization} \end{cases}$$

denotes the *electric displacement* or *electric induction* and

$$\langle H \rangle = \langle B \rangle - 4\pi \langle M \rangle \ , \ \langle M \rangle \begin{cases} = \dfrac{\text{mean magnetic dipole moment}}{\text{volume}} \\ = \text{mean magnetization} \end{cases}$$

denotes the *magnetic field strength*. $\langle P \rangle$ and $\langle M \rangle$ are related to the neutral charge and current density distributions $\langle \rho_{\mathrm{nt}} \rangle$ and $\langle j_{\mathrm{nt}} \rangle$ via

$$\langle \rho_{\mathrm{nt}}(x,t) \rangle = -\nabla \langle P(x,t) \rangle$$
$$\langle j_{\mathrm{nt}}(x,t) \rangle = \frac{\partial \langle P(x,t) \rangle}{\partial t} + c\nabla \times \langle M(x,t) \rangle \ .$$

The macroscopic Maxwell equations (I') to (IV') imply the macroscopic continuity equation

$$\frac{\partial \langle \rho_{\mathrm{fr}}(x,t) \rangle}{\partial t} + \nabla \langle j_{\mathrm{fr}}(x,t) \rangle = 0 \ .$$

For a better understanding of this theorem, consider the following points:

• The macroscopic Maxwell equations are written solely in terms of macroscopically averaged quantities.

• The electric induction vector $\langle D \rangle$ is composed of two parts. The first contribution is the mean electric field $\langle E \rangle$ (externally applied, as well as induced by the free charges). The second contribution is due to the mean polarization vector $\langle P \rangle$ of the medium, which generally depends on the external field $\langle E \rangle$.

• The magnetic field vector $\langle H \rangle$ is described by the mean magnetic field $\langle B \rangle$ (applied externally, as well as induced by the free currents) minus the mean magnetization $\langle M \rangle$ of the medium, which generally depends on the external field $\langle B \rangle$.[15]

• The macroscopic Maxwell equations have a similar structure to their microscopic counterparts (I) to (IV). However, in contrast to (I) to (IV), they are phenomenologically inspired rather than of a fundamental nature. In particular, they cannot be formulated Lorentz-covariantly but are only valid in the specific inertial system in which the medium is at rest on average.

[15] Note that there is a semantic inconsistency here: the fundamental magnetic field B is given the unfortunate name of "magnetic induction", whereas the matter field H is called "magnetic field". This is contrary to the naming conventions for E and D. It would have been better to have the reverse terminology for B and H.

Macroscopic Lorentz force. Of course, our averaging procedure can also be applied to the Lorentz force. This yields the mean force $\langle \boldsymbol{F}_{\mathrm{L}} \rangle$, which is exerted on a charge q by the mean fields $\langle \boldsymbol{E} \rangle$ and $\langle \boldsymbol{B} \rangle$:

$$\langle \boldsymbol{F}_{\mathrm{L}}(\boldsymbol{x},t) \rangle = q \left(\langle \boldsymbol{E}(\boldsymbol{x},t) \rangle + \frac{\dot{\boldsymbol{x}}}{c} \times \langle \boldsymbol{B}(\boldsymbol{x},t) \rangle \right) .$$

Macroscopic energy. As the macroscopic Lorentz force $\langle \boldsymbol{F}_{\mathrm{L}} \rangle$ is structurally similar to the microscopic force $\boldsymbol{F}_{\mathrm{L}}$, we can use the same derivation as in Subsection 2.1.3 to obtain the macroscopic energy balance equation

$$\frac{\partial \langle \epsilon_{\mathrm{mech}} \rangle}{\partial t} + \frac{\partial \langle \epsilon_{\mathrm{em}} \rangle}{\partial t} = -\boldsymbol{\nabla} \langle \boldsymbol{S} \rangle ,$$

with

$$\frac{\partial \langle \epsilon_{\mathrm{mech}} \rangle}{\partial t} = \langle \boldsymbol{j}_{\mathrm{fr}} \rangle \langle \boldsymbol{E} \rangle$$

$$\left. \begin{aligned} \frac{\partial \langle \epsilon_{\mathrm{em}} \rangle}{\partial t} &= \frac{\langle \boldsymbol{E} \rangle \langle \dot{\boldsymbol{D}} \rangle + \langle \boldsymbol{H} \rangle \langle \dot{\boldsymbol{B}} \rangle}{4\pi} \\ \langle \boldsymbol{S} \rangle &= \frac{c}{4\pi} \langle \boldsymbol{E} \rangle \times \langle \boldsymbol{H} \rangle . \end{aligned} \right\} \tag{2.99}$$

Notation. For macroscopic relationships, it is common practice to drop the averaging brackets as well as the index fr. In the following, we shall adopt this convention, keeping in mind the fact that the microscopic and macroscopic expressions have very different interpretations.

2.6.2 Material Equations

In order to be able to apply Maxwell's equations in matter, we need explicit expressions for the polarization and the magnetization in terms of the macroscopic electromagnetic fields. Clearly, these relations will depend mainly on the microscopic structure of the material under consideration and are therefore found in the framework of quantum mechanics and quantum statistics. However, even within electrodynamics, we can make some phenomenologically motivated attempts, whose validity can be verified experimentally. If we restrict ourselves to isotropic (no preferred direction in space) and homogeneous (no preferred location in space) materials, a meaningful ansatz is[16]

$$\boldsymbol{P}(\boldsymbol{x},t) = \boldsymbol{P}_0 + \int \mathrm{d}^3 x' \mathrm{d}t' \alpha(|\boldsymbol{x} - \boldsymbol{x}'|, t - t') \boldsymbol{E}(\boldsymbol{x}',t')$$

$$\boldsymbol{M}(\boldsymbol{x},t) = \boldsymbol{M}_0 + \int \mathrm{d}^3 x' \mathrm{d}t' \beta(|\boldsymbol{x} - \boldsymbol{x}'|, t - t') \boldsymbol{B}(\boldsymbol{x}',t') ,$$

[16] For \boldsymbol{P} (or \boldsymbol{M}, respectively), a term proportional to \boldsymbol{B} (or \boldsymbol{E}, respectively) is not possible due to their conflicting behavior under a parity transformation.

where α and β are scalar functions of time and space. In most cases, the space dependency of α and β is relevant only on microscopic length scales (a few atomic length units) and, therefore, is much smaller than the macroscopic scale considered here. Neglecting this dependency, we have

$$
\left.
\begin{aligned}
\boldsymbol{P}(\boldsymbol{x},t) &= \boldsymbol{P}_0 + \int \mathrm{d}t'\,\alpha(t-t')\boldsymbol{E}(\boldsymbol{x},t') \\
\boldsymbol{M}(\boldsymbol{x},t) &= \boldsymbol{M}_0 + \int \mathrm{d}t'\,\beta(t-t')\boldsymbol{B}(\boldsymbol{x},t')\ .
\end{aligned}
\right\}
\tag{2.100}
$$

This ansatz honors the phenomenon of *hysteresis*, which is observed in ferro-electric and ferromagnetic materials, where the polarization and magnetization depend on the "prehistory" of the applied \boldsymbol{E} and \boldsymbol{B} fields. If we disregard these types of material, it turns out that for a large range of field strengths and even for alternating fields of not-too-high frequency, the ansatz (2.100) can be replaced by the following linear relationships between \boldsymbol{P} and \boldsymbol{E} and between \boldsymbol{M} and \boldsymbol{B}:

$$
\boldsymbol{P} = \chi_{\mathrm{e}}\boldsymbol{E}\ ,\quad \boldsymbol{M} = \chi_{\mathrm{m}}\boldsymbol{B}\ .
\tag{2.101}
$$

Here, χ_{e} and χ_{m} respectively denote the *electric susceptibility* and the *magnetic susceptibility* of the material. Using this, we find simple relations between the fields \boldsymbol{E} and \boldsymbol{D} as well as \boldsymbol{B} and \boldsymbol{H}:

$$
\left.
\begin{aligned}
\boldsymbol{D} &= \epsilon\boldsymbol{E}\ ,\quad \epsilon = 1 + 4\pi\chi_{\mathrm{e}} \\
\boldsymbol{B} &= \mu\boldsymbol{H}\ ,\quad \mu = 1 + 4\pi\chi_{\mathrm{m}}\ .
\end{aligned}
\right\}
\tag{2.102}
$$

The material constants ϵ and μ are respectively called the *dielectric constant* and the *permeability constant*. In the following, we disregard more complicated functional dependencies and concentrate solely on the linear relations (2.101) and (2.102). Generally, one distinguishes between the following electric materials:

- *Dielectric:* electric dipoles are induced by applying an external electric field. They result from displacements of the electron clouds relative to the atomic nuclei, as well as from the displacement of positive ions relative to the negative ones.

- *Paraelectric:* these materials have permanent electric dipoles. In the absence of electric fields, these dipoles are distributed randomly due to the heat-induced motion. They become aligned by applying an external electric field. Contrary to dielectric materials, the electric susceptibility χ_{e} is temperature-dependent.

- *Ferroelectric:* these materials also have permanent electric dipoles, which become aligned by applying an external electric field. With growing field strength, the polarization grows very rapidly, reaches a saturation value, and does not vanish when the field is switched off. The polarization vanishes completely only when an opposite field is applied (electric hysteresis).

Above a critical temperature this behavior disappears and the material becomes paraelectric.

For dielectric and paramagnetic substances, we always have $\chi_e > 0$ and $\epsilon > 0$. Similar to electric materials, magnetic materials are classified as follows:

- *Diamagnetic:* these substances are characterized by $\chi_m < 0$. Magnetic dipoles are induced by applying an external magnetic field. According to Lenz's rule, these dipoles are oriented in such a way that they oppose the external magnetic field (therefore, $\chi_m < 0$).

- *Paramagnetic:* paramagnets have permanent magnetic dipoles, which are aligned in the presence of an external magnetic field. In the absence of magnetic fields, the dipoles are randomly distributed. Contrary to diamagnetic materials, the magnetic susceptibility χ_m is positive and temperature-dependent.

- *Ferromagnetic:* ferromagnetic substances have permanent magnetic dipoles, which are also aligned when an external magnetic field is applied. The relationship between magnetization and magnetic field is not linear; similarly to ferroelectric materials, there exist magnetic hysteresis effects. Above a critical temperature the ferromagnetic properties disappear and the material becomes paramagnetic.

2.6.3 Continuity Conditions at Boundaries

We shall now investigate the behavior of the macroscopic fields E, B, D, and H at a boundary surface that separates two different media 1 and 2. To this end, we can adopt the arguments of Subsection 2.5.3 by placing a volume element ΔV or surface element ΔF of height h into the separating boundary plane (Fig. 2.15). By applying Gauss's law to (I') and (III'), and taking the limit $h \to 0$, we obtain

$$(D_2 - D_1)n = 4\pi\sigma \, , \, \sigma = \frac{\delta q}{\delta F}$$
$$(B_2 - B_1)n = 0 \, ,$$

Fig. 2.15. Integration domains at a surface separating two different media.

where n is the normal vector pointing into medium 2 and σ is the surface charge density of the boundary plane. Applying Stokes's law to (IV') yields

$$\int_{\Delta F} \mathrm{d}\boldsymbol{F} \nabla \times \boldsymbol{H} = \oint_C \mathrm{d}\boldsymbol{s}\boldsymbol{H} = \frac{1}{c} \int_{\Delta F} \mathrm{d}\boldsymbol{F} \frac{\partial \boldsymbol{D}}{\partial t} + \frac{4\pi}{c} \int_{\Delta F} \mathrm{d}\boldsymbol{F}\boldsymbol{j} \ .$$

Assuming that $\partial \boldsymbol{D}/\partial t$ is finite at the boundary plane, this contribution vanishes as h tends to zero, so that

$$\boldsymbol{t}(\boldsymbol{H}_2 - \boldsymbol{H}_1) = \frac{4\pi}{c}J \ , \ J = \frac{i}{\delta l} \ .$$

Here, \boldsymbol{t} is a tangential vector of the boundary surface, i the current flowing perpendicularly to \boldsymbol{t} through ΔF, and J the corresponding surface current density [current/length=(charge/surface)×velocity]. The corresponding calculation for (II') leads to

$$\boldsymbol{t}(\boldsymbol{E}_2 - \boldsymbol{E}_1) = 0 \ .$$

Theorem 2.25: Boundary conditions at surfaces

At a surface separating two different materials, the following continuity conditions for the macroscopic electromagnetic fields hold:

$$\boldsymbol{n}(\boldsymbol{D}_2 - \boldsymbol{D}_1) = 4\pi\sigma \ , \ \sigma = \text{surface charge density}$$
$$\boldsymbol{n}(\boldsymbol{B}_2 - \boldsymbol{B}_1) = 0$$
$$\boldsymbol{t}(\boldsymbol{H}_2 - \boldsymbol{H}_1) = \frac{4\pi}{c}J \ , \ J = \text{surface current (perpendicular to } \boldsymbol{n} \text{ and } \boldsymbol{t})$$
$$\boldsymbol{t}(\boldsymbol{E}_2 - \boldsymbol{E}_1) = 0 \ ,$$

where \boldsymbol{n} is the normal vector pointing into medium 2 and \boldsymbol{t} is a tangential vector of the boundary plane.

Summary

- Macroscopic electrodynamic phenomena in matter are described by the **macroscopic Maxwell equations**. These are given in terms of the fields $\boldsymbol{E} = \langle \boldsymbol{E} \rangle$, $\boldsymbol{B} = \langle \boldsymbol{B} \rangle$, $\boldsymbol{D} = \langle \boldsymbol{D} \rangle$, and $\boldsymbol{H} = \langle \boldsymbol{H} \rangle$ as well as the free charge and current density distributions $\rho = \langle \rho_{\mathrm{fr}} \rangle$ and $\boldsymbol{j} = \langle \boldsymbol{j}_{\mathrm{fr}} \rangle$, which are obtained by averaging over sufficiently many atomic length units.

- Contrary to Maxwell's microscopic equations (I) to (IV), these equations are not fundamental but motivated phenomenologically.

- The neutral charge and current density distributions bound in the material are described by the **polarization** $\boldsymbol{P} = \langle \boldsymbol{P} \rangle$ and the **magnetization** $\boldsymbol{M} = \langle \boldsymbol{M} \rangle$. They are related to the fields \boldsymbol{D} and \boldsymbol{H}.

\triangleright

- For a large class of materials, approximate linear relationships exist between P and E, as well as between M and B. Using the **dielectric constant** ϵ and the **permeability** μ, they can be expressed as $D = \epsilon E$ and $B = \mu H$.

- Application of Gauss's and Stokes's law to the macroscopic Maxwell equations yields boundary conditions for the macroscopic electromagnetic fields at a boundary surface.

Applications

31. Dielectric sphere in a homogeneous electric field. Consider a charge-free sphere, of radius R and dielectric constant ϵ, that is placed in a homogeneous electric field $E_0 = E_0 e_z$. Calculate the electrostatic potential inside and outside the sphere, as well as the corresponding electric field E.

Solution. By assumption, there are no free charges inside (region I) and outside (region II) the sphere. Therefore, we need to solve the Laplace equation in both regions, taking into account the relevant boundary conditions. Putting the origin of our coordinate system at the sphere's center, our problem becomes azimuthally symmetric, so that, according to Theorem 2.23, we can make the following ansatz:

$$\phi_{\mathrm{I}}(\boldsymbol{x}) = \sum_{l=0}^{\infty} A_l r^l P_l(\cos\theta)$$

$$\phi_{\mathrm{II}}(\boldsymbol{x}) = \sum_{l=0}^{\infty} \left[B_l r^l + C_l r^{-l-1} \right] P_l(\cos\theta) \ .$$

(Due to the absence of charges inside the sphere, the inner potential cannot become singular at the origin.) The coefficients A_l, B_l, and C_l are obtained from the following boundary conditions:

- In the limit $|\boldsymbol{x}| \to \infty$, the E field is equal to the initial field:

$$\nabla\phi_{\mathrm{II}}\big|_{|\boldsymbol{x}|\to\infty} = -E_0 \Longrightarrow B_1 = -E_0 \ , \ B_{l\neq 1} = 0 \ .$$

- The tangential components of E are continuous at the sphere's surface:

$$\frac{\partial\phi_{\mathrm{I}}}{\partial\theta}\bigg|_{r=R} = \frac{\partial\phi_{\mathrm{II}}}{\partial\theta}\bigg|_{r=R} \Longrightarrow A_1 = -E_0 + \frac{C_1}{R^3} \ , \ A_l = \frac{C_l}{R^{2l+1}} \ , \ l \neq 1 \ .$$

- The normal components of D are continuous at the sphere's surface:

$$\epsilon\frac{\partial\phi_{\mathrm{I}}}{\partial r}\bigg|_{r=R} = \frac{\partial\phi_{\mathrm{II}}}{\partial r}\bigg|_{r=R} \Longrightarrow \begin{cases} \epsilon A_1 = -E_0 - \dfrac{2C_1}{R^3} \\[2mm] \epsilon l A_l = -(l+1)\dfrac{C_l}{R^{2l+1}} \ , \ l \neq 1 \ . \end{cases}$$

From this, it follows that

$$A_1 = -E_0 \frac{3}{\epsilon + 2} \ , \quad C_1 = E_0 R^3 \frac{\epsilon - 1}{\epsilon + 2} \ , \quad A_{l \neq 1} = C_{l \neq 1} = 0 \ ,$$

and the potential reads

$$\phi_{\mathrm{I}}(\boldsymbol{x}) = -E_0 \frac{3}{\epsilon + 2} r \cos\theta$$

$$\phi_{\mathrm{II}}(\boldsymbol{x}) = -E_0 r \cos\theta + E_0 \frac{\epsilon - 1}{\epsilon + 2} \frac{R^3}{r^2} \cos\theta \ .$$

The corresponding \boldsymbol{E} field is shown in Fig. 2.16. Inside the sphere is a constant electric field of strength $E_{\mathrm{I}} = \frac{3}{\epsilon + 2} E_0 < E_0$ in the z-direction. Outside the sphere we have the initial field plus the field of an electric dipole $p = E_0 R^3 (\epsilon - 1)/(\epsilon + 2)$ pointing in the z-direction.

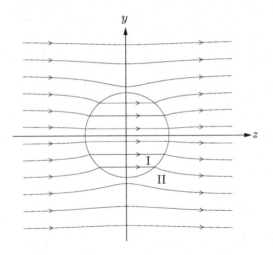

Fig. 2.16. Electric field resulting from a dielectric sphere placed in a field that was originally homogeneous in the z-direction.

32. Permeable hollow sphere in a homogeneous magnetic field. Consider a hollow sphere with inner radius a, outer radius b, and permeability constant μ, which is placed in an initially homogeneous magnetic field $\boldsymbol{B}_0 = B_0 \boldsymbol{e}_z$ (see left-hand side of Fig. 2.17). What is the magnetic field \boldsymbol{B} in the regions I, II, and III?

Solution. Since there are no currents, we have $\nabla \times \boldsymbol{B} = \boldsymbol{0}$ in all three areas, so that \boldsymbol{B} can be written as the gradient of a scalar field ψ:

$$\boldsymbol{B}(\boldsymbol{x}) = -\nabla \psi(\boldsymbol{x}) \ .$$

Because $\nabla \boldsymbol{B} = 0$, this leads to the Laplace equation

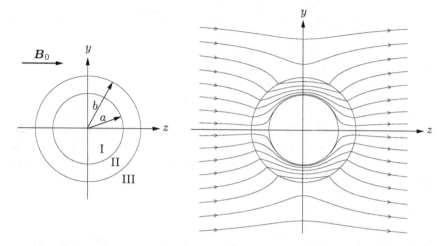

Fig. 2.17. Magnetic field resulting from a highly permeable sphere placed in a field that was originally homogeneous in the z-direction.

$$\nabla^2 \psi(\boldsymbol{x}) = 0 \ .$$

Therefore, we make the following ansatz for the scalar potential ψ:

$$\psi_{\mathrm{I}}(\boldsymbol{x}) = \sum_{l=0}^{\infty} \alpha_l r^l P_l(\cos\theta)$$

$$\psi_{\mathrm{II}}(\boldsymbol{x}) = \sum_{l=0}^{\infty} \left[\beta_l r^l + \gamma_l r^{-l-1} \right] P_l(\cos\theta)$$

$$\psi_{\mathrm{III}}(\boldsymbol{x}) = -B_0 r \cos\theta + \sum_{l=0}^{\infty} \delta_l r^{-l-1} P_l(\cos\theta) \ ,$$

where the condition $\boldsymbol{B}(|\boldsymbol{x}| \to \infty) = \boldsymbol{B}_0$ is already satisfied. The constants α_l, β_l, γ_l, and δ_l are obtained from the following continuity conditions:

- The tangential components of \boldsymbol{H} are continuous at $r = a$ and $r = b$:

$$\left.\frac{\partial\psi_{\mathrm{I}}}{\partial\theta}\right|_{r=a} = \frac{1}{\mu}\left.\frac{\partial\psi_{\mathrm{II}}}{\partial\theta}\right|_{r=a} \ , \quad \frac{1}{\mu}\left.\frac{\partial\psi_{\mathrm{II}}}{\partial\theta}\right|_{r=b} = \left.\frac{\partial\psi_{\mathrm{III}}}{\partial\theta}\right|_{r=b} \ .$$

- The normal components of \boldsymbol{B} are continuous at $r = a$ and $r = b$:

$$\left.\frac{\partial\psi_{\mathrm{I}}}{\partial r}\right|_{r=a} = \left.\frac{\partial\psi_{\mathrm{II}}}{\partial r}\right|_{r=a} \ , \quad \left.\frac{\partial\psi_{\mathrm{II}}}{\partial r}\right|_{r=b} = \left.\frac{\partial\psi_{\mathrm{III}}}{\partial r}\right|_{r=b} \ .$$

After some manipulation, we find

$$\alpha_1 = -\frac{9\mu B_0}{(2\mu+1)(\mu+2) - 2\frac{a^3}{b^3}(\mu-1)^2} \ , \quad \alpha_{l\neq1} = 0$$

$$\beta_1 = -\frac{3(2\mu + 1)B_0}{(2\mu + 1)(\mu + 2) - 2\frac{a^3}{b^3}(\mu - 1)^2} \;,\; \beta_{l \neq 1} = 0$$

$$\gamma_1 = -\frac{3a^3(\mu - 1)B_0}{(2\mu + 1)(\mu + 2) - 2\frac{a^3}{b^3}(\mu - 1)^2} \;,\; \gamma_{l \neq 1} = 0$$

$$\delta_1 = \frac{(2\mu + 1)(\mu - 1)(b^3 - a^3)B_0}{(2\mu + 1)(\mu + 2) - 2\frac{a^3}{b^3}(\mu - 1)^2} \;,\; \delta_{l \neq 1} = 0 \;.$$

Outside the sphere the potential is equivalent to that of the initial field B_0 plus the field of a magnetic dipole of size δ_1 pointing in the z-direction. In the inner region is a constant magnetic field of size $-\alpha_1$ in the z-direction. In particular, for $\mu \gg 1$, we have

$$\alpha_1 = -\frac{9B_0}{2\mu \left(1 - \frac{a^3}{b^3}\right)} \;.$$

According to this, inside a highly permeable hollow sphere the magnetic field is strongly reduced, even if the width of the sphere's shell is relatively small. This magnetic screening effect shown on the right-hand side of Fig. 2.17 is analogous to the electric screening within a Faraday cage.

2.7 Electromagnetic Waves

This section deals with the propagation of electromagnetic waves. Starting from the macroscopic Maxwell equations, the properties of plane monochromatic waves in nonconducting media are derived. Taking into account the continuity relations at boundary planes of different types of media, we derive the *reflection* and *refraction laws*. Furthermore, we consider *wave packets* consisting of a superposition of plane waves and their "widening" in *dispersive media*. Next, we expand our discussion to include the wave propagation in conducting media. Here, we will encounter complex wave vectors leading to damping phenomena. At the end of this section we discuss the wave propagation in a cylindrical *hollow conductor*.

2.7.1 Plane Waves in Nonconducting Media

Consider a nonconducting homogeneous isotropic medium characterized by the dielectric constant ϵ and the permeability constant μ. In the following we assume that both quantities are real and positive. They may depend on the wave frequency ω. With

$$\boldsymbol{D} = \epsilon \boldsymbol{E} \;,\; \boldsymbol{B} = \mu \boldsymbol{H} \;,$$

the source-free ($\rho = 0$, $\boldsymbol{j} = \boldsymbol{0}$) macroscopic Maxwell equations are

$$\left.\begin{array}{l} \epsilon \boldsymbol{\nabla} \boldsymbol{E} = 0 \ , \quad \boldsymbol{\nabla} \times \boldsymbol{E} + \dfrac{1}{c}\dfrac{\partial \boldsymbol{B}}{\partial t} = 0 \\[3mm] \boldsymbol{\nabla} \boldsymbol{B} = 0 \ , \quad \boldsymbol{\nabla} \times \dfrac{\boldsymbol{B}}{\mu} - \dfrac{\epsilon}{c}\dfrac{\partial \boldsymbol{E}}{\partial t} = 0 \ . \end{array}\right\} \tag{2.103}$$

For $\epsilon = \mu = 1$, they also formally describe the vacuum case. However, as we have noted previously, in a vacuum the microscopic fields \boldsymbol{E} and \boldsymbol{B} have a different interpretation than the averaged macroscopic quantities. Combining the curl equations while honoring the divergence equations, we obtain the following wave equations symmetric in \boldsymbol{E} and \boldsymbol{B}:

$$\boldsymbol{\nabla}^2 \boldsymbol{E} - \frac{1}{v^2}\frac{\partial^2 \boldsymbol{E}}{\partial t^2} = 0 \ , \ \boldsymbol{\nabla}^2 \boldsymbol{B} - \frac{1}{v^2}\frac{\partial^2 \boldsymbol{B}}{\partial t^2} = 0 \ , \ v = \frac{c}{\sqrt{\epsilon\mu}} \ , \tag{2.104}$$

where the frequency-dependent quantity v has the dimension of velocity. To solve these equations for a particular frequency ω, we make the ansatz

$$\boldsymbol{E}(\boldsymbol{x}, t) = \boldsymbol{E}_0 \mathrm{e}^{\mathrm{i}(\boldsymbol{kx}-\omega t)} \ , \quad \boldsymbol{B}(\boldsymbol{x}, t) = \boldsymbol{B}_0 \mathrm{e}^{\mathrm{i}(\boldsymbol{kx}-\omega t)} \ . \tag{2.105}$$

These fields describe plane monochromatic, i.e., unifrequent, waves propagating in the direction of the wave vector \boldsymbol{k}. Note that the physical fields are given by their real parts. Inserting these equations into (2.104), we obtain a relation between ω and k called the *dispersion relation*:

$$\omega^2 = \frac{c^2 k^2}{\epsilon\mu} = \frac{c^2 k^2}{n^2} \ , \ n = \sqrt{\epsilon\mu} \ , \tag{2.106}$$

where n is the frequency-dependent *refraction index* of the medium. By assumption, n, as well as ω, and \boldsymbol{k} are real,[17] whereas the amplitude vectors $\boldsymbol{E}_0, \boldsymbol{B}_0$ are generally complex-valued. Equation (2.106) does not yet define the solutions (2.105) uniquely. If we insert (2.105) into the Maxwell equations (2.103), we obtain the additional constraints

$$\boldsymbol{k}\boldsymbol{E}_0 = 0 \ , \quad \boldsymbol{k} \times \boldsymbol{E}_0 = \frac{\omega}{c}\boldsymbol{B}_0$$

$$\boldsymbol{k}\boldsymbol{B}_0 = 0 \ , \quad \boldsymbol{k} \times \boldsymbol{B}_0 = -\epsilon\mu\frac{\omega}{c}\boldsymbol{E}_0 \ ,$$

implying that

$$\boldsymbol{k} \perp \boldsymbol{E}_0, \boldsymbol{B}_0 \ , \quad \boldsymbol{B}_0 = \frac{c}{\omega}\boldsymbol{k} \times \boldsymbol{E}_0 \ . \tag{2.107}$$

This means that $\boldsymbol{k}, \boldsymbol{E}_0, \boldsymbol{B}_0$ form an orthogonal trihedral (in this order). Therefore, (2.105) describes *transverse waves* whose directions of oscillation, $\boldsymbol{E}_0, \boldsymbol{B}_0$, are perpendicular to their direction of propagation, \boldsymbol{k}.

Energy and momentum. Taking into account (2.99), the equations for the time-averaged quantities $\overline{\boldsymbol{S}}$ and $\overline{\epsilon_{\mathrm{em}}}$, analogous to those of (2.11) and (2.12), are

[17] A complex n (and, consequently, a complex k) leads to exponentially damped waves (*dissipation*).

$$\overline{\boldsymbol{S}} = \frac{c}{8\pi\mu}\mathrm{Re}\left[\boldsymbol{E}_0 \times \boldsymbol{B}_0^*\right] = \frac{c}{8\pi}\sqrt{\frac{\epsilon}{\mu}}|\boldsymbol{E}_0|^2\hat{\boldsymbol{k}} \ , \ \ \hat{\boldsymbol{k}} = \frac{\boldsymbol{k}}{|\boldsymbol{k}|}$$

$$\overline{\epsilon_{\mathrm{em}}} = \frac{1}{16\pi}\left[\epsilon|\boldsymbol{E}_0|^2 + \frac{1}{\mu}|\boldsymbol{B}_0|^2\right] = \frac{\epsilon}{8\pi}|\boldsymbol{E}_0|^2 \ .$$

The ratio of these two quantities determines the velocity of the energy flux in the direction of propagation:

$$\frac{\overline{\boldsymbol{S}}}{\overline{\epsilon_{\mathrm{em}}}} = \frac{c}{\sqrt{\epsilon\mu}}\hat{\boldsymbol{k}} = v\hat{\boldsymbol{k}} \ .$$

Therefore, we can identify v as the *phase velocity* v_φ that describes the velocity of wave trains with constant phase in the direction of \boldsymbol{k}:

$$\boldsymbol{k}\boldsymbol{x} - \omega t = \mathrm{const} \implies |\boldsymbol{k}|(|\boldsymbol{x}| - v_\varphi t) = \mathrm{const} \ , \ \ v_\varphi = \frac{\omega}{|\boldsymbol{k}|} \ .$$

In addition to frequency ω and phase velocity v_φ, another characteristic of plane monochromatic waves is the *wavelength* λ. This is the length between points of equal phase (in the direction of propagation) at fixed time:

$$|\boldsymbol{k}|(|\boldsymbol{x}| + \lambda) = |\boldsymbol{k}||\boldsymbol{x}| + 2\pi \implies \lambda = \frac{2\pi}{|\boldsymbol{k}|} \ .$$

Polarization. Due to the linearity of the wave equations, a linear combination (superposition) of their solutions yields a new solution. We wish to consider the following two specific solutions with the same frequency ω:

$$\boldsymbol{E}_1(\boldsymbol{x}, t) = E_1\boldsymbol{e}_1\mathrm{e}^{\mathrm{i}(\boldsymbol{k}\boldsymbol{x}-\omega t)} \ , \ \ \boldsymbol{B}_1(\boldsymbol{x}, t) = \frac{c}{\omega}\boldsymbol{k} \times \boldsymbol{E}_1(\boldsymbol{x}, t)$$

$$\boldsymbol{E}_2(\boldsymbol{x}, t) = E_2\boldsymbol{e}_2\mathrm{e}^{\mathrm{i}(\boldsymbol{k}\boldsymbol{x}-\omega t)} \ , \ \ \boldsymbol{B}_2(\boldsymbol{x}, t) = \frac{c}{\omega}\boldsymbol{k} \times \boldsymbol{E}_2(\boldsymbol{x}, t) \ ,$$

where \boldsymbol{e}_1 and \boldsymbol{e}_2 are real orthogonal unit vectors. These fields describe *linearly polarized* waves because their direction of oscillation is along the direction of the *polarization vectors* \boldsymbol{e}_i or $\hat{\boldsymbol{k}} \times \boldsymbol{e}_i$ for all times. Superimposing these two \boldsymbol{E} fields yields the new solution

$$\boldsymbol{E}(\boldsymbol{x}, t) = (\boldsymbol{e}_1 E_1 + \boldsymbol{e}_2 E_2)\mathrm{e}^{\mathrm{i}(\boldsymbol{k}\boldsymbol{x}-\omega t)} = (\boldsymbol{e}_1|E_1|\mathrm{e}^{\mathrm{i}\delta_1} + \boldsymbol{e}_2|E_2|\mathrm{e}^{\mathrm{i}\delta_2})\mathrm{e}^{\mathrm{i}(\boldsymbol{k}\boldsymbol{x}-\omega t)} \ ,$$

whose physical part is given by

$$\mathrm{Re}\boldsymbol{E}(\boldsymbol{x}, t) = \boldsymbol{e}_1|E_1|\cos(\boldsymbol{k}\boldsymbol{x} - \omega t + \delta_1) + \boldsymbol{e}_2|E_2|\cos(\boldsymbol{k}\boldsymbol{x} - \omega t + \delta_2). (2.108)$$

(Similar considerations hold for the \boldsymbol{B} fields.) Depending on the values of the phases δ_1, δ_2, and of the amplitudes $|E_1|, |E_2|$, the last expression can be classified by three polarization states:

- *Elliptical polarization:* $\delta_1 \neq \delta_2$. This is the most general case. Considering a fixed location, the electrical field vector $\mathrm{Re}\boldsymbol{E}(\boldsymbol{x} = \mathrm{const}, t)$ rotates on an ellipse in a plane spanned by \boldsymbol{e}_1 and \boldsymbol{e}_2 and with period $T = 2\pi/\omega$.

- *Circular polarization:* $\delta_2 = \delta_1 \pm \pi/2$, $|E_1| = |E_2|$. In this special case, (2.108) becomes

$$\mathrm{Re}\boldsymbol{E}(\boldsymbol{x},t) = |E_1| \left[\boldsymbol{e}_1 \cos(\boldsymbol{k}\boldsymbol{x} - \omega t + \delta_1) \mp \boldsymbol{e}_2 \sin(\boldsymbol{k}\boldsymbol{x} - \omega t + \delta_1) \right] .$$

This means that $\mathrm{Re}\boldsymbol{E}(\boldsymbol{x} = \mathrm{const}, t)$ describes a circle of radius $|E_1|$. Depending on the sign of $\delta_2 - \delta_1 = \pm\pi/2$, the direction of circulation is

(+): positive (*left circular polarization, positive helicity*),

(−): negative (*right circular polarization, negative helicity*).

- *Linear polarization:* $\delta_1 = \delta_2$. Equation (2.108) now reads

$$\mathrm{Re}\boldsymbol{E}(\boldsymbol{x},t) = (\boldsymbol{e}_1|E_1| + \boldsymbol{e}_2|E_2|) \cos(\boldsymbol{k}\boldsymbol{x} - \omega t + \delta_1) ,$$

and $\mathrm{Re}\boldsymbol{E}(\boldsymbol{x} = \mathrm{const}, t)$ moves along a straight line in the $\boldsymbol{e}_1\boldsymbol{e}_2$-plane.

Theorem 2.26: Plane waves in nonconducting media

Plane monochromatic waves in nonconducting media (ϵ, μ real) are given by the real parts of the transverse fields

$$\boldsymbol{E}(\boldsymbol{x},t) = \boldsymbol{E}_0 \mathrm{e}^{\mathrm{i}(\boldsymbol{k}\boldsymbol{x} - \omega t)} , \quad \boldsymbol{B}(\boldsymbol{x},t) = \boldsymbol{B}_0 \mathrm{e}^{\mathrm{i}(\boldsymbol{k}\boldsymbol{x} - \omega t)} ,$$

with

$$\omega^2 = \frac{c^2 k^2}{n^2} , \quad n = \sqrt{\epsilon\mu} , \quad \boldsymbol{k} \perp \boldsymbol{E}_0\boldsymbol{B}_0 , \quad \boldsymbol{B}_0 = \frac{c}{\omega}\boldsymbol{k} \times \boldsymbol{E}_0 ,$$

$\omega, \boldsymbol{k}, n$ real , $\boldsymbol{E}_0, \boldsymbol{B}_0$ complex .

Characteristics of these waves are: the frequency ω (related to the wave vector \boldsymbol{k} via the dispersion relation of the medium), the phase velocity $v_\varphi = \omega/|\boldsymbol{k}|$, and the wavelength $\lambda = 2\pi/|\boldsymbol{k}|$. The superposition of two linearly polarized waves yields elliptically polarized waves. Special cases of elliptical polarization are linear and circular polarization.

2.7.2 Reflection and Refraction

In this subsection we investigate the behavior of plane monochromatic waves at a boundary plane aligned in the xy-plane and separating two media with different refraction indices $n = \sqrt{\epsilon\mu}$ and $n' = \sqrt{\epsilon'\mu'}$ (see Fig. 2.18). We make the following simple ansatz for the \boldsymbol{E} and \boldsymbol{B} fields of the incoming, reflected, and refracted waves:

- **Incoming waves:**

$$\boldsymbol{E}(\boldsymbol{x},t) = \boldsymbol{E}_0 \mathrm{e}^{\mathrm{i}(\boldsymbol{k}\boldsymbol{x} - \omega t)} , \quad \boldsymbol{B}(\boldsymbol{x},t) = \frac{c}{\omega}\boldsymbol{k} \times \boldsymbol{E}(\boldsymbol{x},t)$$

$$\boldsymbol{k} \perp \boldsymbol{E}_0 , \quad \omega^2 = \frac{c^2 k^2}{n^2} .$$

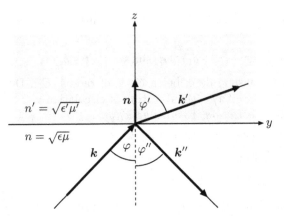

Fig. 2.18. Reflection and refraction of electromagnetic waves at a boundary plane separating two different media.

- **Refracted waves:**

$$E'(x,t) = E_0 e^{i(k'x-\omega't)} \ , \quad B'(x,t) = \frac{c}{\omega'}k' \times E'(x,t)$$

$$k' \perp E'_0 \ , \quad \omega'^2 = \frac{c^2 k'^2}{n'^2} \ .$$

- **Reflected waves:**

$$E''(x,t) = E_0'' e^{i(k''x-\omega''t)} \ , \quad B''(x,t) = \frac{c}{\omega''}k'' \times E''(x,t)$$

$$k'' \perp E_0'' \ , \quad \omega''^2 = \frac{c^2 k''^2}{n^2} \ .$$

This ansatz satisfies the wave equations (2.104), as well as the orthogonality conditions (2.107) for each medium. Additionally, according to Theorem 2.25, the tangential (t) and normal (n) components of the fields must obey certain continuity conditions at the boundary plane ($z = 0$). In the absence of surface charges and currents, these conditions are

$$[E_{\rm t}, \ H_{\rm t}, \ B_{\rm n}, \ D_{\rm n}]_{z=0} \quad \text{continuous} \ . \tag{2.109}$$

Is is easily seen that these conditions can only be satisfied for all times (up to integer multiples of 2π, which can be neglected without restricting generality) if

$$[kx - \omega t = k'x - \omega't = k''x - \omega''t]_{z=0} \ .$$

For $x = 0$ this implies

$$\omega = \omega' = \omega'' \implies |k| = |k''| \ ,$$

and for $t = 0$

$$[kx = k'x = k''x]_{z=0} \ .$$

This last relation means that all wave vectors must lie in a plane spanned by the wave vector \mathbf{k} of the incoming wave and the normal vector \mathbf{n} of the boundary plane. Defining the angles φ, φ' and φ'' as shown in Fig. 2.18,

$$k_y = |k| \sin \varphi , \ k_y' = |k'| \sin \varphi' , \ k_y'' = |k| \sin \varphi'' ,$$

we find

Theorem 2.27: Reflection and refraction laws

If a unifrequent plane wave hits a boundary separating two different media characterized by the refraction coefficients n and n', the following holds:

- Incoming, reflected, and refracted waves have the same frequency.

- The absolute values of the wave vectors of incoming and reflected waves are equal.

- The angles of incidence and reflection are identical:

 $\varphi = \varphi''$ *(reflection law)* .

- The angles of incidence and refraction are related to the media's refraction indices via

 $$\frac{\sin \varphi'}{\sin \varphi} = \frac{n}{n'} \ \text{(refraction law)} .$$

Total reflection. If the electromagnetic wave is moving from an optically denser medium to an optically thinner one $(n > n')$, the refraction law tells us that

$$\sin \varphi = \frac{n'}{n} \sin \varphi' < 1 .$$

This means there exists an angle of incidence,

$$\varphi_{\text{TR}} = \arcsin \frac{n'}{n} ,$$

for which the refraction angle is $\pi/2$, so that the refracted wave travels along the boundary plane. For larger angles, $\varphi > \varphi_{\text{TR}}$, the incoming wave is totally reflected (see Application 33). By experimentally determining the limiting angle φ_{TR} at which total refraction begins to occur, the refraction index of an unknown medium can be obtained.

Intensity relations. In contrast to the reflection and refraction laws, the intensity and polarization of the reflected and refracted waves depend critically on the vector character of the waves. To see this, we write the continuity conditions (2.109) once again in explicit form:

$$
\left.
\begin{aligned}
(\boldsymbol{E}_0 + \boldsymbol{E}_0'' - \boldsymbol{E}_0')\boldsymbol{t} &= 0 \\
\left[\frac{1}{\mu}(\boldsymbol{k} \times \boldsymbol{E}_0 + \boldsymbol{k}'' \times \boldsymbol{E}_0'') - \frac{1}{\mu'}\boldsymbol{k}' \times \boldsymbol{E}_0'\right]\boldsymbol{t} &= 0 \\
(\boldsymbol{k} \times \boldsymbol{E}_0 + \boldsymbol{k}'' \times \boldsymbol{E}_0'' - \boldsymbol{k}' \times \boldsymbol{E}_0')\boldsymbol{n} &= 0 \\
[\epsilon(\boldsymbol{E}_0 + \boldsymbol{E}_0'') - \epsilon' \boldsymbol{E}_0']\boldsymbol{n} &= 0\,,
\end{aligned}
\right\}
\tag{2.110}
$$

where $\boldsymbol{n} = \begin{pmatrix} 0 \\ 0 \\ 1 \end{pmatrix}$ is the normal vector and $\boldsymbol{t} = \begin{pmatrix} t_1 \\ t_2 \\ 0 \end{pmatrix}$ a tangential vector of

the boundary plane. For further evaluation of these equations, it is instructive to consider two special cases in which the incoming waves are linearly polarized in different directions:

a. Perpendicular polarization: the polarization vector of the incoming electric field is perpendicular to the plane of incidence spanned by \boldsymbol{n} and \boldsymbol{k} (see Fig. 2.19). In this case our ansatz for the vector quantities of the \boldsymbol{E} fields is

$$
\boldsymbol{k} = \begin{pmatrix} 0 \\ k_2 \\ k_3 \end{pmatrix} , \quad \boldsymbol{k}' = \begin{pmatrix} 0 \\ k_2 \\ k_3' \end{pmatrix} , \quad \boldsymbol{k}'' = \begin{pmatrix} 0 \\ k_2 \\ -k_3 \end{pmatrix}
$$

$$
\boldsymbol{E}_0 = \begin{pmatrix} E_1 \\ 0 \\ 0 \end{pmatrix} , \quad \boldsymbol{E}_0' = \begin{pmatrix} E_1' \\ E_2' \\ E_3' \end{pmatrix} \quad \boldsymbol{E}_0'' = \begin{pmatrix} E_1'' \\ E_2'' \\ E_3'' \end{pmatrix} ,
$$

with

$$
\left.
\begin{aligned}
0 &= k_2 E_2' + k_3' E_3' \quad (\boldsymbol{k}' \boldsymbol{E}_0' = 0) \\
0 &= k_2 E_2'' - k_3 E_3'' \quad (\boldsymbol{k}'' \boldsymbol{E}_0'' = 0)\,.
\end{aligned}
\right\}
\tag{2.111}
$$

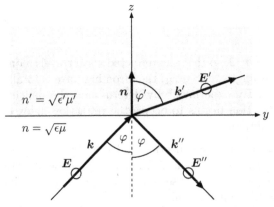

Fig. 2.19. Reflection and refraction of electromagnetic waves whose \boldsymbol{E} field is polarized perpendicular to the plane of incidence.

The first and last relations in (2.110) imply that

$$0 = E_2'' - E_2'$$
$$0 = \epsilon E_3'' - \epsilon' E_3'$$

which, in conjunction with (2.111), forms a homogeneous system of equations for the four unknowns E_2', E_3', E_2'', E_3''. The only solution is the trivial one:

$$E_2' = E_3' = E_2'' = E_3'' = 0 \;.$$

Consequently, reflected and refracted waves are polarized in the same way as the incoming wave. The remaining equations in (2.110) form an inhomogeneous system of equations in E_1', E_1'', and k_3', which yield the *Fresnel formulae* for electromagnetic waves polarized perpendicular to the plane of incidence:

$$\frac{E_1'}{E_1} = \frac{2n\cos\varphi}{n\cos\varphi + \frac{\mu}{\mu'}\sqrt{n'^2 - n^2\sin^2\varphi}}$$

$$\frac{E_1''}{E_1} = \frac{n\cos\varphi - \frac{\mu}{\mu'}\sqrt{n'^2 - n^2\sin^2\varphi}}{n\cos\varphi + \frac{\mu}{\mu'}\sqrt{n'^2 - n^2\sin^2\varphi}} \;. \tag{2.112}$$

b. Parallel polarization: the polarization vector of the incoming electric field is parallel to the plane of incidence spanned by \boldsymbol{n} and \boldsymbol{k} (see Fig. 2.20). Here, we choose

$$\boldsymbol{k} = k\begin{pmatrix} 0 \\ \sin\varphi \\ \cos\varphi \end{pmatrix} \;, \quad \boldsymbol{k}' = k'\begin{pmatrix} 0 \\ \sin\varphi' \\ \cos\varphi' \end{pmatrix} \;, \quad \boldsymbol{k}'' = k\begin{pmatrix} 0 \\ \sin\varphi \\ -\cos\varphi \end{pmatrix} \;,$$

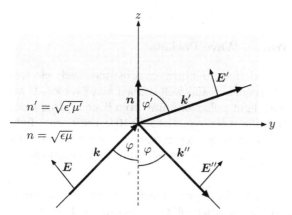

Fig. 2.20. Reflection and refraction of electromagnetic waves whose \boldsymbol{E} field is polarized parallel to the plane of incidence.

$$\boldsymbol{E}_0 = E_0 \begin{pmatrix} 0 \\ -\cos\varphi \\ \sin\varphi \end{pmatrix} , \quad \boldsymbol{E}_0' = E_0' \begin{pmatrix} 0 \\ -\cos\varphi' \\ \sin\varphi' \end{pmatrix} , \quad \boldsymbol{E}_0'' = E_0'' \begin{pmatrix} 0 \\ \cos\varphi \\ \sin\varphi \end{pmatrix} .$$

The tangential equations in (2.110) then yield an inhomogeneous system for E_0' and E_0'',

$$\cos\varphi' E_0' + \cos\varphi E_0'' = \cos\varphi E_0$$
$$\frac{k'}{\mu'} E_0' - \frac{k}{\mu} E_0'' = \frac{k}{\mu} E_0 ,$$

whose solution is given by

$$\frac{E_0'}{E_0} = \frac{2nn'\cos\varphi}{\frac{\mu}{\mu'}n'^2 \cos\varphi + n\sqrt{n'^2 - n^2 \sin^2\varphi}}$$
$$\frac{E_0''}{E_0} = \frac{\frac{\mu}{\mu'}n'^2 \cos\varphi - n\sqrt{n'^2 - n^2 \sin^2\varphi}}{\frac{\mu}{\mu'}n'^2 \cos\varphi + n\sqrt{n'^2 - n^2 \sin^2\varphi}} . \tag{2.113}$$

These are the Fresnel formulae for the case of parallel polarization.

Brewster angle. In case b, there is an angle, called the *Brewster angle*, for which the amplitude of the reflected wave in (2.113) vanishes. If $\mu = \mu'$, this angle is

$$\varphi_B = \arctan\frac{n'}{n} .$$

Note that in case a, (2.112) can never be zero. Since nonpolarized waves can be split into both polarizations a and b, there are no reflected waves with \boldsymbol{E}_0'' parallel to the incident plane if the angle of incidence is equal to the Brewster angle. In this case the reflected waves are linearly polarized perpendicular to the plane of incidence. This effect can be used to create linearly polarized light.

2.7.3 Superposition of Waves, Wave Packets

The previous subsection handled the unrealistic case of (idealized) electromagnetic waves with distinct frequency and well-defined wave vector. However, in practice, one deals with light pulses of finite (even if small) frequency and wavelength ranges. Since the wave equations (2.104) are linear, light pulses (*wave packets*) with a given frequency range can be constructed by superposition of monochromatic waves, and the following phenomena can be observed:

- If the medium is *dispersive*, i.e., the (real) refraction index is frequency-dependent, the phase velocities of each of the superimposed waves are different. This implies that the individual wave components propagate at different speeds, thereby changing their relative phases. This leads to a

deformation of the wave packet that grows in time. In addition, the wave components are refracted differently at a boundary plane of two different media.[18]

- In a dispersive medium, the wave packet propagation velocity, called the *group velocity* v_g, is generally different to the phase velocities v_φ of the individual waves. The group velocity determines the energy transport and the velocity of the signal of the wave packet.

- In a *dissipative medium*, i.e., the refraction index is complex, damping effects occur. It is then possible for the group velocity to become even larger than the velocity of light, c. However, since the propagation of the wave packet is bounded by absorption, the inequality $v_g > c$ does not contradict special relativity, which states that c is the maximal velocity of any kind of signals.

To better understand the concept of "group velocity", we consider the propagation of a wave packet in an arbitrary medium. For simplification, we set the direction of propagation along the z-axis and consider just one component, ψ, of the \boldsymbol{E} field. Its most general form is then given by

$$\psi(z,t) = \frac{1}{\sqrt{2\pi}} \int\limits_{-\infty}^{\infty} \mathrm{d}k A(k)\mathrm{e}^{\mathrm{i}[kz - \omega(k)t]} . \tag{2.114}$$

Additionally, we assume the following, very general, dispersion relation between ω and k:

$$\omega = \frac{ck}{n(\omega)} \Longleftrightarrow \omega = \omega(k) .$$

For $t = 0$, (2.114) is simply the Fourier representation of the function $\psi(z,0)$, with Fourier components

$$A(k) = \frac{1}{\sqrt{2\pi}} \int\limits_{-\infty}^{\infty} \mathrm{d}z \psi(z,0)\mathrm{e}^{-\mathrm{i}kz} . \tag{2.115}$$

If $A(k)$ has a sharp maximum around k_0, we can expand $\omega(k)$ in a Taylor expansion around k_0:

$$\omega(k) = \omega(k_0) + \left.\frac{\mathrm{d}\omega}{\mathrm{d}k}\right|_{k_0} (k - k_0) + \frac{1}{2}\left.\frac{\mathrm{d}^2\omega}{\mathrm{d}k^2}\right|_{k_0} (k - k_0)^2 + \ldots$$

$$= \omega_0 + v_g(k - k_0) + \gamma(k - k_0)^2 + \ldots , \tag{2.116}$$

with

[18] This effect is responsible for the spectral decomposition of light within a prism.

$$\omega_0 = \omega(k_0)$$

$$v_g = \left.\frac{d\omega}{dk}\right|_{k_0} \qquad \text{(group velocity)}$$

$$\gamma = \left.\frac{1}{2}\frac{d^2\omega}{dk^2}\right|_{k_0} \qquad (\textit{dispersion parameter}) .$$

Inserting the linear approximation of (2.116) into (2.114), we obtain

$$\psi(z,t) \approx \frac{1}{\sqrt{2\pi}} e^{i(v_g k_0 - \omega_0)t} \int_{-\infty}^{\infty} dk\, A(k) e^{i(kz - v_g kt)}$$

$$\approx e^{i(v_g k_0 - \omega_0)t}\psi(z - v_g t, 0) .$$

The intensity of this wave is

$$|\psi(z,t)|^2 = |\psi(z - v_g t, 0)|^2 .$$

Thus, in the linear approximation, the wave packet moves with group velocity v_g in the z-direction without changing its shape, i.e., without dispersion. For linear dispersion relations $\omega \sim |k|$ (nondispersive media) such as in a vacuum, this approximation is exact. In the dispersive case, beyond the linear approximation, one finds that the wave packet changes its shape, i.e., it flows apart. This effect is described to lowest order by the dispersion parameter γ, while the group velocity determines the velocity of the wave packet's center of mass.

Wave packet in a dispersive medium. To illustrate dispersion effects, we now calculate (without any approximation) the z-directional propagation of a one-dimensional light pulse in a medium governed by the following dispersion relation:

$$\omega(k) = \nu\left(1 + \frac{a^2 k^2}{2}\right) .$$

Here, ν denotes a constant frequency and a a constant length. As we shall see below, a can be regarded as a characteristic wavelength for the occurrence of dispersion effects. For the light pulse's initial shape, we assume the following Gaussian distribution:

$$\psi(z,0) = e^{-\frac{z^2}{2\Delta^2}} e^{ik_0 z} .$$

According to (2.115), the distribution of the corresponding Fourier amplitudes is given by

$$A(k) = \frac{1}{\sqrt{2\pi}} \int_{-\infty}^{\infty} dz\, e^{-\frac{z^2}{2\Delta^2}} e^{-i(k-k_0)z} = \Delta e^{-\frac{\Delta^2(k-k_0)^2}{2}} .$$

Using (2.114), the shape of the light pulse at a later time t evaluates to

$$\psi(z,t) = \frac{1}{\sqrt{2\pi}} \int_{-\infty}^{\infty} dk A(k) e^{i[kz-\omega(k)t]}$$

$$= \frac{\Delta}{\sqrt{2\pi}} \int_{-\infty}^{\infty} dk e^{-\frac{\Delta^2(k-k_0)^2}{2}} e^{i\left[kz-\nu\left(1+\frac{a^2 k^2}{2}\right)t\right]}$$

$$= \frac{\Delta}{\sqrt{2\pi}} e^{-\left(\frac{\Delta^2 k_0^2}{2}+i\nu t\right)} \int_{-\infty}^{\infty} dk e^{-\frac{\alpha(t)}{2}k^2+k(\Delta^2 k_0+iz)} , \tag{2.117}$$

with

$$\alpha(t) = \Delta^2 + i\nu a^2 t .$$

The integration in (2.117) can be performed using quadratic completion and yields

$$\psi(z,t) = \frac{\Delta}{\sqrt{\alpha(t)}} \exp\left(-\frac{(z-\nu a^2 t k_0)^2}{2\alpha(t)}\right) \exp\left[ik_0 z - i\nu\left(1+\frac{a^2 k_0^2}{2}\right)t\right] .$$

The intensity of the wave packet is

$$|\psi(z,t)|^2 = \psi^*(z,t)\psi(z,t) = \frac{\Delta}{\sqrt{\beta(t)}} \exp\left(-\frac{(z-\nu a^2 t k_0)^2}{\beta(t)}\right) ,$$

with

$$\beta(t) = \Delta^2 + \frac{\nu^2 a^4 t^2}{\Delta^2} .$$

The width of the wave packet is described by the time-dependent quantity β, which grows in time (Fig. 2.21). Its rate of change per time unit is found by differentiation of β with respect to t:

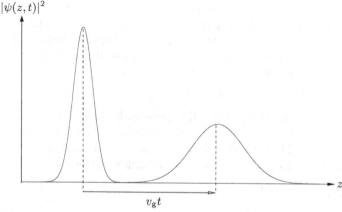

Fig. 2.21. Widening of a wave packet in a dispersive medium.

$$\beta'(t) = \frac{2\nu^2 a^4 t}{\Delta^2} .$$

From this, one sees that $\Delta \gg a$ is a necessary criterion for a small deformation and an (almost) dispersion-free propagation of the wave packet in the medium.

2.7.4 Plane Waves in Conducting Media

The propagation of electromagnetic waves in conducting media differs somewhat from the nonconducting case. For conductors, in addition to the macroscopic Maxwell equations, one must also take into account *Ohm's law*

$$\boldsymbol{j} = \sigma \boldsymbol{E} ,$$

where σ denotes the conductibility of the conductor. Therefore, we have

$$\epsilon \boldsymbol{\nabla} \boldsymbol{E} = 0 \ , \quad \boldsymbol{\nabla} \times \boldsymbol{E} + \frac{\mu}{c} \frac{\partial \boldsymbol{H}}{\partial t} = 0$$

$$\mu \boldsymbol{\nabla} \boldsymbol{H} = 0 \ , \quad \boldsymbol{\nabla} \times \boldsymbol{H} - \frac{\epsilon}{c} \frac{\partial \boldsymbol{E}}{\partial t} = \frac{4\pi\sigma}{c} \boldsymbol{E} .$$

By combining the two curl equations, we find the *telegraph equations*, which are symmetric in \boldsymbol{E} and \boldsymbol{H}:

$$\boldsymbol{\nabla}^2 \boldsymbol{E} - \frac{\epsilon\mu}{c} \frac{\partial^2 \boldsymbol{E}}{\partial t^2} - \frac{4\pi\mu\sigma}{c^2} \frac{\partial \boldsymbol{E}}{\partial t} = 0 \ , \quad \boldsymbol{\nabla}^2 \boldsymbol{H} - \frac{\epsilon\mu}{c} \frac{\partial^2 \boldsymbol{H}}{\partial t^2} - \frac{4\pi\mu\sigma}{c^2} \frac{\partial \boldsymbol{H}}{\partial t} = 0 .$$

To solve these equations we proceed along similar lines to Subsection 2.7.1 and make the ansatz

$$\boldsymbol{E}(\boldsymbol{x}, t) = \boldsymbol{E}_0 e^{i(\boldsymbol{k}\boldsymbol{x} - \omega t)} \ , \quad \boldsymbol{H}(\boldsymbol{x}, t) = \boldsymbol{H}_0 e^{i(\boldsymbol{k}\boldsymbol{x} - \omega t)} . \tag{2.118}$$

This gives the dispersion relation

$$k^2 = \frac{\omega^2}{c^2} \mu\eta = \frac{\omega^2}{c^2} p^2 \ , \quad \eta = \epsilon + \frac{4\pi i \sigma}{\omega} \ , \quad p = \sqrt{\mu\eta} ,$$

where η is the *generalized dielectric constant* and p the *generalized refraction index*. Thus, in the case of conducting media, the wave vector is complex, so that we expect damped propagation. For example, consider a wave propagating in the x-direction and set $k = \alpha + i\beta$. The exponential part of the wave function can then be written as

$$e^{i(kx - \omega t)} = e^{-\beta x} e^{i(\alpha x - \omega t)} \ , \quad \beta > 0 .$$

This means the wave's amplitude falls off exponentially in the direction of propagation and β can be regarded as a measure for the wave's *penetration depth* into the medium. The orthogonality relations corresponding to (2.107) are found by inserting (2.118) into the above Maxwell equations:

$$\boldsymbol{k}\boldsymbol{E}_0 = 0 \ , \quad \boldsymbol{k} \times \boldsymbol{E}_0 = \frac{\omega\mu}{c} \boldsymbol{H}_0$$

$$\boldsymbol{k}\boldsymbol{H}_0 = 0 \ , \quad \boldsymbol{k} \times \boldsymbol{H}_0 = -\frac{\omega\eta}{c} \boldsymbol{E}_0$$

$$\Longrightarrow \boldsymbol{k} \perp \boldsymbol{E}_0, \boldsymbol{H}_0 \ , \ \boldsymbol{H}_0 = \frac{c}{\omega\mu}\boldsymbol{k} \times \boldsymbol{E}_0 \ .$$

Accordingly, the electromagnetic waves in conducting media are also transverse and \boldsymbol{k}, \boldsymbol{E}_0, and \boldsymbol{H}_0 form, as before, an orthogonal trihedral (in this order).

2.7.5 Cylindrical Hollow Conductor

By *hollow conductors* we mean long metal bodies with open ends, whereas bodies with closed ends are called *cavity resonators*. In the following we discuss the propagation of electromagnetic waves in a cylindrical hollow conductor with constant cross section along the z-axis (Fig. 2.22). The surface S of

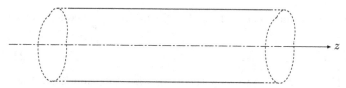

Fig. 2.22. Hollow conductor with cylindrical symmetry.

this body is assumed to be an ideal conductor. The internal space is filled by a dispersive medium with dielectric constant ϵ and permeability constant μ. Due to the problem's cylindrical geometry, we only expect waves propagating along the positive and negative z-axis. Therefore, our ansatz is

$$\boldsymbol{E}(\boldsymbol{x},t) = \boldsymbol{E}(x,y)e^{\pm ikz-i\omega t} \ , \ \boldsymbol{B}(\boldsymbol{x},t) = \boldsymbol{B}(x,y)e^{\pm ikz-i\omega t} \ . \qquad (2.119)$$

Before inserting these fields into the wave equations (2.104), it is advantageous to decompose them into components parallel and perpendicular to the z-axis:

$$\boldsymbol{E}(\boldsymbol{x},t) = \left[\boldsymbol{E}_z(\boldsymbol{x}) + \boldsymbol{E}_t(\boldsymbol{x})\right]e^{-i\omega t} \ ,$$

with

$$\boldsymbol{E}_t = (\boldsymbol{e}_z\boldsymbol{E})\boldsymbol{e}_z = \begin{pmatrix} 0 \\ 0 \\ E_z \end{pmatrix} \ , \ \boldsymbol{E}_t = (\boldsymbol{e}_z \times \boldsymbol{E}) \times \boldsymbol{e}_z = \begin{pmatrix} E_x \\ E_y \\ 0 \end{pmatrix} \ .$$

Corresponding considerations hold for \boldsymbol{B}. After a few intermediate calculations and taking into account the explicit z-dependency (2.119), we find the following expressions for the transverse fields from the curl equations in (2.103):

$$\left.\begin{aligned} \boldsymbol{E}_t &= \frac{1}{\gamma^2}\left[\boldsymbol{\nabla}_t\left(\frac{\partial E_z}{\partial z}\right) - i\frac{\omega}{c}(\boldsymbol{e}_z \times \boldsymbol{\nabla}_t)B_z\right] \\ \boldsymbol{B}_t &= \frac{1}{\gamma^2}\left[\boldsymbol{\nabla}_t\left(\frac{\partial B_z}{\partial z}\right) + i\frac{\omega}{c}\epsilon\mu(\boldsymbol{e}_z \times \boldsymbol{\nabla}_t)E_z\right] \ , \end{aligned}\right\} \qquad (2.120)$$

with

$$\gamma^2 = \frac{\omega^2}{c^2}\epsilon\mu - k^2 \ , \ \boldsymbol{\nabla}_t = \boldsymbol{\nabla} - \frac{\partial}{\partial z} \ .$$

This means the transverse fields are solely determined by the longitudinal ones, so that it is sufficient to solve the wave equations (2.104) for the longitudinal fields \boldsymbol{E}_z and \boldsymbol{B}_z:

$$\left(\frac{\partial^2}{\partial x^2} + \frac{\partial^2}{\partial y^2} + \gamma^2\right)\left(\begin{matrix} E_z \\ B_z \end{matrix}\right) = 0 \ . \tag{2.121}$$

Since the cylinder's surface is assumed to be an ideal conductor, we have the boundary conditions

$$\boldsymbol{n}\boldsymbol{B} = 0 \ , \ \boldsymbol{n} \times \boldsymbol{E} = \boldsymbol{0} \ ,$$

where \boldsymbol{n} denotes the normal vector of S. These conditions are equivalent to the constraints

$$E_z|_{\boldsymbol{x} \in S} = 0 \ , \ \boldsymbol{n}\boldsymbol{\nabla}B_z|_{\boldsymbol{x} \in S} = 0 \ . \tag{2.122}$$

As one can see, E_z and B_z have to obey the same wave equation (2.121) but with different boundary conditions (2.122). In general, this cannot be fulfilled by both fields simultaneously. Consequently, one distinguishes among three types of solutions:

- *Transverse magnetic modes (TM):* the eigenvalue equation (2.121) is solved for E_z with the boundary condition $E_z|_{\boldsymbol{x} \in S} = 0$, leading to a specific eigenvalue spectrum γ^2_{TM}. The solution of the eigenvalue equation in B_z is chosen to be the trivial one: $B_z = 0 \ \forall \ \boldsymbol{x}$.

- *Transverse electric modes (TE):* equation (2.121) is solved for B_z with the boundary condition $\boldsymbol{n}\boldsymbol{\nabla}B_z|_{\boldsymbol{x} \in S} = 0$, leading to another specific eigenvalue spectrum γ^2_{TE}. The solution in E_z is chosen to be the trivial one: $E_z = 0 \ \forall \ \boldsymbol{x}$.

- *Transverse electric and magnetic modes (TEM):* these modes are characterized by vanishing longitudinal components of the electric and magnetic field: $E_z = B_z = 0 \ \forall \ \boldsymbol{x}$. From (2.120), we see that nonvanishing transverse fields must obey the constraint

$$\gamma^2_{\mathrm{TEM}} = \frac{\omega^2}{c^2}\epsilon\mu - k^2 = 0 \ .$$

However, this is just the dispersion relation for a dispersive medium. In other words: TEM-waves propagate as if they were in an unconfined medium.

Summary

- The wave propagation in nonconducting media is described by wave equations that are symmetric in E and B. Their solutions are composed of **plane monochromatic waves**, whose wave vectors and frequencies are related to one another via a medium-dependent **dispersion relation**.

- By superimposing two differently polarized waves, one generally obtains **elliptically polarized** light. Special cases of elliptical polarization are **circular** and **linear polarization**.

- If a monochromatic wave hits a boundary plane separating two different optical media, the incoming, reflected, and refracted parts are related via the **reflection** and **refraction laws**.

- In **dispersive media**, the various monochromatic waves of a wave packet have different **phase velocities**, resulting in an increasing deformation of the wave packet in time. The overall movement, i.e., the center of mass movement of the wave packet is governed by its **group velocity**.

- The **telegraph equations** are symmetric in E and H and describe the propagation of electromagnetic waves in conducting media. Here, the wave vectors are complex and lead to damped propagation (**dissipation**).

- The propagation of electromagnetic waves in **cylindrical hollow conductors** is uniquely determined by their longitudinal E and B components in conjunction with the corresponding continuity conditions. One distinguishes among three types of solutions, namely **transverse magnetic modes** (TM), **transverse electric modes** (TE), and **transverse electric and magnetic modes** (TEM).

Applications

33. Total reflection. Show that the time-averaged energy flux \overline{nS} through a boundary plane vanishes in the case of total reflection, $\varphi > \varphi_{\mathrm{TR}}$.

Solution. Referring to Fig. 2.18, we have

$$k' = k' \begin{pmatrix} 0 \\ \sin\varphi' \\ \cos\varphi' \end{pmatrix} , \; n = \begin{pmatrix} 0 \\ 0 \\ 1 \end{pmatrix} .$$

In the case of total reflection, it holds that

$$\sin\varphi' = \alpha > 1 ,$$

i.e., φ' is complex and $\cos\varphi'$ is purely imaginary:

$$\cos \varphi' = \sqrt{1 - \sin^2 \varphi'} = i\sqrt{\sin^2 \varphi' - 1} \ .$$

The time-averaged normal component of the energy flux follows as

$$\overline{nS} = \frac{c}{8\pi\mu}\text{Re}[n(E' \times B'^*)] = \frac{c^2}{8\pi\omega\mu}\text{Re}[nk'|E'_0|^2] \ .$$

Since $nk' = k' \cos \varphi'$ is purely imaginary, we arrive at $\overline{nS} = 0$.

34. Cavity resonator with circular cross section. Determine the oscillation modes of an ideal hollow metal volume with circular cross section. The cross-sectional radius is R, the length of the resonator is L, and the internal space is filled by a dispersive medium (ϵ, μ).

Solution. Due to the closed ends at $z = 0$ and $z = L$, electromagnetic waves are reflected at the ends and result in *standing waves* along the z-axis. Therefore, in the case of TM-waves, we make the following ansatz for the longitudinal parts:

$$B_z(x) = 0 \ , \ E_z(x) = \psi(x,y)\cos\left(\frac{p\pi z}{L}\right) \ , \ p = 1, 2, \dots \ .$$

The transverse parts are obtained as

$$E_t(x) = -\frac{p\pi}{L\gamma^2}\sin\left(\frac{p\pi z}{L}\right)\nabla_t\psi \ , \ B_t(x) = \frac{i\epsilon\mu\omega}{c\gamma^2}\cos\left(\frac{p\pi z}{L}\right)e_z \times \nabla_t\psi \ ,$$

with

$$\gamma^2 = \frac{\omega^2}{c^2}\epsilon\mu - \left(\frac{p\pi}{L}\right)^2 \ .$$

Note that this ansatz automatically honors the boundary conditions $E_t(z = 0) = E_t(z = L)$ at the closed ends. For TE-modes, we choose

$$E_z(x) = 0 \ , \ B_z(x) = \psi(x,y)\sin\left(\frac{p\pi z}{L}\right) \ , \ p = 1, 2, \dots \ ,$$

from which the transverse components follow as

$$E_t(x) = -\frac{i\omega}{c\gamma^2}\sin\left(\frac{p\pi z}{L}\right)e_z \times \nabla_t\psi \ , \ B_t(x) = \frac{p\pi}{L\gamma^2}\cos\left(\frac{p\pi z}{L}\right)\nabla_t\psi \ .$$

Again, the corresponding boundary conditions at the closed ends, $B_z(z = 0) = B_z(z = L)$, are already taken into account. To solve the wave equation

$$\left(\frac{\partial^2}{\partial x^2} + \frac{\partial^2}{\partial y^2} + \gamma^2\right)\psi(x,y) = 0 \ , \tag{2.123}$$

we use cylindrical coordinates as suggested by the problem's geometry,

$$x = r\cos\varphi \ , \ y = r\sin\varphi \ ,$$

so that (2.123) turns into

$$\left(\frac{\partial^2}{\partial r^2} + \frac{1}{r}\frac{\partial}{\partial r} + \frac{1}{r^2}\frac{\partial^2}{\partial \varphi^2} + \gamma^2\right)\psi(r,\varphi) = 0 \ .$$

The ansatz

$$\psi(r,\varphi) = J(r)\mathrm{e}^{\mathrm{i}m\varphi} \ , \quad m = 0,\pm 1,\pm 2,\dots \ ,$$

leads to

$$\frac{\mathrm{d}^2 J}{\mathrm{d}r^2} + \frac{1}{r}\frac{\mathrm{d}J}{\mathrm{d}r} + \left(\gamma^2 - \frac{m^2}{r^2}\right)J = 0 \ .$$

Using the substitution $x = \gamma r$, this equation can be cast into the *Bessel differential equation* (see Section A.5)

$$\frac{\mathrm{d}^2 J}{\mathrm{d}x^2} + \frac{1}{x}\frac{\mathrm{d}J}{\mathrm{d}x} + \left(1 - \frac{m^2}{x^2}\right)J = 0 \ ,$$

which is solved by the *Bessel functions*

$$J_m(x) = \left(\frac{x}{2}\right)^m \sum_{i=0}^{\infty} \frac{(-1)^i}{i!(i+m)!}\left(\frac{x}{2}\right)^{2i} \ .$$

For TM-modes, we therefore obtain the longitudinal components

$$E_z(\boldsymbol{x}) = J_m(\gamma r)\mathrm{e}^{\mathrm{i}m\varphi}\cos\left(\frac{p\pi z}{L}\right) \ , \quad B_z = 0 \ .$$

The corresponding boundary condition at the surface shell yields the constraint

$$E_z(\boldsymbol{x})\big|_{\sqrt{x^2+y^2}=R} = 0 \Longrightarrow J_m(\gamma R) = 0 \ .$$

Therefore, the allowed eigenvalues γ of the wave equation are determined by the zeros of the Bessel functions (indicated by n). For rotational symmetric eigenoscillations ($m = 0$), they are given by

$$\gamma_{m=0,n=0}R = 2.405 \ , \quad \gamma_{m=0,n=1}R = 5.520 \ , \quad \gamma_{m=0,n=2}R = 8.654 \ ,\dots \ .$$

The eigenfrequencies belonging to γ_{mn} are

$$\omega_{mnp}^2 = \frac{c}{\sqrt{\epsilon\mu}}\sqrt{\gamma_{mn}^2 + \left(\frac{p\pi}{L}\right)^2} \ .$$

For TE-modes, we have the longitudinal fields

$$E_z(\boldsymbol{x}) = 0 \ , \quad B_z(\boldsymbol{x}) = J_m(\gamma r)\mathrm{e}^{\mathrm{i}m\varphi}\sin\left(\frac{p\pi z}{L}\right) \ ,$$

with the following constraint:

$$\boldsymbol{n}\boldsymbol{\nabla} B_z(x,y)\big|_{\sqrt{x^2+y^2}=R} = 0 \Longrightarrow \frac{\mathrm{d}J_m(x)}{\mathrm{d}x} = 0\bigg|_{x=\gamma R} = 0 \ .$$

It relates the allowed γ with the zeros of the first derivative of the Bessel functions. For $m = 0$, the first few zeros read

$$x_{0,n} = \gamma_{0,n}R = 3.832 \ , \quad 7.016 \ , \quad 10.173 \ , \quad \dots \ .$$

In general, the x_{mn} lead to TE-eigenfrequencies ω_{mnp}, which are different from the TM-eigenfrequencies.

2.8 Lagrange Formalism in Electrodynamics

The Lagrange and Hamilton formalisms play a less dominant role in electro-dynamics than, for example, in classical mechanics. In classical mechanics, the Lagrange function provides a direct route to the equations of motion. However, in electrodynamics, these equations are already given in the form of Maxwell's equations. Nevertheless, the Lagrange formalism yields an alternative approach to electrodynamics, which is of particular interest with a view to the construction of other field theories. In this section we there-fore present the most important results of Lagrangian electrodynamics. First, the Lagrange formulation of the equations of motion of a particle within a given electromagnetic field are considered. This formalism is then extended to the fields themselves. Furthermore, we discuss the conservation laws ema-nating from the formalism and close by considering the relationship between symmetries and the principle of *gauge invariance*.

2.8.1 Lagrange and Hamilton Functions of a Charged Particle

According to Theorem 2.10, we have the following relativistic equation of motion for a particle of rest mass m_0 and charge q moving through the electromagnetic fields \boldsymbol{E} and \boldsymbol{B}:

$$\frac{\mathrm{d}}{\mathrm{d}t} \frac{m_0\dot{\boldsymbol{x}}}{\sqrt{1 - \frac{\dot{\boldsymbol{x}}^2}{c^2}}} = \dot{\boldsymbol{p}} = \boldsymbol{F}_\mathrm{L} = q\left(\boldsymbol{E} + \frac{\dot{\boldsymbol{x}}}{c} \times \boldsymbol{B}\right) \ .$$

Expressing the fields in terms of scalar and vector potentials leads to

$$\frac{\mathrm{d}}{\mathrm{d}t} \frac{m_0\dot{\boldsymbol{x}}}{\sqrt{1 - \frac{\dot{\boldsymbol{x}}^2}{c^2}}} = q\left[-\boldsymbol{\nabla}\phi - \frac{1}{c}\frac{\partial \boldsymbol{A}}{\partial t} + \frac{\dot{\boldsymbol{x}}}{c} \times (\boldsymbol{\nabla} \times \boldsymbol{A})\right]$$

$$= q\left[-\boldsymbol{\nabla}\phi - \frac{1}{c}\frac{\partial \boldsymbol{A}}{\partial t} + \frac{1}{c}\boldsymbol{\nabla}(\boldsymbol{A}\dot{\boldsymbol{x}}) - \frac{1}{c}(\dot{\boldsymbol{x}}\boldsymbol{\nabla})\boldsymbol{A}\right]$$

$$= q\boldsymbol{\nabla}\left(\frac{\boldsymbol{A}\dot{\boldsymbol{x}}}{c} - \phi\right) - \frac{q}{c}\frac{\mathrm{d}\boldsymbol{A}}{\mathrm{d}t} \ . \tag{2.124}$$

Comparing this expression with the Lagrange equation in vector form,

$$\boldsymbol{\nabla}_{\boldsymbol{x}}L - \frac{\mathrm{d}}{\mathrm{d}t}\boldsymbol{\nabla}_{\dot{\boldsymbol{x}}}L = \boldsymbol{0} \ ,$$

we see that (2.124) is obtained from the Lagrange function

$$L = -m_0c^2\sqrt{1 - \frac{\dot{\boldsymbol{x}}^2}{c^2}} - q\phi + \frac{q}{c}\boldsymbol{A}\dot{\boldsymbol{x}} \ . \tag{2.125}$$

From this, the generalized momentum \boldsymbol{P} of the particle is found to be

$$\boldsymbol{P} = \boldsymbol{\nabla}_{\dot{\boldsymbol{x}}}L = \boldsymbol{p} + \frac{q}{c}\boldsymbol{A} = \frac{m_0\dot{\boldsymbol{x}}}{\sqrt{1 - \frac{\dot{\boldsymbol{x}}^2}{c^2}}} + \frac{q}{c}\boldsymbol{A} \ . \tag{2.126}$$

Therefore, the corresponding Hamilton function is given by

$$H = \boldsymbol{P}\dot{\boldsymbol{x}} - L = \frac{m_0 c^2}{\sqrt{1 - \frac{\dot{x}^2}{c^2}}} + q\phi .$$

With the help of (2.126), we eliminate $\dot{\boldsymbol{x}}$ in favor of \boldsymbol{P} and finally obtain

$$H = \sqrt{m_0^2 c^4 + c^2 \left(\boldsymbol{P} - \frac{q}{c}\boldsymbol{A}\right)^2} + q\phi .$$

Theorem 2.28: Lagrange and Hamilton functions of a charge q in an electromagnetic field

$$L = -m_0 c^2 \sqrt{1 - \frac{\dot{x}^2}{c^2}} - q\phi + \frac{q}{c}\boldsymbol{A}\dot{\boldsymbol{x}}$$

$$H = \sqrt{m_0^2 c^4 + c^2 \left(\boldsymbol{P} - \frac{q}{c}\boldsymbol{A}\right)^2} + q\phi .$$

From Subsection 1.6.5, we know that the relativistic Lagrange function of a free particle is given by

$$L_{\text{free}} = -m_0 c^2 \sqrt{1 - \frac{\dot{x}^2}{c^2}} .$$

Therefore, we conclude that the Lagrange function

$$L' = L - L_{\text{free}} = -q\phi + \frac{q}{c}\boldsymbol{A}\dot{\boldsymbol{x}} = -\frac{q}{\gamma c} u_\mu A^\mu , \quad \gamma = \frac{1}{\sqrt{1 - \frac{\dot{x}^2}{c^2}}}$$

describes the interaction of the particle with an electromagnetic field.

2.8.2 Lagrange Density of the Electromagnetic Field

We now show how the relativistic Lagrange formalism of point particles can be applied to continuous fields. To this end, consider initially a system of fields $\phi_i(x)$ that depend on the four-vectors x^μ. Our starting point is the Lagrange function L, which we generalize to the *Lagrange density* $\mathcal{L}(\phi_i, \partial_\mu \phi_i, x)$ of the fields ϕ_i. The action S is obtained by integrating over the four-dimensional space-time:

$$S = \int \mathrm{d}^4 x \mathcal{L}(\phi_i, \partial_\mu \phi_i, x) .$$

At this stage it is reasonable to introduce some constraints on the form of the Lagrange density. First, we need to guarantee that our formalism leads to the right equations of motion – in the case of electrodynamics these are Maxwell's

equations. This implies that the Lagrange density can contain derivatives of
the fields only up to first order. Second, we wish to consider only local field
theories so that the Lagrange density can depend on the fields at only one
single location x. Finally, \mathcal{L} should be altered at most by a total divergence
under gauge transformations, so that the action remains invariant.

Lagrange equations. Let us first consider the variation of the action by
the variation $\delta\phi$ of an arbitrary field component $\phi(x)$:

$$\delta S = \int d^4x \left[\frac{\partial \mathcal{L}}{\partial \phi} \delta\phi + \frac{\partial \mathcal{L}}{\partial(\partial_\mu \phi)} \delta(\partial_\mu \phi) \right]$$

$$= \int d^4x \left[\frac{\partial \mathcal{L}}{\partial \phi} - \partial_\mu \frac{\partial \mathcal{L}}{\partial(\partial_\mu \phi)} \right] \delta\phi + \int d^4x \partial_\mu \left[\frac{\partial \mathcal{L}}{\partial(\partial_\mu \phi)} \delta\phi \right] . \qquad (2.127)$$

In analogy to Subsection 1.2.3, $\phi(x)$ is assumed to be fixed on the space-time
hyperplane at infinity ("fixed endpoints") or to vanish there:

$$\delta\phi(x)|_{|x^\mu|\to\infty} = [\phi'(x) - \phi(x)]_{|x^\mu|\to\infty} = 0 .$$

Since the last term of (2.127) is a total divergence, it can be transformed,
using the four-dimensional Gauss theorem, into the surface integral

$$\int d^4x \partial_\mu \left[\frac{\partial \mathcal{L}}{\partial(\partial_\mu \phi)} \delta\phi \right] = \oint d\sigma_\mu \left[\frac{\partial \mathcal{L}}{\partial(\partial_\mu \phi)} \delta\phi \right] ,$$

which, according to our assumption, vanishes. Therefore, demanding station-
arity of the action yields the following *Lagrange equation* for the field ϕ:

$$\delta S = 0 \implies \frac{\partial \mathcal{L}}{\partial \phi} - \partial_\mu \frac{\partial \mathcal{L}}{\partial(\partial_\mu \phi)} = 0 . \qquad (2.128)$$

Lagrange density of the electromagnetic field. Due to the above con-
straints on the Lagrange density, one can show that \mathcal{L} can contain only
Lorentz scalars of the form

$$\partial_\mu A^\nu \partial^\mu A_\nu , \quad \partial_\mu A^\nu \partial_\nu A^\mu , \quad (\partial_\mu A^\mu)^2 , \quad A_\mu A^\mu , \quad j_\mu A^\mu .$$

These terms must be combined in the Lagrange function such that (2.128)
yields the inhomogeneous Maxwell equations (see Theorem 2.9)[19]

$$\partial^\mu F_{\mu\nu} = \frac{4\pi}{c} j_\nu .$$

In Application 35, we show that the following Lagrange density satisfies this
condition:

Theorem 2.29: Lagrange density of the electromagnetic field

$$\mathcal{L} = -\frac{1}{16\pi} F_{\mu\nu} F^{\mu\nu} - \frac{1}{c} j_\mu A^\mu + \text{total divergence} .$$

[19] Due to the definition of $F^{\mu\nu}$, the homogeneous Maxwell equations $\partial_\mu G^{\mu\nu} = 0$
are satisfied automatically.

Gauge invariance. Applying the gauge transformation

$$A^\mu \longrightarrow A'^\mu = A^\mu - \partial^\mu \chi \, ,$$

we find that \mathcal{L}' contains an additional term $j_\mu \partial^\mu \chi / c$ compared to \mathcal{L}. However, this term can be expressed as a total divergence, provided that the current is conserved:

$$j_\mu \partial^\mu \chi = \partial^\mu (j_\mu \chi) \, .$$

In other words, conservation of the four-current j^μ is necessary and sufficient for the theory to be gauge-invariant.

Hamilton formalism. Having found a formal analogy between the Lagrange formalism of a particle and that of electromagnetic fields, we now look for a generalization of the Hamilton function

$$H = \sum_i p_i \dot{q}_i - L$$

for fields. We consider the *Hamilton density*

$$\mathcal{H} = \frac{\partial \mathcal{L}}{\partial \left(\frac{\partial \phi}{\partial t}\right)} \frac{\partial \phi}{\partial t} - \mathcal{L}$$

and generalize it to a contravariant tensor of rank 2, called the *energy-momentum tensor:*

$$\Theta^{\mu\nu} = \frac{\partial \mathcal{L}}{\partial (\partial_\mu \phi)} \partial^\nu \phi - g^{\mu\nu} \mathcal{L} \, . \tag{2.129}$$

Note that in the present form, and in conjunction with Theorem 2.29, this tensor is not identical with $T^{\mu\nu}$ from (2.46) in Subsection 2.3.5. In particular, $\Theta^{\mu\nu}$ is not necessarily symmetric, as required by angular momentum conservation. However, we can always add to $\Theta^{\mu\nu}$ a term

$$\bar{\Theta}^{\mu\nu} = \partial_\kappa \varphi^{\kappa\mu\nu} \, ,$$

with

$$\varphi^{\kappa\mu\nu} = -\varphi^{\mu\kappa\nu} \, , \quad \partial_\mu \partial_\kappa \varphi^{\kappa\mu\nu} = 0 \, .$$

In the case of electrodynamics, this term is given by

$$\bar{\Theta}^{\mu\nu} = \frac{1}{4\pi} \partial_\kappa (F^{\kappa\mu} A^\nu) \, ,$$

so that the symmetrized energy-momentum tensor of electrodynamics,

$$\Theta^{\mu\nu} + \bar{\Theta}^{\mu\nu} = T^{\mu\nu} = \frac{1}{4\pi} \left(g^{\mu\kappa} F_{\kappa\lambda} F^{\lambda\nu} + \frac{1}{4} g^{\mu\nu} F_{\kappa\lambda} F^{\kappa\lambda} \right) \, ,$$

is now indeed identical with the stress tensor (2.46).

2.8.3 Conservation Laws and the Noether Theorem

Proceeding in a similar manner to Subsection 1.2.2, we shall now show the relation between symmetries and conservation laws. For this purpose, we initially consider a single scalar field $\phi(x)$ and calculate the variation of the action under a very general simultaneous variation of x and ϕ:

$$x^\mu \longrightarrow x'^\mu = x^\mu + \delta x^\mu \ , \ \phi(x) \longrightarrow \phi'(x') = \phi(x) + \Delta\phi(x) \ .$$

Here, we stress that $\Delta\phi$ is the total variation of ϕ at two different space-time points. It is connected via

$$\left.\begin{aligned}
\Delta\phi(x) &= \phi'(x') - \phi(x') + \phi(x') - \phi(x) = \delta\phi + (\partial_\nu\phi)\delta x^\nu \\
\delta\phi(x)|_{|x^\mu|\to\infty} &= 0 \ ,
\end{aligned}\right\} \tag{2.130}$$

with the variation $\delta\phi$ of ϕ at the same space-time point. The last relation again reflects that ϕ is fixed at the space-time boundary. To evaluate the variation

$$\delta S = \int \delta(\mathrm{d}^4 x)\mathcal{L} + \int \mathrm{d}^4 x \delta\mathcal{L} \ ,$$

we still need to know the functional determinant

$$\mathrm{d}^4 x' = \left|\det\left[\frac{\partial x'^\mu}{\partial x^\nu}\right]\right| \mathrm{d}^4 x = |\det[\delta^\mu_\nu + \partial_\nu(\delta x^\mu)]| \, \mathrm{d}^4 x = [1 + \partial_\mu(\delta x^\mu)]\mathrm{d}^4 x$$

$$\Longrightarrow \delta(\mathrm{d}^4 x) = \partial_\mu(\delta x^\mu)\mathrm{d}^4 x.$$

Overall, we find

$$\begin{aligned}
\delta S &= \int \mathrm{d}^4 x \mathcal{L}\partial_\mu(\delta x^\mu) + \int \mathrm{d}^4 x \left[(\partial_\mu\mathcal{L})\delta x^\mu + \frac{\partial\mathcal{L}}{\partial\phi}\delta\phi + \frac{\partial\mathcal{L}}{\partial(\partial_\mu\phi)}\partial_\mu(\delta\phi)\right] \\
&= \int \mathrm{d}^4 x \left[\frac{\partial\mathcal{L}}{\partial\phi} - \partial_\mu\frac{\partial\mathcal{L}}{\partial(\partial_\mu\phi)}\right]\delta\phi + \delta S_\sigma \ ,
\end{aligned}$$

with

$$\begin{aligned}
\delta S_\sigma &= \int \mathrm{d}^4 x \partial_\mu(\mathcal{L}\delta x^\mu) + \int \mathrm{d}^4 x \partial_\mu\left[\frac{\partial\mathcal{L}}{\partial(\partial_\mu\phi)}\delta\phi\right] \\
&= \int \mathrm{d}^4 x \partial_\mu(\mathcal{L}\delta x^\mu) + \int \mathrm{d}^4 x \partial_\mu\left[\frac{\partial\mathcal{L}}{\partial(\partial_\mu\phi)}(\Delta\phi - (\partial_\nu\phi)\delta x^\nu)\right] \\
&= \int \mathrm{d}^4 x \partial_\mu\left\{\frac{\partial\mathcal{L}}{\partial(\partial_\mu\phi)}\Delta\phi + \left[\mathcal{L}g^{\mu\nu} - \frac{\partial\mathcal{L}}{\partial(\partial_\mu\phi)}(\partial^\nu\phi)\right]\delta x_\nu\right\} \\
&= \int \mathrm{d}^4 x \partial_\mu\left\{\frac{\partial\mathcal{L}}{\partial(\partial_\mu\phi)}\Delta\phi - \Theta^{\mu\nu}\delta x_\nu\right\} \ ,
\end{aligned}$$

where partial integration, (2.130), and the energy-momentum tensor (2.129) have been used. Now, let us consider an infinitesimal transformation of the kind[20]

[20] In the presence of more than one field, the second relation must be replaced by $\Delta\phi_i = \Psi_{ia}\delta\epsilon^a$.

$$\delta x_\nu = X_{\nu a}(x)\delta\epsilon^a \ , \quad \Delta\phi = \Psi_a(x)\delta\epsilon^a \ ,$$

with a matrix X, a vector Ψ, and the infinitesimal parameters ϵ^a. Then, assuming that the transformed field satisfies the Lagrange equation (2.128), the invariance condition of the action leads to the current conservation law

$$\partial_\mu j_a^\mu = 0 \iff \partial_0 j_a^0 = -\boldsymbol{\nabla}\boldsymbol{j}_a \ , \quad j_a^\mu = \frac{\partial\mathcal{L}}{\partial(\partial_\mu\phi)}\Psi_a - \Theta^{\mu\nu}X_{\nu a} \ ,$$

where j_a^μ represents the *Noether current* for each index a. Furthermore, integration of this relation over a sufficiently large spatial volume yields the corresponding charge conservation law:

$$\frac{1}{c}\frac{\mathrm{d}}{\mathrm{d}t}\int\mathrm{d}^3x\, j_a^0 = -\int\mathrm{d}^3x\,\boldsymbol{\nabla}\boldsymbol{j}_a = -\oint\mathrm{d}\boldsymbol{F}\boldsymbol{j}_a = 0$$

$$\implies Q_a = \int\mathrm{d}^3x\, j_a^0 = \text{const} \ .$$

Theorem 2.30: Noether's theorem

The invariance of the action $S = \int\mathrm{d}^4x\mathcal{L}(\phi,\partial_\mu\phi,x)$ under the transformations

$$x^\mu \longrightarrow x'^\mu = x^\mu + \delta x^\mu \quad , \quad \delta x_\nu = X_{\nu a}(x)\delta\epsilon^a$$

$$\phi(x) \longrightarrow \phi'(x') = \phi(x) + \Delta\phi(x) \ , \quad \Delta\phi(x) = \Psi_a(x)\delta\epsilon^a$$

implies conservation of the Noether currents

$$j_a^\mu = \frac{\partial\mathcal{L}}{\partial(\partial_\mu\phi)}\Psi_a - \Theta^{\mu\nu}X_{\nu a} \ , \quad \partial_\mu j_a^\mu = 0$$

and of the charges

$$Q_a = \int\mathrm{d}^3x\, j_a^0 \ , \quad \frac{\mathrm{d}}{\mathrm{d}t}Q_a = 0 \ .$$

As an example of this theorem, consider a transformation that shifts the origin of time and space:

$$X_{\nu a} = \delta_{\nu a} \ , \quad \Psi_a = 0 \implies \delta x_\nu = \epsilon_\nu \ , \quad \Delta\phi = 0 \ .$$

This immediately leads to

$$j_a^\mu = -\Theta_a^\mu \ , \quad \int\Theta_a^0\mathrm{d}^3x = \text{const} \ .$$

From our previous discussion, we know that $\int\Theta_\nu^0\mathrm{d}^3x$ is the four-momentum of the field. This means that the invariance of the action integral under space-time translations leads to conservation of momentum and energy of the electromagnetic field. This is totally analogous to our findings in classical

mechanics. In a similar manner, we could derive the conservation of angular momentum and the center of mass law from the invariance of the action under rotations of x^μ and under translations of x^μ with constant velocity.

2.8.4 Internal Symmetries and Gauge Principle

The 10 classical conservation laws follow from the invariance of the action under transformations concerning space and time. If one is dealing with more than one field component, then additional symmetries under internal transformations with $\delta x^\mu = 0$ are possible (*internal symmetries*). In this case, the invariance of the action is tantamount to the invariance of the Lagrange density, up to a possible total divergence.

In the following, we consider the simple example of a two-component field and discuss the invariance of the Lagrange density under rotations of these two components. We start with a *global gauge transformation*, which transforms both components at all locations in the same way. Following this, we shall see what implications arise if we demand that the theory is invariant even under *local gauge transformations*. This will lead us to the principle of local gauge invariance, which is of great importance, particularly in high-energy physics.

Global gauge invariance. As our starting point we choose the Lagrange density

$$\mathcal{L} = (\partial_\mu \phi)(\partial^\mu \phi^*) - m^2 \phi \phi^* , \qquad (2.131)$$

which depends on the two-component scalar fields

$$\phi = \frac{1}{\sqrt{2}}(\phi_1 + i\phi_2) , \quad \phi^* = \frac{1}{\sqrt{2}}(\phi_1 - i\phi_2) .$$

This describes, for example, the relativistic motion of a free electron with two internal states (spin up and spin down). The Lagrange equations yield the *Klein-Gordon equations*

$$(\partial_\mu \partial^\mu + m^2)\phi = 0 , \quad (\partial_\mu \partial^\mu + m^2)\phi^* = 0 ,$$

which are relativistic generalizations of the Schrödinger equation. In addition to the space-time symmetries mentioned above, the Lagrange density (2.131) has an additional symmetry in that it is invariant under the internal global transformations

$$\left.\begin{array}{l} \phi(x) \longrightarrow \phi'(x) = e^{-iq\Lambda}\phi(x) \approx \phi(x)(1 - iq\Lambda) \\[2mm] \phi^*(x) \longrightarrow \phi'^*(x) = e^{iq\Lambda}\phi^*(x) \approx \phi^*(x)(1 + iq\Lambda) \end{array}\right\} , \quad \delta x^\mu = 0 ,$$

where q and Λ are real constants. In the above notation we have

$$X_{\nu a} = 0 , \quad \Psi = -iq\phi , \quad \Psi^* = iq\phi^* , \quad \delta\epsilon = \Lambda ,$$

from which the corresponding conserved Noether current follows:

$$j^\mu = \frac{\partial\mathcal{L}}{\partial(\partial_\mu \phi)}\Psi + \frac{\partial\mathcal{L}}{\partial(\partial_\mu \phi^*)}\Psi^* = iq(\phi^* \partial^\mu \phi - \phi \partial^\mu \phi^*) .$$

Local gauge invariance. Next, we demand that the theory also remain invariant if we perform a local instead of a global gauge transformation. To this end, we replace the constant Λ by a function $\Lambda(x)$. Using

$$\phi'(x) = [1 - iq\Lambda(x)]\phi(x) , \quad \phi'^*(x) = [1 + iq\Lambda(x)]\phi^*(x) ,$$

and

$$\partial_\mu \phi' = (1 - iq\Lambda)\partial_\mu \phi - iq\phi\partial_\mu\Lambda , \quad \partial^\mu \phi'^* = (1 + iq\Lambda)\partial^\mu \phi^* + iq\phi^*\partial^\mu\Lambda ,$$

we see that the terms $\partial_\mu \phi$ and $\partial^\mu \phi^*$ transform differently to ϕ and ϕ^*. Consequently, the Lagrange density (2.131) can no longer be invariant under these transformations. Inserting in the primed quantities and exploiting the Lagrange equations for ϕ and ϕ^* lead to the change

$$\delta\mathcal{L} = -iq\Lambda\partial_\mu \left[\frac{\partial\mathcal{L}}{\partial(\partial_\mu\phi)}\phi \right] - iq\frac{\partial\mathcal{L}}{\partial(\partial_\mu\phi)}\phi\partial_\mu\Lambda - (\phi \to \phi^*)$$

in the Lagrange density. The first term is a total divergence and causes no change in the action. Therefore, we are left with

$$\delta\mathcal{L} = -iq\left[\phi(\partial^\mu\phi^*)\partial_\mu\Lambda - \phi^*(\partial^\mu\phi)\partial_\mu\Lambda\right] = j^\mu\partial_\mu\Lambda .$$

However, a way to rescue the local gauge invariance of this theory is to add two additional terms

$$\mathcal{L}_1 = -j^\mu A_\mu = -iq(\phi^*\partial^\mu\phi - \phi\partial^\mu\phi^*)A_\mu$$
$$\mathcal{L}_2 = q^2 A_\mu A^\mu \phi^* \phi$$

to \mathcal{L} and to define a rule for the transformational behavior of the four-vector A_μ under local gauge transformations:

$$A_\mu(x) \longrightarrow A'_\mu(x) = A_\mu(x) + \partial_\mu\Lambda(x) .$$

Using these directives, we find, as desired,

$$\delta\mathcal{L} + \delta\mathcal{L}_1 + \delta\mathcal{L}_2 = 0 .$$

We see that local gauge invariance of the theory is ensured only if the current j^μ is coupled to a new field A^μ. The strength of the coupling is given by the *coupling constant* q, which, in turn, is identified with the charge q of the field ϕ. Our theory is completed by allowing a kinetic quadratic term for the field A^μ, which, of course, must also satisfy local gauge invariance. From previous considerations we already know that

$$\mathcal{L}_3 = -\frac{1}{16\pi}F_{\mu\nu}F^{\mu\nu} , \quad F_{\mu\nu} = \partial_\mu A_\nu - \partial_\nu A_\mu$$

satisfies this condition. The complete locally gauge-invariant Lagrange density is therefore given by

$$\mathcal{L} = (\partial_\mu + iqA_\mu)\phi(\partial^\mu - iqA^\mu)\phi^* - m^2\phi^*\phi - \frac{1}{16\pi}F_{\mu\nu}F^{\mu\nu} . \tag{2.132}$$

In summary, we obtain the following remarkable result: to extend global gauge invariance of our theory to local gauge invariance, we must introduce an electromagnetic field A^μ called the *gauge potential*, with a particular trans-formational behavior, and couple it to the Noether current j^μ. This is the content of the *gauge principle*. By adding a quadratic term in A^μ, the gauge potential itself becomes a dynamical object of the theory.

Introducing the *gauge-covariant derivatives*

$$D_\mu = \partial_\mu + iqA_\mu \ , \ D^{*\mu} = \partial^\mu - iqA^\mu \ ,$$

with

$$D'_\mu \phi' = (1 - iq\Lambda)D_\mu \phi \ , \ D'^{*\mu}\phi'^* = (1 + iq\Lambda)D^\mu \phi^* \ ,$$

we can write (2.132) as

$$\mathcal{L} = (D_\mu \phi)(D^{*\mu}\phi^*) - m^2 \phi^* \phi - \frac{1}{16\pi}F_{\mu\nu}F^{\mu\nu} \ .$$

The Lagrange equations for the fields A^μ lead to

$$\partial_\nu F^{\mu\nu} = -4\pi j^\mu \ ,$$

where

$$j^\mu = iq(\phi^* D^\mu \phi - \phi D^{*\mu}\phi^*)$$

is the conserved Noether current of the locally gauge-invariant theory. Finally, note that adding a term of the type

$$m^2 A_\mu A^\mu$$

breaks gauge invariance. Therefore, we conclude that the electromagnetic field is massless.

Theorem 2.31: Gauge principle

The Lagrange density of a complex scalar field

$$\mathcal{L} = (\partial_\mu \phi)(\partial^\mu \phi^*) - m^2 \phi^* \phi$$

is invariant under the local gauge transformations

$$\phi'(x) = e^{-iq\Lambda(x)}\phi(x) \ , \ \phi'^*(x) = e^{iq\Lambda(x)}\phi^*(x)$$

if

- a vector field $A^\mu(x)$ is introduced, which transforms as

$$A_\mu \longrightarrow A'_\mu = A_\mu + \partial_\mu \Lambda \ ,$$

- and the derivatives ∂_μ and ∂^*_μ are replaced by the covariant derivatives

$$D_\mu = \partial_\mu + iqA_\mu \quad \text{and} \quad D^*_\mu = \partial_\mu - iqA_\mu \ .$$

\triangleright

The locally gauge-invariant Lagrange density is then given by

$$\mathcal{L} = D_\mu \phi D^{*\mu} \phi^* - m^2 \phi^* \phi - \frac{1}{16\pi} F_{\mu\nu} F^{\mu\nu} .$$

Demanding local gauge invariance leads necessarily to the existence of the electromagnetic field.

Summary

- The dynamics of a charged particle in an electromagnetic field can be described using a Lagrange or a Hamilton function.

- The Lagrange formalism (with few degrees of freedom) can be extended to continuous systems (with infinitely many degrees of freedom) by replacing the Lagrange function with the **Lagrange density**. In analogy to the example of point mechanics, the corresponding Lagrange equations describe the dynamics of continuous systems. In the case of electrodynamics, these lead to the Maxwell equations.

- The invariance of the action under transformations of the space-time four-vector and the fields in the Lagrange density yields a set of conservation laws. This is the content of the **Noether theorem**, which also holds for internal symmetries.

- Demanding invariance of the Lagrange theory under local gauge transforms leads to the existence of **gauge fields**.

Applications

35. Lagrange equations. Calculate the Lagrangian equations of motions following from the Lagrange density of the electromagnetic field, as well as from the Lagrange density of a real scalar particle.

Solution. The Lagrange density of the electromagnetic field is

$$\mathcal{L} = -\frac{1}{16\pi} F_{\alpha\beta} F^{\alpha\beta} - \frac{1}{c} A^\alpha j_\alpha$$

$$= -\frac{1}{16\pi} g_{\alpha\gamma} g_{\beta\rho} (\partial^\gamma A^\rho - \partial^\rho A^\gamma)(\partial^\alpha A^\beta - \partial^\beta A^\alpha) - \frac{1}{c} A^\alpha j_\alpha .$$

From this, we calculate

$$\frac{\partial \mathcal{L}}{\partial A^\nu} = -\frac{1}{c} j_\nu$$

and

$$\frac{\partial \mathcal{L}}{\partial(\partial^\mu A^\nu)} = -\frac{1}{16\pi} g_{\alpha\gamma} g_{\beta\rho} \left[(\delta_\mu^\gamma \delta_\nu^\rho - \delta_\mu^\rho \delta_\nu^\gamma) F^{\alpha\beta} + (\delta_\mu^\alpha \delta_\nu^\beta - \delta_\mu^\beta \delta_\nu^\alpha) F^{\gamma\rho} \right]$$

$$= -\frac{1}{16\pi} (F_{\mu\nu} - F_{\nu\mu} + F_{\mu\nu} - F_{\nu\mu}) = -\frac{1}{4\pi} F_{\mu\nu} \ .$$

It follows that

$$\partial^\mu F_{\mu\nu} = \frac{4\pi}{c} j_\nu \ .$$

The Lagrange density of a real scalar field is

$$\mathcal{L} = (\partial_\alpha \phi)(\partial^\alpha \phi) - m^2 \phi^2 = g^{\alpha\beta}(\partial_\alpha \phi)(\partial_\beta \phi) - m^2 \phi^2 \ .$$

Evaluating

$$\frac{\partial \mathcal{L}}{\partial \phi} = -2m^2 \phi$$

and

$$\frac{\partial \mathcal{L}}{\partial(\partial_\mu \phi)} = g^{\alpha\beta}(\delta_\alpha^\mu \partial_\beta \phi + \delta_\beta^\mu \partial_\alpha \phi) = 2\partial^\mu \phi \ ,$$

we arrive at

$$(\partial_\mu \partial^\mu + m^2)\phi = 0 \ .$$

3. Quantum Mechanics

Quantum mechanics describes microscopic phenomena on atomic length scales ($\sim 10^{-10}$ m). It was developed in the 1920s by Max Planck, Niels Bohr, Erwin Schrödinger, Werner Heisenberg, and others, when it was revealed that the behavior of the microcosmos differs fundamentally from that of the macroscopic world. Two of the main characteristics of the microscopic world are, first, the indistinguishability of wave-like and particle-like behavior of corpuscular phenomena. Depending on the experimental setup, particles, such as electrons, can be interpreted as point-like bodies or as waves. And, second, dynamical quantities of a bound system such as energy or angular momentum can take on only certain quantized values that are discrete multiples of a particular constant. Through the introduction of novel mathematical concepts (novel to physics), and by leaving behind the classical conceptions of the end of the 19th century, quantum theory could be formulated in such a way as to describe the new discoveries, and it was able to make predictions that were subsequently corroborated by experiment. As a nonrelativistic theory – it was only in the 1930s that Paul Dirac successfully attempted a formulation of relativistic quantum mechanics – its range of validity is confined to one of velocities small compared to that of light, and of binding energies small compared to the constituent masses of bound systems. Nevertheless, even today, this theory still remains the physical-mathematical foundation for the development of atomic and elementary particle theories, as well as for quantum field theories that describe the interaction of elementary particles through the exchange of virtual field quanta.

Contrary to all classical theories describing particle behavior, quantum mechanics is of probabilistic nature and makes predictions about probabilities of physical systems. Each state is described by an abstract vector in a Hilbert space that represents the probability amplitude for all possible configurations of the state. The time development of these vectors, on the other hand, is completely deterministic and follows a partial differential equation called the Schrödinger equation. This is the fundamental equation of nonrelativistic quantum mechanics. Quantization of the theory, in short, is achieved by representing the erstwhile classical dynamical quantities such as energy and momentum by operators. As a consequence of the probabilistic concept,

quantum mechanical measurements differ fundamentally from classical ones. They are closely connected to the eigenvalue problem of Hermitean operators.

In the first section of this chapter we introduce the mathematical formalism of the theory, due mostly to Paul Dirac. It consists mainly of elements of linear algebra although sometimes using a new notation. We discuss the representation of quantum mechanical states as bra and ket vectors, as well as the eigenvalue problem of linear operators.

Section 3.2 deals with the physical foundation of quantum theory. We introduce the quantum mechanical postulates and discuss them in a general frame. We will see that quantum theory can be written in infinitely many abstract representations called pictures. These are all related by unitary transformations. Choosing a particular basis within such a picture yields a specific representation of the theory. Within pictures these representations are also connected by unitary transformations.

In Section 3.3 we consider, as first concrete applications, some one-dimensional systems for which the Schrödinger equation is exactly soluble, and in which typical quantum effects can be studied easily. We will encounter the effect of a wave packet dispersing with time. This effect has already been studied in the context of electrodynamics and will now be reinterpreted quantum mechanically.

Angular momentum plays a particularly special role in three-dimensional quantum mechanics and is the topic of Section 3.4. It is generally defined via commutator relations. Accordingly, the quantum mechanical counterpart of the ordinary classical angular momentum is a special case of a generic type of angular momentum quantum operator. We introduce a further operator of this kind, called spin, which has no classical counterpart and can be interpreted as the intrinsic rotation of a particle. Finally, we discuss the addition (also called coupling) of quantum mechanical angular momenta.

In Section 3.5 we discuss three-dimensional quantum systems with central symmetry. For such systems the angular part of the Schrödinger equation can be separated and leads to eigensolutions of the angular momentum operator. It remains to solve the radial Schrödinger equation, and we shall do so for some simple systems such as the naive hydrogen atom.

In Section 3.6 we discuss quantum mechanical implications of electromagnetic phenomena. From the laws of electrodynamics we shall deduce the quantum description of electron movement in external electromagnetic fields. Expanding the electrodynamical gauge principle to quantum mechanics leads to interesting new phenomena that can be verified experimentally. Additionally, we discuss the Stern-Gerlach experiment, which provides a belated justification for the spin-dynamical degree of freedom.

Since most quantum mechanical systems cannot be solved exactly, one generally has to resort to approximative methods. For static bound systems the time-independent perturbation theory is a suitable approximation, and we introduce it in Section 3.7. It will allow us to discuss adequately a more

realistic version of the hydrogen atom by taking into account the coupling between angular momentum and spin of the electron (fine structure splitting), as well as the coupling between electron and nucleus spin (hyperfine structure splitting).

Section 3.8 deals with atomic transitions. They are a consequence of the interaction between atoms and electromagnetic radiation (emission and absorption), which causes atomic electrons to change their quantum state. These systems are also not exactly soluble, and they lead us to consider time-dependent perturbation theory. With its help, we calculate matrix elements (transition rates) of some atomic transitions in the dipole approximation.

In contrast to classical mechanics, where identical particles can always be distinguished by their different trajectories, quantum mechanics knows of two types of particles: distinguishable and identical ones. This has implications for quantum mechanical many-particle systems, which are studied in Section 3.9.

In the last section of this chapter we consider the scattering of particles. The idea is to describe the scattering process by an asymptotic wave function that splits into an incoming and a scattered part. The amplitude of the scattered wave then allows us to calculate the differential cross section of the process.

3.1 Mathematical Foundations of Quantum Mechanics

In this section we introduce some mathematical concepts of relevance to quantum mechanics. In general, these are taken from linear algebra, and we therefore assume most readers to be familiar with the terrain. Consequently, we review only the bare minimum, at the same time introducing Dirac's notation of *bras* and *kets*. In addition to the central concept of *Hilbert spaces*, we discuss *linear operators* and their *eigenvalue problems*. At the end of this section we take a look at vectors and operators in particular representations (basis systems).

3.1.1 Hilbert Space

In general, each quantum mechanical state is assigned a particular kind of vector that, following Dirac's notation, is called the *ket vector* or simply *ket* and is denoted by the symbol $|\cdot\rangle$. To distinguish among different kets one inserts into this symbol one or more indices that can take on discrete or continuous values. The set of all kets forms a vector space, the so-called *Hilbert space*. With its help, it is possible to formulate quantum mechanics in a general, representation-independent manner.

Definition: Hilbert space \mathcal{H}

A Hilbert space \mathcal{H} is a linear vector space defined on the body of complex numbers with a Hermitean scalar product. Two relations are defined on Hilbert spaces:

- If $|v\rangle \in \mathcal{H}$ and $\lambda \in \mathbf{C}$, then $\lambda |v\rangle$ is also an element of the Hilbert space.

- If $|v\rangle, |u\rangle \in \mathcal{H}$, then $|v\rangle + |u\rangle \in \mathcal{H}$ (principle of superposition).

Metric of the Hilbert space

The scalar product of a ket $|u\rangle$ with a ket $|v\rangle$ is generally a complex number $\langle v||u\rangle = \langle v|u\rangle$ with the following attributes:

- The scalar product of $|v\rangle$ and $|u\rangle$ is the complex conjugate of the scalar product of $|u\rangle$ with $|v\rangle$:

$$\langle v|u\rangle = \langle u|v\rangle^* \, . \tag{3.1}$$

- Each vector has a real and positive square norm:

$$\langle u|u\rangle \geq 0 \qquad (= 0 \text{ if, and only if, } |u\rangle = 0) \, .$$

- The scalar product $\langle v|u\rangle$ is linear in $|u\rangle$ and antilinear in $|v\rangle$:

$$\langle v|\, (\lambda_1 |u_1\rangle + \lambda_2 |u_2\rangle) = \lambda_1 \langle v|u_1\rangle + \lambda_2 \langle v|u_2\rangle$$
$$((\langle v_1| \lambda_1 + \langle v_2| \lambda_2)|u\rangle = \lambda_1^* \langle v_1|u\rangle + \lambda_2^* \langle v_2|u\rangle \, , \ \lambda_{1,2} \in \mathbf{C} \, .$$

The above implies the Schwarz inequality

$$|\langle v|u\rangle|^2 \leq \langle v|v\rangle \langle u|u\rangle \, .$$

The identity holds if $|v\rangle$ and $|u\rangle$ are proportional to one another, i.e., $|v\rangle = \lambda |u\rangle$.

In quantum mechanics one deals with finite- and infinite-dimensional Hilbert spaces, depending on the system under consideration. In the following we shall generally assume, without giving a proof, that the space we are handling is complete with regard to the norm $|\langle u|u\rangle|$ defined by the scalar product.

Dual space. It is well known from linear algebra that each vector space can be assigned a dual vector space whose elements are all linear functions $\chi(|u\rangle)$ of the kets $|u\rangle$, and for which the superposition principle also holds. Such vectors are called *bra vectors*, or simply *bras*, and they are represented by the symbol $\langle \cdot |$. The scalar product of two ket vectors $|v\rangle$ and $|u\rangle$ can therefore be interpreted as the value of the linear function v applied to the ket $|u\rangle$:

$$\langle v|u\rangle = v(|u\rangle) \, .$$

A unique antilinear relation called *conjugation* defines a ket vector for each bra, and vice versa:

$$|u\rangle = \sum_i \lambda_i |i\rangle \quad \longleftrightarrow \quad \langle u| = \sum_i \langle i| \lambda_i^* . \tag{3.2}$$

To understand the following sections it will be sufficient to regard the scalar product as a functional relation between two vectors of the Hilbert space: $\langle \cdot | \cdot \rangle : \mathcal{H} \times \mathcal{H} \to \mathbf{C}$.

Definition: Orthonormal basis, projection operator

- Two vectors $|n\rangle$ and $|m\rangle$ are *orthonormal* if $\langle n| m\rangle = \delta_{nm}$.

- A minimal set of orthonormal vectors $\{|n\rangle, n = 1, 2, \ldots\}$ is called an *orthonormal basis* of a subspace of \mathcal{H} if any vector $|u\rangle$ of this space can be written as a linear combination of the form

$$|u\rangle = \sum_n |n\rangle c_n , \ c_n = \langle n| u\rangle$$

$$= \sum_n |n\rangle \langle n| u\rangle$$

$$= \mathcal{P}_{\{n\}} |u\rangle , \ \mathcal{P}_{\{n\}} = \sum_n |n\rangle \langle n| = 1 .$$

A similar relation holds for bra vectors:

$$\langle v| = \sum_n \langle v| n\rangle \langle n| = \langle v| \mathcal{P}_{\{n\}} .$$

$\mathcal{P}_{\{n\}}$ defines the *unit operator*. It consists of the sum of *projection operators* $\mathcal{P}_n = |n\rangle \langle n|$ that project onto the basis vector $|n\rangle$.

Using the unit operator, the scalar product of two vectors $|v\rangle$ and $|u\rangle$ can be written as

$$\langle v| u\rangle = \langle v| \mathcal{P}_{\{n\}} \mathcal{P}_{\{n\}} |u\rangle = \sum_{n,m} \langle v| n\rangle \langle n| m\rangle \langle m| u\rangle$$

$$= \sum_n \langle v| n\rangle \langle n| u\rangle .$$

This leads to the *completeness relation*

$$\langle u| u\rangle = \sum_n |\langle n| u\rangle|^2 .$$

With the help of this relation, one can verify explicitly whether a particular set of orthonormal vectors forms a basis set.

Improper Hilbert vectors. We shall see in later sections that it is also necessary to consider vectors whose norm is not finite and which depend on at least one continuous index. These *improper Hilbert vectors* do not, in all stringency, belong to \mathcal{H}. However, linear combinations of the form

$$|\omega\rangle = \int_{\nu_1}^{\nu_2} \lambda(\nu) \, |\nu\rangle \, d\nu$$

are indeed elements of the Hilbert space, satisfying all the corresponding attributes. For these types of vectors, the above definitions and relations need to be extended in the following way:

Definition: Orthonormal basis, projection operator (for continuous indices)

- Two improper Hilbert vectors $|\mu\rangle , |\nu\rangle$ are orthonormal if $\langle \mu | \nu \rangle = \delta(\mu - \nu)$.

- A minimal set of orthonormal improper Hilbert vectors $\{|\nu\rangle , \nu \in \mathbb{R}\}$ is called orthonormal basis of a subspace of \mathcal{H} if any proper vector $|u\rangle$ of this subspace can be written as an integral over ν such that

$$|u\rangle = \int d\nu \, |\nu\rangle \, \lambda(\nu) \, , \quad \lambda(\nu) = \langle \nu | u \rangle$$

$$= \int d\nu \, |\nu\rangle \, \langle \nu | u \rangle$$

$$= \mathcal{P}_{\{\nu\}} |u\rangle \, , \quad \mathcal{P}_{\{\nu\}} = \int d\nu \, |\nu\rangle \, \langle \nu | = \mathbf{1} \, .$$

Using this definition, the scalar product $\langle v | u \rangle$ is obtained in complete analogy to the finite discrete case:

$$\langle v | u \rangle = \langle v | \, \mathcal{P}_{\{\nu\}} \mathcal{P}_{\{\nu\}} \, | u \rangle$$

$$= \int d\nu \int d\nu' \, \langle v | \nu \rangle \, \langle \nu | \nu' \rangle \, \langle \nu' | u \rangle$$

$$= \int d\nu \int d\nu' \, \langle v | \nu \rangle \, \delta(\nu - \nu') \, \langle \nu | \nu' \rangle \, \langle \nu' | u \rangle$$

$$= \int d\nu \, \langle v | \nu \rangle \, \langle \nu | u \rangle \, .$$

The completeness relation is

$$\langle u | u \rangle = \int d\nu | \langle \nu | u \rangle |^2 \, .$$

Combined systems. In the context of N-particle systems, we will need the combination of particles from different vector spaces. To this end, we define

Definition: Tensor product of two vector spaces

Let ϵ_1 and ϵ_2 be two vector spaces and $|u^{(1)}\rangle \in \epsilon_1$, $|u^{(2)}\rangle \in \epsilon_2$. The kets $|u^{(1)}\rangle \otimes |u^{(2)}\rangle = |u^{(1)}; u^{(2)}\rangle$ span the *tensor space* $\epsilon_1 \otimes \epsilon_2$ that, per definition, has the following properties:

\triangleright

- $\dim(\epsilon_1 \otimes \epsilon_2) = \dim(\epsilon_1) \cdot \dim(\epsilon_2)$ (as long as $\epsilon_{1,2}$ are finite-dimensional).
- Commutativity: $\left| u^{(1)}; u^{(2)} \right\rangle = \left| u^{(2)}; u^{(1)} \right\rangle$.
- Distributivity of addition:

$$\left| u^{(1)} \right\rangle = \lambda \left| v^{(1)} \right\rangle + \mu \left| w^{(1)} \right\rangle$$

$$\implies \left| u^{(1)}; u^{(2)} \right\rangle = \lambda \left| v^{(1)}; u^{(2)} \right\rangle + \mu \left| w^{(1)}; u^{(2)} \right\rangle$$

and

$$\left| u^{(2)} \right\rangle = \lambda \left| v^{(2)} \right\rangle + \mu \left| w^{(2)} \right\rangle$$

$$\implies \left| u^{(1)}; u^{(2)} \right\rangle = \lambda \left| u^{(1)}; v^{(2)} \right\rangle + \mu \left| u^{(1)}; w^{(2)} \right\rangle .$$

Let $\boldsymbol{A}^{(1)}$ and $\boldsymbol{A}^{(2)}$ denote two *operators* (see next subsection) on vector spaces ϵ_1 and ϵ_2, respectively. We have:

$$\boldsymbol{A}^{(1)} \left| u^{(1)} \right\rangle = \left| v^{(1)} \right\rangle \implies \boldsymbol{A}^{(1)} \left| u^{(1)}; u^{(2)} \right\rangle = \left| v^{(1)}; u^{(2)} \right\rangle$$

$$\boldsymbol{A}^{(2)} \left| u^{(2)} \right\rangle = \left| v^{(2)} \right\rangle \implies \boldsymbol{A}^{(2)} \left| u^{(1)}; u^{(2)} \right\rangle = \left| u^{(1)}; v^{(2)} \right\rangle .$$

For the *commutator* (see page 234), we find

$$[\boldsymbol{A}^{(1)}, \boldsymbol{A}^{(2)}] = \boldsymbol{A}^{(1)} \boldsymbol{A}^{(2)} - \boldsymbol{A}^{(2)} \boldsymbol{A}^{(1)} = 0 ,$$

which states that the operators on ϵ_1 commute with those on ϵ_2. Similar relations hold for the dual tensor space.

3.1.2 Linear Operators[1]

Linear operators play a central role in quantum mechanics, and they are closely related to physical quantities. In addition, changes of the state of a system, particularly those caused by measurements, are represented by linear operators.

Definition: Linear operator A

A linear operator \boldsymbol{A} assigns each ket (bra) of a subspace of the Hilbert space \mathcal{H}, the domain, a ket (bra) of a subspace of \mathcal{H}, the co-domain:

$$\boldsymbol{A} \left| u \right\rangle = \left| v \right\rangle , \quad \left\langle u' \right| \boldsymbol{A} = \left\langle v' \right| .$$

\triangleright

[1] To avoid confusion with classical quantities, quantum mechanical operators are always represented by boldfaced capital letters. This holds equally for vector-like operators (e.g., angular momentum \boldsymbol{L}), and for scalar-like operators (e.g., the three Cartesian components of angular momentum: $\boldsymbol{L_x}$, $\boldsymbol{L_y}$, $\boldsymbol{L_z}$).

It has the following properties:

$$A(\lambda_1 |u_1\rangle + \lambda_2 |u_2\rangle) = \lambda_1 A |u_1\rangle + \lambda_2 A |u_2\rangle$$
$$(\langle u_1| \lambda_1 + \langle u_2'| \lambda_2)A = (\langle u_1'| A)\lambda_1 + (\langle u_2'| A)\lambda_2 \ .$$

Two operators are identical if they are defined on the same domain and if each state is projected by these operators onto the same state of the co-domain.

Thus, $u(A |v\rangle)$ is a linear function of $|v\rangle$ since both u and A are linear. It follows that

$$u(A |v\rangle) = (\langle u| A) |v\rangle = \langle u| A |v\rangle \ .$$

The order of applying u and A on $|v\rangle$ is not important, and the brackets can be dropped. The most important algebraic operations that can be performed with operators are

- Multiplication with a constant c:

$$(cA) |u\rangle = c(A |u\rangle) \ , \ \langle u'| (cA) = (\langle u'| A)c \ .$$

- Operator sum $S = A + B$:

$$S |u\rangle = A |u\rangle + B |u\rangle \ , \ \langle u'| S = \langle u'| A + \langle u'| B \ .$$

- Operator product $P = AB$:

$$P |u\rangle = A(B |u\rangle) \ , \ \langle u'| P = (\langle u'| A)B \ .$$

Here, the domain of A must include the co-domain of B.

Commutator. Contrary to the sum, the product of two linear operators A and B is not, in general, commutative, and the *commutator*

$$[A, B] = AB - BA$$

does not necessarily vanish. Some useful rules obtained from the definition of the commutator are as follows:

$$[A, B] = -[B, A]$$
$$[A, B + C] = [A, B] + [A, C]$$
$$[A, BC] = [A, B]C + B[A, C]$$
$$0 = [A, [B, C]] + [B, [C, A]] + [C, [A, B]]$$
$$[A, B^n] = \sum_{s=0}^{n-1} B^s [A, B] B^{n-s-1} \ .$$

Hermitean and unitary operators. From the one-to-one relation (3.2) between bras and kets, we can find an analogous relation, called *adjunction*, between operators of the Hilbert space and the space dual to it:

$$|v\rangle = A\,|u\rangle \quad \longleftrightarrow \quad \langle v| = \langle u|\,A^\dagger\;.$$

This relation defines the linear operator A^\dagger which is the *Hermitean conjugate*, or *adjoint*, operator to A. Using the definition of the scalar product (3.1), we find the conjugation

$$\langle u|\,A\,|v\rangle = \langle v|\,A^\dagger\,|u\rangle^*$$

and therefore

$$(A^\dagger)^\dagger = A \qquad , \quad (cA)^\dagger = c^*A^\dagger\,,\ c \in \mathbf{C}\,,$$

$$(A + B)^\dagger = A^\dagger + B^\dagger \quad , \quad (AB)^\dagger = B^\dagger A^\dagger\;.$$

Forming the adjoint of an operator corresponds to performing the conjugation between bras and kets, or finding the complex conjugate of a complex number.

Definition: Hermitean, anti-Hermitean operators

- A linear operator H is called *Hermitean* if it is identical to its adjoint: $H = H^\dagger$.

- A linear operator I is called *anti-Hermitean* if it is identical to the negative of its adjoint: $I = -I^\dagger$.

This definition implies that

- Any linear operator A can be written uniquely as the sum of a Hermitean and an anti-Hermitean operator:

$$A = H_A + I_A\ ,\ H_A = \frac{A + A^\dagger}{2}\ ,\ I_A = \frac{A - A^\dagger}{2}\;.$$

- Any linear combination of Hermitean operators with real coefficients is Hermitean.

Another important class of operators is given by

Definition: Unitary operators

An operator U is called *unitary* if it is equal to the inverse of its adjoint:

$$UU^\dagger = U^\dagger U = 1\;.$$

It follows that

- The product of two unitary operators is a unitary operator.

- Unitary operators leave the scalar product of vectors to which they are applied invariant.

- If U is the infinitesimal unitary transformation

$$U = 1 + i\epsilon F \; , \; |\epsilon| \ll 1 \; ,$$

then

$$UU^\dagger = 1 = (1 + i\epsilon F)(1 - i\epsilon F^\dagger) \Longrightarrow F = F^\dagger \; ,$$

i.e., the operator F is Hermitean.

- If F is Hermitean, then e^{iF} is unitary.

3.1.3 Eigenvalue Problems

Many problems of quantum mechanics can be formulated as *eigenvalue problems*.

Definition: Eigenvectors, eigenvalues

Let A be a linear operator. The complex number a for which

$$A|u\rangle = a|u\rangle$$

is called the *eigenvalue* of A, and the ket $|u\rangle$ is called the *eigenket* or *eigenstate* to the eigenvalue a. Similarly, $\langle u'|$ is the eigenbra to the eigenvalue a' if

$$\langle u'| A = \langle u'| a' \; .$$

Some consequences are as follows:

- If $|u\rangle$ is an eigenket of A, any multiple $c|u\rangle$ of this vector is an eigenket of A to the same eigenvalue.

- If there are several linearly independent eigenkets to the same eigenvalue a, any linear combination of these kets is an eigenvector of A to this eigenvalue. Thus, the set of eigenkets of A to a particular eigenvalue a forms a vector space, the *subspace to the eigenvalue* a. The level of degeneracy of a is given by the dimension of this subspace.

- Any eigenvector of A with eigenvalue a is also an eigenvector of $f(A)$ with eigenvalue $f(a)$: $A|u\rangle = a|u\rangle \Longrightarrow f(A)|u\rangle = f(a)|u\rangle$.

Similar properties hold for the eigenbras of A. The totality of all eigenvalues of an operator is called the *eigenvalue spectrum* of the operator.

The eigenvalue problem of Hermitean operators is particularly important since, in quantum mechanics, physically measurable quantities are generally identified with Hermitean operators.

Theorem 3.1: Eigenvalue problem of Hermitean operators

If A is a Hermitean operator on the Hilbert space \mathcal{H}, then

1. The ket- and bra-eigenvalue spectra are identical.

2. All eigenvalues are real.

3. Any bra vector that is the conjugate of an eigenket of A is an eigenbra with the same eigenvalue, and vice versa. In other words: the subspace of the eigenbras to a particular eigenvalue is the dual space of the subspace of the eigenkets with the same eigenvalue.

4. Eigenvectors to different eigenvalues are orthogonal.

Proof.

To 2. If

$$A = A^\dagger \quad \text{and} \quad A\,|u\rangle = a\,|u\rangle\,,$$

it follows that

$$\langle u|\,A\,|u\rangle = \langle u|\,A\,|u\rangle^* = a\,\langle u|\,u\rangle\,.$$

Since $\langle u|\,u\rangle$ is real, a is also real. The proof for the eigenvalues of the eigenbras is completely analogous.

To 1 and 3. Since all eigenvalues are real, it follows from $A\,|u\rangle = a\,|u\rangle$ that $\langle u|\,A = \langle u|\,a$. Both statements follow from this.

To 4. Let

$$A\,|u\rangle = a\,|u\rangle\,\,,\quad A\,|v\rangle = b\,|v\rangle\,\,,\quad a \neq b\,.$$

By scalar multiplication of the first equation with $\langle v|$, of the second equation with $|u\rangle$, and subsequent subtraction of the resulting relations, we find

$$(a - b)\,\langle v|\,u\rangle = 0 \Longrightarrow \langle v|\,u\rangle = 0\,.$$

Continuous spectrum. So far we have assumed that the eigenvectors belong to the Hilbert space or, equivalently, that the eigenvalue spectrum is discrete. Generally, however, the spectrum consists of a discrete and a continuous part, where the latter corresponds to eigenvectors that do not have a finite norm, and therefore do not belong to the Hilbert space. Nevertheless, by normalizing these improper Hilbert vectors to the δ-function, they can be smoothly included into the eigenvalue problem so that all statements of the previous theorem remain valid.

Observables, completeness. Consider, as an example, the following very general eigenvalue spectrum of a Hermitean operator A:

$$n, r \in \mathbb{Z}$$

$$\boldsymbol{A} \,|\, nr \rangle = a_n \,|\, nr \rangle \quad, \quad \boldsymbol{A} \,|\, \nu, \rho, r \rangle = a(\nu, \rho) \,|\, \nu, \rho, r \rangle \quad, \quad \nu, \rho \in \mathbb{R}$$

$$a_n, a(\nu, \rho) \in \mathbb{R}.$$

It consists of a discrete part $\{a_n\}$ and a continuous part $\{a(\nu, \rho)\}$. The eigenkets to the eigenvalue a_n are $|\, nr \rangle$, whereas the eigenkets to the eigenvalue $a(\nu, \rho)$ are given by $|\, \nu, \rho, r \rangle$. The corresponding degeneracies are given by the running index r. The eigenvectors can be normalized such that the following orthonormality relations hold:

$$\langle n, r | \, n', r' \rangle = \delta_{nn'} \delta_{rr'}$$

$$\langle n, r | \, \nu', \rho', r' \rangle = 0$$

$$\langle \nu, \rho, r | \, \nu', \rho', r' \rangle = \delta(\nu - \nu') \delta(\rho - \rho') \delta_{rr'} \,.$$

If these vectors span the complete space, we can expand (as a series or an integral) any vector with finite norm using this basis. In this case, the vectors are said to form a *complete system*, and the Hermitean operator is called an *observable*. Often, it can only be shown for the case of Hermitean operators with a discrete spectrum that the basis system is complete. The proof for operators with a mixed or purely continuous spectrum is usually quite complex. In the following we shall always take it for granted that the eigenket system of Hermitean operators with a discrete or continuous spectrum is a complete system. The unit operator to the above basis system is

$$\boldsymbol{P} = \boldsymbol{P}_{\{n,r\}} + \boldsymbol{P}_{\{\nu,\rho,r\}}$$

$$= \sum_{n,r} |\, nr \rangle \, \langle nr | + \sum_r \int \mathrm{d}\nu \int \mathrm{d}\rho \, |\, \nu, \rho, r \rangle \, \langle \nu, \rho, r | \,,$$

and the expansion of a ket $|\, u \rangle$ is given by

$$|\, u \rangle = \boldsymbol{P} \,|\, u \rangle$$

$$= \sum_{n,r} |\, nr \rangle \, \langle nr | \, u \rangle + \sum_r \int \mathrm{d}\nu \int \mathrm{d}\rho \, |\, \nu, \rho, r \rangle \, \langle \nu, \rho, r | \, u \rangle \,.$$

For the basis system at hand, the completeness relation reads

$$\langle u | \, u \rangle = \sum_{n,r} |\, \langle nr | \, u \rangle \,|^2 + \sum_r \int \mathrm{d}\nu \int \mathrm{d}\rho \, |\, \langle \nu, \rho, r | \, u \rangle \,|^2 \,.$$

In the next section we shall identify observables with measurable physical quantities. For this reason the following theorem is particularly important:

Theorem 3.2: Commuting observables

Two observables \boldsymbol{A} and \boldsymbol{B} commute if, and only if, they have at least one common orthonormal basis.

Proof. Let $\{|n\rangle\}$ be a complete orthonormal system of \boldsymbol{A} with $\boldsymbol{A}|n\rangle = a_n|n\rangle$. It follows that

$$\boldsymbol{A} = \boldsymbol{\mathcal{P}}_{\{n\}}\boldsymbol{A}\boldsymbol{\mathcal{P}}_{\{n\}} = \sum_{n,n'}|n\rangle\langle n|\boldsymbol{A}|n'\rangle\langle n'| = \sum_n|n\rangle\,a_n\,\langle n|$$

$$\boldsymbol{B} = \boldsymbol{\mathcal{P}}_{\{n\}}\boldsymbol{B}\boldsymbol{\mathcal{P}}_{\{n\}} = \sum_{n,n'}|n\rangle\langle n|\boldsymbol{B}|n'\rangle\langle n'|\ .$$

Therefore,

$$\boldsymbol{AB} = \sum_{n,n',n''}|n\rangle\,a_n\,\langle n|n'\rangle\langle n'|\boldsymbol{B}|n''\rangle\langle n''|$$

$$= \sum_{n,n'}|n\rangle\,a_n\,\langle n|\boldsymbol{B}|n'\rangle\langle n'|$$

$$\boldsymbol{BA} = \sum_{n,n',n''}|n\rangle\langle n|\boldsymbol{B}|n'\rangle\langle n'|n''\rangle\,a_{n''}\,\langle n''|$$

$$= \sum_{n,n'}|n\rangle\langle n|\boldsymbol{B}|n'\rangle\,a_{n'}\,\langle n'|\ .$$

Using $[\boldsymbol{A},\boldsymbol{B}] = \boldsymbol{0}$, we find

$$\langle n|\boldsymbol{B}|n'\rangle\,(a_n - a_{n'}) = 0\ .$$

If there is no degeneracy ($a_n \neq a_{n'}$ for $n \neq n'$), then

$$\langle n|\boldsymbol{B}|n'\rangle = 0 \quad \text{for} \quad n \neq n'\ , \tag{3.3}$$

so that each eigenvector of \boldsymbol{A} is also eigenvector of \boldsymbol{B}. If some eigenvalues a_n are degenerate, one can always choose basis vectors such that (3.3) is satisfied. This completes the proof in forward direction. On the other hand, if $|n\rangle$ are also eigenvectors of \boldsymbol{B}, we have

$$\boldsymbol{B}|n\rangle = b_n|n\rangle \implies \langle n|\boldsymbol{B}|n'\rangle = b_n\delta_{nn'} \implies [\boldsymbol{A},\boldsymbol{B}] = \boldsymbol{0}\ .$$

3.1.4 Representation of Vectors and Linear Operators

Let us consider an arbitrary complete, discrete and nondegenerate eigenvalue spectrum $\{|n\rangle\}$ of a Hermitean operator \boldsymbol{Q}:

$$\boldsymbol{Q}|n\rangle = q_n|n\rangle\ ,\quad \langle m|n\rangle = \delta_{mn}\ ,\quad \boldsymbol{\mathcal{P}}_Q = \sum_n|n\rangle\langle n| = 1\ .$$

Then, any kets, bras, and operators of the Hilbert space can be expanded in the eigenkets $\{|n\rangle\}$ as follows:

$$|u\rangle = \boldsymbol{\mathcal{P}}_Q|u\rangle = \sum_n|n\rangle\langle n|u\rangle$$

$$\langle v| = \langle v|\boldsymbol{\mathcal{P}}_Q = \sum_n\langle v|n\rangle\langle n| = \sum_n\langle n|v\rangle^*\langle n|$$

$$\boldsymbol{A} = \boldsymbol{\mathcal{P}}_Q\boldsymbol{A}\boldsymbol{\mathcal{P}}_Q = \sum_{n,m}|n\rangle\langle n|\boldsymbol{A}|m\rangle\langle m|\ .$$

The projections $\langle n | u \rangle$ and $\langle v | n \rangle$ onto the basis vectors, as well as $\langle n | A | m \rangle$, can be interpreted as elements of coordinate vectors $(| u \rangle)^Q$, $(\langle v |)^Q$ and matrices $(A)^Q$ that represent the abstract quantities $| u \rangle$, $\langle v |$, and A in the Q-representation:

$$(| u \rangle)^Q = \begin{pmatrix} \langle 1 | u \rangle \\ \langle 2 | u \rangle \\ \vdots \end{pmatrix} \ , \ (\langle v |)^Q = (\langle v | 1 \rangle , \langle v | 2 \rangle , \ldots)$$

$$(A)^Q = \begin{pmatrix} \langle 1 | A | 1 \rangle & \langle 1 | A | 2 \rangle & \cdots \\ \langle 2 | A | 1 \rangle & \langle 2 | A | 2 \rangle & \cdots \\ \vdots & \vdots & \vdots \end{pmatrix} .$$

We have the following properties:

- The bra-row matrix $(\langle u |)^Q$, which is the conjugate of the ket-column matrix $(| u \rangle)^Q$, is the complex conjugate of the transposed ket matrix: $\langle u | n \rangle = \langle n | u \rangle^*$.

- The adjoint matrix $(A^\dagger)^Q$ of the square operator matrix $(A)^Q$ is obtained by complex conjugation and transposition of the original matrix:

$$\left(A^\dagger \right)^Q_{mn} = \langle m | A^\dagger | n \rangle = \langle n | A | m \rangle^* = \left[(A)^Q_{nm} \right]^* .$$

It is easy to see that the algebraic operations of vectors and operators are the same as those of the matrices that represent them. The extension to the continuous case is unproblematic. The matrices then have continuous indices, and the sums in the matrix operations must be replaced by integrals.

Note that the observable Q has a particularly simple form in the Q-representation since there it is given by a diagonal matrix in which all off-diagonal elements vanish. The same is true for any function $f(Q)$, and, following Theorem 3.2, for any observable commuting with Q (whereby the degeneracy of eigenvalues may have to be used to choose a suitable basis system).

Change of representation. We consider two basis systems consisting of the eigenvectors $\{| n \rangle , n = 1, 2, \ldots\}$ of an observable Q and $\{| \chi \rangle , \chi \in \mathbb{R}\}$ of an observable Θ:

$$Q | n \rangle = q_n | n \rangle \ , \ \langle n | m \rangle = \delta_{nm} \ , \ P_Q = \sum_n | n \rangle \langle n | = 1$$

$$\Theta | \chi \rangle = \theta(\chi) | \chi \rangle \ , \ \langle \chi | \chi' \rangle = \delta(\chi - \chi') \ , \ P_\Theta = \int d\chi | \chi \rangle \langle \chi | = 1 .$$

The basis vectors of one representation can be expanded in the basis of the other representation as

$$| n \rangle = P_\Theta | n \rangle = \int d\chi | \chi \rangle \langle \chi | n \rangle \ , \ | \chi \rangle = P_Q | \chi \rangle = \sum_n | n \rangle \langle n | \chi \rangle .$$

The expansion coefficients $\langle \chi | n \rangle$ and $\langle n | \chi \rangle$ can be interpreted as the elements of matrices $S(\chi, n)$ and $T(n, \chi)$, which, in this particular case, have a continuous as well as a discrete index. Due to $\langle \chi | n \rangle = \langle n | \chi \rangle^*$, we find $T = S^\dagger$. In addition,

$$\langle \chi | \chi' \rangle = \sum_n \langle \chi | n \rangle \langle n | \chi' \rangle = \delta(\chi - \chi')$$

$$\langle n | n' \rangle = \int \mathrm{d}\chi \, \langle n | \chi \rangle \langle \chi | n' \rangle = \delta_{nn'}$$

$$\Longrightarrow SS^\dagger = 1 \ , \ TT^\dagger = S^\dagger S = 1 \ ,$$

which means that S is unitary. Now, suppose a ket $|u\rangle$ and an operator A are given in the Q-representation:

$$|u\rangle = \sum_n |n\rangle \langle n|u\rangle \ , \quad A = \sum_{n,n'} |n\rangle \langle n| A |n'\rangle \langle n'| \ .$$

It follows that in the Θ-representation

$$|u\rangle = \sum_n \int \mathrm{d}\chi \, |\chi\rangle \langle \chi|n\rangle \langle n|u\rangle$$

$$\Longleftrightarrow (|u\rangle)^\Theta = S(|u\rangle)^Q \quad \text{and accordingly} \quad (\langle v|)^\Theta = (\langle v|)^Q S^\dagger$$

and

$$A = \sum_{n,n'} \int \mathrm{d}\chi \int \mathrm{d}\chi' \, |\chi\rangle \langle \chi|n\rangle \langle n| A |n'\rangle \langle n'|\chi'\rangle \langle \chi'|$$

$$\Longleftrightarrow (A)^\Theta = S(A)^Q S^\dagger \ .$$

Thus, the transition from matrices in the Q-representation to those in the Θ-representation is given by a unitary transformation S. The elements of this matrix have the following properties:

- As a function of the column index n, the elements $\langle \chi|n\rangle$ of the χth row are the components of the row vector $(\langle \chi|)^Q$, which is the eigenbra $\langle \chi|$ of Θ in the Q-representation.

- As a function of the row index χ, the elements of the the nth column are the components of the column vector $(|n\rangle)^\Theta$, which is the eigenket $|n\rangle$ of Q in the Θ-representation.

The solution of the eigenvalue problem of an operator Θ in the Q-representation is therefore mathematically equivalent to determining the transformation S that diagonalizes the matrix $(\Theta)^Q$.

Summary

- **Proper ket vectors** are elements of the **Hilbert space**, a linear vector space of at most countable infinite dimension. They possess a finite norm with regard to the Hilbert metric. **Bra vectors** are conjugate to the kets and are elements of the **dual Hilbert space**.

- **Improper Hilbert vectors** that do not have a finite norm can be normalized to the δ-function. They satisfy an analogous set of relations as proper Hilbert vectors.

- The notion of **operators** plays a central role in quantum mechanics. Of particular interest are **Hermitean operators**, which always have real eigenvalues. Their eigenvectors are mutually orthogonal. If the eigenbasis of a Hermitean operator is complete, the operator is called an **observable**.

- Two observables commute if, and only if, they possess a common orthonormal basis.

- Expanding kets and operators in the same basis of the Hilbert space, the expansion coefficients describe these quantities in the **representation** of this particular basis.

Applications

36. Properties of projection operators. Consider a finite-dimensional subspace of the Hilbert space that is spanned by the orthonormal system $|a_1\rangle, \ldots, |a_n\rangle$. Show that the operator

$$\mathcal{P}_{\{a\}} = \sum_{i=1}^{n} |a_i\rangle \langle a_i|$$

satisfies all typical properties of projection operators:

a. $\mathcal{P}_{\{a\}}$ is linear,

b. $\mathcal{P}_{\{a\}}^2 = \mathcal{P}_{\{a\}}$,

c. $\mathcal{P}_{\{a\}}$ is Hermitean: $\mathcal{P}_{\{a\}}^\dagger = \mathcal{P}_{\{a\}}$.

What are the eigenvalues and eigenvectors of $\mathcal{P}_{\{a\}}$?

Solution.

To a. We have

$$\boldsymbol{P}_{\{a\}}(\lambda_1 \, |\psi_1\rangle + \lambda_2 \, |\psi_2\rangle) = \sum_{i=1}^{n} |a_i\rangle \, \langle a_i| \, \lambda_1\psi_1 + \lambda_2\psi_2\rangle$$

$$= \sum_{i=1}^{n} |a_i\rangle \, \langle a_i| \, \lambda_1\psi_1\rangle + \sum_{i=1}^{n} |a_i\rangle \, \langle a_i| \, \lambda_2\psi_2\rangle$$

$$= \lambda_1 \sum_{i=1}^{n} |a_i\rangle \, \langle a_i| \, \psi_1\rangle + \lambda_2 \sum_{i=1}^{n} |a_i\rangle \, \langle a_i| \, \psi_2\rangle$$

$$= \lambda_1 \boldsymbol{P}_{\{a\}} \, |\psi_1\rangle + \lambda_2 \boldsymbol{P}_{\{a\}} \, |\psi_2\rangle \ .$$

Therefore, $\boldsymbol{P}_{\{a\}}$ is linear.

To b.

$$\boldsymbol{P}^2_{\{a\}} = \sum_{i,j=1}^{n} |a_i\rangle \, \langle a_i| \, a_j\rangle \, \langle a_j| = \sum_{i,j=1}^{n} |a_i\rangle \, \delta_{ij} \, \langle a_j| = \sum_{i=1}^{n} |a_i\rangle \, \langle a_i| = \boldsymbol{P}_{\{a\}}.$$

To c.

$$\left\langle \psi_1 \left| \sum_{i=1}^{n} |a_i\rangle \, \langle a_i| \right| \psi_2 \right\rangle = \sum_{i=1}^{n} \langle \psi_1| \, a_i\rangle \, \langle a_i| \, \psi_2\rangle = \sum_{i=1}^{n} \langle \psi_2| \, a_i\rangle^* \, \langle a_i| \, \psi_1\rangle^*$$

$$= \left\langle \psi_2 \left| \sum_{i=1}^{n} |a_i\rangle \, \langle a_i| \right| \psi_1 \right\rangle^*$$

$$\Longrightarrow \boldsymbol{P}_{\{a\}} = \boldsymbol{P}^\dagger_{\{a\}} \ .$$

Clearly,

$$\boldsymbol{P}_{\{a\}} \, |a_j\rangle = \sum_{i=1}^{n} |a_i\rangle \, \langle a_i| \, a_j\rangle = \sum_{i=1}^{n} \delta_{ij} \, |a_i\rangle = |a_j\rangle \ .$$

Thus, $\boldsymbol{P}_{\{a\}}$ has the eigenvectors $|a_i\rangle$ with eigenvalues one.

37. Commuting operators. Verify that the following Hermitean operators commute and specify a simultaneous eigenbasis:

$$\Omega = \begin{pmatrix} 1 & 0 & 1 \\ 0 & 0 & 0 \\ 1 & 0 & 1 \end{pmatrix} , \ \Lambda = \begin{pmatrix} 2 & 1 & 1 \\ 1 & 0 & -1 \\ 1 & -1 & 2 \end{pmatrix} .$$

In addition, show that both matrices are diagonal in this basis representation.

Solution. We immediately see that

$$\Omega\Lambda = \Lambda\Omega = \begin{pmatrix} 3 & 0 & 3 \\ 0 & 0 & 0 \\ 3 & 0 & 3 \end{pmatrix} \Longrightarrow [\Omega, \Lambda] = \boldsymbol{0} \ .$$

To determine the eigenbases, we need to solve the eigenvalue problems

$$\boldsymbol{\Omega x} = \omega \boldsymbol{x} \ , \quad \boldsymbol{\Lambda x} = \lambda \boldsymbol{x} \ . \tag{3.4}$$

For $\boldsymbol{\Omega}$ we find

$$\begin{vmatrix} 1-\omega & 0 & 1 \\ 0 & -\omega & 0 \\ 1 & 0 & 1-\omega \end{vmatrix} = -\omega^2(\omega - 2) = 0$$

$$\implies \begin{cases} \omega_1 = 0: \ \boldsymbol{e}_1 = \dfrac{1}{\sqrt{3}} \begin{pmatrix} 1 \\ 1 \\ -1 \end{pmatrix} \ , \quad \omega_2 = 0: \ \boldsymbol{e}_2 = \dfrac{1}{\sqrt{3}} \begin{pmatrix} 1 \\ -1 \\ -1 \end{pmatrix} \\[3mm] \omega_3 = 2: \ \boldsymbol{e}_3 = \dfrac{1}{\sqrt{2}} \begin{pmatrix} 1 \\ 0 \\ 1 \end{pmatrix} \ , \end{cases}$$

whereas for $\boldsymbol{\Lambda}$

$$\begin{vmatrix} 2-\lambda & 1 & 1 \\ 1 & -\lambda & -1 \\ 1 & -1 & 2-\lambda \end{vmatrix} = -(\lambda - 2)(\lambda + 1)(\lambda - 3) = 0$$

$$\implies \begin{cases} \lambda_1 = 2: \ \boldsymbol{f}_1 = \dfrac{1}{\sqrt{3}} \begin{pmatrix} 1 \\ 1 \\ -1 \end{pmatrix} \ , \quad \lambda_2 = -1: \ \boldsymbol{f}_2 = \dfrac{1}{\sqrt{6}} \begin{pmatrix} 1 \\ -2 \\ -1 \end{pmatrix} \\[3mm] \lambda_3 = 3: \ \boldsymbol{f}_3 = \dfrac{1}{\sqrt{2}} \begin{pmatrix} 1 \\ 0 \\ 1 \end{pmatrix} \ . \end{cases}$$

As a simultaneous basis system $\{\boldsymbol{g}_i\}$ of $\boldsymbol{\Omega}$ and $\boldsymbol{\Lambda}$ we can take

$$\boldsymbol{g}_1 = \boldsymbol{e}_1 = \boldsymbol{f}_1 = \frac{1}{\sqrt{3}} \begin{pmatrix} 1 \\ 1 \\ -1 \end{pmatrix}$$

$$\boldsymbol{g}_2 = \boldsymbol{f}_2 = \frac{1}{2\sqrt{2}} (-\boldsymbol{e}_1 + 3\boldsymbol{e}_2) = \frac{1}{\sqrt{6}} \begin{pmatrix} 1 \\ -2 \\ -1 \end{pmatrix}$$

$$\boldsymbol{g}_3 = \boldsymbol{e}_3 = \boldsymbol{f}_3 = \frac{1}{\sqrt{2}} \begin{pmatrix} 1 \\ 0 \\ 1 \end{pmatrix} \ ,$$

where we note that \boldsymbol{g}_2 is a linear combination of the eigenvectors \boldsymbol{e}_1 and \boldsymbol{e}_2 of $\boldsymbol{\Omega}$ to the same eigenvalue. Transforming the eigenvalue equations (3.4) given in the canonical basis into the g-basis, we have

$$\boldsymbol{\Omega}' \boldsymbol{x}' = \omega \boldsymbol{x}' \ , \quad \boldsymbol{\Lambda}' \boldsymbol{x}' = \boldsymbol{\Lambda} \boldsymbol{x}' \ , \quad \text{with } \boldsymbol{A}' = \boldsymbol{R} \boldsymbol{A} \boldsymbol{R}^{-1} \ , \quad \boldsymbol{x}' = \boldsymbol{R} \boldsymbol{x} \ ,$$

where the transformation matrix \boldsymbol{R} is given by

$$\boldsymbol{R}^{-1} = \frac{1}{\sqrt{6}} \begin{pmatrix} \sqrt{2} & 1 & \sqrt{3} \\ \sqrt{2} & -2 & 0 \\ -\sqrt{2} & -1 & \sqrt{3} \end{pmatrix} \;,\quad \boldsymbol{R} = \frac{1}{\sqrt{6}} \begin{pmatrix} \sqrt{2} & \sqrt{2} & -\sqrt{2} \\ 1 & -2 & -1 \\ \sqrt{3} & 0 & \sqrt{3} \end{pmatrix}\,.$$

Thus, we find for $\boldsymbol{\Omega}'$ and $\boldsymbol{\Lambda}'$

$$\boldsymbol{\Omega}' = \begin{pmatrix} 0 & 0 & 0 \\ 0 & 0 & 0 \\ 0 & 0 & 2 \end{pmatrix} \;,\quad \boldsymbol{\Lambda}' = \begin{pmatrix} 2 & 0 & 0 \\ 0 & -1 & 0 \\ 0 & 0 & 3 \end{pmatrix}\,.$$

3.2 Formulation of Quantum Theory

Using the mathematical formalism developed in the previous section, we now turn to the formulation of quantum theory. Following a brief motivation based on some experiments indicating the inability of classical mechanics to describe certain microscopic phenomena, we present the postulates of quantum mechanics in axiomatic form. In this context, the concept of *measurement* plays a central role as quantum mechanical measurements differ fundamentally from classical ones. Furthermore, we investigate some general aspects concerning the temporal evolution of quantum systems (Schrödinger equation). This section ends with considerations about representations and pictures of quantum theory.

3.2.1 Limits of Classical Physics

Two properties of the microscopic world that fundamentally contradict the classical view of nature can be viewed as the primary reason for the development of quantum mechanics toward the end of the 19th century:

- the quantized nature of physical quantities such as energy and momentum,

- the wave-particle duality.

Representative for many experimental manifestations of these two phenomena (black-body radiation, Franck-Hertz experiment, Stern-Gerlach experiment, Zeeman effect, atomic spectroscopy, Compton effect, etc.), we pick out two specific experiments:

Photoelectric effect. When an alkali metal placed in a vacuum is radiated with ultraviolet light, it emits electrons if the frequency of the light is above a particular (material-dependent) threshold frequency. The strength of the induced electric current is proportional to the intensity of the radiation absorbed by the metal. However, the energy of these electrons is independent of the radiation intensity and, in contradiction to classical physics, depends only on the frequency. Einstein found an explanation for this observation

by postulating that the radiation is formed by light quanta (*photons*) with energy

$$E = h\nu \quad \text{or} \quad E = \hbar\omega \ , \ \omega = 2\pi\nu \ ,$$

where ν is the frequency of the light and

$$h = 6.62 \cdot 10^{-27} \, \text{erg} \cdot \text{s} \ , \ \hbar = \frac{h}{2\pi}$$

is *Planck's action quantum*. Accordingly, an electron absorbing a light quantum receives an amount of energy $h\nu$ that is used partially to detach the electron from the metal: *separation work W*. The remaining energy is transformed into kinetic energy of the electron:

$$E_e = \frac{1}{2} m_e v^2 = h\nu - W \ .$$

Bending of matter particles. When a homogeneous electron beam that has passed through a crystal is captured on a screen, one sees an interference pattern with a sequence of intensity minima and maxima, quite analogous to the deflection of electromagnetic waves. The same holds for the deflection of mono-energetic beams of helium atoms, hydrogen molecules, neutrons, and so on. Obviously, and in contradiction to classical expectations, particles exhibit attributes that leave room for a wave interpretation. On the other hand, when one tries to detect the particles at the location of the screen, one can only ever detect proper particles hitting the screen one by one. Accordingly, at a sufficiently small radiation time, the interference pattern is discontinuous and indeed results from the bounces of each single particle on the screen.

A satisfying noncontradictory interpretation of these and many other phenomena is obtained if the strictly classical distinction between corpuscular and wave-like phenomena is abandoned and replaced by the postulate that each particle is described by a wave function $\psi(\boldsymbol{x}, t)$ whose intensity $|\psi(\boldsymbol{x}, t)|^2$ is a measure of the probability of finding the particle at time t at location \boldsymbol{x}. This is the content of the statistical interpretation of the wave-particle duality. One important consequence of this wave-mechanical view is the superposition principle, according to which different states can be linearly superimposed as in wave optics. As we shall see in the next sections, this result is fundamental to the theoretical foundation of quantum mechanics where states exist in an abstract vector space, the Hilbert space. Wave mechanics then arises as a special representation of quantum mechanics, called *coordinate representation*.

Another principle that quantum theory has to obey is the *correspondence principle*. It states that quantum mechanics must, in some sense, converge to the classical theory when the extent of matter waves becomes small. In this case, we can regard matter waves as being point-like so that they behave like classical particles following exact trajectories. This corresponds to the limit $\hbar \to 0$ and is analogous to the limiting case of wave optics $\xrightarrow{\lambda \to 0}$ geometrical optics.

3.2.2 Postulates of Quantum Mechanics

In the following, we present the postulates of nonrelativistic quantum mechanics. To highlight differences between quantum and classical mechanics, these postulates are opposed by the corresponding classical laws in the Hamilton formalism.

Theorem 3.3: Postulates of quantum mechanics

<u>Classical mechanics</u>

<u>Quantum mechanics</u>

The classical state of a system at time t is given by the generalized coordinates $q_1(t), \ldots, q_N(t)$ and momenta $p_1(t), \ldots, p_N(t)$ defined on a real state space.

I. The quantum mechanical state of a system is given by a non-null vector, the *state vector* $|\psi(t)\rangle$, defined on a complex unitary Hilbert space. Vectors differing only by a constant factor describe the same state.

Each classical dynamical variable ω is a function of q_i and p_i on the state space:
$$\omega = \omega(q_1, \ldots, q_N, p_1, \ldots, p_N).$$

II. Physical observables are quantities that can be measured experimentally. They are generally described by linear Hermitean operators so that, in particular, their eigenvalues are real.

The generalized coordinates and momenta obey the Poisson brackets
$$\{q_i, q_k\} = \{p_i, p_k\} = 0 \ ,$$
$$\{q_i, p_k\} = \delta_{ik} \ .$$

III. The independent classical quantities x_i and p_i are represented by Hermitean operators[2] \boldsymbol{X}_i and \boldsymbol{P}_i for which the following commutation rules hold:

$$[\boldsymbol{X}_i, \boldsymbol{X}_j] = [\boldsymbol{P}_i, \boldsymbol{P}_j] = \boldsymbol{0} \ ,$$
$$[\boldsymbol{X}_i, \boldsymbol{P}_j] = i\hbar\delta_{ij} \ , \ i,j = 1,2,3 \ .$$

The Hermitean operators corresponding to the classical variables $\omega(x_i, p_i)$ are

$$\Omega(\boldsymbol{X}_i, \boldsymbol{P}_i) = \omega(x_i \to \boldsymbol{X}_i, p_i \to \boldsymbol{P}_i).$$

However, there are observables with no classical counterpart.

\triangleright

[2] See footnote 1 on page 233.

A measurement of a dynamical variable ω yields: $\omega(q_1, \ldots, q_N, p_i, \ldots, p_N)$. The measurement does not necessarily influence the state of the particle.

IV. If the system is in the state $|\psi\rangle$, a measurement of the dynamical variable represented by Ω yields one of the eigenvalues ω of Ω with the probability

$$W(\omega) = \frac{\langle\psi|\,\boldsymbol{\mathcal{P}}_\omega\,|\psi\rangle}{\langle\psi|\,\psi\rangle} \,,$$

where $\boldsymbol{\mathcal{P}}_\omega$ projects onto the eigenspace corresponding to the eigenvalue ω. A necessary consequence of this (ideal) measurement is that the state changes from $|\psi\rangle$ to $\boldsymbol{\mathcal{P}}_\omega|\psi\rangle$.

The state variables q_i and p_i obey Hamilton's equations:

$$\dot{q}_i = \frac{\partial H}{\partial p_i} \,,\quad \dot{p}_i = -\frac{\partial H}{\partial q_i} \,.$$

V. The state vector $|\psi(t)\rangle$ obeys the *Schrödinger equation*

$$i\hbar \frac{\mathrm{d}}{\mathrm{d}t}|\psi(t)\rangle = \boldsymbol{H}\,|\psi(t)\rangle \,.$$

\boldsymbol{H} is the operator of total energy, called the *Hamilton operator*. In the simplest case, it can be obtained via the correspondence principle

$$\boldsymbol{H} = H(x \to \boldsymbol{X}_i, p_i \to \boldsymbol{P}_i)$$

from the Hamilton function of the corresponding classical system.

To I. This postulate emphasizes the importance of Hilbert spaces in quantum mechanics. Since Hilbert spaces are closed with respect to addition, adding two vectors results in another possible state. This is the principle of superposition, which is related to quantum mechanical interference effects. Vectors differing only by a multiplicative constant are equivalent. Each state $|\psi\rangle$ is therefore associated with a one-dimensional subspace of the Hilbert space, called *beam*. It is generated by the projection operator $\boldsymbol{\mathcal{P}}_\psi = |\psi\rangle\langle\psi|$. Here, $|\psi\rangle$ is a *pure state* as opposed to *mixed states*, that we will discuss in the context of statistical physics. In general, a quantum physical state is described by a vector $|\psi(t)\rangle$ that is a linear combination of proper and improper Hilbert vectors. Proper vectors are normalizable to unity, whereas improper vectors can only by normalized to the δ-function. However, $|\psi(t)\rangle$ can always be normalized to unity so that it is a proper Hilbert vector in any case.

To II. Axiom II shows the importance of the theory of linear operators discussed in Subsection 3.1.2. They constitute the quantum mechanical objects

that act on state vectors. Note, however, that these vectors are not themselves observables. It is the real eigenvalues of the operators that we measure in experiments (see Subsection 3.2.3).

To III. This commutation rule has already been motivated in classical mechanics. The Poisson bracket obeys the same algebra as the commutator. Hamiltonian mechanics and quantum mechanics can therefore be interpreted as two realizations of the same algebraic structure. The replacement $x_i \to \boldsymbol{X}_i$, $p_i \to \boldsymbol{P}_i$ can be ambiguous. This is partially due to the fact that the coordinate operator (position operator) does not commute with the momentum operator. Consider, for example, the classical quantity $\omega = x_i p_i = p_i x_i$. We can derive two operators from it: $\boldsymbol{\Omega} = \boldsymbol{X}_i \boldsymbol{P}_i$ and $\boldsymbol{\Omega} = \boldsymbol{P}_i \boldsymbol{X}_i \neq \boldsymbol{X}_i \boldsymbol{P}_i$. In most cases the mixed terms in \boldsymbol{X}_i and \boldsymbol{P}_i can be symmetrized, which here yields $\boldsymbol{\Omega} = (\boldsymbol{X}_i \boldsymbol{P}_i + \boldsymbol{P}_i \boldsymbol{X}_i)/2$. The symmetry is also necessary to make $\boldsymbol{\Omega}$ Hermitean. Another ambiguity is due to the fact that the above replacement is not invariant under coordinate transformations of the classical quantities. Therefore, it is agreed that operator replacements are made based on Cartesian classical quantities.

To IV. In quantum mechanics, we can no longer speak of particle "trajectories". Statements about quantum systems are purely statistical and are interpreted as the result of measurements of many identical systems. In addition, a measurement of a system leads to a noncausal change of it such that immediately after the measurement the system is in a (more or less) well-defined state. This is called *state reduction*. We shall discuss this axiom further in Subsection 3.2.3.

To V. The Schrödinger equation is the dynamical foundation of quantum mechanics in the *Schrödinger picture*. In this picture, the operators are generally time-independent, whereas the states depend on time. A thorough discussion of the Schrödinger equation follows in Subsection 3.2.4. Given an initial state $|\psi(t_0)\rangle$, this equation fully describes the temporal evolution $|\psi(t)\rangle$ of a system as long as no measurement is performed. In this form, the equation is more general than the Schrödinger equation of wave mechanics as it does not refer to a special representation (i.e., a special basis system).

Extension of the postulates. The above postulates are not sufficient to fully develop nonrelativistic quantum mechanics. In Section 3.4 we add a postulate by introducing a purely quantum mechanical degree of freedom called *spin*. In Section 3.9 we extend the postulates to N-particle systems.

3.2.3 Quantum Mechanical Measurement

To predict the result of a measurement of a quantity corresponding to the operator $\boldsymbol{\Omega}$ in a statistical sense, postulate IV tells us to calculate its orthonormal eigenvectors $|\omega_i\rangle$ and its eigenvalues ω_i, as well as the corresponding

projection operators $\boldsymbol{\mathcal{P}}_{\omega_i}$. The probability of finding the eigenvalue ω_i from this measurement is then given by

$$W(\omega_i) = \frac{\langle\psi|\,\boldsymbol{\mathcal{P}}_{\omega_i}\,|\psi\rangle}{\langle\psi|\,\psi\rangle} = \frac{|\langle\omega_i|\,\psi\rangle|^2}{\langle\psi|\,\psi\rangle} \ , \quad \boldsymbol{\mathcal{P}}_{\omega_i} = |\omega_i\rangle\,\langle\omega_i|\ . \tag{3.5}$$

If the state vector $|\psi\rangle$ is a linear combination of the eigenvectors of $\boldsymbol{\Omega}$, i.e., $|\psi\rangle = \sum_j \alpha_j\,|\omega_j\rangle$, it follows that

$$W(\omega_i) = \frac{|\sum\limits_j \alpha_j\,\langle\omega_i|\,\omega_j\rangle|^2}{\sum\limits_j |\alpha_j|^2} = \frac{|\alpha_i|^2}{\sum\limits_j |\alpha_j|^2}\ .$$

In particular, we see that for $|\psi\rangle = |\omega_j\rangle$: $W(\omega_i) = \delta_{ij}$, which means that the measurement always (with probability one) yields the value ω_j. In case of degeneracy, we have analogously to (3.5)

$$W(\omega_i) = \frac{\langle\psi|\,\boldsymbol{\mathcal{P}}_{\omega_i}\,|\psi\rangle}{\langle\psi|\,\psi\rangle} = \frac{\sum\limits_r |\langle\omega_i,r|\,\psi\rangle|^2}{\langle\psi|\,\psi\rangle}\ , \quad \boldsymbol{\mathcal{P}}_{\omega_i} = \sum_r |\omega_i,r\rangle\,\langle\omega_i,r|\ . \tag{3.6}$$

If the spectrum of $\boldsymbol{\Omega}$ is degenerate and continuous, the quantity $W(\omega)$ is interpreted as a *probability density*. The probability $W[\omega_1,\omega_2]$ of measuring a value in the interval $[\omega_1 : \omega_2]$ is then given by

$$W[\omega_1,\omega_2] = \int\limits_{\omega_1}^{\omega_2} d\omega W(\omega) = \frac{\int\limits_{\omega_1}^{\omega_2} d\omega\,\langle\psi|\,\boldsymbol{\mathcal{P}}_\omega\,|\psi\rangle}{\langle\psi|\,\psi\rangle} = \frac{\sum\limits_r \int\limits_{\omega_1}^{\omega_2} d\omega\,|\langle\omega,r|\,\psi\rangle|^2}{\langle\psi|\,\psi\rangle}\ ,$$

with $\boldsymbol{\mathcal{P}}_\omega$ and $W(\omega)$ from (3.6).

Ideal measurement and state reduction. The statement of postulate IV, i.e., the collapse of the state vector from $|\psi\rangle$ to $\boldsymbol{\mathcal{P}}_\omega\,|\psi\rangle$, is valid only for *ideal measurements*. In fact, one can use this to define an ideal measurement as follows:

Definition: Ideal measurement and reduction of the state vector

If the result of an ideal measurement on a system with the state vector $|\psi\rangle$ is given by the eigenvalue ω, then, immediately after the experiment, the system is found in the state $\boldsymbol{\mathcal{P}}_\omega\,|\psi\rangle$. This is called the *projection postulate*.

Thus, the ideal measurement functions like a perfect filter, letting pass only that part of $|\psi\rangle$ that corresponds to the eigenstate of the eigenvalue ω. If the measurement is not ideal, the filter will not be perfect, and a certain amount of distortion will occur. An ideal measurement can be used to prepare a state such that a following measurement of the same operator on the state yields a sharp measurement (with probability one).

Normalization. In the following, we shall assume that physical state vectors are always normalized to unity: $\langle \psi | \psi \rangle = 1$.

Statistical properties of observables. Often we are not so much interested in the probability density $W(\omega)$ of an operator Ω, but we wish to determine its statistical average $\langle \Omega \rangle$. This is achieved by performing a large number of equal measurements on identically prepared systems. Its calculation proceeds as follows:

$$\langle \Omega \rangle = \sum_i W(\omega_i)\omega_i = \sum_i \langle \psi | \omega_i \rangle \langle \omega_i | \psi \rangle \omega_i = \sum_i \langle \psi | \Omega | \omega_i \rangle \langle \omega_i | \psi \rangle \ ,$$

where $\Omega | \omega_i \rangle = \omega_i | \omega_i \rangle$ has been used. Noting that $\sum_i | \omega_i \rangle \langle \omega_i | = 1$, we find

Theorem 3.4: Expectation value $\langle \Omega \rangle$ of an operator Ω

The *expectation value* of an observable Ω in a state $| \psi \rangle$ is given by

$$\langle \Omega \rangle = \langle \psi | \Omega | \psi \rangle \ .$$

To calculate $\langle \Omega \rangle$ we just need to know the state vector $| \psi \rangle$, as well as the operator Ω, in a particular basis (for example, as column vectors and matrices), but not its eigenvectors and eigenvalues. Generally, the expectation value $\langle \Omega \rangle$ of an operator Ω is accompanied by an amount of uncertainty. This is usually quantified by the mean-square deviation, defined as

$$\Delta\Omega = \left[\langle \Omega^2 \rangle - \langle \Omega \rangle^2 \right]^{1/2} \ .$$

For eigenstates of Ω we clearly have $\Delta\Omega = 0$.

Compatibility of observables. The question naturally arises whether it is possible to prepare a state such that it yields sharp measurement results for two different observables Ω and Λ.[3] This is only possible if the operators possess a common set of eigenvectors that, according to Theorem 3.2, is equivalent to $[\Omega, \Lambda] = 0$. To characterize or to prepare the state of a quantum system most precisely, one performs measurements of as many simultaneously commuting observables (*compatible observables*) as possible. If the common orthonormal system of these observables is not unique, one keeps adding a further observable that commutes with all others, and so on.

Definition: Complete set of observables

A set of observables A, B, C, \ldots forms a *complete set of commuting observables* if all the observables commute among one another and if their basis system is uniquely determined. To each set of eigenvalues a, b, c, \ldots there exists (up to a constant) a unique common eigenvector $| a, b, c, \ldots \rangle$.

[3] In the following sense: three measurements are performed: initially, a measurement of Ω and then one of Λ. A repeat measurement of Ω should yield the same result as the first one.

Heisenberg's uncertainty principle. In the following, we show that the product of the uncertainties of two observables is bounded from below.[4] Let A and B denote two Hermitean operators with

$$\Delta A = \left[\langle A^2 \rangle - \langle A \rangle^2 \right]^{1/2} \ , \ \Delta B = \left[\langle B^2 \rangle - \langle B \rangle^2 \right]^{1/2} \ .$$

For the new variables

$$\hat{A} = A - \langle A \rangle \ , \ \hat{B} = B - \langle B \rangle \ ,$$

we find

$$\Delta A = \Delta \hat{A} = \left\langle \hat{A}^2 \right\rangle^{1/2} \ , \ \Delta B = \Delta \hat{B} = \left\langle \hat{B}^2 \right\rangle^{1/2}$$

and, using the Schwarz inequality,

$$(\Delta A)^2 (\Delta B)^2 = \left\langle \psi | \hat{A}^2 | \psi \right\rangle \left\langle \psi | \hat{B}^2 | \psi \right\rangle \geq \left| \left\langle \psi | \hat{A}\hat{B} | \psi \right\rangle \right|^2 \ .$$

Splitting the operator $\hat{A}\hat{B}$ into its Hermitean and its anti-Hermitean parts yields

$$(\Delta A)^2 (\Delta B)^2 \geq \frac{1}{4} \left| \left\langle \psi | \{\hat{A}, \hat{B}\} | \psi \right\rangle + \left\langle \psi | [\hat{A}, \hat{B}] | \psi \right\rangle \right|^2 \ ,$$

where $\{\hat{A}, \hat{B}\} = \hat{A}\hat{B} + \hat{B}\hat{A}$ is the *anticommutator*. Since $\{\hat{A}, \hat{B}\}$ is Hermitean, its expectation value is real, whereas $[\hat{A}, \hat{B}] = [A, B]$ is anti-Hermitean and has a purely imaginary expectation value. This means

$$(\Delta A)^2 (\Delta B)^2 \geq \frac{1}{4} \left\langle \psi | \{\hat{A}, \hat{B}\} | \psi \right\rangle^2 + \frac{1}{4} \left| \left\langle \psi | [A, B] | \psi \right\rangle \right|^2 \ .$$

Theorem 3.5: Heisenberg's uncertainty principle

Let A and B denote two Hermitean operators. Heisenberg's uncertainty principle states that

$$\Delta A \cdot \Delta B \geq \frac{1}{2} \left| \left\langle [A, B] \right\rangle \right| \ .$$

Note that the equality holds only if $\left\langle \psi | \{\hat{A}, \hat{B}\} | \psi \right\rangle = 0$ and $\hat{A} | \psi \rangle = c\hat{B} | \psi \rangle$. An interesting special case of the uncertainty principle is that of *canonical conjugate* operators such as coordinates and momenta, for which $[X_i, P_i] = i\hbar$. For these variables, Theorem 3.5 is independent of the state:

$$\Delta X_i \cdot \Delta P_i \geq \frac{\hbar}{2} \ .$$

[4] Recall that the state vector $| \psi \rangle$ is normalized to unity.

3.2.4 Schrödinger Picture and Schrödinger Equation

Postulate V describes the temporal evolution of a quantum system within a particular *picture* of quantum theory, called the *Schrödinger picture*. Since physical observables such as expectation values are invariant under unitary transformations, we can define an infinite number of pictures, all of which are connected by unitary transformations.

In this subsection we discuss the Schrödinger picture in which states are time-dependent, whereas operators can at most be explicitly dependent on time. Two more pictures that are also commonly used are the *Heisenberg picture* and the *Dirac picture*. These pictures are considered in the next subsection.

Schrödinger equation and temporal evolution operator. If the state $|\psi\rangle$ of a quantum system is known at a particular time (prepared in the above sense), the temporal evolution of the system in the Schrödinger picture is uniquely determined by the Schrödinger equation

$$i\hbar\frac{\mathrm{d}}{\mathrm{d}t}|\psi(t)\rangle = \boldsymbol{H}(t)|\psi(t)\rangle \ , \tag{3.7}$$

as long as the system is not perturbed in the relevant time interval (for example, by a measurement). The Hamilton operator may depend explicitly on time, i.e., $\mathrm{d}\boldsymbol{H}/\mathrm{d}t = \partial\boldsymbol{H}/\partial t$, say due to the presence of time-dependent fields. The general solution of this equation can be written in the form

$$|\psi(t)\rangle = \boldsymbol{U}(t,t_0)|\psi(t_0)\rangle \ , \ \boldsymbol{U}(t_0,t_0) = \boldsymbol{1} \ ,$$

where $\boldsymbol{U}(t,t_0)$ is the *temporal evolution operator*. Inserting this expression into (3.7), we obtain an operator equation that is equivalent to the Schrödinger equation,

$$i\hbar\frac{\mathrm{d}}{\mathrm{d}t}\boldsymbol{U}(t,t_0) = \boldsymbol{H}(t)\boldsymbol{U}(t,t_0) \ , \tag{3.8}$$

from which it follows that

$$\boldsymbol{U}(t+\Delta t,t) = \boldsymbol{1} - \frac{i}{\hbar}\Delta t\boldsymbol{H}(t) \ , \ \Delta t \ll 1 \ .$$

Since the Hamilton operator \boldsymbol{H} is Hermitean, $\boldsymbol{U}(t+\Delta t,t)$ is an infinitesimal unitary operator. The full operator $\boldsymbol{U}(t,t_0)$ is obtained as a series of infinitesimal unitary transformations

$$\boldsymbol{U}(t,t_0) = \lim_{\Delta t \to 0}\boldsymbol{U}(t,t-\Delta t)\boldsymbol{U}(t-\Delta t,t-2\Delta t)\cdots\boldsymbol{U}(t_0+\Delta t,t_0)$$

and is therefore also unitary. The unitarity of the evolution operator \boldsymbol{U} or, equally, the Hermitecity of the Hamilton operator \boldsymbol{H} follows necessarily from the condition that the norm of the state vector $|\psi\rangle$ is constant in time. This, in turn, is necessary to formulate probability statements.

Theorem 3.6: Conservation of the norm

Due to the Hermitecity of the Hamilton operator H and, consequently, the unitarity of the time evolution operator U, the norm of a state vector $|\psi(t)\rangle$ is constant in time:

$$\langle \psi(t)|\psi(t)\rangle = \langle \psi(t_0)|U^\dagger(t,t_0)U(t,t_0)|\psi(t_0)\rangle = \langle \psi(t_0)|\psi(t_0)\rangle \ .$$

Formally, (3.8) is an ordinary differential equation of the first order. Note, however, that the naive solution ansatz

$$U(t,t_0) = \exp\left(-\frac{i}{\hbar}\int_{t_0}^{t} H(t')dt'\right)$$

is not correct since the exponential function contains an arbitrary number of powers of $\int H(t')dt'$ that generally do not commute with one another: $[H(t), H(t')] \neq 0$.[5] In the following, we consider an important exception to this rule.

Conservative Hamilton operator, time-independent Schrödinger equation. Due to (3.8), for the special case of a closed conservative system for which $\partial H/\partial t = 0$, we can write the temporal evolution operator U as

$$U(t,t_0) = e^{-iH(t-t_0)/\hbar} \ .$$

We would like to find an explicit expression for U and $|\psi\rangle$ in terms of the energy eigenstates $|E_n\rangle$, i.e., the normalized eigenkets of H that obey the *time-independent Schrödinger equation*

$$H|E_n\rangle = E_n|E_n\rangle \ .$$

Expanding $|\psi(t)\rangle$ in these eigenkets,

$$|\psi(t)\rangle = \sum_n |E_n\rangle\langle E_n|\psi(t)\rangle = \sum_n a_n(t)|E_n\rangle \ , \quad a_n(t) = \langle E_n|\psi(t)\rangle \ ,$$

and inserting this expression into (3.7), we obtain the equation

$$i\hbar\dot{a}_n(t) = E_n a_n(t) \ ,$$

which is solved by

[5] With the help of the *time-ordering operator* T it is, in fact, possible to find a formal solution of this type, namely

$$U(t,t_0) = T\left\{\exp\left(-\frac{i}{\hbar}\int_{t_0}^{t} H(t')dt'\right)\right\} \ .$$

Here, the various expressions are arranged in chronological order.

$$a_n(t) = a_n(t_0)e^{-iE_n(t-t_0)/\hbar} \ .$$

It follows that

$$\langle E_n|\,\psi(t)\rangle = \langle E_n|\,\psi(t_0)\rangle\, e^{-iE_n(t-t_0)/\hbar}$$

$$\implies |\psi(t)\rangle = \sum_n |E_n\rangle\,\langle E_n|\,\psi(t_0)\rangle\, e^{-iE_n(t-t_0)/\hbar} \ .$$

Theorem 3.7: General solution of the Schrödinger equation for conservative systems

$$i\hbar\frac{\mathrm{d}}{\mathrm{d}t}|\psi(t)\rangle = \boldsymbol{H}\,|\psi(t)\rangle \implies |\psi(t)\rangle = \boldsymbol{U}(t,t_0)\,|\psi(t_0)\rangle \ ,$$

with

$$\boldsymbol{U}(t,t_0) = \sum_n |E_n\rangle\,\langle E_n|\, e^{-iE_n(t-t_0)/\hbar} \ , \quad \boldsymbol{H}\,|E_n\rangle = E_n\,|E_n\rangle \ .$$

Accordingly, in case of degeneracy, we have

$$\boldsymbol{U}(t,t_0) = \sum_{n,r} |E_n,r\rangle\,\langle E_n,r|\, e^{-iE_n(t-t_0)/\hbar} \ ,$$

and in the continuous case the sums have to be replaced with integrals. The states $|E_n\rangle\, e^{-iE_n(t-t_0)/\hbar}$ are the *stationary solutions* of the time-dependent Schrödinger equation. They vary periodically in time, and their frequency ω satisfies Einstein's relation $E = \hbar\omega$. For these states the probability distribution $W(\lambda)$ of a variable Λ is independent of time since

$$W(\lambda,t) = |\langle\lambda|\,E(t)\rangle\,|^2 = \left|\langle\lambda|\,E\rangle\, e^{-iE(t-t_0)/\hbar}\right|^2 = |\langle\lambda|\,E\rangle\,|^2 = W(\lambda,t_0) \ .$$

3.2.5 Other Pictures of Quantum Theory

In the previous subsection we have been using a particular approach to quantum mechanics, namely the Schrödinger picture. In this picture the dynamics of the system is described by rotations of the state vector in the Hilbert space, while keeping the basis vectors fixed. Since the scalar product is invariant under unitary transformations, we can use other pictures, for example, one where the state vector remains fixed in time, but the basis system is rotating. This is called the *Heisenberg picture*. Herein, observables are described by time-dependent operators, whereas states appear fixed in time. Yet another picture is the *Dirac picture* in which state vectors and basis vectors are both time-dependent in a particular way. This turns out to be very useful in perturbation theory. In the following, we shall take a closer look at the Heisenberg and the Dirac pictures.

Heisenberg picture. The Heisenberg picture is defined as follows:

Definition: Heisenberg picture

Let $|\psi_S(t)\rangle = U(t, t_0)|\psi_S(t_0)\rangle$ denote a state vector, and A_S an operator in the Schrödinger picture. The corresponding quantities $|\psi_H\rangle$ and A_H in the Heisenberg picture are then defined by

$$|\psi_H\rangle = U^\dagger(t, t_0)|\psi_S(t)\rangle = |\psi_S(t_0)\rangle$$

$$A_H(t) = U^\dagger(t, t_0)A_S U(t, t_0) \ .$$

Thus, in the Heisenberg picture, states are independent of time, whereas operators are time-dependent even if A_S is not explicitly a function of time. The temporal evolution of A_H is easily found:

$$
\begin{aligned}
i\hbar\frac{\mathrm{d}}{\mathrm{d}t}A_H(t) &= i\hbar\frac{\mathrm{d}U^\dagger}{\mathrm{d}t}A_S U + i\hbar U^\dagger\frac{\partial A_S}{\partial t}U + i\hbar U^\dagger A_S\frac{\mathrm{d}U}{\mathrm{d}t} \\
&= -U^\dagger H_S A_S U + i\hbar U^\dagger\frac{\partial A_S}{\partial t}U + U^\dagger A_S H_S U \\
&= U^\dagger[A_S, H_S]U + i\hbar U^\dagger\frac{\partial A_S}{\partial t}U \ .
\end{aligned}
$$

Here, we have used (3.8) and its adjoint. Because of $\frac{\partial A_H}{\partial t} = U^\dagger\frac{\partial A_S}{\partial t}U$ and[6] $U^\dagger A_S H_S U = U^\dagger A_S U U^\dagger H_S U = A_H H_H$, we obtain

Theorem 3.8: Heisenberg equation and conserved quantities

In the Heisenberg picture, the relation corresponding to the Schrödinger equation is given by

$$i\hbar\frac{\mathrm{d}A_H}{\mathrm{d}t} = [A_H, H_H] + i\hbar\frac{\partial A_H}{\partial t} \qquad (\textit{Heisenberg equation}) \ .$$

Using

$$\frac{\mathrm{d}\langle A_H\rangle}{\mathrm{d}t} = \left\langle\psi_H\left|\frac{\mathrm{d}A_H}{\mathrm{d}t}\right|\psi_H\right\rangle \ , \quad \frac{\mathrm{d}|\psi_H\rangle}{\mathrm{d}t} = 0 \ ,$$

we obtain the *Heisenberg equation for expectation values*

$$\frac{\mathrm{d}\langle A\rangle}{\mathrm{d}t} = \frac{1}{i\hbar}\langle[A, H]\rangle + \left\langle\frac{\partial A}{\partial t}\right\rangle \ . \tag{3.9}$$

Due to the invariance of the scalar product, this equation is independent of the chosen picture so that the index H can be dropped.

Furthermore, we see a simple rule for conserved quantities: an observable A that is not explicitly time-dependent, and that commutes with the Hamilton operator H, is a conserved quantity.

[6] This relation holds since $\frac{\partial A_S}{\partial t}$ is a function of observables, and therefore transforms as any other operator.

If the operator A is not explicitly time-dependent, (3.9) reduces to the *Ehrenfest theorem*

$$\frac{\mathrm{d}\langle A \rangle}{\mathrm{d}t} = \frac{1}{\mathrm{i}\hbar}\langle [A, H] \rangle \ . \tag{3.10}$$

We can use this equation to show an uncertainty principle relating energy and time. Consider, to this end, a system with a time-independent Hamilton operator H, as well as a time-independent observable A. Denoting by ΔA and ΔE the mean-square deviations of their expectation values, Theorem 3.5 yields

$$\Delta A \cdot \Delta E \geq \frac{1}{2} | \langle [A, H] \rangle | \ ,$$

and therefore,

$$\Delta A \cdot \Delta E \geq \frac{\hbar}{2} \frac{\mathrm{d}\langle A \rangle}{\mathrm{d}t} \ .$$

Rearranging this equation, we obtain

Theorem 3.9: Energy-time uncertainty

$$\Delta \tau \cdot \Delta E \geq \frac{\hbar}{2} \ , \quad \Delta \tau = \frac{\Delta A}{\frac{\mathrm{d}\langle A \rangle}{\mathrm{d}t}} \ .$$

Here, $\Delta \tau$ is the time interval in which the expectation value of A changes by the amount ΔA. It represents the minimal time span required for a noticeable change of $\langle A \rangle$ and can be regarded as a characteristic time for the evolution of the system. The above relation tells us that the energy of a system that is in a fixed state over a time $\Delta \tau$ has an uncertainty $\Delta E \geq \hbar/(2\Delta \tau)$. Another way of looking at this is that violations ΔE of classical energy conservation are possible within the time interval $\Delta \tau \sim \hbar/\Delta E$. Note that the time-energy uncertainty is qualitatively different from the uncertainty relations of Theorem 3.5 since $\Delta \tau$ is not a dynamic variable, but just an external evolution parameter.

Dirac picture. Another picture that is often used when the Hamilton operator shows a time dependency is the Dirac or interaction picture. Here, the Hamilton operator H_S of the Schrödinger picture is split into two Hermitean parts,

$$H_\mathrm{S}(t) = H_\mathrm{S}^{(0)} + H_\mathrm{S}'(t) \ ,$$

where the *unperturbed* part $H_\mathrm{S}^{(0)}$ is time-independent, and the *perturbation* $H_\mathrm{S}'(t)$ is time-dependent. To solve the Schrödinger equation

$$\mathrm{i}\hbar\frac{\mathrm{d}}{\mathrm{d}t}U(t, t_0) = H_\mathrm{S}(t)U(t, t_0) \ ,$$

the unitary operator U is written as the product of two unitary operators,

$$U(t, t_0) = U^{(0)}(t, t_0) U'(t, t_0) \, , \quad U^{(0)}(t_0, t_0) = U'(t_0, t_0) = 1 \, ,$$

where $U^{(0)}$ is the solution of the unperturbed Schrödinger equation

$$i\hbar \dot{U}^{(0)} = H_{\mathrm{S}}^{(0)} U^{(0)} \, .$$

The Dirac picture is now defined as follows:

Definition: Dirac picture

Let the Hamilton operator, a state vector, and an operator be given by $H_{\mathrm{S}}(t) = H_{\mathrm{S}}^{(0)} + H_{\mathrm{S}}'(t)$, $|\psi_{\mathrm{S}}(t)\rangle$ and A_{S} in the Schrödinger picture. If $U^{(0)}(t, t_0)$ is a unitary operator, for which $i\hbar \dot{U}^{(0)} = H_{\mathrm{S}}^{(0)} U^{(0)}$, the corresponding quantities in the Dirac picture are given by

$$|\psi_{\mathrm{I}}(t)\rangle = U^{(0)\dagger}(t, t_0) |\psi_{\mathrm{S}}(t)\rangle$$

$$A_{\mathrm{I}}(t) = U^{(0)\dagger}(t, t_0) A_{\mathrm{S}} U^{(0)}(t, t_0) \, .$$

With part of the time dependency in the state vector, and part of it in the operator, the Dirac picture is halfway between the Schrödinger and the Heisenberg one. We also have

$$\begin{aligned}
i\hbar \frac{\mathrm{d}}{\mathrm{d}t} |\psi_{\mathrm{I}}(t)\rangle &= i\hbar \left(\dot{U}^{(0)\dagger} |\psi_{\mathrm{S}}(t)\rangle + U^{(0)\dagger} \frac{\mathrm{d}}{\mathrm{d}t} |\psi_{\mathrm{S}}(t)\rangle \right) \\
&= -U^{(0)\dagger} H_{\mathrm{S}}^{(0)} |\psi_{\mathrm{S}}(t)\rangle + U^{(0)\dagger} H_{\mathrm{S}} |\psi_{\mathrm{S}}(t)\rangle \\
&= U^{(0)\dagger} H_{\mathrm{S}}' |\psi_{\mathrm{S}}(t)\rangle = U^{(0)\dagger} H_{\mathrm{S}}' U^{(0)} |\psi_{\mathrm{I}}\rangle = H_{\mathrm{I}}' |\psi_{\mathrm{I}}\rangle
\end{aligned}$$

and therefore,

Theorem 3.10: State and operator equations in the Dirac picture

In the Dirac picture the evolution equations are

$$i\hbar \frac{\mathrm{d}}{\mathrm{d}t} |\psi_{\mathrm{I}}(t)\rangle = H_{\mathrm{I}}' |\psi_{\mathrm{I}}(t)\rangle \qquad \textit{(state equation)}$$

or, using $|\psi_{\mathrm{I}}(t)\rangle = U'(t, t_0) |\psi_{\mathrm{I}}(t_0)\rangle$,

$$i\hbar \frac{\mathrm{d}}{\mathrm{d}t} U' = H_{\mathrm{I}}' U'$$

and

$$i\hbar \frac{\mathrm{d}A_{\mathrm{I}}}{\mathrm{d}t} = [A_{\mathrm{I}}, H_{\mathrm{I}}^{(0)}] + i\hbar \frac{\partial A_{\mathrm{I}}}{\partial t} \qquad \textit{(operator equation)} \, .$$

The last equation can be derived in similar fashion as we have done for the Heisenberg equation.

Accordingly, the time-dependent states $|\psi_{\mathrm{I}}\rangle$ obey a Schrödinger equation that

contains the perturbation H'_{I}, whereas the time-dependent operators obey a Heisenberg equation with the unperturbed Hamilton operator $H_{\mathrm{I}}^{(0)}$.

3.2.6 Representations in Quantum Mechanics

Using vectors in an abstract Hilbert space, we have, up to now, developed quantum mechanics in a basis- or representation-independent manner. However, if we wish to obtain components of the vector, we must choose a basis system that spans the complete vector space. Such a *representation* is given by the eigenvectors of a complete set of commuting observables. The projections of the state vector $|\psi\rangle$ onto a system $\{|q\rangle\}$ of basis vectors yield a component vector that quantifies the state in the q-representation, whereas operators are represented by the elements of the Hermitean matrix $\langle q| A |q'\rangle$. By projecting the vector $|\psi\rangle$ onto a particular basis we transform the abstract Schrödinger equation (or the operator equations of motion) into differential equations and algebraic relations, which, when solved, yield quantitative information verifiable by experiment.

In any picture of quantum mechanics there is an infinite set of possible representations related via unitary transformations.[7] However, representations in which the operator matrices are diagonal are clearly particularly useful since the Schrödinger equation is much easier to solve in these. If, for example, the Hamilton operator is a function only of coordinates and momenta, $H = H(X,P)$, we have a reasonable choice between the *coordinate representation* and the *momentum representation*. Which of these we choose will depend on the form of the potential V. Both representations are allowed as both, the coordinate operators (position operators) X_i, as well as the momentum operators P_i, $i = 1,2,3$, each define a complete set of commuting observables (disregarding spin), according to postulate III.

Coordinate representation (wave mechanics). The *coordinate representation* is defined as the representation in which the position operator is diagonal. Since it has a continuous spectrum, the one-dimensional case of this condition is given by

Definition: Coordinate representation

$$\langle x| X |x'\rangle = x\delta(x - x') , \quad \mathcal{P}_X = \int \mathrm{d}x\, |x\rangle \langle x| , \quad \psi(x,t) = \langle x|\psi(t)\rangle .$$

\mathcal{P}_X denotes the unit operator in this representation. In conjunction with the commutation rules $[X,P] = \mathrm{i}\hbar$, this leads to a corresponding relation for the momentum operator:

$$\mathrm{i}\hbar\delta(x - x') = \langle x| XP - PX |x'\rangle = (x - x')\langle x| P |x'\rangle$$

[7] These are not to be confused with transformations relating different pictures of the theory.

$$\implies \langle x| \, \boldsymbol{P} \, |x'\rangle = i\hbar \frac{\mathrm{d}}{\mathrm{d}x'} \delta(x - x') = -i\hbar \frac{\mathrm{d}}{\mathrm{d}x} \delta(x - x') \ .$$

Now, we have

$$|\psi(t)\rangle = \boldsymbol{\mathcal{P}}_X \, |\psi(t)\rangle = \int \mathrm{d}x \, |x\rangle \, \langle x| \, \psi(t)\rangle = \int \mathrm{d}x \, |x\rangle \, \psi(x, t) \ ,$$

where $\psi(x, t) = \langle x| \, \psi(t)\rangle$ are the components of the state vector $|\psi(t)\rangle$ in the coordinate representation. This expression is called the *coordinate wave function* or simply *wave function*. Applying \boldsymbol{X} and \boldsymbol{P} to $|\psi\rangle$, we find

$$\begin{aligned} \boldsymbol{X} \, |\psi(t)\rangle &= \boldsymbol{\mathcal{P}}_X \boldsymbol{X} \boldsymbol{\mathcal{P}}_X \, |\psi(t)\rangle \\ &= \int \mathrm{d}x \int \mathrm{d}x' \, |x\rangle \, \langle x| \, \boldsymbol{X} \, |x'\rangle \, \langle x'| \, \psi(t)\rangle \\ &= \int \mathrm{d}x \, |x\rangle \, x\psi(x, t) \end{aligned}$$

$$\begin{aligned} \boldsymbol{P} \, |\psi(t)\rangle &= \boldsymbol{\mathcal{P}}_X \boldsymbol{P} \boldsymbol{\mathcal{P}}_X \, |\psi(t)\rangle \\ &= \int \mathrm{d}x \int \mathrm{d}x' \, |x\rangle \, \langle x| \, \boldsymbol{P} \, |x'\rangle \, \langle x'| \, \psi(t)\rangle \\ &= i\hbar \int \mathrm{d}x \, |x\rangle \int \mathrm{d}x' \left(\frac{\mathrm{d}}{\mathrm{d}x'} \delta(x - x') \right) \psi(x', t) \\ &= -i\hbar \int \mathrm{d}x \, |x\rangle \int \mathrm{d}x' \delta(x - x') \frac{\mathrm{d}}{\mathrm{d}x'} \psi(x', t) \\ &= -i\hbar \int \mathrm{d}x \, |x\rangle \, \frac{\mathrm{d}}{\mathrm{d}x} \psi(x, t) \ , \end{aligned}$$

which yields the following correspondence relations for operators in the coordinate representation:

> **Theorem 3.11: Operators in coordinate representation**
>
> $$\boldsymbol{X} \longrightarrow \boldsymbol{X}_X = x \ , \ \boldsymbol{P} \longrightarrow \boldsymbol{P}_X = -i\hbar \frac{\mathrm{d}}{\mathrm{d}x} \ ,$$
>
> $$\Omega(\boldsymbol{X}, \boldsymbol{P}) \longrightarrow \Omega_X = \Omega \left(\boldsymbol{X} \to x, \boldsymbol{P} \to -i\hbar \frac{\mathrm{d}}{\mathrm{d}x} \right) \ .$$

Accordingly, the Schrödinger equation and expectation values in the coordinate representation are given by

$$i\hbar \frac{\mathrm{d}}{\mathrm{d}t} \psi(x, t) = \boldsymbol{H}_X \psi(x, t) \tag{3.11}$$

$$\langle \boldsymbol{\Omega} \rangle = \int \mathrm{d}x \psi^*(x, t) \boldsymbol{\Omega}_X \psi(x, t) = \langle \psi| \, \boldsymbol{\Omega}_X \, |\psi\rangle = \langle \boldsymbol{\Omega}_X \rangle \ . \tag{3.12}$$

Proper Hilbert vectors with a finite norm correspond to wave functions for which

$$\langle\psi|\psi\rangle = \int dx |\psi(x,t)|^2 = \int dx |\psi(x,0)|^2 < \infty .$$

They span the *Hilbert space of square-integrable functions.*

N-particle systems. The generalization of the above to systems with N degrees of freedom is unproblematic.[8] In this case we have the corresponding relations

$$\langle x_1,\ldots,x_N| \, \boldsymbol{X}_i \, |x_1',\ldots,x_N'\rangle = x_i\delta(x_1 - x_1')\cdots\delta(x_N - x_N')$$

$$\langle x_1,\ldots,x_N| \, \boldsymbol{P}_i \, |x_1',\ldots,x_N'\rangle = i\hbar\delta(x_1 - x_1')\cdots\frac{d}{dx_i'}\delta(x_i - x_i')$$

$$\times\cdots\delta(x_N - x_N')$$

$$\boldsymbol{P}_X = \int dx_1\cdots dx_N \, |x_1,\ldots,x_N\rangle \, \langle x_1,\ldots,x_N|$$

and

$$|\psi(t)\rangle = \boldsymbol{P}_X \, |\psi(t)\rangle = \int dx_1\cdots dx_N \, |x_1,\ldots,x_N\rangle \, \psi(x_1,\ldots,x_N,t) ,$$

with

$$\psi(x_1,\ldots,x_N,t) = \langle x_1,\ldots,x_N| \, \psi(t)\rangle .$$

In particular, for expectation values, we have

$$\langle\boldsymbol{\Omega}\rangle = \int dx_1\cdots dx_N \psi^*(x_1,\ldots,x_N,t)\boldsymbol{\Omega}_X\psi(x_1,\ldots,x_N) .$$

Momentum representation. In the *momentum representation* the matrix describing momentum is diagonal (as above, we consider the one-dimensional case):

Definition: Momentum representation

$$\langle p| \, \boldsymbol{P} \, |p'\rangle = p\delta(p - p') , \quad \boldsymbol{P}_P = \int dp \, |p\rangle \, \langle p| \, , \quad \varphi(p,t) = \langle p| \, \psi(t)\rangle .$$

In this representation the matrix elements of the position operator are

$$i\hbar\delta(p - p') = \langle p| \, \boldsymbol{X}\boldsymbol{P} - \boldsymbol{P}\boldsymbol{X} \, |p'\rangle = (p' - p) \, \langle p| \, \boldsymbol{X} \, |p'\rangle$$

$$\Longrightarrow \langle p| \, \boldsymbol{X} \, |p'\rangle = -i\hbar\frac{d}{dp'}\delta(p - p') = i\hbar\frac{d}{dp}\delta(p - p') .$$

In analogy to the coordinate representation we have

$$|\psi(t)\rangle = \boldsymbol{P}_P \, |\psi(t)\rangle = \int dp \, |p\rangle \, \langle p| \, \psi(t)\rangle = \int dp \, |p\rangle \, \varphi(p,t) .$$

[8] Note that the description of a one-dimensional N-particle system is mathematically equivalent to N one-dimensional systems.

The expression $\varphi(p,t) = \langle p | \psi(t) \rangle$ describes the component of $| \psi(t) \rangle$ in the momentum basis and is therefore called the *momentum wave function*. The operator assignments corresponding to those of the coordinate representation follow from

$$
\begin{aligned}
\boldsymbol{X} | \psi(t) \rangle &= \boldsymbol{\mathcal{P}}_P \boldsymbol{X} \boldsymbol{\mathcal{P}}_P | \psi(t) \rangle \\
&= \int \mathrm{d}p \int \mathrm{d}p' \, | p \rangle \, \langle p | \, \boldsymbol{X} \, | p' \rangle \, \langle p' | \psi(t) \rangle \\
&= \mathrm{i}\hbar \int \mathrm{d}p \, | p \rangle \, \frac{\mathrm{d}}{\mathrm{d}p} \varphi(p,t) \\
\boldsymbol{P} | \psi(t) \rangle &= \boldsymbol{\mathcal{P}}_P \boldsymbol{P} \boldsymbol{\mathcal{P}}_P | \psi(t) \rangle \\
&= \int \mathrm{d}p \int \mathrm{d}p' \, | p \rangle \, \langle p | \, \boldsymbol{P} \, | p' \rangle \, \langle p' | \psi(t) \rangle \\
&= \int \mathrm{d}p \, | p \rangle \, p \varphi(p,t) \ ,
\end{aligned}
$$

and we find

Theorem 3.12: Operators in momentum representation

$$
\boldsymbol{X} \longrightarrow \boldsymbol{X}_P = \mathrm{i}\hbar \frac{\mathrm{d}}{\mathrm{d}p} \ , \quad \boldsymbol{P} \longrightarrow \boldsymbol{P}_P = p \ ,
$$

$$
\Omega(\boldsymbol{X}, \boldsymbol{P}) \longrightarrow \boldsymbol{\Omega}_P = \Omega \left(\boldsymbol{X} \to \mathrm{i}\hbar \frac{\mathrm{d}}{\mathrm{d}p}, \boldsymbol{P} \to p \right) \ .
$$

Furthermore, in momentum representation, we see that (3.11) and (3.12) are replaced by

$$
\mathrm{i}\hbar \frac{\mathrm{d}}{\mathrm{d}t} \varphi(p,t) = \boldsymbol{H}_P \varphi(p,t)
$$

and

$$
\langle \boldsymbol{\Omega} \rangle = \int \mathrm{d}p \varphi^*(p,t) \boldsymbol{\Omega}_P \varphi(p,t) = \langle \varphi | \, \boldsymbol{\Omega}_P \, | \varphi \rangle = \langle \boldsymbol{\Omega}_P \rangle \ .
$$

The generalization to the N-dimensional case in this representation follows along the same line as in coordinate representation and boils down to the replacements

$$
\begin{aligned}
\int \mathrm{d}p &\longrightarrow \int \mathrm{d}p_1 \cdots \mathrm{d}p_N \\
| p \rangle &\longrightarrow | p_1, \ldots, p_N \rangle \\
\varphi(p,t) &\longrightarrow \varphi(p_1, \ldots, p_N, t) \ .
\end{aligned}
$$

Transformation from coordinate to momentum representation. We have a special relation between the coordinate and momentum representations: the wave functions of one representation are just Fourier transforms of the other. To see this, we write

$$\psi(x,t) = \langle x|\,\psi(t)\rangle = \int \mathrm{d}p\,\langle x|\,p\rangle\,\langle p|\,\psi(t)\rangle = \int \mathrm{d}p\,\langle x|\,p\rangle\,\varphi(p,t)\ ,$$

where the expansion coefficients $\langle x|\,p\rangle$ satisfy

$$
\begin{aligned}
p\,\langle x|\,p\rangle &= \langle x|\,\boldsymbol{P}\,|p\rangle \\
&= \int \mathrm{d}x'\,\langle x|\,\boldsymbol{P}\,|x'\rangle\,\langle x'|\,p\rangle \\
&= \mathrm{i}\hbar \int \mathrm{d}x'\left(\frac{\mathrm{d}}{\mathrm{d}x'}\delta'(x-x')\right)\langle x'|\,p\rangle = -\mathrm{i}\hbar\frac{\mathrm{d}}{\mathrm{d}x}\,\langle x|\,p\rangle\ .
\end{aligned}
$$

This is a differential equation in $\langle x|\,p\rangle$ that is solved by $\langle x|\,p\rangle \sim \mathrm{e}^{\mathrm{i}px/\hbar}$. Therefore, one finds

Theorem 3.13: Relation between coordinate and momentum representations, De Broglie relation

The wave functions in the coordinate and the momentum representation are related by Fourier transformations:

$$\psi(x,t) = \sqrt{\frac{\hbar}{2\pi}}\int \mathrm{d}k\,\mathrm{e}^{\mathrm{i}kx}\varphi(k,t) = \frac{1}{\sqrt{2\pi\hbar}}\int \mathrm{d}p\,\mathrm{e}^{\mathrm{i}px/\hbar}\varphi(p,t)$$

$$\varphi(k,t) = \frac{1}{\sqrt{2\pi\hbar}}\int \mathrm{d}x\,\mathrm{e}^{-\mathrm{i}kx}\psi(x,t)\ .$$

The *wave number* k and the momentum p obey the *De Broglie relation*

$$p = \hbar k\ .$$

Summary

- Any quantum mechanical state is described by a proper Hilbert vector whose temporal evolution is governed by the **Schrödinger equation**.

- An **ideal quantum mechanical measurement** of an observable Ω yields one of its eigenvalues ω. The measurement process reduces the quantum system to an eigenstate of Ω that belongs to ω (state reduction). The simultaneous measurement of two noncommuting observables leads to uncertainties governed by the **Heisenberg uncertainty principle**.

- Apart from the most commonly used **Schrödinger picture**, there are infinitely many equivalent pictures of quantum mechanics that are all connected by unitary transformations. Of particular interest are the **Heisenberg picture** and the **Dirac** or **interaction picture**.

\triangleright

- Within a given picture the choice of a complete basis system defines a particular **representation** in which the physical quantities (state vectors and observables) are described by column vectors and quadratic matrices. All representations are equivalent and also connected by unitary transformations. Two commonly used representations are the **coordinate** and the **momentum representations**. Since position and momentum operator are canonical conjugates, there is a special relationship between **coordinate wave functions** and **momentum wave functions**: they are Fourier transform pairs.

Applications

38. Ehrenfest equations. Prove the validity of the *Ehrenfest equations*

$$\frac{d\langle X \rangle}{dt} = \left\langle \frac{\partial H}{\partial P} \right\rangle \; , \; \frac{d\langle P \rangle}{dt} = -\left\langle \frac{\partial H}{\partial X} \right\rangle$$

for a quantum mechanical particle in a scalar, space-dependent potential by using the Ehrenfest theorem.

Solution. Following (3.10), we have for

$$H = \frac{P^2}{2m} + V(X)$$

the representation-independent relations

$$\frac{d\langle X \rangle}{dt} = \frac{1}{2mi\hbar} \langle [X, P^2] \rangle \tag{3.13}$$

$$\frac{d\langle P \rangle}{dt} = \frac{1}{i\hbar} \langle [P, V(X)] \rangle \; . \tag{3.14}$$

If we use

$$[X, P^2] = P[X, P] + [X, P]P = 2i\hbar P \; ,$$

(3.13) becomes

$$\frac{d\langle X \rangle}{dt} = \frac{\langle P \rangle}{m} = \left\langle \frac{\partial H}{\partial P} \right\rangle \; .$$

To solve (3.14), we choose the coordinate representation:

$$P \longrightarrow -i\hbar \boldsymbol{\nabla} \; , \; V(X) \longrightarrow V(\boldsymbol{x}) \; .$$

In this representation it is easy to check that

$$[\boldsymbol{\nabla}, V(\boldsymbol{x})]\psi(\boldsymbol{x}, t) = \boldsymbol{\nabla}(V(\boldsymbol{x})\psi(\boldsymbol{x}, t)) - V(\boldsymbol{x})\boldsymbol{\nabla}\psi(\boldsymbol{x}, t) = (\boldsymbol{\nabla}V(\boldsymbol{x}))\psi(\boldsymbol{x}, t) \; ,$$

so that, independently of the representation, we can conclude that

$$[\boldsymbol{\nabla}, V(\boldsymbol{X})] = \boldsymbol{\nabla} V(\boldsymbol{X})$$

$$\implies \frac{d\langle \boldsymbol{P} \rangle}{dt} = -\langle \boldsymbol{\nabla} V(\boldsymbol{X}) \rangle = -\left\langle \frac{\partial H}{\partial \boldsymbol{X}} \right\rangle .$$

Note the formal analogy of the Ehrenfest equations with the Hamilton equations of classical mechanics.

39. Measurement probabilities. Consider the following Hermitean operators:

$$\boldsymbol{L}_x = \frac{1}{\sqrt{2}} \begin{pmatrix} 0 & 1 & 0 \\ 1 & 0 & 1 \\ 0 & 1 & 0 \end{pmatrix} , \quad \boldsymbol{L}_y = \frac{1}{\sqrt{2}} \begin{pmatrix} 0 & -i & 0 \\ i & 0 & -i \\ 0 & i & 0 \end{pmatrix} , \quad \boldsymbol{L}_z = \begin{pmatrix} 1 & 0 & 0 \\ 0 & 0 & 0 \\ 0 & 0 & -1 \end{pmatrix} .$$

a. What are the possible eigenvalues l_z and eigenvectors of \boldsymbol{L}_z?

b. Verify Heisenberg's uncertainty principle of the operators \boldsymbol{L}_x and \boldsymbol{L}_y for the eigenstate of \boldsymbol{L}_z with $l_z = 1$.

c. What are the possible results of measurements and their probabilities for \boldsymbol{L}_x if the system is in the eigenstate of \boldsymbol{L}_z with $l_z = -1$?

Solution.

To a. From the form of \boldsymbol{L}_z we see immediately that the eigenvalues and eigenvectors are given by

$$l_z = +1 : e_+ = \begin{pmatrix} 1 \\ 0 \\ 0 \end{pmatrix} , \quad l_z = 0 : e_0 = \begin{pmatrix} 0 \\ 1 \\ 0 \end{pmatrix} , \quad l_z = -1 : e_- = \begin{pmatrix} 0 \\ 0 \\ 1 \end{pmatrix} .$$

To b. For the state e_+ we first need to calculate

$$\langle \boldsymbol{L}_x \rangle = e_+^\dagger \boldsymbol{L}_x e_+ = 0 , \quad \langle \boldsymbol{L}_x^2 \rangle = e_+^\dagger \boldsymbol{L}_x^2 e_+ = \frac{1}{2}$$

$$\implies \Delta L_x = \sqrt{\langle \boldsymbol{L}_x^2 \rangle - \langle \boldsymbol{L}_x \rangle^2} = \frac{1}{\sqrt{2}}$$

$$\langle \boldsymbol{L}_y \rangle = e_+^\dagger \boldsymbol{L}_y e_+ = 0 , \quad \langle \boldsymbol{L}_y^2 \rangle = e_+^\dagger \boldsymbol{L}_y^2 e_+ = \frac{1}{2}$$

$$\implies \Delta L_y = \sqrt{\langle \boldsymbol{L}_y^2 \rangle - \langle \boldsymbol{L}_y \rangle^2} = \frac{1}{\sqrt{2}}$$

$$|\langle [\boldsymbol{L}_x, \boldsymbol{L}_y] \rangle| = |e_+^\dagger [\boldsymbol{L}_x, \boldsymbol{L}_y] e_+| = e_+^\dagger \boldsymbol{L}_z e_+ = 1 .$$

Thus, we have

$$\Delta L_x \cdot \Delta L_y = \frac{1}{2} = \frac{1}{2} |\langle [\boldsymbol{L}_x, \boldsymbol{L}_y] \rangle| .$$

To c. To find the possible results of a measurement of L_x, we evaluate the eigenvalues and eigenvectors of L_x:

$$L_x x = l_x x \iff (L_x - l_x I) x = 0$$

$$\implies \begin{vmatrix} -l_x & \frac{1}{\sqrt{2}} & 0 \\ \frac{1}{\sqrt{2}} & -l_x & \frac{1}{\sqrt{2}} \\ 0 & \frac{1}{\sqrt{2}} & -l_x \end{vmatrix} = -l_x(l_x + 1)(l_x - 1) = 0 \ .$$

From this, we find

$$l_x = 1: \ f_+ = \frac{1}{2}\begin{pmatrix} 1 \\ \sqrt{2} \\ 1 \end{pmatrix} \ , \ l_x = 0: \ f_0 = \frac{1}{\sqrt{2}}\begin{pmatrix} 1 \\ 0 \\ -1 \end{pmatrix} \ ,$$

$$l_x = -1: \ f_- = \frac{1}{2}\begin{pmatrix} 1 \\ -\sqrt{2} \\ 1 \end{pmatrix} \ .$$

Next, we need to expand the state e_- in the eigenbasis of L_x:

$$e_- = f_+\langle f_+ | e_- \rangle + f_0 \langle f_0 | e_- \rangle + f_-\langle f_- | e_- \rangle \ .$$

The absolute values of the expansion coefficients are the relative probabilities of the measurement:

$$W(l_x = 1) = |\langle f_+ | e_- \rangle|^2 = \frac{1}{4}$$

$$Wl_x = 0) = |\langle f_0 | e_- \rangle|^2 = \frac{1}{2}$$

$$W(l_x = -1) = |\langle f_- | e_- \rangle|^2 = \frac{1}{4} \ .$$

And, as expected, the sum of probabilities is one:

$$W(l_x = 1) + W(l_x = 0) + W(l_x = -1) = 1 \ .$$

3.3 One-Dimensional Systems

Having established the general structure of quantum theory, we now turn our attention to the simplest type of problems: a single particle in one dimension. Although this type of setup might seem somewhat oversimplified, such systems are instructive with a view to more complex, three-dimensional problems, and, of course, they are the easiest to solve.

We have already seen that it is advantageous, in solving the Schrödinger equation, to choose a representation in which the structure of the problem is simplified as much as possible. In some cases it might also be possible to proceed purely algebraically. In the following we discuss three examples of one-dimensional systems:

- Free particle and wave packets: solution in momentum space.

- Potential step and potential well: solution in coordinate space.

- Harmonic oscillator: algebraic solution.

The most common method for solving the Schrödinger equation proceeds by setting up the differential equation in coordinate space and using wave mechanics. Therefore, we start with a general discussion of this method.

3.3.1 Aspects of the Schrödinger Equation in Coordinate Space

In Subsection 3.2.6 we found that the dynamics of a one-dimensional non-relativistic system in the absence of particle spin is given by the differential equation

$$i\hbar \frac{\mathrm{d}}{\mathrm{d}t}\psi(x,t) = \boldsymbol{H}_X \psi(x,t) \ , \tag{3.15}$$

where \boldsymbol{H}_X is of the form

$$\boldsymbol{H}_X = \frac{\boldsymbol{P}_X^2}{2m} + V(\boldsymbol{X}_X, t) = -\frac{\hbar^2}{2m}\frac{\mathrm{d}^2}{\mathrm{d}x^2} + V(x,t) \ , \ V \in \mathbb{R} \ , \tag{3.16}$$

with an interaction potential V. In this representation, we interpret $|\psi(x,t)|^2$ as the probability density of finding the particle at time t and at position x if we perform a measurement on the system. In other words: measuring the position of N identically prepared, noninteracting particles each described by ψ, $N|\psi(\boldsymbol{x},t)|^2\mathrm{d}x$ equals the number of particles that are found in the range $[x : x + \mathrm{d}x]$ at time t. This is the *Born interpretation* of quantum mechanics.

An important theorem, of which we will make use later, is

Theorem 3.14: Continuity equation in coordinate space

The *continuity equation* follows from the Schrödinger equation (3.15), (3.16), and its adjoint:

$$\frac{\mathrm{d}}{\mathrm{d}t}|\psi(x,t)|^2 + \frac{\mathrm{d}}{\mathrm{d}x}j(x,t) = 0 \ ,$$

where

$$j(x,t) = \frac{\hbar}{2im}\left(\psi^*\frac{\mathrm{d}\psi}{\mathrm{d}x} - \psi\frac{\mathrm{d}\psi^*}{\mathrm{d}x}\right) \tag{3.17}$$

is the *probability current density* or *particle current density*.

According to this theorem, a change of the probability density within a certain x-range induces a particle flux:

$$\frac{\mathrm{d}}{\mathrm{d}t}\int_a^b \mathrm{d}x|\psi(x,t)|^2 = j(a,t) - j(b,t) \ .$$

Time-independent Schrödinger equation. Following Subsection 3.2.4, the time dependency of the Schrödinger equation for a conservative Hamilton operator can be separated in coordinate space by the ansatz $\psi(x,t) = \Psi(x)e^{-i\omega t}$, $\omega = E/\hbar$, which leads from (3.15) and (3.16) to the time-independent eigenvalue equation

$$\left(-\frac{\hbar^2}{2m}\frac{d^2}{dx^2} + V(x)\right)\Psi(x) = E\Psi(x)$$

or

$$\frac{d^2\Psi(x)}{dx^2} = [U(x) - \epsilon]\,\Psi(x)\ ,\ \ U(x) = \frac{2mV(x)}{\hbar^2}\ ,\ \ \epsilon = \frac{2mE}{\hbar^2}\ . \tag{3.18}$$

Continuity conditions. In general, we also wish to consider potentials with finite discontinuities (steps). Let $x = a$ be the location of such a step of the potential $U(x)$. The solution $\Psi(x)$ is then still continuously differentiable in a since we see from (3.18) that in the surrounding $[a - \delta : a + \delta], \delta \ll 1$

$$\Psi'(a + \delta) - \Psi'(a - \delta) = \int_{a-\delta}^{a+\delta} dx\frac{d}{dx}\Psi'(x) = \int_{a-\delta}^{a+\delta} dx\,[U(x) - \epsilon]\,\Psi(x) = 0\ ,$$

where we have used the continuity of Ψ at $x = a$. Thus, Ψ and Ψ' are both continuous in a.[9]

Symmetry considerations. Symmetries can often be used to simplify the problem at hand. To see how this works, we define the *parity operator* \boldsymbol{P}:

$$\boldsymbol{P}q(x) = q(-x)\ .$$

Its eigenvalues are $+1$ for even, and -1 for odd functions q. Now, let $\boldsymbol{P}V(x) = V(x)$. Since the kinetic part of the Hamilton operator contains only the second derivative with respect to the coordinate, the operators \boldsymbol{H} and \boldsymbol{P} commute. Applying the parity operator on

$$\boldsymbol{H}\Psi(x) = E\Psi(x)\ , \tag{3.19}$$

we have

$$\boldsymbol{H}\Psi(-x) = E\Psi(-x)\ .$$

Given a solution $\Psi(x)$, we see that $\Psi(-x)$ is also a solution of (3.19). These two functions can be combined to give two new solutions,

$$\Psi_\pm(x) = \Psi(x) \pm \Psi(-x)\ ,\ \ \boldsymbol{P}\Psi_\pm = \pm\Psi_\pm\ ,$$

which are simultaneous eigenstates of \boldsymbol{H} and \boldsymbol{P}. Thus, for symmetric potentials, the basis states can always be split into even and odd functions.

[9] For infinite steps (for example, a δ-potential), this argument no longer holds (see Application 41).

Qualitative discussion of the spectrum. In our subsequent investigations of the time-independent Schrödinger equation we will come across a set of characteristics of the solutions that are generally valid. For a given potential we should always ask whether it allows *bound states* as well as *scattered states*. Just as in classical mechanics, both types of solutions are possible, depending on the type of interaction and the energy of the particle (think of the Kepler problem with its bound elliptical states and its hyperbolic scattered states). However, for bound states, quantum mechanical solutions display a phenomenon unknown to classical mechanics: quantum mechanical bound states are necessarily discrete. Without proof, this behavior can be understood by the following consideration: for a box-shaped potential V with $V(x \in [-a : a]) = -V_0 < 0$, $V(x \notin [-a : a]) = 0$, there exist two points of return, namely the walls of the box, between which a classical particle can oscillate with arbitrary energy. However, quantum mechanically, the particle can also be found behind the walls of the box with nonvanishing probability, i.e., in the classically forbidden zone. In these regions the wave function of the time-independent Schrödinger equation has to fall off exponentially. The conditions of continuity for the wave function at the boundaries of classically allowed (oscillatory) and forbidden (exponential) zones imply that solutions only exist for very specific, discrete energies.

Using the previously introduced simple examples, we shall now verify the following points explicitly:

- The bound spectrum is discrete, and there always exists at least one bound state called the *ground state*. Its wave function has no zero crossing.

- In the one-dimensional case, bound states are not degenerate.

- If, in case of a symmetric potential, there exist more than one bound state, they are given intermittently by even and odd wave functions, and the number of zero crossings increases by one for each new solution.

3.3.2 Widening of a Free Wave Packet

The most simple one-dimensional problem is that of a free particle. Its classical Hamilton function is

$$H(x, p) = \frac{p^2}{2m} ,$$

where m denotes the particle mass. Since the only dynamic variable is momentum, we choose the momentum representation for the solution of this problem. Herein, the Schrödinger equation has the form

$$i\hbar \frac{d}{dt} \varphi(p, t) = \frac{p^2}{2m} \varphi(p, t)$$

or

$$i\hbar \frac{d}{dt}\varphi(k,t) = \frac{\hbar^2 k^2}{2m}\varphi(k,t) \ .$$

Its general solution is given by

$$\varphi(k,t) = \phi(k)e^{-i\omega t} \ , \quad \omega = \frac{\hbar k^2}{2m} \ ,$$

where $\phi(k)$ is an arbitrary function of k obeying the normalization condition

$$\langle \phi(k)| \phi(k)\rangle = \langle \varphi(k,t)| \varphi(k,t)\rangle = \frac{1}{\hbar}$$

for physical particles with finite momentum width. The corresponding solution in coordinate space is found using Theorem 3.13 as

$$\psi(x,t) = \sqrt{\frac{\hbar}{2\pi}} \int dk \varphi(k,t)e^{ikx} = \sqrt{\frac{\hbar}{2\pi}} \int dk \phi(k)e^{i(kx-\omega t)} \ .$$

We now wish to discuss the temporal evolution of the wave function ψ. We assume that, at time $t = 0$, it is given by the normalized Gauss function

$$\psi(x,0) = \frac{1}{\sqrt{\Delta}\pi^{1/4}}e^{-\frac{x^2}{2\Delta^2}}e^{ik_0 x} \ , \quad k_0 > 0 \ , \tag{3.20}$$

describing a particle with average momentum $\langle P\rangle = \hbar k_0$. Following Theorem 3.13, we have

$$\phi(k) = \frac{1}{\sqrt{2\hbar\Delta}\pi^{3/4}} \int dx e^{-\frac{x^2}{2\Delta^2}}e^{-i(k-k_0)x} = \frac{\sqrt{\Delta}}{\sqrt{\hbar}\pi^{1/4}}e^{-\frac{\Delta^2(k-k_0)^2}{2}} \ .$$

For the coordinate wave function we find

$$\psi(x,t) = \frac{\sqrt{\Delta}}{\sqrt{2}\pi^{3/4}} \int dk e^{-\frac{\Delta^2}{2}(k-k_0)^2}e^{ikx}e^{-\frac{i\hbar k^2 t}{2m}}$$

$$= \frac{\sqrt{\Delta}}{\sqrt{\alpha(t)}\pi^{1/4}} \exp\left(-\frac{\left(x - \frac{\hbar k_0 t}{m}\right)^2}{2\alpha(t)}\right) \exp\left[ik_0\left(x - \frac{\hbar k_0 t}{2m}\right)\right] \ , \tag{3.21}$$

with

$$\alpha(t) = \Delta^2 + i\frac{\hbar t}{m} \ .$$

Finally, the probability density $|\psi(x,t)|^2$ is obtained as

$$|\psi(x,t)|^2 = \psi^*(x,t)\psi(x,t) = \frac{1}{\sqrt{\pi\beta(t)}} \exp\left(-\frac{\left(x - \frac{\hbar k_0 t}{m}\right)^2}{\beta(t)}\right) \ ,$$

with

$$\beta(t) = \Delta^2 + \frac{\hbar^2 t^2}{\Delta^2 m^2} \ .$$

Obviously, $|\psi(x,t)|^2$ is also given by a Gaussian distribution whose width $\beta(t)$ increases with time. The center of mass of the wave packet moves with

velocity $\hbar k_0/m = p_0/m$. Using this example, we can verify explicitly the Heisenberg uncertainty principle for position and momentum:

$$\langle \boldsymbol{X} \rangle = \frac{1}{\sqrt{\pi \beta}} \int \mathrm{d}x \, x \exp\left(-\frac{\left(x - \frac{\hbar k_0 t}{m}\right)^2}{\beta}\right) = \frac{\hbar k_0 t}{m}$$

$$\langle \boldsymbol{X}^2 \rangle = \frac{1}{\sqrt{\pi \beta}} \int \mathrm{d}x \, x^2 \exp\left(-\frac{\left(x - \frac{\hbar k_0 t}{m}\right)^2}{\beta}\right) = \frac{\beta}{2} + \frac{\hbar^2 k_0^2 t^2}{m^2}$$

$$\langle \boldsymbol{P} \rangle = \frac{\Delta}{\sqrt{\pi}} \int \mathrm{d}k \, k \, e^{-\Delta^2 (k - k_0)^2} = \hbar k_0$$

$$\langle \boldsymbol{P}^2 \rangle = \frac{\Delta \hbar}{\sqrt{\pi}} \int \mathrm{d}k \, k^2 e^{-\Delta^2 (k - k_0)^2} = \frac{\hbar^2}{2\Delta^2} + \hbar^2 k_0^2 \ .$$

$$\Longrightarrow \Delta X \cdot \Delta P = \frac{\hbar \sqrt{\beta(t)}}{2\Delta} \geq \frac{\hbar}{2} \ .$$

3.3.3 Potential Step

Consider a particle moving under the influence of the step potential

$$V(x) = V_0 \Theta(x) \ , \quad \Theta(x) = \left\{ \begin{array}{l} 0 \text{ for } x < 0 \\ 1 \text{ for } x \geq 0 \end{array} \right\} \ , \quad V_0 > 0$$

(see Fig. 3.1). To solve this problem, we initially determine the eigenfunctions

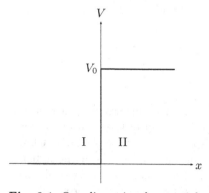

Fig. 3.1. One-dimensional potential step.

and eigenvalues for a particle with fixed momentum that are the solutions of the stationary Schrödinger equation (3.18) in coordinate representation. Subsequently, we turn to the more realistic case of the scattering of a wave packet at this potential step.

Solution of the Schrödinger equation for given momentum. For the stationary solutions in zones I and II, we need to distinguish between the following two cases:

Case 1: $\epsilon > U_0$. In both zones, the general solution is given as a superposition of sine and cosine oscillations, and we make the following ansatz:

$$\Psi_I(x) = Ae^{ik_1x} + Be^{-ik_1x} \quad , \quad k_1 = \sqrt{\epsilon}$$

$$\Psi_{II}(x) = Ce^{ik_2x} \quad , \quad k_2 = \sqrt{\epsilon - U_0} = \sqrt{k_1^2 - U_0} \, .$$

In principle, Ψ_{II} could also contain an additional term proportional to e^{-ik_2x} that corresponds to an incoming wave from $+\infty$ in the negative x-direction. But restricting ourselves to the case of an incoming wave only from the left, and a transmitted wave in zone II only to the right, this term does not contribute. The constants A, B, and C can be found using the continuity conditions

$$\left. \begin{array}{c} \Psi_I(0) = \Psi_{II}(0) \\[2mm] \Psi'_I(0) = \Psi'_{II}(0) \end{array} \right\} \implies \left\{ \begin{array}{c} B = A\dfrac{k_1 - \sqrt{k_1^2 - U_0}}{k_1 + \sqrt{k_1^2 - U_0}} \\[4mm] C = A\dfrac{2k_1}{k_1 + \sqrt{k_1^2 - U_0}} \, . \end{array} \right.$$

Case 2: $0 \le \epsilon \le U_0$. In zone II, we expect the solution to decay exponentially, so we put

$$\Psi_I(x) = Ae^{ik_1x} + Be^{-ik_1x} \quad , \quad k_1 = \sqrt{\epsilon}$$

$$\Psi_{II}(x) = Ce^{-k_2x} \quad , \quad k_2 = \sqrt{U_0 - \epsilon} = \sqrt{U_0 - k_1^2} \, .$$

An exponentially growing term e^{k_2x} in zone II can be eliminated as unphysical since its norm diverges. For the constants, we find

$$B = A\frac{ik_1 + k_2}{ik_1 - k_2} = A\frac{ik_1 + \sqrt{U_0 - k_1^2}}{ik_1 - \sqrt{U_0 - k_1^2}}$$

$$C = A\frac{2ik_1}{ik_1 - k_2} = A\frac{2ik_1}{ik_1 - \sqrt{U_0 - k_1^2}} \, .$$

In both cases, the solution is composed of an incoming wave ψ_I, a reflected wave ψ_R, and a transmitted wave ψ_T. The character of the solution is therefore significantly different from the corresponding classical scenario. Classically, in the first case, a particle would continue on its way without being reflected, but slower from $x = 0$ onward. In the second case, a classical particle would be totally reflected. The fact that there is a nonzero probability of finding a quantum mechanical particle behind the potential wall when $\epsilon < U_0$ is called the *tunnel effect*.

Two interesting quantities characterizing a scattering process are the reflection and transmission coefficients. They are defined as follows:

Definition: Reflection and transmission coefficients R, T

Let j_I, j_R, j_T be the current densities of incoming, reflected, and transmitted wave functions, respectively. We define

\triangleright

$$R = \left| \frac{j_R}{j_I} \right| \ , \ T = \left| \frac{j_T}{j_I} \right| \ , \ T = 1 - R \ .$$

The third relation is a statement of *global current conservation* and follows from Theorem 3.14. The reflection and transmission coefficients describe the proportion of the reflected and the transmitted particle flux relative to the incoming flux. For the potential step, they are given by

Case 1:

$$\left. \begin{aligned} R &= \left| \frac{B}{A} \right|^2 = \frac{2k_1^2 - U_0 - 2k_1\sqrt{k_1^2 - U_0}}{2k_1^2 - U_0 + 2k_1\sqrt{k_1^2 - U_0}} \\ T &= \left| \frac{C}{A} \right|^2 \frac{\sqrt{k_1^2 - U_0}}{k_1} = \frac{4k_1\sqrt{k_1^2 - U_0}}{2k_1^2 - U_0 + 2k_1\sqrt{k_1^2 - U_0}} \end{aligned} \right\} \quad (3.22)$$

Case 2: $R = 1$, $T = 0$.

If, in the first case, ϵ is much larger than U_0, we have practically no reflection, so that the incoming wave from the left propagates without disturbance to the right. In the second case, note that, although there is a nonzero probability of finding the particle in zone II, there is no overall movement in the positive x-direction.

Scattering of a wave packet. To describe the motion of a physical particle, we need to take the solutions of the Schrödinger equation, combine them to a wave packet, and trace its evolution in time. Often, in experiments, the wave packet has a relatively narrow momentum distribution corresponding to a large uncertainty in the particle's localization. The case we have been looking at so far, with solutions spread out over all space, can therefore be regarded as the limit of vanishing momentum uncertainty and is also a good approximation for small momentum distributions. These considerations hold independently of the shape of the potential so that we have

Theorem 3.15: Reflection and transmission coefficients of a wave packet

The reflected and transmitted parts of a wave packet with small momentum distribution depend only on the average momentum value.

We shall verify the correctness of this statement by showing that, in case of a sharp momentum, the reflection and transmission coefficients are identical to the probabilities

$$R' = \int\limits_{-\infty}^{0} dx |\psi(x, t \to \infty)|^2 \ , \ T' = \int\limits_{0}^{\infty} dx |\psi(x, t \to \infty)|^2$$

of finding the particle to the left of (zone I) and to the right of (zone II) the potential step in the limit $t \to \infty$. To this end, we assume the particle is initially described by the following Gaussian distribution [see (3.20)]:

$$\psi(x,0) = \Psi_{\mathrm{I}}(x) = \frac{1}{\sqrt{\Delta}\pi^{1/4}} e^{-\frac{(x+a)^2}{2\Delta^2}} e^{ik_0(x+a)} \ , \quad a \gg \Delta \gg 1 \ .$$

Here, we choose $a \gg \Delta$, so that to start all the wave packet is far left of the potential step. Assuming that $k_0^2 > U_0$ (1. case), the general solution of (3.18) with fixed momentum is of the form

$$\Psi_{k_1}(x) = A\left\{ \left[e^{ik_1 x} + \left(\frac{B(k_1)}{A} \right) e^{-ik_1 x} \right] \Theta(-x) \right.$$
$$\left. + \left(\frac{C(k_1)}{A} \right) e^{ik_2(k_1)x} \Theta(x) \right\} \ .$$

Bar a constant, the projection $\langle \Psi_{k_1} | \Psi_{\mathrm{I}} \rangle$ is given by

$$\phi(k_1) = \langle \Psi_{k_1} | \Psi_{\mathrm{I}} \rangle$$
$$= A^* \left\{ \int \mathrm{d}x \left[e^{-ik_1 x} + \left(\frac{B}{A} \right)^* e^{ik_1 x} \right] \Theta(-x) \Psi_{\mathrm{I}}(x) \right.$$
$$\left. + \int \mathrm{d}x \left(\frac{C}{A} \right)^* e^{-ik_2 x} \Theta(x) \Psi_{\mathrm{I}}(x) \right\} \ .$$

The third integral can be neglected to a very good approximation since, by assumption, $\Psi_{\mathrm{I}}(x)$ vanishes for $x \geq 0$. The second integral also vanishes as Ψ_{I} has a distinct maximum in momentum space around $k_0 > 0$ and is therefore orthogonal to negative momentum states. Thus, ϕ is just the Fourier transform of Ψ_{I}:

$$\phi(k_1) \approx A^* \int \mathrm{d}x e^{-ik_1 x} \Psi_{\mathrm{I}}(x)$$
$$= \frac{A^*}{\sqrt{\Delta}\pi^{1/4}} \int \mathrm{d}x e^{-ik_1 x} e^{ik_0(x+a)} e^{-\frac{(x+a)^2}{2\Delta^2}}$$
$$= \frac{A^*\sqrt{2\pi\Delta}}{\pi^{1/4}} e^{ik_1 a} e^{-\frac{\Delta^2}{2}(k_1 - k_0)^2} \ .$$

As we shall see,

$$A = A^* = \frac{1}{\sqrt{2\pi}}$$

provides the correct normalization. For $\psi(x,t)$, we now find

$$\psi(x,t) \approx \int \mathrm{d}k_1 \phi(k_1) \Psi_{k_1}(x) e^{-i\frac{E(k_1)t}{\hbar}}$$
$$= \frac{\sqrt{\Delta}}{\sqrt{2\pi}^{3/4}} \int \mathrm{d}k_1 e^{ik_1 a} e^{-\frac{\Delta^2}{2}(k_1 - k_0)^2} e^{-\frac{i\hbar k_1^2 t}{2m}}$$
$$\times \left[\left(e^{ik_1 x} + \frac{B}{A} e^{-ik_1 x} \right) \Theta(-x) + \frac{C}{A} e^{ik_2 x} \Theta(x) \right] \ . \tag{3.23}$$

If we use (3.21), the first term yields

$$\Theta(-x)G(-a, k_0, t) = \psi_I(x, t) \, ,$$

with

$$G(-a, k_0, t) = \frac{\sqrt{\Delta}}{\sqrt{\alpha(t)}\pi^{1/4}} \exp\left(-\frac{\left(x + a - \frac{\hbar k_0 t}{m}\right)^2}{2\alpha(t)}\right)$$

$$\times \exp\left[ik_0\left(x + a - \frac{\hbar k_0 t}{2m}\right)\right] .$$

It describes a wave packet coming from $x = -a$ and moving in positive direction toward the potential. For large times, its center of mass is given by $\langle X \rangle = -a + \hbar k_0 t/m \approx \hbar k_0 t/m > 0$, so that the product $\Theta(-x)G(-a, k_0, t)$ vanishes in the limit $t \to \infty$. For $t = 0$, we have $\psi_I(x, 0) = \Psi_I(x)$, which justifies our choice of the normalization constant A. The second term yields

$$\Theta(-x)\frac{B(k_0)}{A}G(a, -k_0, t) = \psi_R(x, t) \, , \qquad (3.24)$$

where the ratio B/A could be pulled out of the integral due to the vanishing dispersion of the momentum $\phi(k_1)$ around k_0. $\psi_R(x, t)$ describes the reflected wave packet, which comes originally from $x = +a$ and moves in the negative direction. For large t, we have $\langle X \rangle = a - \hbar k_0 t/m \approx -\hbar k_0 t/m < 0$, so that the factor $\Theta(-x)$ in (3.24) can be dropped. Thus,

$$R' \approx \int_{-\infty}^{0} dx |\psi(x, t \to \infty)|^2 \approx \int_{-\infty}^{\infty} dx |\psi_R(x, t \to \infty)|^2$$

$$\approx \left|\frac{B(k_0)}{A}\right|^2 \int_{-\infty}^{\infty} dx |G(a, -k_0, t \to \infty)|^2 = \left|\frac{B(k_0)}{A}\right|^2 .$$

The third term in (3.23) describes the transmitted part ψ_T of ψ. Its calculation is not necessary in order to determine T' as we can use the conservation of the norm as well as the orthonormality of ψ_R and ψ_T to find

$$\langle \psi_I | \psi_I \rangle = \langle \psi_R + \psi_T | \psi_R + \psi_T \rangle = R' + T' = 1$$

$$\Longrightarrow T' \approx \left|\frac{C(k_0)}{A}\right|^2 \frac{\sqrt{k_0^2 - U_0}}{k_0} .$$

So, R' and T' are indeed equal to the reflection and transmission coefficients R and T from (3.22) for a particle with fixed momentum.

3.3.4 Potential Well

Consider a particle moving in the square potential well

$$V(x) = \left\{ \begin{array}{ll} 0 & \text{for } -a \leq x \leq a \\ V_0 & \text{else} \end{array} \right\} , \quad V_0 > 0$$

(see Fig. 3.2). To solve this problem, we again start with the time-independent

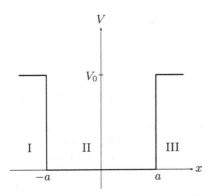

Fig. 3.2. One-dimensional potential well.

Schrödinger equation (3.18) in coordinate representation. Depending on the size of ϵ, two cases need to be considered:

Bound states: $\epsilon < U_0$.

$$\Psi_{\text{I}}(x) = Ae^{k_1 x} \qquad\qquad\qquad , \quad k_1 = \sqrt{U_0 - \epsilon} = \sqrt{U_0 - k_2^2}$$

$$\Psi_{\text{II}}(x) = B\cos(k_2 x) + C\sin(k_2 x) \quad , \quad k_2 = \sqrt{\epsilon}$$

$$\Psi_{\text{III}}(x) = De^{-k_1 x} .$$

Here, Ψ_{II} is chosen in trigonometric form for convenience. The constants are obtained using the continuity conditions for the solution and its first derivative at the boundaries:

$$\begin{aligned} Ae^{-k_1 a} &= B\cos(k_2 a) - C\sin(k_2 a) \\ k_1 Ae^{-ik_1 a} &= k_2\left[B\sin(k_2 a) + C\cos(k_2 a)\right] \\ De^{-k_1 a} &= B\cos(k_2 a) + C\sin(k_2 a) \\ -k_1 De^{-k_1 a} &= k_2\left[-B\sin(k_2 a) + C\cos(k_2 a)\right] . \end{aligned}$$

Combining the two first and the two last equations, we find

$$k_1 = k_2 \frac{B\sin(k_2 a) + C\cos(k_2 a)}{B\cos(k_2 a) - C\sin(k_2 a)} = k_2 \frac{B\sin(k_2 a) - C\cos(k_2 a)}{B\cos(k_2 a) + C\sin(k_2 a)} ,$$

which results in the condition $BC = 0$. This means either $C = 0$, and the solutions are even in x (positive parity), or we have $B = 0$, and the solutions are odd in x (negative parity). Clearly, this is a direct consequence of the symmetry of the potential. For even solutions, we have the condition

$$\tan k_2 a = \frac{k_1}{k_2} = \frac{\sqrt{U_0 - k_2^2}}{k_2}$$

or

$$\tan y = \frac{\sqrt{U_0' - y^2}}{y} \; , \quad y = k_2 a \; , \quad U_0' = a^2 U_0 \; .$$

In Fig. 3.3, the functions $\tan y$ and $\sqrt{U_0' - y^2}/y$ are plotted versus y. The allowed values of y are the intersections of the two curves. We see that there are more and more bound states the higher the walls of the potential well are. But, in any case, there exists always at least one bound state, the ground state. For odd solutions, the corresponding constraint is

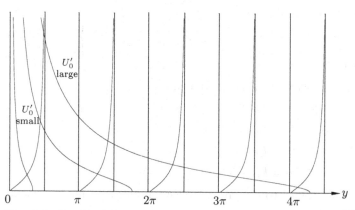

Fig. 3.3. Graphical determination of intersections of $\tan y$ and $\sqrt{U_0' - y^2}/y$ for the case of even bound solutions.

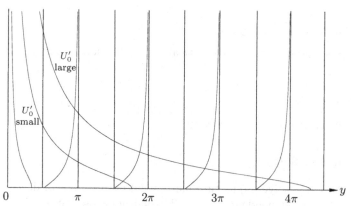

Fig. 3.4. Graphical determination of intersections of $\tan(y + \pi/2)$ and $\sqrt{U_0' - y^2}/y$ for the case of odd bound solutions.

$$- \cot y = \tan\left(y + \frac{\pi}{2}\right) = \frac{\sqrt{U_0' - y^2}}{y} \, ,$$

which is shown graphically in Fig. 3.4. Here, it may happen that there is no bound state solution at all if the potential well is too flat. Considering the solutions we have found, as well as Figs. 3.3 and 3.4, the following can be ascertained: the wave function of the ground state is even and has no zero crossing. At higher excitations the symmetry of the wave function alternates between even and odd, while the number of zero crossings increases by one with each energy level.

Unbound states: $\epsilon \geq U_0$. Here, we shall consider only incoming waves moving from left to right. The general solution then reads

$$\left. \begin{array}{l} \Psi_{\mathrm{I}}(x) = A e^{ik_1 x} + B e^{-ik_1 x} \, , \quad k_1 = \sqrt{\epsilon - U_0} = \sqrt{k_2^2 - U_0} \\[2mm] \Psi_{\mathrm{II}}(x) = C e^{ik_2 x} + D e^{-ik_2 x} \, , \quad k_2 = \sqrt{\epsilon} \\[2mm] \Psi_{\mathrm{III}}(x) = E e^{ik_1 x} \, , \end{array} \right\} \qquad (3.25)$$

where the constants follow from the continuity conditions

$$A e^{-ik_1 a} + B e^{ik_1 a} = C e^{-ik_2 a} + D e^{ik_2 a}$$
$$k_1\left(A e^{-ik_1 a} - B e^{ik_1 a}\right) = k_2\left(C e^{-ik_2 a} - D e^{ik_2 a}\right)$$
$$C e^{ik_2 a} + D e^{-ik_2 a} = E e^{ik_1 a}$$
$$k_2\left(C e^{ik_2 a} - D e^{-ik_2 a}\right) = k_1 E e^{ik_1 a} \, .$$

This system can be solved to obtain the reflection and transmission coefficients as

$$R = \frac{(k_1^2 - k_2^2)^2 \sin^2 2k_2 a}{4k_1^2 k_2^2 + (k_1^2 - k_2^2)^2 \sin^2 2k_2 a}$$
$$T = \frac{4k_1^2 k_2^2}{4k_1^2 k_2^2 + (k_1^2 - k_2^2)^2 \sin^2 2k_2 a} \, .$$

As in the previous example, there is practically no reflection if $\epsilon \gg U_0$. If $\sin 2k_2 a = 0$, i.e.,

$$\epsilon = \left(\frac{n\pi}{2a}\right)^2 \, , \quad n = 1, 2, \dots \, ,$$

the reflection coefficient even vanishes exactly.

3.3.5 Harmonic Oscillator

Sometimes it is easier to solve a quantum mechanical problem algebraically, i.e., in a representation-independent manner. The harmonic oscillator, whose classical Hamilton function is

$$H(x, p) = \frac{p^2}{2m} + V(x) \, , \quad V(x) = \frac{m\omega^2}{2} x^2 \, ,$$

provides an example for this approach. Its representation-independent Schrö-dinger equation reads

$$i\hbar\frac{d}{dt}\,|\psi(t)\rangle = H(\boldsymbol{X},\boldsymbol{P})\,|\psi(t)\rangle$$

$$H(\boldsymbol{X},\boldsymbol{P}) = \boldsymbol{H} = \frac{\boldsymbol{P}^2}{2m} + \frac{m\omega^2}{2}\boldsymbol{X}^2 \ .$$

Introducing the lowering and raising operators

$$\boldsymbol{A} = \frac{1}{\sqrt{2\hbar}}\left(\sqrt{m\omega}\boldsymbol{X} + \frac{i}{\sqrt{m\omega}}\boldsymbol{P}\right) \qquad (lowering\ operator)$$

$$\boldsymbol{A}^\dagger = \frac{1}{\sqrt{2\hbar}}\left(\sqrt{m\omega}\boldsymbol{X} - \frac{i}{\sqrt{m\omega}}\boldsymbol{P}\right) \qquad (raising\ operator),$$

it follows that

$$\boldsymbol{H} = \hbar\omega\left(\boldsymbol{A}^\dagger\boldsymbol{A} + \frac{1}{2}\right)$$

and

$$[\boldsymbol{A},\boldsymbol{A}^\dagger] = 1 \ , \ [\boldsymbol{H},\boldsymbol{A}] = -\hbar\omega\boldsymbol{A} \ , \ [\boldsymbol{H},\boldsymbol{A}^\dagger] = \hbar\omega\boldsymbol{A}^\dagger \ .$$

Obviously, the evaluation of the spectrum of the Hermitean Hamilton oper-ator \boldsymbol{H} from the eigenvalue equation

$$\boldsymbol{H}\,|n\rangle = E_n\,|n\rangle \ , \ E_n = \hbar\omega\left(n + \frac{1}{2}\right)$$

is equivalent to the determination of the spectrum of the Hermitean operator $\boldsymbol{N} = \boldsymbol{A}^\dagger\boldsymbol{A}$ from

$$\boldsymbol{N}\,|n\rangle = n\,|n\rangle \ ,$$

where \boldsymbol{N} denotes the *occupation number operator*. The eigenvector and eigen-values of \boldsymbol{N} are obtained from the following observations.

- The above commutation rules imply that

$$\boldsymbol{N}\boldsymbol{A}\,|n\rangle = \boldsymbol{A}(\boldsymbol{N}-1)\,|n\rangle = (n-1)\boldsymbol{A}\,|n\rangle$$
$$\boldsymbol{N}\boldsymbol{A}^\dagger\,|n\rangle = \boldsymbol{A}^\dagger(\boldsymbol{N}+1)\,|n\rangle = (n+1)\boldsymbol{A}^\dagger\,|n\rangle \ .$$

So, if $|n\rangle$ is an eigenstate of \boldsymbol{N} with eigenvalue n, then $\boldsymbol{A}\,|n\rangle$ and $\boldsymbol{A}^\dagger\,|n\rangle$ are also eigenstates but with eigenvalues $n-1$ and $n+1$, respectively. Thus, using the raising and lowering operators, one can construct neighboring states and energies of $|n\rangle$ in ascending and descending directions. This finally results in a complete discrete set of eigenstates and corresponding energies.

- All energy eigenvalues are positive:

$$E_n = \langle n|\, \boldsymbol{H}\, |n\rangle = \hbar\omega \left(\frac{1}{2} + \langle n|\, \boldsymbol{A}^\dagger \boldsymbol{A}\, |n\rangle \right)$$

$$= \hbar\omega \left(\frac{1}{2} + \sum_m \langle n|\, \boldsymbol{A}^\dagger\, |m\rangle \langle m|\, \boldsymbol{A}\, |n\rangle \right)$$

$$= \hbar\omega \left(\frac{1}{2} + \sum_m |\langle m|\, \boldsymbol{A}\, |n\rangle|^2 \right) > 0 \ .$$

- If $|0\rangle$ is the state with the smallest energy eigenvalue $E_0 = \hbar\omega/2$ (*ground-state energy*), it must be true that

$$\boldsymbol{A}\, |0\rangle = 0 \ .$$

In coordinate representation this yields the differential equation

$$\frac{1}{\sqrt{2\hbar}} \left(\sqrt{m\omega}\, x + \frac{\hbar}{\sqrt{m\omega}} \frac{\mathrm{d}}{\mathrm{d}x} \right) \Psi_0(x) = 0 \ ,$$

whose normalized solution is

$$\Psi_0(x) = \frac{\sqrt{b}}{\pi^{1/4}} \mathrm{e}^{-\frac{b^2}{2} x^2} \ , \quad b^2 = \frac{m\omega}{\hbar} \ .$$

- From

$$\boldsymbol{A}^\dagger\, |n\rangle = \alpha_{n+1}\, |n+1\rangle \ , \quad \boldsymbol{A}\, |n\rangle = \beta_{n-1}\, |n-1\rangle \ ,$$

we obtain the coefficients

$$|\alpha_{n+1}|^2 = \langle n|\, \boldsymbol{A}\boldsymbol{A}^\dagger\, |n\rangle = \langle n|\, \boldsymbol{N} + 1\, |n\rangle = n + 1$$
$$|\beta_{n-1}|^2 = \langle n|\, \boldsymbol{A}^\dagger \boldsymbol{A}\, |n\rangle = \langle n|\, \boldsymbol{N}\, |n\rangle = n \ .$$

Here, any phases can be neglected, so that

$$|n+1\rangle = \frac{1}{\sqrt{n+1}} \boldsymbol{A}^\dagger\, |n\rangle \ , \quad |n-1\rangle = \frac{1}{\sqrt{n}} \boldsymbol{A}\, |n\rangle \ .$$

Iterating the first relation leads to

$$|n\rangle = \frac{1}{\sqrt{n!}} (\boldsymbol{A}^\dagger)^n\, |0\rangle \ , \quad n \in \mathbb{N}.$$

In coordinate representation this becomes

$$\Psi_n(\hat{x}) = \frac{1}{\sqrt{2^n n!}} \left(\hat{x} - \frac{\mathrm{d}}{\mathrm{d}\hat{x}} \right)^n \Psi_0(\hat{x}) \ , \quad \hat{x} = \sqrt{\frac{m\omega}{\hbar}}\, x \ .$$

Summary

- Due to their simplicity, one-dimensional systems are well-suited to explore the main characteristics of quantum theory. For example, we saw that the width of a particle wave packet changes with time and spreads out.

- The behavior of a wave packet moving toward a potential step depends on its energy in the following manner: if the energy is larger than the potential step, part of the packet is reflected (**reflection coefficient** R) and the remainder is transmitted (**transmission coefficient** T). If the energy of the incoming wave packet is smaller than the potential, no transmission occurs ($T = 0$). Nevertheless, the wave packet also penetrates the classically forbidden zone. To calculate R and T, it is often sufficient to consider **static solutions** with infinite spatial extension, as they provide a good approximation of realistic experimental setups (small momentum uncertainty).

- The solutions for particles in a potential well can be classified as follows: for particle energies larger than the well, we find unbound states (**continuous spectrum**) that contain both reflected and transmitted parts. If the energy is smaller, there exist bound states, but only for specific (quantized) values of the particle energy (**discrete spectrum**).

- The harmonic oscillator is a good example for problems that can be solved virtually without recourse to a particular representation.

Applications

40. Potential barrier. Consider a one-dimensional particle of mass m and energy E moving from $x = -\infty$ toward the potential barrier

$$V(x) = \left\{ \begin{array}{ll} V_0 \text{ for } -a \leq x \leq a \\ 0 \qquad\quad \text{else} \end{array} \right\} \, , \; V_0 > 0$$

as shown in Fig. 3.5. Evaluate the transmission coefficient T for both cases, $0 \leq E \leq V_0$ and $E > V_0$.

Solution. The stationary Schrödinger equation for this problem is

$$\frac{\mathrm{d}^2\Psi}{\mathrm{d}x^2} = [U(x) - \epsilon]\Psi(x) \, , \; U(x) = \frac{2mV(x)}{\hbar^2} \, , \; \epsilon = \frac{2mE}{\hbar^2} \, .$$

It is solved by the ansatz

$$\Psi_{\mathrm{I}}(x) = Ae^{i\kappa_1 x} + Be^{-i\kappa_1 x} \, , \; \kappa_1 = \sqrt{\epsilon}$$

$$\Psi_{\mathrm{II}}(x) = Ce^{i\kappa_2 x} + De^{-i\kappa_2 x} \, , \; \kappa_2 = \sqrt{\epsilon - U_0}$$

$$\Psi_{\mathrm{III}}(x) = Ee^{i\kappa_1 x} \, .$$

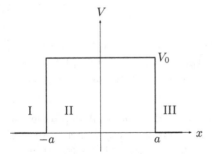

Fig. 3.5. One-dimensional potential barrier.

This is identical in form to the ansatz (3.25) of unbound states for the potential well if we make the substitutions

$$k_1 \longrightarrow \kappa_1 \ , \ k_2 \longrightarrow \kappa_2 \ .$$

The transmission coefficient can therefore be obtained from the same problem, and we obtain

1. case: $\epsilon \geq U_0$.

$$T = \frac{4\epsilon(\epsilon - U_0)}{4\epsilon(\epsilon - U_0) + U_0^2 \sin^2(2a\sqrt{\epsilon - U_0})} \ .$$

2. case: $0 \leq \epsilon < U_0$.

$$T = \frac{4\epsilon(U_0 - \epsilon)}{4\epsilon(U_0 - \epsilon) + U_0^2 \sinh^2(2a\sqrt{U_0 - \epsilon})} \ .$$

In the first case, the transmission coefficient approaches unity for $\epsilon \gg U_0$. For the specific eigenvalues

$$\epsilon = U_0 + \left(\frac{n\pi}{2a}\right)^2 \ , \ n = 1, 2, \ldots \ ,$$

T is exactly one, and there is no reflection at all. In the second case, and for fixed energy ϵ, T decreases with increasing potential height U_0 and potential width a.

41. δ-potential. Consider a particle of mass m and energy E, coming from $x = -\infty$ and moving toward the δ-potential

$$V(x) = V_0 \delta(x - x_0) \ .$$

What are the solutions of the Schrödinger equation? Show that for $V_0 < 0$, there is exactly one bound state.

Solution. The Schrödinger equation is

$$\frac{\mathrm{d}^2 \Psi}{\mathrm{d}x^2} = [U(x) - \epsilon]\Psi(x) \ , \ U(x) = U_0\delta(x - x_0) \ , \tag{3.26}$$

with

$$U_0 = \frac{2mV_0}{\hbar^2} \ , \ \epsilon = \frac{2mE}{\hbar^2} \ .$$

Since we are dealing with an infinite discontinuity at x_0, the first derivatives of the wave functions in zones I ($x < x_0$) and II ($x \geq x_0$) must satisfy the following condition resulting from (3.26):

$$\Psi'(x_0 + \delta) - \Psi'(x_0 - \delta) = \int_{x_0-\delta}^{x_0+\delta} dx \frac{d}{dx} \Psi'(x)$$

$$= U_0 \int_{x_0-\delta}^{x_0+\delta} dx\delta(x-x_0)\Psi(x) - \epsilon \int_{x_0-\delta}^{x_0+\delta} dx\Psi(x)$$

$$\implies \Psi'(x_0 + \delta) - \Psi'(x_0 - \delta) = U_0\Psi(x_0) \ .$$

For $\epsilon \geq 0$, our ansatz for the solution is

$$\Psi_I(x) = Ae^{ikx} + Be^{-ikx} \ , \ k = \sqrt{\epsilon}$$
$$\Psi_{II}(x) = Ce^{ikx} \ .$$

The constants follow from the continuity conditions

$$Ae^{ikx_0} + Be^{-ikx_0} = Ce^{ikx_0}$$
$$ik(Ce^{ikx_0} - Ae^{ikx_0} + Be^{-ikx_0}) = U_0Ce^{ikx_0}$$

as

$$B = A\frac{U_0e^{2ikx_0}}{2ik - U_0} \ , \ C = A\frac{2ik}{2ik - U_0} \ .$$

The transmission and reflection coefficients are therefore

$$T = \left|\frac{C}{A}\right|^2 = \frac{4k^2}{4k^2 + U_0^2} \ , \ R = 1 - T = \frac{U_0^2}{4k^2 + U_0^2} \ ,$$

respectively. For $U_0 < 0$, one can also have bound states ($\epsilon < 0$). A physically meaningful ansatz for this is given by

$$\Psi_I(x) = Ae^{kx} \ , \ k = \sqrt{-\epsilon}$$
$$\Psi_{II}(x) = Be^{-kx} \ .$$

The corresponding continuity conditions yield the equations

$$B = Ae^{2kx_0} \ , \ B = -A\frac{ke^{2kx_0}}{U_0 + k} \ .$$

Obviously, they can simultaneously be satisfied only if $k = -U_0/2$, which means that there exists only one bound state.

3.4 Quantum Mechanical Angular Momenta

Many three-dimensional problems in quantum mechanics have central symmetry. They are therefore most easily solved by replacing the Cartesian coordinates by spherical ones. In this context, quantum mechanical angular momentum operators play a very important role. In preparation for subsequent sections where we will deal with the three-dimensional Schrödinger equation, we consider in this section the properties of angular momentum operators. In particular, we solve its eigenvalue problem and discuss the addition (coupling) of angular momentum.

3.4.1 General Properties of Angular Momentum Operators

In analogy to classical mechanics, the quantum mechanical angular momentum is given by

$$\boldsymbol{L} = \boldsymbol{X} \times \boldsymbol{P} \; .$$

In Cartesian coordinates, it has the following components:

$$\left.\begin{aligned}
L_x &= -i\hbar \left(y\frac{\partial}{\partial z} - z\frac{\partial}{\partial y} \right) \\
L_y &= -i\hbar \left(z\frac{\partial}{\partial x} - x\frac{\partial}{\partial z} \right) \\
L_z &= -i\hbar \left(x\frac{\partial}{\partial y} - y\frac{\partial}{\partial x} \right)
\end{aligned}\right\} \tag{3.27}$$

This leads to commutation rules such as

$$[L_x, L_y] = i\hbar L_z, \; , \; [L_z, L_x] = i\hbar L_y, \; , \; [L_y, L_z] = i\hbar L_x$$

and, more generally,

$$[L_i, L_j] = i\hbar \epsilon_{ijk} L_k \; .$$

As we will show later, there exist other operators obeying the same commutative laws. Therefore, we shall treat the term "angular momentum" more generally by the following definition:

Definition: Angular momentum \boldsymbol{J}

A Hermitean vector operator \boldsymbol{J} is called *angular momentum* if its components obey the commutator algebra

$$[J_i, J_j] = i\hbar \epsilon_{ijk} J_k \; , \quad [J_i, \boldsymbol{J}^2] = 0 \; . \tag{3.28}$$

The last relation follows automatically from the first.

Accordingly, the angular momentum \boldsymbol{L}, which we defined in analogy to classical mechanics, should be viewed as a special case. In the following, we call this particular operator *orbital angular momentum*.

Due to the Hermitecity of J (and therefore also of its components), the norm of the state vector $J_i |\psi\rangle$ obeys

$$0 \leq \langle \psi J_i | J_i \psi \rangle = \langle \psi | J_i^2 | \psi \rangle \ .$$

This means that the operators J_i^2 and J^2 have nonnegative eigenvalues. Since the components J_i commute with J^2, there exists a system of common eigenvectors of J^2 and of one of the angular momentum components, say, for example, J_z. Without giving a proof, we assume that this basis system is complete (if this wasn't so, J would not qualify as an observable). Let us summarize the solution of the eigenvalue problem of J^2 and J_z by the following theorem and then proceed to a proof of the statements:

Theorem 3.16: Eigenvalue problem of the angular momentum operator J

Let $|j, m\rangle$ denote the vectors of the common eigenbasis of J^2 and J_z. The eigenvalue equations are given by

$$J^2 |j, m\rangle = \hbar^2 j(j+1) |j, m\rangle \ , \quad J_z |j, m\rangle = \hbar m |j, m\rangle \ .$$

Then, the following holds:

1. The possible values of the quantum number j of J^2 are

$$j = 0, \frac{1}{2}, 1, \frac{3}{2}, 2, \ldots \ .$$

2. The possible values of the quantum number m of J_z are

$$m = -j, -j+1, \ldots, j-1, j \ .$$

This means that j has $(2j+1)$-fold degeneracy.

3. Supposing that the states $|j, j\rangle$ and $|j, -j\rangle$ are normalized to unity (they describe states with angular momenta in direction of and in opposite direction to the z-axis), the normalized states with quanta (j, m) are given by

$$|j, m\rangle = \hbar^{m-j} \sqrt{\frac{(j+m)!}{(2j)!(j-m)!}} J_-^{j-m} |j, j\rangle$$

$$= \hbar^{-m-j} \sqrt{\frac{(j-m)!}{(2j)!(j+m)!}} J_+^{j+m} |j, -j\rangle \ ,$$

with

$$J_+ = J_x + iJ_y \ , \quad J_- = J_+^\dagger = J_x - iJ_y \ .$$

Proof.

To 2. The operators J_+ and J_- obey the relations

$$[J_z, J_\pm] = \pm\hbar J_\pm \ , \ [J_+, J_-] = 2\hbar J_z \ , \ [J^2, J_\pm] = 0$$

and

$$J^2 = \frac{1}{2}(J_+J_- + J_-J_+) + J_z^2$$
$$J_+J_- = J^2 - J_z^2 + \hbar J_z$$
$$J_-J_+ = J^2 - J_z^2 - \hbar J_z \ .$$

This implies that

$$\left.\begin{array}{r}\langle j,m|\, J_-J_+\,|j,m\rangle = \hbar^2(j-m)(j+m+1)\,\langle j,m|\,j,m\rangle \\ \langle j,m|\, J_+J_-\,|j,m\rangle = \hbar^2(j+m)(j-m+1)\,\langle j,m|\,j,m\rangle \ . \end{array}\right\} \qquad (3.29)$$

These expressions are the squares of the norms of the states $J_+|j,m\rangle$ and $J_-|j,m\rangle$ and are therefore nonnegative. It follows that m is restricted to

$$-j \le m \le j \ . \qquad (3.30)$$

Applying the operators J^2 and J_z to the states $J_+|j,m\rangle$ and $J_-|j,m\rangle$ yields

$$J^2 J_+|j,m\rangle = J_+J^2|j,m\rangle = \hbar^2 j(j+1)J_+|j,m\rangle$$
$$J_z J_+|j,m\rangle = (J_+J_z + \hbar J_+)|j,m\rangle = \hbar(m+1)J_+|j,m\rangle$$

and

$$J^2 J_-|j,m\rangle = \hbar^2 j(j+1)J_-|j,m\rangle$$
$$J_z J_-|j,m\rangle = \hbar(m-1)J_-|j,m\rangle \ .$$

So, if $|j,m\rangle$ is an eigenstate of J^2 and J_z with eigenvalues $\hbar^2 j(j+1)$ and $\hbar m$, we see that $J_\pm|j,m\rangle$ are themselves eigenstates with eigenvalues $\hbar^2 j(j+1)$ and $\hbar(m\pm 1)$. Therefore, analogously to the harmonic oscillator formalism in Subsection 3.3.5, the operators J_+ and J_- are ladder operators that, by repeated application to a given state $|j,m\rangle$, generate all other states $|j,-j\rangle, |j,-j+1\rangle, \ldots, |j,j-1\rangle, |j,j\rangle$ belonging to the same j. In so doing, we have to consider the additional constraints

$$J_+|j,j\rangle = 0 \ , \ J_-|j,-j\rangle = 0$$

following from (3.30).

To 1. The possible values of j are found by the following reasoning: repeated p-fold application of J_+ to the state $|j,m\rangle$ takes us to a state $|j,j\rangle$, so that $m+p = j$. On the other hand, q-fold application of J_- to $|j,m\rangle$ yields the state $|j,-j\rangle$, so that $m-q = -j$. Thus, the sum of the nonnegative integers p and q,

$$p+q = j-m+j+m = 2j \ ,$$

is also nonnegative and an integer. This implies the constraint of point 1.

To 3. Equation (3.29) yields

$$\left.\begin{aligned}
J_+ |j, m\rangle &= \hbar\sqrt{(j - m)(j + m + 1)}\,|j, m + 1\rangle \\
J_- |j, m\rangle &= \hbar\sqrt{(j + m)(j - m + 1)}\,|j, m - 1\rangle \ .
\end{aligned}\right\} \tag{3.31}$$

Iterating these relations, we find the third conjecture.

3.4.2 Orbital Angular Momentum

Using the results from the previous subsection, we now turn to the eigenvalue problem of the orbital angular momentum operator L in coordinate representation. Introducing spherical coordinates r, φ, θ,

$$x = r \cos\varphi \sin\theta \ , \quad y = r \sin\varphi \sin\theta \ , \quad z = r \cos\theta \ ,$$

the corresponding components of L are derived from (3.27) as

$$L_x = i\hbar \left(\sin\varphi \frac{\partial}{\partial\theta} + \cos\varphi \cot\theta \frac{\partial}{\partial\varphi} \right)$$

$$L_y = i\hbar \left(-\cos\varphi \frac{\partial}{\partial\theta} + \sin\varphi \cot\theta \frac{\partial}{\partial\varphi} \right)$$

$$L_z = -i\hbar \frac{\partial}{\partial\varphi} \ .$$

From this, we find

$$L^2 = -\hbar^2 \left(\frac{1}{\sin\theta} \frac{\partial}{\partial\theta} \sin\theta \frac{\partial}{\partial\theta} + \frac{1}{\sin^2\theta} \frac{\partial^2}{\partial\varphi^2} \right)$$

$$L_\pm = L_x \pm iL_y = \hbar e^{\pm i\varphi} \left(\pm\frac{\partial}{\partial\theta} + i\cot\theta \frac{\partial}{\partial\varphi} \right) \ . \tag{3.32}$$

As before, and due to the importance of the orbital angular momentum operator, we first summarize the results for the eigenvalue problem of L^2 and L_z in concise form, before going into more detail:

Theorem 3.17: Eigenvalue problem of the orbital angular momentum L in spherical coordinate representation

The solutions of the eigenvalue problem

$$L^2 Y_{l,m}(\theta, \varphi) = \hbar^2 l(l + 1) Y_{l,m}(\theta, \varphi)$$

$$L_z Y_{l,m}(\theta, \varphi) = \hbar m Y_{l,m}(\theta, \varphi)$$

are given by the *spherical harmonics*

\triangleright

$$Y_{l,m}(\theta,\varphi) = \frac{(-1)^l}{2^l l!}\sqrt{\frac{(2l+1)!}{4\pi}}\sqrt{\frac{(l+m)!}{(2l)!(l-m)!}}$$

$$\times e^{im\varphi}\sin^{-m}\theta\frac{d^{l-m}}{d(\cos\theta)^{l-m}}\sin^{2l}\theta\;.$$

They form a complete orthonormal system of square-integrable functions on the unit sphere,

$$\int Y_{l,m}^*(\theta,\varphi)Y_{l',m'}(\theta,\varphi)d\Omega = \delta_{ll'}\delta_{mm'}\;,$$

where the scalar product contains the integration measure

$$d\Omega = d\varphi\sin\theta d\theta$$

denoting the spherical surface element. The quantum numbers l (*orbital quantum number*) and m (*magnetic quantum number*) are restricted to the integer values

$$l = 0,1,2,\dots\;,\quad m = -l,\dots,l\;.$$

Proof. Because of the form of L_z, the function $Y_{l,m}$ has to be of the type

$$Y_{l,m}(\theta,\varphi) = f_{l,m}(\theta)e^{im\varphi}\;,$$

where m and therefore also l must be integers, so that $Y_{l,m}(\theta,\varphi)$ is unique under the replacement $\varphi \to \varphi + 2\pi$. For the eigenfunction with the largest value of m, $m = l$, it clearly must hold that

$$L_+ Y_{l,l}(\theta,\varphi) = 0\;.$$

If we use (3.32), this yields the differential equation

$$\left(\frac{\partial}{\partial\theta} - l\cot\theta\right)f_{l,l}(\theta) = 0\;.$$

Its solution is given by

$$f_{l,l}(\theta) = c_l\sin^l\theta\;.$$

The value of the constant c_l follows from the normalization condition

$$1 = \int_0^{2\pi}d\varphi\int_0^{\pi}d\theta\sin\theta Y_{l,l}^*(\theta,\varphi)Y_{l,l}(\theta,\varphi) = 2\pi|c_l|^2\int_0^{\pi}d\theta\sin^{2l+1}\theta$$

$$\Longrightarrow |c_l| = \frac{1}{\sqrt{4\pi}}\frac{\sqrt{(2l+1)!}}{2^l l!}\;.$$

All other eigenfunctions belonging to the orbital quantum number l are found using Theorem 3.16:

$$Y_{l,m}(\theta, \varphi) = \hbar^{m-l} \sqrt{\frac{(l+m)!}{(2l)!(l-m)!}} L_{-}^{l-m} Y_{l,l}(\theta, \varphi) \ .$$

Performing the calculations, we obtain the spherical functions as stated in Theorem 3.17 if the phase of c_l is chosen conventionally as

$$\frac{c_l}{|c_l|} = (-1)^l \ .$$

Some properties of the spherical harmonics are given in Section A.6.

3.4.3 Spin

As we will show in Subsection 3.6.3, it is necessary to qualify most quantum mechanical particles by a new degree of freedom called *spin*. Since it has no classical analogue, it is not possible to deduce its form from the third quantum mechanical postulate. Rather, one proceeds quite intuitively using the available information from experiments. It turns out that spin has to be thought of as an intrinsic angular momentum of a particle, also governed by the angular momentum algebra (3.28). Furthermore, spin is decoupled from all other degrees of freedom of a particle, so that the spin operator commutes with all other dynamical quantities.

In the following, we restrict our discussion to the eigenvalue problem of an electron spin operator \boldsymbol{S} whose quantum numbers s and m_s are given by $s = 1/2$ and $m_s = \pm 1/2$. The results for particles with higher spin follow quite analogously, as shown in Application 42. The electron spin operator has two basis states corresponding to the quantum numbers $(s = 1/2, m_s = 1/2)$ and $(s = 1/2, m_s = -1/2)$. For simplicity, we denote these states with $|+\rangle$ (spin up) and $|-\rangle$ (spin down):

$$\left| \frac{1}{2}, \frac{1}{2} \right\rangle = |+\rangle \ , \quad \left| \frac{1}{2}, -\frac{1}{2} \right\rangle = |-\rangle \ .$$

By definition, these vectors satisfy the eigenvalue equations

$$\boldsymbol{S}^2 |\pm\rangle = \frac{3\hbar^2}{4} |\pm\rangle \ , \quad \boldsymbol{S}_z |\pm\rangle = \pm \frac{\hbar}{2} |\pm\rangle \ .$$

Using a matrix representation, in which the spin states $|+\rangle$ and $|-\rangle$ are represented by the column vectors (*spinors*)

$$\chi(+) = \begin{pmatrix} 1 \\ 0 \end{pmatrix} \ , \quad \chi(-) = \begin{pmatrix} 0 \\ 1 \end{pmatrix} \ ,$$

the matrices for \boldsymbol{S}^2 and \boldsymbol{S}_z are given by

$$\boldsymbol{S}^2 = \frac{3\hbar^2}{4} \boldsymbol{I} \ , \quad \boldsymbol{S}_z = \frac{\hbar}{2} \begin{pmatrix} 1 & 0 \\ 0 & -1 \end{pmatrix} \ ,$$

where I denotes the (2×2)-unit matrix. The components S_x and S_y are found using the ladder operators S_\pm, which, from (3.31), have the following effect on the basis states:

$$S_+\chi(+) = 0 \ , \ S_+\chi(-) = \hbar\chi(+) \Longrightarrow S_+ = \hbar \begin{pmatrix} 0 & 1 \\ 0 & 0 \end{pmatrix}$$

$$S_-\chi(+) = \hbar\chi(-) \ , \ S_-\chi(-) = 0 \Longrightarrow S_- = \hbar \begin{pmatrix} 0 & 0 \\ 1 & 0 \end{pmatrix} \ .$$

This yields

$$S_x = \frac{1}{2}(S_+ + S_-) = \frac{\hbar}{2}\begin{pmatrix} 0 & 1 \\ 1 & 0 \end{pmatrix} \ , \ S_y = \frac{1}{2\mathrm{i}}(S_+ - S_-) = \frac{\hbar}{2}\begin{pmatrix} 0 & -\mathrm{i} \\ \mathrm{i} & 0 \end{pmatrix} \ .$$

In the matrix representation, it is conventional to write the electron spin operator as

$$S = \frac{\hbar}{2}\sigma \ ,$$

where σ is composed of the *Pauli matrices*

$$\sigma_x = \begin{pmatrix} 0 & 1 \\ 1 & 0 \end{pmatrix} \ , \ \sigma_y = \begin{pmatrix} 0 & -\mathrm{i} \\ \mathrm{i} & 0 \end{pmatrix} \ , \ \sigma_z = \begin{pmatrix} 1 & 0 \\ 0 & -1 \end{pmatrix} \ .$$

Some of these matrices' properties are

$$[\sigma_i, \sigma_j] = 2\mathrm{i}\epsilon_{ijk}\sigma_k \ , \ \{\sigma_i, \sigma_j\} = 2I\delta_{ij} \ , \ \sigma_i^2 = I \ .$$

3.4.4 Addition of Angular Momenta

Consider the eigenvalue problem of the sum of two angular momentum operators

$$J = J_1 + J_2 \ , \ J_z = J_{1z} + J_{2z} \ ,$$

where J_1 and J_2 respectively denote the angular momenta of systems 1 and 2 that compose the combined system. The common eigenbasis of the simultaneously commuting operators $J_1^2, J_{1z}, J_2^2, J_{2z}$ is composed of the tensor products of the eigenstates of J_1^2, J_{1z} and J_2^2, J_{2z}:

$$|j_1 m_1\rangle \otimes |j_2, m_2\rangle = |j_1, m_1; j_2, m_2\rangle \ .$$

On the other hand, we know that the operators J_1^2, J_2^2, J^2, J_z also form a complete set of commuting observables, so that we can build an eigenbasis of these operators, whose elements we will denote by $|j_1, j_2, J, M\rangle$.[10] They obey the eigenvalue equations

[10] Note the use of the semicolon in the product basis states in order to help distinguish between the product basis and the total angular momentum basis.

$$J_1^2 |j_1, j_2, J, M\rangle = \hbar^2 j_1(j_1 + 1) |j_1, j_2, J, M\rangle$$
$$J_2^2 |j_1, j_2, J, M\rangle = \hbar^2 j_2(j_2 + 1) |j_1, j_2, J, M\rangle$$
$$J^2 |j_1, j_2, J, M\rangle = \hbar^2 J(J + 1) |j_1, j_2, J, M\rangle$$
$$J_z |j_1, j_2, J, M\rangle = \hbar M |j_1, j_2, J, M\rangle \ , \quad M = m_1 + m_2 \ .$$

Using our knowledge from the preceding subsection, we can conclude that, given j_1 and j_2, the quantum number J can only take on the values

$$J = |j_1 - j_2|, \ldots, j_1 + j_2 - 1, j_1 + j_2 \ ,$$

whereas M must be one of

$$M = m_1 + m_2 = -J, \ldots, J - 1, J \ .$$

The following dimensional argument underpins this reasoning: the number of eigenproduct states is $(2j_1 + 1)(2j_2 + 1)$, and this is equal to the number of eigenstates of the total angular momentum:[11]

$$\sum_{J=j_1-j_2}^{j_1+j_2} (2J + 1) = \sum_{n=0}^{2j_2} [2(j_1 - j_2 + n) + 1] = (2j_1 + 1)(2j_2 + 1) \ .$$

The completeness of the product basis allows for an expansion of the total angular momentum eigenstates in the states of this basis:

$$|j_1, j_2, J, M\rangle = \sum_{m_1, m_2} |j_1, m_1; j_2, m_2\rangle \langle j_1, m_1; j_2, m_2 | j_1, j_2, J, M\rangle . \quad (3.33)$$

The expansion coefficients

$$\langle j_1, m_1; j_2, m_2 | j_1, j_2, J, M\rangle = \langle j_1, m_1; j_2, m_2 | J, M\rangle$$

are called *Clebsch-Gordan-(CG-)coefficients* or *vector addition coefficients*.[12] Some of their properties are

1. $\langle j_1, m_1; j_2, m_2 | J, M\rangle \neq 0 \implies |j_1 - j_2| \leq J \leq j_1 + j_2$.

2. $\langle j_1, m_1; j_2, m_2 | J, M\rangle \neq 0 \implies M = m_1 + m_2$.

3. By convention, the CG-coefficients are taken to be real.

4. $\langle j_1, m_1; j_2, m_2 | J, J\rangle$ is positive by convention.

5. $\langle j_1, m_1; j_2, m_2 | J, M\rangle = (-1)^{j_1+j_2-J} \langle j_1, -m_1; j_2, -m_2 | J, -M\rangle$.

6. As coefficients of a unitary transformation, the CG-coefficients obey the following orthogonality relations:

$$\sum_{m_1, m_2} \langle j_1, m_1; j_2, m_2 | J, M\rangle \langle j_1, m_1; j_2, m_2 | J', M'\rangle = \delta_{JJ'}\delta_{MM'}$$

[11] Without restricting generality, we assume that $j_1 \geq j_2$.
[12] Since the indices j_1 and j_2 occur in the bra vector, they can be suppressed in the ket.

$$\sum_{J,M} \langle j_1, m_1; j_2, m_2 | J, M \rangle \langle j_1, m_1', j_2, m_2' | J, M \rangle = \delta_{m_1 m_1'} \delta_{m_2 m_2'} \ .$$

The explicit calculation of the CG-coefficients can be quite cumbersome. In some simple cases, however, the linear combination (3.33) can be written down directly (see next subsection). For this, note that the state with quantum numbers $J = j_1 + j_2$, $M = J$ is given by

$$| j_1, j_2, j_1 + j_2, j_1 + j_2 \rangle = | j_1, j_1; j_2, j_2 \rangle \ ,$$

since, with the help of

$$\boldsymbol{J}^2 = \boldsymbol{J}_1^2 + \boldsymbol{J}_2^2 + 2\boldsymbol{J}_{1z}\boldsymbol{J}_{2z} + \boldsymbol{J}_{1+}\boldsymbol{J}_{2-} + \boldsymbol{J}_{1-}\boldsymbol{J}_{2+} \ ,$$

we find

$$\begin{aligned}
\boldsymbol{J}^2 | j_1, j_2, J, J \rangle &= \hbar^2 (j_1(j_1 + 1) + j_2(j_2 + 1) + 2j_1 j_2) | j_1, j_1; j_2, j_2 \rangle \\
&= \hbar^2 J(J + 1) | j_1, j_2, J, J \rangle
\end{aligned}$$

and

$$\boldsymbol{J}_z | j_1, j_2, J, J \rangle = \hbar (j_1 + j_2) | j_1, j_1; j_2, j_2 \rangle = \hbar J | j_1, j_2, J, J \rangle \ .$$

Using the lowering operator $\boldsymbol{J}_- = \boldsymbol{J}_{1-} + \boldsymbol{J}_{2-}$, we obtain the other states $| j_1, j_2, J, J - 1 \rangle, \ldots, | j_1, j_2, J, -J \rangle$. The state $| j_1, j_2, j_1 + j_2 - 1, j_1 + j_2 - 1 \rangle$ is now uniquely determined from its orthogonality to $| j_1, j_2, j_1 + j_2, j_1 + j_2 - 1 \rangle$ in conjunction with the CG-phase constraint No. 4. If we use \boldsymbol{J}_-, all other states with $J = j_1 + j_2 - 1$ can then be constructed. And so on.

Theorem 3.18: Addition of angular momenta

Let \boldsymbol{J}_1 and \boldsymbol{J}_2 be two angular momentum operators of a particle, $\boldsymbol{J} = \boldsymbol{J}_1 + \boldsymbol{J}_2$ and $\boldsymbol{J}_z = \boldsymbol{J}_{1z} + \boldsymbol{J}_{2z}$ the total angular momentum and its z-component. Then, the operators $\boldsymbol{J}_1^2, \boldsymbol{J}_2^2, \boldsymbol{J}^2, \boldsymbol{J}_z$ have a common complete basis system. Its elements $| j_1, j_2, J, M \rangle$ can be expanded in the product basis states $| j_1, m_1; j_2, m_2 \rangle$ of $\boldsymbol{J}_1^2, \boldsymbol{J}_{1z}$ and $\boldsymbol{J}_2^2, \boldsymbol{J}_{2z}$ as follows:

$$| j_1, j_2, J, M \rangle = \sum_{m_1, m_2} | j_1, m_1; j_2, m_2 \rangle \underbrace{\langle j_1, m_1; j_2, m_2 | J, M \rangle}_{\text{CG-coefficients}} \ .$$

For given j_1 and j_2, the eigenvalues of the total angular momentum operator are confined to

$$J = |j_1 - j_2|, \ldots, j_1 + j_2 \ , \quad M = m_1 + m_2 = -J, \ldots, J \ .$$

3.4.5 Spin-Orbit and Spin-Spin Coupling

Two useful examples for the addition of angular momenta, which will be discussed now, are the coupling of the orbital angular momentum with the

electron spin and the coupling of two electron spins. We will return to the results obtained here in our discussion of the hydrogen atom in Subsections 3.7.3 to 3.7.5.

Spin-orbit coupling. Consider the eigenvalue problem of the total angular momentum

$$J = L + S \;,\; J_z = L_z + S_z \;,$$

which is composed of the orbital angular momentum L and the electron spin S of a particle. The quantum numbers of L, S, and J are denoted by (l, m), $(s = 1/2, m_s = \pm 1/2)$, and (J, M), respectively. For each $l > 0$, there are only two possible values of J: $J = l \pm 1/2$. Since m_s can take on only the values $\pm 1/2$, each vector of the total angular momentum basis is composed of exactly two product basis vectors. Using the spin basis notation $\left| \frac{1}{2}, \pm\frac{1}{2} \right\rangle = |\pm\rangle$, they are

$$\left| l, \frac{1}{2}, l + \frac{1}{2}, M \right\rangle = \alpha \left| l, M - \frac{1}{2}; + \right\rangle + \beta \left| l, M + \frac{1}{2}; - \right\rangle$$

$$\left| l, \frac{1}{2}, l - \frac{1}{2}, M \right\rangle = \alpha' \left| l, M - \frac{1}{2}; + \right\rangle + \beta' \left| l, M + \frac{1}{2}; - \right\rangle \;,$$

where M is a half-integer. The orthonormality condition of these states yields three equations for the expansion coefficients:

$$\alpha^2 + \beta^2 = 1$$
$$\alpha'^2 + \beta'^2 = 1$$
$$\alpha\alpha' + \beta\beta' = 0 \;.$$

A fourth condition is, for example, given by

$$J^2 \left| l, \frac{1}{2}, l + \frac{1}{2}, M \right\rangle = \hbar^2 \left(l + \frac{1}{2} \right) \left(l + \frac{3}{2} \right) \left| l, \frac{1}{2}, l + \frac{1}{2}, M \right\rangle \;,$$

such that

$$\frac{\beta}{\alpha} = \sqrt{\frac{l + \frac{1}{2} - M}{l + \frac{1}{2} + M}} \;,$$

where

$$J^2 = L^2 + S^2 + 2L_z S_z + L_+ S_- + L_- S_+ \;.$$

Using the CG-phase convention, we finally obtain

$$\left| l, \frac{1}{2}, l \pm \frac{1}{2}, M \right\rangle = \frac{1}{\sqrt{2l + 1}} \left(\pm \sqrt{l + \frac{1}{2} \pm M} \left| l, M - \frac{1}{2}; + \right\rangle \right.$$

$$\left. + \sqrt{l + \frac{1}{2} \mp M} \left| l, M + \frac{1}{2}; - \right\rangle \right) \;. \tag{3.34}$$

This equation also holds for $l = 0 \implies J = 1/2$. In this case, we have

$$\left|0,\frac{1}{2},\frac{1}{2},\frac{1}{2}\right\rangle = |0,0;+\rangle \ , \quad \left|0,\frac{1}{2},\frac{1}{2},-\frac{1}{2}\right\rangle = |0,0;-\rangle \ .$$

Spin-spin coupling. The problem simplifies significantly if we consider the total spin

$$\boldsymbol{S} = \boldsymbol{S}_1 + \boldsymbol{S}_2 \ , \quad S_z = S_{1z} + S_{2z}$$

of two electron spins \boldsymbol{S}_1 and \boldsymbol{S}_2 ($s_{1,2} = 1/2, m_{1,2} = \pm1/2$). The total angular momentum quantum numbers S and M can then have only the values

$$S = 0 \Longrightarrow M = 0 \quad \text{or} \quad S = 1 \Longrightarrow M = -1,0,1 \ ,$$

and we can write down the expansion in product spin states as follows:

$$\left|\frac{1}{2},\frac{1}{2},1,1\right\rangle = |+;+\rangle$$

$$\left|\frac{1}{2},\frac{1}{2},1,0\right\rangle = \frac{1}{\sqrt{2}}\left(|+;-\rangle + |-;+\rangle\right)$$

$$\left|\frac{1}{2},\frac{1}{2},1,-1\right\rangle = |-;-\rangle$$

$$\left|\frac{1}{2},\frac{1}{2},0,0\right\rangle = \frac{1}{\sqrt{2}}\left(|+;-\rangle - |-;+\rangle\right) \ .$$

Summary

- Any Hermitean operator \boldsymbol{J} satisfying a particular commutator algebra is called **angular momentum**. The operators \boldsymbol{J}^2 and \boldsymbol{J}_z have a common set of basis vectors $|j,m\rangle$, with $j = 0, 1/2, 1, 3/2, \ldots$ and $m = -j, -j+1, \ldots, j-1, j$.

- The **orbital angular momentum** \boldsymbol{L} is an example of an operator satisfying this algebra. In coordinate representation, the eigenstates $|l,m\rangle$ of \boldsymbol{L}^2 and L_z are the **spherical harmonics** $Y_{l,m}$, l an integer. The electron **spin** \boldsymbol{S} is another example. It allows for only two eigenvalues $s = 1/2, m_s = \pm1/2$. In matrix representation, the eigenstates of \boldsymbol{S}^2 and S_z are called **spinors**.

- For the addition of two angular momenta, $\boldsymbol{J} = \boldsymbol{J}_1 + \boldsymbol{J}_2$, one basis is given by the tensor product of eigenstates of $\boldsymbol{J}_1^2, \boldsymbol{J}_{1z}$ and $\boldsymbol{J}_2^2, \boldsymbol{J}_{2z}$. Another basis (**total angular momentum basis**) is given by the eigenstates of $\boldsymbol{J}_1^2, \boldsymbol{J}_2^2, \boldsymbol{J}^2, \boldsymbol{J}_z^2$. Both basis systems are related by a unitary transformation (**Clebsch-Gordan coefficients**). The quantum numbers J, M of \boldsymbol{J}^2 and \boldsymbol{J}_z are restricted to the values $J = |j_1 - j_2|, \ldots, j_1 + j_2$, $M = m_1 + m_2 = -J, \ldots, J$.

Applications

42. Spin-1 algebra. What are the components of an $s=1$ spin operator S in matrix representation if its eigenstates $|1,1\rangle$, $|1,0\rangle$, and $|1,-1\rangle$ are given by the vectors

$$\chi(+) = \begin{pmatrix} 1 \\ 0 \\ 0 \end{pmatrix} \ , \ \chi(0) = \begin{pmatrix} 0 \\ 1 \\ 0 \end{pmatrix} \ , \ \chi(-) = \begin{pmatrix} 0 \\ 0 \\ 1 \end{pmatrix} .$$

Solution. The components of S are given by Hermitean (3×3)-matrices. Per definition, the actions of S^2 and S_z on the basis states are

$$S^2\chi(+) = 2\hbar^2\chi(+) \ , \ S^2\chi(0) = 2\hbar^2\chi(0) \ , \ S^2\chi(-) = 2\hbar^2\chi(-)$$

$$S_z\chi(+) = \hbar\chi(+) \ , \ S_z\chi(0) = 0 \ , \ S_z\chi(-) = -\hbar\chi(-) \ .$$

This gives

$$S^2 = 2\hbar^2 I \ , \ S_z = \hbar \begin{pmatrix} 1 & 0 & 0 \\ 0 & 0 & 0 \\ 0 & 0 & -1 \end{pmatrix} \ ,$$

where I denotes the (3×3)-unit matrix. The corresponding matrices for S_x and S_y are obtained by using the ladder operators S_+ and S_-, which, according to (3.31), operate on the basis states in the following way:

$$S_+\chi(+) = 0 \ , \ S_+\chi(0) = \sqrt{2}\hbar\chi(+) \ , \ S_+\chi(-) = \sqrt{2}\hbar\chi(0)$$

$$\Longrightarrow S_+ = \sqrt{2}\hbar \begin{pmatrix} 0 & 1 & 0 \\ 0 & 0 & 1 \\ 0 & 0 & 0 \end{pmatrix}$$

and

$$S_-\chi(+) = \sqrt{2}\hbar\chi(0) \ , \ S_-\chi(0) = \sqrt{2}\hbar\chi(-) \ , \ S_-\chi(-) = 0$$

$$\Longrightarrow S_- = \sqrt{2}\hbar \begin{pmatrix} 0 & 0 & 0 \\ 1 & 0 & 0 \\ 0 & 1 & 0 \end{pmatrix} \ .$$

Thus, we find

$$S_x = \frac{1}{2}(S_+ + S_-) = \frac{\hbar}{\sqrt{2}} \begin{pmatrix} 0 & 1 & 0 \\ 1 & 0 & 1 \\ 0 & 1 & 0 \end{pmatrix}$$

$$S_y = \frac{1}{2i}(S_+ - S_-) = \frac{\hbar}{\sqrt{2}} \begin{pmatrix} 0 & -i & 0 \\ i & 0 & -i \\ 0 & i & 0 \end{pmatrix} \ .$$

43. Temporal evolution of a spin-1/2 system. As we will show in Sub-section 3.6.3, the Hamilton operator describing the interaction between the spin S of an electron with charge e and mass m_e, and an external magnetic field B is given by

$$H = -\frac{e}{m_e c} SB .$$

Consider a localized electron whose only degree of freedom is its spin. Let the particle be in an eigenstate of S_x with eigenvalue $\hbar/2$ at time $t = 0$. At this time, we switch on a magnetic field $\boldsymbol{B} = Be_z$ that affects the particle for a duration T, whence the field is rotated instantaneously to the y-axis: $\boldsymbol{B} = Be_y$. After a second time unit T, a measurement of S_x is performed. What is the probability of finding the value $\hbar/2$?

Solution. In matrix representation, the temporal evolution of the electron can be described by the spinor

$$\psi(t) = \psi_+(t)\chi(+) + \psi_-(t)\chi(-) = \begin{pmatrix} \psi_+(t) \\ \psi_-(t) \end{pmatrix} ,$$

with

$$\chi(+) = \begin{pmatrix} 1 \\ 0 \end{pmatrix} , \quad \chi(-) = \begin{pmatrix} 0 \\ 1 \end{pmatrix} .$$

Thus, within the time interval $0 \le t \le T$, the Schrödinger equation is given by

$$i\hbar \frac{d}{dt} \begin{pmatrix} \psi_+ \\ \psi_- \end{pmatrix} = \hbar\omega \begin{pmatrix} 1 & 0 \\ 0 & -1 \end{pmatrix} \begin{pmatrix} \psi_+ \\ \psi_- \end{pmatrix} , \quad \omega = -\frac{eB}{2m_e c}$$

and has the solution

$$\begin{pmatrix} \psi_+(t) \\ \psi_-(t) \end{pmatrix}_1 = \begin{pmatrix} a_0 e^{-i\omega t} \\ b_0 e^{i\omega t} \end{pmatrix} .$$

The constants a_0 and b_0 are determined from the normalization

$$a_0^2 + b_0^2 = 1$$

and the initial condition

$$S_x \psi(0) = \frac{\hbar}{2}\psi(0)$$

as

$$a_0 = b_0 = \frac{1}{\sqrt{2}} .$$

Within the time interval $T \le t \le 2T$, the Schrödinger equation reads

$$i\hbar \frac{d}{dt} \begin{pmatrix} \psi_+ \\ \psi_- \end{pmatrix} = \hbar\omega \begin{pmatrix} 0 & -i \\ i & 0 \end{pmatrix} \begin{pmatrix} \psi_+ \\ \psi_- \end{pmatrix} .$$

Its solution is given by

$$\begin{pmatrix} \psi_+(t) \\ \psi_-(t) \end{pmatrix}_2 = \begin{pmatrix} ae^{i\omega t} + be^{-i\omega t} \\ -iae^{i\omega t} + ibe^{-i\omega t} \end{pmatrix} ,$$

where the continuity condition

$$\begin{pmatrix} \psi_+(T) \\ \psi_-(T) \end{pmatrix}_1 = \begin{pmatrix} \psi_+(T) \\ \psi_-(T) \end{pmatrix}_2$$

constrains the constants a and b to

$$a = \frac{1}{2\sqrt{2}} \left(e^{-2i\omega T} + i \right) , \quad b = \frac{1}{2\sqrt{2}} \left(1 - ie^{2i\omega T} \right) .$$

At time $2T$, the electron state is therefore given by

$$\begin{pmatrix} \psi_+(2T) \\ \psi_-(2T) \end{pmatrix}_2 = \frac{1}{2\sqrt{2}} \begin{pmatrix} 1 - i + e^{-2i\omega T} + ie^{2i\omega T} \\ 1 - i + e^{2i\omega T} + ie^{-2i\omega T} \end{pmatrix} . \tag{3.35}$$

The probability of now finding the eigenvalue $\hbar/2$ of S_x is given by the projection of (3.35) onto the corresponding eigenstate of S_x:

$$W\left(s_x = \frac{1}{2} \right) = \left| \frac{1}{\sqrt{2}}(1,1)\frac{1}{2\sqrt{2}} \begin{pmatrix} 1 - i + e^{-2i\omega T} + ie^{2i\omega T} \\ 1 - i + e^{2i\omega T} + ie^{-2i\omega T} \end{pmatrix} \right|^2$$

$$= \frac{1}{2}(1 + \cos^2 2\omega T) .$$

Accordingly,

$$W\left(s_x = -\frac{1}{2} \right) = \left| \frac{1}{\sqrt{2}}(1,-1)\frac{1}{2\sqrt{2}} \begin{pmatrix} 1 - i + e^{-2i\omega T} + ie^{2i\omega T} \\ 1 - i + e^{2i\omega T} + ie^{-2i\omega T} \end{pmatrix} \right|^2$$

$$= \frac{1}{2} \sin^2 2\omega T$$

$$= 1 - W\left(s_x = \frac{1}{2} \right)$$

is the probability of finding the value $-\hbar/2$.

3.5 Schrödinger Equation in Three Dimensions

In Cartesian coordinates, the three-dimensional Schrödinger equation for a particle in a time-independent scalar potential is given by

$$i\hbar\frac{d}{dt}\psi(\boldsymbol{x},t) = H\psi(\boldsymbol{x},t) , \quad H = -\frac{\hbar^2}{2m}\boldsymbol{\nabla}^2 + V(\boldsymbol{x}) .$$

Using the ansatz

$$\psi(\boldsymbol{x},t) = \Psi(\boldsymbol{x})e^{-i\omega t} ,$$

we find, analogously to the one-dimensional case, the time-independent Schrödinger equation

$$H\Psi(\boldsymbol{x}) = E\Psi(\boldsymbol{x}) \ , \ E = \hbar\omega \ .$$

If the potential can be written in the form

$$V(\boldsymbol{x}) = V_1(x) + V_2(y) + V_3(z) \ ,$$

the solution of this problem becomes quite simple, since the coordinates x, y, z are then separable and the general solution can be factorized,

$$\Psi(\boldsymbol{x}) = u_1(x)u_2(y)u_3(z) \ ,$$

where the functions u_i are each solutions to the one-dimensional Schrödinger equation

$$\left(-\frac{\hbar^2}{2m}\frac{\mathrm{d}^2}{\mathrm{d}\xi^2} + V_i(\xi) \right) u_i(\xi) = E_i u_i(\xi) \ , \ i = 1,2,3 \ ,$$

with

$$E = E_1 + E_2 + E_3 \ .$$

If two or more of the potentials V_i are identical, the corresponding equations lead to the same eigenvalues, and we have degeneracy.

In the following, we consider systems with central symmetry. For this case, we work in the polar coordinate representation, as this allows a separation of the angular from the radial part, which greatly simplifies the problem. Before looking at this, we show briefly how two-particle problems can be reformulated as effective one-particle problems.

3.5.1 Two-Particle Systems and Separation of the Center of Mass Dynamics

Let the Hamilton operator of a two-particle system in Cartesian coordinate representation be given by

$$H = \frac{\boldsymbol{P}_1^2}{2m_1} + \frac{\boldsymbol{P}_2^2}{2m_2} + V(\boldsymbol{x}_1, \boldsymbol{x}_2) \ ,$$

where

$$\boldsymbol{P}_1 = -\mathrm{i}\hbar\boldsymbol{\nabla}_1 \ , \ \boldsymbol{P}_2 = -\mathrm{i}\hbar\boldsymbol{\nabla}_2 \ , \ [\boldsymbol{P}_1, \boldsymbol{P}_2] = 0$$

are the momenta of the two particles. If the potential depends only on the relative distance of the two particles,

$$V(\boldsymbol{x}_1, \boldsymbol{x}_2) = V(\boldsymbol{x}_1 - \boldsymbol{x}_2) \ ,$$

the six-dimensional problem can be reduced to two three-dimensional ones. The first of these represents the constant movement of the center of mass, and the second describes the relative motion of an effective one-particle system. Recall how this is done in classical mechanics (see Subsection 1.5.1). There, we introduced center of mass and relative coordinates

$$x_S = \frac{m_1 x_1 + m_2 x_2}{M} \ , \quad x_R = x_1 - x_2 \ ,$$

as well as center of mass- and relative momenta

$$p_S = p_1 + p_2 = M\dot{x}_S \ , \quad p_R = \frac{m_2 p_1 - m_1 p_2}{M} = \mu \dot{x}_R \ ,$$

with

$$M = m_1 + m_2 \ (\text{total mass}) \ , \quad \mu = \frac{m_1 m_2}{m_1 + m_2} \ (\text{reduced mass}) \ .$$

This decouples Newton's equations of motion into two equations, one for the center of mass, the other for the relative motion. For the corresponding quantum mechanical case, we need to make the substitutions

$$\boldsymbol{P}_S = \boldsymbol{P}_1 + \boldsymbol{P}_2 = -i\hbar \boldsymbol{\nabla}_S \ , \quad \boldsymbol{P}_R = \frac{m_2 \boldsymbol{P}_1 - m_1 \boldsymbol{P}_2}{M} = -i\hbar \boldsymbol{\nabla}_R \ ,$$

for which we have the commutation rules

$$[\boldsymbol{P}_S, \boldsymbol{P}_R] = \boldsymbol{0} \ , \quad [x_{S_i}, \boldsymbol{P}_{S_j}] = [x_{R_i}, \boldsymbol{P}_{R_j}] = i\hbar \delta_{ij}$$

as well as

$$\frac{\boldsymbol{P}_1^2}{2m_1} + \frac{\boldsymbol{P}_2^2}{2m_2} = \frac{\boldsymbol{P}_S^2}{2M} + \frac{\boldsymbol{P}_R^2}{2\mu} \ .$$

The time-independent Schrödinger equation is then of the following form:

$$\left(\frac{\boldsymbol{P}_S^2}{2M} + \frac{\boldsymbol{P}_R^2}{2\mu} + V(x_R) \right) \Psi(x_S, x_R) = E\Psi(x_S, x_R) \ .$$

The ansatz

$$\Psi(x_S, x_R) = \Psi_S(x_S)\Psi_R(x_R)$$

decouples this equation into two separate Schrödinger equations for the center of mass and the relative motion:

$$H_S \Psi_S(x_S) = E_S \Psi_S(x_S) \ , \quad H_S = \frac{\boldsymbol{P}_S^2}{2M}$$

$$H_R \Psi_R(x_R) = E_R \Psi_R(x_R) \ , \quad H_R = \frac{\boldsymbol{P}_R^2}{2\mu} + V(x_R) \ , \quad [H_S, H_R] = 0 \ .$$

The first of these equations describes the motion of a particle with total mass M, whereas the second corresponds to the dynamics of a particle with reduced mass μ in a potential V. Just as in classical mechanics, we see that, up to the replacements $m \leftrightarrow \mu$ and $x \leftrightarrow x_R$, there exists no difference between describing the relative motion of a two-particle system and the (absolute) motion of a one-particle system.

3.5.2 Radial Schrödinger Equation

If the potential V is centrally symmetric, $V(\boldsymbol{x}) = V(|\boldsymbol{x}|)$, the Hamilton operator displays spherical symmetry (i.e., it is invariant under spatial rotations). This makes it advantageous to use spherical coordinates:

$$x = r \cos\varphi \sin\theta \ , \ y = r \sin\varphi \sin\theta \ , \ z = r \cos\theta \ .$$

It is straightforward to see that the square of the momentum operator \boldsymbol{P}, in spherical coordinates, is given by

$$\boldsymbol{P}^2 = -\hbar^2 \boldsymbol{\nabla}^2 = \boldsymbol{P}_r^2 + \frac{\boldsymbol{L}^2}{r^2} \ ,$$

where

$$\boldsymbol{P}_r = -i\hbar \frac{1}{r}\frac{\partial}{\partial r} r = -i\hbar \left(\frac{\partial}{\partial r} + \frac{1}{r} \right)$$

is the *radial momentum* (not to be confused with the relative momentum \boldsymbol{P}_R), and \boldsymbol{L} is the particle's orbital momentum, which we discussed in Subsection 3.4.2. The time-independent Schrödinger equation of the one-particle system then becomes

$$H\Psi(r,\theta,\varphi) = E\Psi(r,\theta,\varphi) \ , \ \boldsymbol{H} = \left(\frac{\boldsymbol{P}_r^2}{2m} + \frac{\boldsymbol{L}^2}{2mr^2} + V(r) \right) \ . \tag{3.36}$$

Before attempting to solve this equation, we need to see under which conditions \boldsymbol{P}_r (and therefore \boldsymbol{H}) is Hermitean. For this to hold, we must have

$$0 = \langle \Psi | \, \boldsymbol{P}_r \, | \Psi \rangle - \langle \Psi | \, \boldsymbol{P}_r \, | \Psi \rangle^*$$

$$= -i\hbar \int_0^{2\pi} d\varphi \int_0^{\pi} d\theta \sin\theta \int_0^{\infty} dr\, r \left(\Psi^* \frac{\partial}{\partial r}(r\Psi) + \Psi \frac{\partial}{\partial r}(r\Psi^*) \right)$$

$$= -i\hbar \int_0^{2\pi} d\varphi \int_0^{\pi} d\theta \sin\theta \int_0^{\infty} dr\, \frac{\partial}{\partial r}|r\Psi|^2 \ .$$

For square-integrable functions, we know that $r\Psi \xrightarrow{r \to \infty} 0$, so the integral over r equals its value at the origin. This means that \boldsymbol{P}_r is Hermitean only if we choose square-integrable functions Ψ for which

$$\lim_{r \to 0} r\Psi = 0 \tag{3.37}$$

is satisfied. Furthermore, we also need to investigate if (3.36) is equivalent to the Schrödinger equation for all possible r, including the origin. One can indeed show that this is true if Ψ satisfies the above Hermiticity condition. Now, we have

$$[\boldsymbol{H},\boldsymbol{L}^2] = [\boldsymbol{H},\boldsymbol{L}_z] = [\boldsymbol{L}^2,\boldsymbol{L}_z] = 0 \ .$$

Since there are no other quantities that commute with these, the operators H, L^2, and L_z constitute a complete set of commuting observables. They must therefore have a common unique set of basis vectors. This is given by the spherical harmonics $Y_{l,m}$, i.e., the basis states of L^2 and L_z, and functions $g_l(r)$ depending only on the radius r:

$$\Psi(r, \theta, \varphi) = g_l(r) Y_{l,m}(\theta, \varphi) .$$

Inserting this expression into (3.36), we find, in conjunction with the Hermitecity condition (3.37),

Theorem 3.19: Radial equation for centrally symmetric potentials

$$\left[-\frac{\hbar^2}{2m} \left(\frac{d^2}{dr^2} + \frac{2}{r}\frac{d}{dr} \right) + \frac{\hbar^2 l(l+1)}{2mr^2} + V(r) \right] g_l(r) = E g_l(r) , \qquad (3.38)$$

where

$$\lim_{r \to 0} r g_l(r) = 0 .$$

Substituting $u_l(r) = r g_l(r)$ yields

$$\left(-\frac{\hbar^2}{2m}\frac{d^2}{dr^2} + \frac{\hbar^2 l(l+1)}{2mr^2} + V(r) \right) u_l(r) = E u_l(r) , \quad u_l(r = 0) = 0. \ (3.39)$$

Some immediate implications are

- Of all solutions u_l only those that are normalizable to unity or the δ-function are physically meaningful.

- If the potential diverges at the origin more slowly than $1/r^2$: $\lim_{r \to 0} r^2 V(r) = 0$ (which is true for most potentials), then, around the origin, u_l satisfies the equation

$$\frac{d^2 u_l}{dr^2} - \frac{l(l+1)}{r^2} u_l = 0 ,$$

whose solutions are $u_l(r) \sim r^{l+1}$ (regular solution) and $u_l(r) \sim r^{-l}$.

- If, for $r \to \infty$, the potential converges to zero faster than $1/r$: $\lim_{r \to \infty} r V(r) = 0$, then we have for large r

$$\frac{d^2 u}{dr^2} + \frac{2mE}{\hbar^2} u = 0 .$$

The solutions to this equation behave asymptotically like

$$E < 0 : \quad u(r) \sim e^{-kr}, e^{kr}$$

$$E > 0 : \quad u(r) \sim e^{ikr}, e^{-ikr} , \quad k^2 = \left| \frac{2mE}{\hbar^2} \right| .$$

3.5.3 Free Particles

As a first application we consider the simplest system, namely that of a free particle. The corresponding radial Schrödinger equation is

$$\left[-\frac{\hbar^2}{2m}\left(\frac{d^2}{dr^2} + \frac{2}{r}\frac{d}{dr} \right) + \frac{\hbar^2 l(l+1)}{2mr^2} - E \right] g_l(r) = 0 \ .$$

Using the dimensionless variables

$$k^2 = \frac{2mE}{\hbar^2} \ , \quad \rho = kr \ ,$$

this becomes the *spherical Bessel differential equation* (see Section A.5)

$$\left(\frac{d^2}{d\rho^2} + \frac{2}{\rho}\frac{d}{d\rho} + 1 - \frac{l(l+1)}{\rho^2} \right) g_l(\rho) = 0 \ . \tag{3.40}$$

Its solutions are the *spherical Bessel functions* whose form and asymptotic behavior are given by

$$j_l(\rho) = (-\rho)^l \left(\frac{1}{\rho}\frac{d}{d\rho} \right)^l \frac{\sin\rho}{\rho} \sim \begin{cases} \dfrac{\rho^l}{(2l+1)!!} & \text{for } \rho \to 0 \\[2mm] \dfrac{\sin(\rho - l\pi/2)}{\rho} & \text{for } \rho \to \infty \end{cases}$$

$$n_l(\rho) = (-\rho)^l \left(\frac{1}{\rho}\frac{d}{d\rho} \right)^l \frac{\cos\rho}{\rho} \sim \begin{cases} \dfrac{(2l-1)!!}{\rho^{l+1}} & \text{for } \rho \to 0 \\[2mm] \dfrac{\cos(\rho - l\pi/2)}{\rho} & \text{for } \rho \to \infty \ . \end{cases}$$

Specific combinations of these functions called *Hankel functions* are of particular interest:

$$h_l^{(+)}(\rho) = n_l(\rho) + ij_l(\rho) \overset{\rho\to\infty}{\longrightarrow} \frac{e^{i\left(\rho - \frac{l\pi}{2}\right)}}{\rho}$$

$$h_l^{(-)}(\rho) = n_l(\rho) - ij_l(\rho) \overset{\rho\to\infty}{\longrightarrow} \frac{e^{-i\left(\rho - \frac{l\pi}{2}\right)}}{\rho} \ .$$

Their asymptotic behavior for $k^2 > 0$ corresponds to outgoing and incoming spherical waves, respectively. Depending on the sign of E, we distinguish the following two cases:

- $E < 0$: here, $h_l^{(+)}$ is the only bounded solution of (3.40). However, at the origin, it has a pole of order $l + 1$. Therefore, the eigenvalue problem has no solution, which means that, in accordance with our expectation, there are no eigenstates of a free particle with negative energy.

- $E \geq 0$: in this case, the Bessel equation has exactly one solution, which is bounded everywhere, namely $j_l(\rho)$. The total solution of the time-independent Schrödinger equation is therefore given by

$$\Psi_{l,m}(r,\theta,\varphi) = j_l(kr)Y_{l,m}(\theta,\varphi) \ . \tag{3.41}$$

Note that the above considerations can also be used for the case of a potential $V(r)$ that can be split into regions of constant potential values V_i. In this case, E is to be replaced by $E - V_i$ in each region.

Expansion of plane waves in spherical harmonics. Other than by spherical waves (3.41), a free particle with (infinitely degenerate) energy $E = \hbar^2 k^2 / 2m$ can also be described by plane waves $e^{i\mathbf{k}\mathbf{r}}$. They represent a particle with momentum $\hbar\mathbf{k}$, whereas (3.41) corresponds to a particle with a particular angular momentum. Since spherical harmonics constitute a complete basis system, the countable number of spherical wave functions to a particular wave number k spans the space of eigenfunctions with energy $E = \hbar^2 k^2 / 2m$. Thus, the plane wave $e^{i\mathbf{k}\mathbf{r}}$ can be expanded in these functions:

$$e^{i\mathbf{k}\mathbf{r}} = \sum_{l=0}^{\infty} \sum_{m=-l}^{l} a_{l,m}(\mathbf{k}) j_l(kr) Y_{l,m}(\theta, \varphi) .$$

Putting the z-axis in the direction of \mathbf{k}, we have

$$e^{i\mathbf{k}\mathbf{r}} = e^{ikr\cos\theta} , \quad L_z e^{ikr\cos\theta} = 0 .$$

This implies that the expansion is independent of φ and restricted to terms with $m = 0$. If we use (A.14) and $a_l = a_{l,0}$, it follows that

$$e^{iu\rho} = \sum_{l=0}^{\infty} a_l j_l(\rho) P_l(u) , \quad u = \cos\theta , \tag{3.42}$$

where $P_l = P_{l,0}$ are the *Legendre polynomials* (see Section A.6). To determine the expansion coefficients a_l, we proceed as follows: differentiation of (3.42) yields

$$iue^{iu\rho} = \sum_{l=0}^{\infty} a_l \frac{dj_l}{d\rho} P_l(u) .$$

On the other hand, according to (A.11), we have

$$iue^{iu\rho} = i\sum_{l=0}^{\infty} a_l j_l u P_l = i\sum_{l=0}^{\infty} \left(\frac{l+1}{2l+3} a_{l+1} j_{l+1} + \frac{l}{2l-1} a_{l-1} j_{l-1} \right) P_l .$$

Comparing the coefficients of P_l in the previous two expansions and using (A.9) and (A.10) we obtain

$$l\left(\frac{1}{2l+1} a_l - \frac{i}{2l-1} a_{l-1} \right) j_{l-1} = (l+1) \left(\frac{1}{2l+1} a_l + \frac{i}{2l+3} a_{l+1} \right) j_{l+1}$$

and, therefore,

$$\frac{1}{2l+3} a_{l+1} = \frac{i}{2l+1} a_l \implies a_l = (2l+1) i^l a_0 .$$

Using $j_l(0) = \delta_{l0}$ and $P_0(u) = 1$, we see from the expansion of $e^{iu\rho}$ for $\rho = 0$ that $a_0 = 1$. This finally leads to

Theorem 3.20: Expansion of a plane wave in spherical harmonics

Let the wave vector \boldsymbol{k} define the z-direction. We then have

$$e^{i\boldsymbol{k}\boldsymbol{r}} = e^{ikr\cos\theta} = \sum_{l=0}^{\infty}(2l+1)i^{l}j_{l}(kr)P_{l}(\cos\theta)$$

$$= \sum_{l=0}^{\infty}\sqrt{4\pi(2l+1)}i^{l}j_{l}(kr)Y_{l,0}(\theta,\varphi) \ .$$

3.5.4 Spherically Symmetric Potential Well

Consider the following spherically symmetric potential function (Fig. 3.6):

$$V(r) = \left\{\begin{array}{cc} -V_0 & \text{for } r < a \\ 0 & \text{for } r \geq a \end{array}\right\} \ , \quad V_0 > 0 \ .$$

In the inner area I, the sole solution of the radial Schrödinger equation (3.38), which is regular at the origin, is given by

$$g_{l}^{(\mathrm{I})}(k_1r) = A_l j_l(k_1r) \ , \quad k_1 = \sqrt{\frac{2m(E+V_0)}{\hbar^2}} \ .$$

In area II, we need to distinguish two cases:

Bound states: $E < 0$. Here, we have

$$g_{l}^{(\mathrm{II})}(ik_2r) = B_l h_{l}^{(+)}(ik_2r) \ , \quad k_2 = \sqrt{\frac{-2mE}{\hbar^2}}$$

as the only solution bounded at infinity. The continuity conditions at $r = a$,

$$g_{l}^{(\mathrm{I})}(k_1a) = g_{l}^{(\mathrm{II})}(ik_2a) \ , \quad \frac{\mathrm{d}}{\mathrm{d}r}g_{l}^{(\mathrm{I})}(k_1r)\bigg|_{r=a} = \frac{\mathrm{d}}{\mathrm{d}r}g_{l}^{(\mathrm{II})}(ik_2r)\bigg|_{r=a} \ ,$$

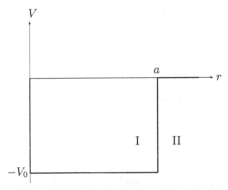

Fig. 3.6. Spherically symmetric potential well.

determine the ratio of the integration constants A_l and B_l. These two conditions can be simultaneously satisfied only for discrete values of E. They determine the energy levels of the bound states of the particle. For $l=0$ states, we obtain the condition

$$\tan k_1 a = -\frac{k_1}{k_2} .$$

Unbounded states: $E > 0$. The general solution is a linear combination of the spherical Bessel functions, for which we can make the following ansatz:

$$g_l^{(II)}(k_2 r) = B_l \left[j_l(k_2 r) \cos \delta_l + n_l(k_2 r) \sin \delta_l \right] , \quad k_2 = \sqrt{\frac{2mE}{\hbar^2}} .$$

For $l = 0$, the corresponding continuity conditions constrain the phase δ_0 to

$$\tan(k_2 a + \delta_0) = \frac{k_2}{k_1} \tan k_1 a . \tag{3.43}$$

3.5.5 Naive Hydrogen Atom

The standard example of a two-particle system is the hydrogen atom. It is built of a positively charged proton and an electron that circles around it. We initially choose to treat the more general case of *hydrogen-like atoms*. They also contain just one electron, whereas the nucleus may consist of more than one proton (and neutron). Furthermore, we disregard effects due to the spins of the nucleus or the electron; they are discussed in Subsections 3.7.3 to 3.7.5. Our starting point is the electrostatic Coulomb potential

$$V(r) = -\frac{Ze^2}{r}$$

acting between electron and nucleus, where e is the elementary charge of the electron and Z is the charge number of the nucleus. The radial Schrödinger equation for the relative motion of electron and nucleus follows from (3.39):

$$\left(-\frac{\hbar^2}{2\mu} \frac{d^2}{dr^2} + \frac{\hbar^2 l(l+1)}{2\mu r^2} - \frac{Ze^2}{r} \right) u_l(r) = E u_l(r) , \quad \mu = \frac{m_e m_k}{m_e + m_k} .$$

Here, m_e denotes the mass of the electron and m_k that of the nucleus.[13] Restricting ourselves to the case of bound states with $E < 0$, it is useful to introduce the dimensionless quantities

$$\rho = \left(-\frac{8\mu E}{\hbar^2} \right)^{1/2} r , \quad \lambda = \frac{Ze^2}{\hbar} \left(-\frac{\mu}{2E} \right)^{1/2} = Z\alpha_e \left(-\frac{\mu c^2}{2E} \right)^{1/2} ,$$

with the *fine structure constant*

[13] Note that the mass of the proton is around 1840 times greater than that of the electron, so that for all hydrogen-like atoms, we can use the approximation $\mu \approx m_e$.

$$\alpha_e = \frac{e^2}{\hbar c} \approx \frac{1}{137} \ .$$

The radial Schrödinger equation then turns into the simple form

$$\left(\frac{d^2}{d\rho^2} - \frac{l(l+1)}{\rho^2} + \frac{\lambda}{\rho} - \frac{1}{4} \right) u_l(\rho) = 0 \ . \tag{3.44}$$

For the case $\rho \to 0$, this is reduced to

$$\left(\frac{d^2}{d\rho^2} - \frac{l(l+1)}{\rho^2} \right) u_l(\rho) = 0 \ ,$$

whose regular solution is proportional to ρ^{l+1}. On the other hand, for $\rho \to \infty$, we have the equation

$$\left(\frac{d^2}{d\rho^2} - \frac{1}{4} \right) u(\rho) = 0 \ .$$

Its normalizable solution, which falls off at infinity, behaves as $e^{-\rho/2}$. Putting it all together, we are led to the following ansatz as a solution of (3.44):

$$u_l(\rho) = e^{-\rho/2} \rho^{l+1} H(\rho) \ .$$

Inserting this in the above, we find the differential equation

$$\rho H'' + (2l + 2 - \rho)H' + (\lambda - l - 1)H = 0 \ . \tag{3.45}$$

The power-type expansion

$$H(\rho) = \sum_{i=0}^{\infty} a_i \rho^i$$

yields

$$\sum_{i=0}^{\infty} \left[(i+1)(i+2l+2)a_{i+1} + (\lambda - l - 1 - i)a_i \right] \rho^i = 0 \ ,$$

and we obtain a recursive formula for the expansion coefficients a_i:

$$a_{i+1} = \frac{i + l + 1 - \lambda}{(i+1)(i+2l+2)} a_i \ .$$

For u_l to display the required asymptotic behavior at infinity, the expansion needs to terminate at some $i = n'$, i.e.,

$$\lambda = n' + l + 1 \ . \tag{3.46}$$

This is just the quantization condition for λ and therefore also for the energy levels of the bound states to a given angular momentum (l, m). Conventionally, one introduces the quantity

$$n = n' + l + 1 \ ,$$

called the *principal quantum number*. With this notation, to each $n > 0$, there corresponds a radial state

$$u_{n,l}(\rho) = e^{-\rho/2}\rho^{l+1} \sum_{i=0}^{n-l-1} (-1)^i \frac{(n-l-1)!(2l+1)!}{(n-l-1-i)!(2l+1+i)!i!}\rho^i$$

$$= e^{-\rho/2}\rho^{l+1} \frac{(n-l-1)!(2l+1)!}{[(n+l)!]^2} L_{n-l-1}^{2l+1}(\rho) , \tag{3.47}$$

where

$$L_p^k(\rho) = \sum_{i=0}^{p} (-1)^i \frac{[(p+k)!]^2}{(p-i)!(k+i)!i!}\rho^i$$

are the *Laguerre polynomials*. The energy levels belonging to (3.47) are given by

$$E_n = -\frac{Z^2 e^4 \mu}{2\hbar^2 n^2} = \frac{E_1}{n^2} , \quad E_1 = -\frac{Z^2 e^4 \mu}{2\hbar^2} = -\frac{1}{2}\mu c^2 Z^2 \alpha_e^2 . \tag{3.48}$$

We see that E_n does not depend on l, so that, for given n, all states with $l < n$ and $-l \leq m \leq l$ have the same energy; they are degenerate. The degree of degeneracy is[14]

$$\sum_{l=0}^{n-1} (2l+1) = n^2 .$$

In atomic spectroscopy, it is conventional to denote the states defined by l by the letters (s, p, d, f, g, \ldots) in ascending order and to precede these with the principal quantum number n. The magnetic quantum number m, which determines the orientation of the system, is usually suppressed. Figure 3.7 shows the energy levels of the hydrogen atom (*term scheme*). $1s$ is the non-degenerate ground state. The first excited level is fourfold degenerate and contains one $2s$ and three $2p$ states. The second excited level contains one $3s$, three $3p$, and five $3d$ states and is ninefold degenerate, and so on. In Subsections 3.7.3 to 3.7.5, we show that if we include the spin degrees of freedom of electron and proton as well as relativistic corrections, the spectrum of this "naive" system splits into further lines, thus breaking the degeneracy seen here.

Transforming back to the original relative coordinate r, (3.47) yields the first few normalized radial wave functions of the hydrogen atom as

$$g_{1,0}(r) = 2\left(\frac{Z}{r_0}\right)^{3/2} e^{-Zr/r_0}$$

$$g_{2,0}(r) = 2\left(\frac{Z}{2r_0}\right)^{3/2} \left(1 - \frac{Zr}{2r_0}\right) e^{-Zr/2r_0}$$

$$g_{2,1}(r) = \frac{1}{\sqrt{3}}\left(\frac{Z}{2r_0}\right)^{3/2} \frac{Zr}{r_0} e^{-Zr/2r_0}$$

[14] Actually, the degree of degeneracy is $2n^2$ due to the two spin directions of the electron, which we have neglected so far.

Fig. 3.7. Term scheme of the "naive hydrogen atom".

$$g_{3,0}(r) = 2 \left(\frac{Z}{3r_0} \right)^{3/2} \left(1 - \frac{2Zr}{3r_0} + \frac{2Z^2r^2}{27r_0^2} \right) e^{-Zr/3r_0}$$

$$g_{3,1}(r) = \frac{4\sqrt{2}}{3} \left(\frac{Z}{3r_0} \right)^{3/2} \frac{Zr}{3r_0} \left(1 - \frac{Zr}{6r_0} \right) e^{-Zr/3r_0}$$

$$g_{3,2}(r) = \frac{2\sqrt{2}}{27\sqrt{5}} \left(\frac{Z}{3r_0} \right)^{3/2} \left(\frac{Zr}{r_0} \right)^2 e^{-Zr/3r_0} \; ,$$

where

$$r_0 = \frac{\hbar}{\mu c \alpha_e}$$

denotes the *Bohr radius*. Some resulting expectation values, whose calculation we present in Application 45, are

$$\langle nlm| \, r \, |nlm \rangle = \langle r \rangle_{nl} = \frac{r_0}{2Z} \left[3n^2 - l(l+1) \right] \tag{3.49}$$

$$\langle nlm| \, r^2 \, |nlm \rangle = \langle r^2 \rangle_{nl} = \frac{r_0^2 n^2}{2Z^2} \left[5n^2 + 1 - 3l(l+1) \right]$$

$$\left. \begin{array}{l} \left\langle nlm \left| \dfrac{1}{r} \right| nlm \right\rangle = \left\langle \dfrac{1}{r} \right\rangle_{nl} = \dfrac{Z}{r_0 n^2} \\[3mm] \left\langle nlm \left| \dfrac{1}{r^2} \right| nlm \right\rangle = \left\langle \dfrac{1}{r^2} \right\rangle_{nl} = \dfrac{Z^2}{r_0^2 n^3 \left(l + \frac{1}{2} \right)} \\[3mm] \left\langle nlm \left| \dfrac{1}{r^3} \right| nlm \right\rangle = \left\langle \dfrac{1}{r^3} \right\rangle_{nl} = \dfrac{Z^3}{r_0^3 n^3 l \left(l + \frac{1}{2} \right) (l+1)} \end{array} \right\} \tag{3.50}$$

If l is maximal, $l = n-1$, the average quadratic radial distance of the electron from the nucleus is given by

$$\Delta r = \sqrt{\langle r^2 \rangle - \langle r \rangle^2} = \frac{\langle r \rangle}{\sqrt{2n+1}} \ .$$

For large n, the quantity $\Delta r / \langle r \rangle$ becomes very small, and the electron is practically confined to the surface of a sphere with radius $R = n^2 r_0 / Z$, where the corresponding energy $E = -Z^2 e^2 / (2 r_0 n^2) = -Z e^2 / 2R$ is equal to the kinetic energy of a classical electron circling around the nucleus with radius R. This is an example of the correspondence principle that we mentioned in Subsection 3.2.1 and that states that the laws of quantum mechanics converge toward the corresponding laws of the classical theory in the limit of large quantum numbers or in the limit $\hbar \to 0$.

Summary

- If, for a three-dimensional two-particle problem, the potential is a function only of the relative distance of the two particles, the Schrödinger equation can be separated into two decoupled equations, just as for the classical mechanical problem of this type. One equation describes the **center of mass motion** and the other the **relative motion** of the particles. The latter can be viewed as an equation for an **effective one-particle problem** within the potential.

- For **centrally symmetric potentials**, we use the polar coordinate representation to separate the radial part of the wave function from the angular part. The solution to the angular part is given by spherical harmonics, the eigenfunctions of the orbital angular momentum. The solutions to the radial part are obtained from the **radial equation for centrally symmetric potentials**.

- For free particles or areas of constant potential, the radial equation becomes the **spherical Bessel differential equation** whose solutions are the **spherical Bessel functions**.

- Using a simplistic (naive) approach, the bound states of hydrogen-like atoms are characterized by three quantum numbers: the **principal quantum number** n, the **orbital quantum number** $l < n$, and the **magnetic quantum number** $m = -l, \ldots, l$. The binding energies depend only on the principal quantum number n and are therefore n^2-fold degenerate if we disregard the two spin orientations of the electron.

Applications

44. Three-dimensional anisotropic oscillator. Determine the energy eigenvalues of a three-dimensional anisotropic oscillator whose stationary Schrödinger equation in coordinate representation is given by

$$H\Psi(\boldsymbol{x}) = E\Psi(\boldsymbol{x}) \ , \quad H = -\frac{\hbar^2}{2m}\boldsymbol{\nabla}^2 + \frac{m}{2}\left(\omega_1^2 x^2 + \omega_2^2 y^2 + \omega_3^2 z^2\right) \ . \quad (3.51)$$

Solution. The product ansatz

$$\Psi(x, y, z) = u_1(x)u_2(y)u_3(z)$$

takes us from (3.51) to

$$\left(-\frac{\hbar^2}{2m}u_1''(x) + \frac{m}{2}\omega_1^2 x^2 u_1(x) - E_1 u_1(x)\right) u_2(y)u_3(z)$$

$$+ \left(-\frac{\hbar^2}{2m}u_2''(y) + \frac{m}{2}\omega_2^2 y^2 u_2(y) - E_2 u_2(y)\right) u_1(x)u_3(z)$$

$$+ \left(-\frac{\hbar^2}{2m}u_3''(z) + \frac{m}{2}\omega_3^2 z^2 u_3(z) - E_3 u_3(z)\right) u_1(x)u_2(y) = 0 \ , \quad (3.52)$$

with

$$E = E_1 + E_2 + E_3 \ .$$

For (3.52) to hold, we see that the brackets must be identically zero:

$$\left(-\frac{\hbar^2}{2m}u_i''(\xi) + \frac{m}{2}\omega_i^2 \xi^2 u_i(\xi)\right) = E_i u_i(\xi) \ , \quad i = 1, 2, 3 \ .$$

Thus, the problem reduces to three equations each of which describe a one-dimensional oscillator. This problem has already been discussed in Subsection 3.3.5, so that we can use those results here to obtain the total energy of the system as

$$E = \hbar\omega_1\left(n_1 + \frac{1}{2}\right) + \hbar\omega_2\left(n_2 + \frac{1}{2}\right) + \hbar\omega_3\left(n_3 + \frac{1}{2}\right) \ .$$

The ground-state wave function $(n_1 = n_2 = n_3 = 0)$ is given by

$$\Psi_{0,0,0}(\boldsymbol{x}) = \frac{\sqrt{b_1 b_2 b_3}}{\pi^{3/4}}e^{-\frac{1}{2}(b_1^2 x^2 + b_2^2 y^2 + b_3^2 z^2)} \ , \quad b_i^2 = \frac{m\omega_i}{\hbar} \ .$$

45. Expectation values for the hydrogen atom. With the help of the radial Schrödinger equation, show that for hydrogen-like atoms we have

$$(s+1)\frac{Z^2}{r_0^2 n^2}\left\langle r^s\right\rangle_{nl} - (2s+1)\frac{Z}{r_0}\left\langle r^{s-1}\right\rangle_{nl}$$

$$+ \frac{s}{4}\left[(2l+1)^2 - s^2\right]\left\langle r^{s-2}\right\rangle_{nl} = 0 \ . \quad (3.53)$$

Solution. For hydrogen-like atoms, we know from (3.39) that

$$u''(r) - \frac{l(l+1)}{r^2}u(r) + \frac{2Z}{r_0 r}u(r) - \frac{Z^2}{r_0^2 n^2}u(r) = 0 \ , \ u(r) = u_{n,l}(r) \ . \ (3.54)$$

Therefore,

$$\int_0^\infty dr r^s u(r) u''(r) = l(l+1) \int_0^\infty dr r^{s-2} u^2(r) - \frac{2Z}{r_0} \int_0^\infty dr r^{s-1} u^2(r)$$

$$+ \frac{Z^2}{r_0^2 n^2} \int_0^\infty dr r^s u^2(r)$$

$$= l(l+1) \left\langle r^{s-2} \right\rangle_{nl} - \frac{2Z}{r_0} \left\langle r^{s-1} \right\rangle_{nl} + \frac{Z^2}{r_0^2 n^2} \left\langle r^s \right\rangle_{nl} . \ (3.55)$$

On the other hand, we can use partial integration to find

$$\int_0^\infty dr r^s u(r) u''(r) = -s \int_0^\infty dr r^{s-1} u(r) u'(r) - \int_0^\infty dr r^s u'^2(r)$$

$$= -s \int_0^\infty dr r^{s-1} u(r) u'(r)$$

$$+ \frac{2}{s+1} \int_0^\infty dr r^{s+1} u'(r) u''(r) \ , \ (3.56)$$

where the last term can be rewritten using (3.54):

$$\int_0^\infty dr r^{s+1} u'(r) u''(r) = l(l+1) \int_0^\infty dr r^{s-1} u(r) u'(r) - \frac{2Z}{r_0} \int_0^\infty dr r^s u(r) u'(r)$$

$$+ \frac{Z^2}{r_0^2 n^2} \int_0^\infty dr r^{s+1} u(r) u'(r) \ . \ (3.57)$$

Taking into account

$$\int_0^\infty dr r^k u(r) u'(r) = -\frac{k}{2} \int_0^\infty dr r^{k-1} u^2(r) = -\frac{k}{2} \left\langle r^{k-1} \right\rangle \ ,$$

the combination of (3.55), (3.56), and (3.57) yields (3.53).

3.6 Electromagnetic Interactions

In Subsection 3.5.5 we discussed the interaction of an electron with an electrostatic Coulomb potential using the hydrogen atom as an example. This

subsection deals more generally with the dynamics of an electron in an electromagnetic field. In this context, the important concept of *gauge invariance* will be introduced, from which some interesting quantum effects can be deduced. At the end of this subsection we provide an a posteriori explanation for the electron spin, which we have already encountered several times previously.

3.6.1 Electron in an Electromagnetic Field

For the following discussion, it may be useful to recall the results of Sections 2.1 and 2.2. Our starting point for the quantum mechanical discussion of the electron motion in an external electromagnetic field is the corresponding classical equation of motion of an electron with mass m_e and charge e:

$$m_e \ddot{\boldsymbol{x}} = e\left(\boldsymbol{E}(\boldsymbol{x},t) + \frac{\dot{\boldsymbol{x}}}{c} \times \boldsymbol{B}(\boldsymbol{x},t)\right) . \tag{3.58}$$

The electric and magnetic fields \boldsymbol{E}, \boldsymbol{B} are related via

$$\boldsymbol{B} = \boldsymbol{\nabla} \times \boldsymbol{A} , \ \boldsymbol{E} = -\frac{1}{c}\frac{\partial \boldsymbol{A}}{\partial t} - \boldsymbol{\nabla}\phi ,$$

with the scalar potential ϕ and the vector potential \boldsymbol{A}. Equation (3.58) can be deduced from the laws of Hamiltonian mechanics if the Hamiltonian is chosen as

$$H = \frac{1}{2m_e}\left(\boldsymbol{p} - \frac{e}{c}\boldsymbol{A}\right)^2 + e\phi = \frac{1}{2m_e}\left(\boldsymbol{p}^2 - \frac{2e}{c}\boldsymbol{p}\boldsymbol{A} + \frac{e^2}{c^2}\boldsymbol{A}^2\right) + e\phi .$$

We then find

$$\boldsymbol{\nabla}_x H = -\frac{e}{m_e c}\boldsymbol{\nabla}_x(\boldsymbol{p}\boldsymbol{A}) + \frac{e^2}{2m_e c^2}\boldsymbol{\nabla}_x(\boldsymbol{A}^2) + e\boldsymbol{\nabla}_x\phi = -\dot{\boldsymbol{p}}$$

$$\boldsymbol{\nabla}_p H = \frac{1}{m_e}\boldsymbol{p} - \frac{e}{m_e c}\boldsymbol{A} = \dot{\boldsymbol{x}}$$

and, therefore,

$$m_e \ddot{\boldsymbol{x}} = \dot{\boldsymbol{p}} - \frac{e}{c}\left((\dot{\boldsymbol{x}}\boldsymbol{\nabla})\boldsymbol{A} + \frac{\partial \boldsymbol{A}}{\partial t}\right) , \ \boldsymbol{\nabla} = \boldsymbol{\nabla}_x$$

$$= \frac{e}{m_e c}\left[(\boldsymbol{p}\boldsymbol{\nabla})\boldsymbol{A} + \boldsymbol{p} \times (\boldsymbol{\nabla} \times \boldsymbol{A})\right]$$

$$- \frac{e^2}{2m_e c^2}\left[2(\boldsymbol{A}\boldsymbol{\nabla})\boldsymbol{A} + 2\boldsymbol{A} \times (\boldsymbol{\nabla} \times \boldsymbol{A})\right]$$

$$- e\boldsymbol{\nabla}\phi - \frac{e}{c}\left[(\dot{\boldsymbol{x}}\boldsymbol{\nabla})\boldsymbol{A} + \frac{\partial \boldsymbol{A}}{\partial t}\right]$$

$$= \frac{e}{c}\left\{\left[\left(\frac{1}{m_e}\boldsymbol{p} - \frac{e}{m_e c}\boldsymbol{A}\right)\boldsymbol{\nabla}\right]\boldsymbol{A} + \left(\frac{1}{m_e}\boldsymbol{p} + \frac{e}{m_e c}\boldsymbol{A}\right) \times (\boldsymbol{\nabla} \times \boldsymbol{A})\right\}$$

$$-e\boldsymbol{\nabla}\phi - \frac{e}{c}\left((\dot{\boldsymbol{x}}\boldsymbol{\nabla})\boldsymbol{A} + \frac{\partial \boldsymbol{A}}{\partial t}\right)$$

$$= \frac{e}{c}\dot{\boldsymbol{x}} \times \boldsymbol{B} + e\boldsymbol{E} \ .$$

The transition to quantum mechanics is performed by the usual operator replacements, and we obtain

Theorem 3.21: Electron in an external electromagnetic field

Let \boldsymbol{A} and ϕ be the vector and the scalar potential of the electromagnetic fields \boldsymbol{E} and \boldsymbol{B}. In coordinate representation, the Hamilton operator for an electron interacting with \boldsymbol{E} and \boldsymbol{B} is given by

$$H = \frac{1}{2m_e}\left(\frac{\hbar}{i}\boldsymbol{\nabla} - \frac{e}{c}\boldsymbol{A}(\boldsymbol{x}, t)\right)^2 + e\phi(\boldsymbol{x}, t) \ .$$

In Coulomb gauge, $\boldsymbol{\nabla}\boldsymbol{A} = 0$, this becomes

$$H = -\frac{\hbar^2}{2m_e}\boldsymbol{\nabla}^2 + \frac{ie\hbar}{m_e c}\boldsymbol{A}\boldsymbol{\nabla} + \frac{e^2}{2m_e c^2}\boldsymbol{A}^2 + e\phi \ . \tag{3.59}$$

If we assume that the magnetic field only has a component in the z-direction,

$$\boldsymbol{B} = \boldsymbol{\nabla} \times \boldsymbol{A} = B\begin{pmatrix} 0 \\ 0 \\ 1 \end{pmatrix} \ , \quad \text{with} \quad \boldsymbol{A} = \frac{B}{2}\begin{pmatrix} -y \\ x \\ 0 \end{pmatrix} = -\frac{1}{2}\boldsymbol{x} \times \boldsymbol{B} \ ,$$

the second and third terms of (3.59) become

$$\frac{ie\hbar}{m_e c}\boldsymbol{A}\boldsymbol{\nabla}\Psi = \frac{ie\hbar}{2m_e c}B(\boldsymbol{x} \times \boldsymbol{\nabla})\Psi = -MB\Psi \ , \quad M = \frac{e}{2m_e c}\boldsymbol{L}$$

$$\frac{e^2}{2m_e c^2}\boldsymbol{A}^2\Psi = \frac{e^2}{8m_e c^2}\left[\boldsymbol{x}^2\boldsymbol{B}^2 - (\boldsymbol{x}\boldsymbol{B})^2\right]\Psi = \frac{e^2 B^2}{8m_e c^2}(x^2 + y^2)\Psi \ ,$$

where M can be interpreted as the magnetic dipole moment of the electron with orbital momentum \boldsymbol{L}.

Given the strengths of magnetic fields B that are usually achieved in the laboratory ($\approx 10^{-4}$ Gauss), the quadratic term in \boldsymbol{A} is several orders of magnitude smaller than the linear term and can safely be dropped.

Normal Zeeman effect. As an application of Theorem 3.21, we consider again the naive hydrogen atom as discussed in Subsection 3.5.5 but now in a constant external magnetic field in the z-direction, $\boldsymbol{B} = B\boldsymbol{e}_z$. If we neglect the \boldsymbol{A}^2-term, the Schrödinger equation is given by

$$(H^{(0)} + H^{(1)})\Psi = (E^{(0)} + E^{(1)})\Psi \ ,$$

with

$$H^{(0)} = -\frac{\hbar^2}{2\mu}\boldsymbol{\nabla}^2 + e\phi \ , \quad \phi(\boldsymbol{x}) = -\frac{Ze}{|\boldsymbol{x}|} \ , \quad H^{(1)} = -\frac{eB}{2\mu c}L_z \ .$$

Since the solutions $\Psi_{n,l,m}$ of the hydrogen equation

$$H^{(0)}\Psi_{n,l,m} = E^{(0)}\Psi_{n,l,m} \ , \ \ E^{(0)} = E_n^{(0)} = -\frac{\mu c^2 Z^2 \alpha_{\mathrm{e}}^2}{2n^2}$$

are also eigenstates of L_z, we obtain $E^{(1)}$ directly from

$$H^{(1)}\Psi_{n,l,m} = E^{(1)}\Psi_{n,l,m}$$

as

$$E^{(1)} = E_{lm}^{(1)} = -\frac{eB}{2\mu c}\hbar m \ .$$

Thus, in the presence of a constant magnetic field, the $(2l+1)$-fold degenerate levels belonging to a fixed value of l split into $2l + 1$ equidistant levels (Fig. 3.8).

Fig. 3.8. Splitting of the naive $l=1$- and $l=2$-hydrogen levels in the presence of an external magnetic field (normal Zeeman effect).

3.6.2 Gauge Invariance of the Schrödinger Equation

Unlike Maxwell's equations, the Schrödinger equation

$$i\hbar\frac{\partial\psi(\boldsymbol{x},t)}{\partial t} = \left[\frac{1}{2m}\left(\frac{\hbar}{i}\boldsymbol{\nabla} - \frac{e}{c}\boldsymbol{A}(\boldsymbol{x},t)\right)^2 + e\phi(\boldsymbol{x},t)\right]\psi(\boldsymbol{x},t) \tag{3.60}$$

is not invariant under gauge transformations of the kind

$$\boldsymbol{A} \longrightarrow \boldsymbol{A}' = \boldsymbol{A} + \boldsymbol{\nabla}\chi \ , \ \ \phi \longrightarrow \phi' = \phi - \frac{1}{c}\frac{\partial\chi}{\partial t} \ .$$

However, its gauge invariance can be restored if the wave function ψ is multiplied with a suitably chosen phase factor:

$$\psi(\boldsymbol{x},t) \longrightarrow \psi'(\boldsymbol{x},t) = e^{i\Lambda(\boldsymbol{x},t)}\psi(\boldsymbol{x},t) \ .$$

Substituting A, ϕ, and ψ by the primed quantities A', ϕ', and ψ' in (3.60), we find

$$i\hbar e^{-i\Lambda}\left(-i\frac{\partial\Lambda}{\partial t}\psi' + \frac{\partial\psi'}{\partial t}\right)$$

$$= \frac{1}{2m}\left(\frac{\hbar}{i}\nabla - \frac{e}{c}A' + \frac{e}{c}\nabla\chi\right)\left(\frac{\hbar}{i}\nabla - \frac{e}{c}A' + \frac{e}{c}\nabla\chi\right)e^{-i\Lambda}\psi'$$

$$+ e\left(\phi' + \frac{1}{c}\frac{\partial\chi}{\partial t}\right)e^{-i\Lambda}\psi'$$

$$= \frac{1}{2m}\left(\frac{\hbar}{i}\nabla - \frac{e}{c}A' + \frac{e}{c}\nabla\chi\right)e^{-i\Lambda}\left(\frac{\hbar}{i}\nabla - \frac{e}{c}A' + \frac{e}{c}\nabla\chi - \hbar\nabla\Lambda\right)\psi'$$

$$+ e\left(\phi' + \frac{1}{c}\frac{\partial\chi}{\partial t}\right)e^{-i\Lambda}\psi'$$

$$= \frac{1}{2m}e^{-i\Lambda}\left(\frac{\hbar}{i}\nabla - \frac{e}{c}A' + \frac{e}{c}\nabla\chi - \hbar\nabla\Lambda\right)^2\psi'$$

$$+ e\left(\phi' + \frac{1}{c}\frac{\partial\chi}{\partial t}\right)e^{-i\Lambda}\psi' \ . \tag{3.61}$$

If we choose

$$\Lambda(\boldsymbol{x},t) = \frac{e}{\hbar c}\chi(\boldsymbol{x},t) \ ,$$

(3.61) transforms into an equation of the same form as the original Schrödinger equation (3.60):

$$i\hbar\frac{\partial\psi'(\boldsymbol{x},t)}{\partial t} = \left[\frac{1}{2m}\left(\frac{\hbar}{i}\nabla - \frac{e}{c}A'(\boldsymbol{x},t)\right)^2 + e\phi'(\boldsymbol{x},t)\right]\psi'(\boldsymbol{x},t) \ . \tag{3.62}$$

Thus, as desired, (3.60) is indeed gauge-invariant under the above gauge transformations.

In a situation without magnetic fields ($\boldsymbol{B} = 0$), there are two choices to describe the motion of an electron in a purely electric potential ϕ. Either one solves the Schrödinger equation

$$i\hbar\frac{\partial\psi}{\partial t} = \left(-\frac{\hbar^2}{2m_e}\nabla^2 + e\phi\right)\psi \ , \tag{3.63}$$

in which the vector potential is not present at all, or one considers gauge invariance and solves the more general equation (3.62), setting

$$A' = \nabla\chi \ . \tag{3.64}$$

The wave function ψ' is then related to ψ from (3.63) via

$$\psi'(\boldsymbol{x},t) = e^{ie/(\hbar c)\chi(\boldsymbol{x},t)}\psi(\boldsymbol{x},t) \ .$$

As only the square of its absolute value is experimentally accessible, it would seem that the phase factor in the wave function is of no importance. However, there are situations in which the wave function itself, and thus its phase, plays an important role. Two examples of that kind are now presented.

Quantization of the magnetic flux. Consider a torus-shaped supercon-
ductor in an external constant magnetic field below its critical temperature
T_c (Fig. 3.9). Due to the *Meissner effect*, the magnetic field is pushed out of

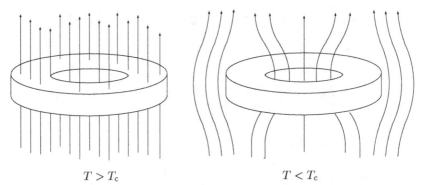

$$T > T_c \qquad\qquad\qquad T < T_c$$

Fig. 3.9. Flux line distribution of an external magnetic field in the presence of
a superconducting medium above its critical temperature T_c (*left*) and below T_c
(*right*).

the torus, so that its inside becomes field-free. According to (3.64), the phase
factor χ of the wave function of an electron in the torus is

$$\chi(\boldsymbol{x},t) = \int_{\boldsymbol{x}_0}^{\boldsymbol{x}} d\boldsymbol{x}' \boldsymbol{A}(\boldsymbol{x}',t) \; ,$$

where \boldsymbol{x}_0 is an arbitrary fixed point inside the torus and $\boldsymbol{A}(=\boldsymbol{A}')$ denotes
the vector potential of the magnetic field outside the torus. However, this
path integral is not unique due to the hole in the torus through which the
magnetic field enters. Consider, for example, two paths 1 and 2 differing
from one another by one extra winding around the torus. The difference in
the integral is then

$$\int_1 d\boldsymbol{x}' \boldsymbol{A}(\boldsymbol{x}',t) - \int_2 d\boldsymbol{x}' \boldsymbol{A}(\boldsymbol{x}',t) = \oint d\boldsymbol{x}' \boldsymbol{A}(\boldsymbol{x}',t) = \int_F d\boldsymbol{F} \boldsymbol{\nabla}' \times \boldsymbol{A}(\boldsymbol{x}',t)$$

$$= \int_F d\boldsymbol{F} \boldsymbol{B}(\boldsymbol{x}',t) = \varPhi_{\mathrm{m}} \; ,$$

where \varPhi_{m} is just the magnetic flux through the area spanned by the paths
1 and 2. For the wave function to be unique, and therefore physically mean-
ingful, it must not feel the difference between the two paths. This leads to a
quantization of the magnetic flux:

$$\varPhi_{\mathrm{m}} = \frac{2\pi\hbar c}{e} n \; , \quad n = 0, \pm 1, \pm 2, \dots \; .$$

Such a quantization is indeed experimentally verifiable, but with a slight modification:

$$\Phi_{\mathrm{m}} = \frac{2\pi\hbar c}{2e} n \ .$$

The factor of 2 is explained by *Cooper's theory*, according to which two electrons in a superconducting metal form correlated states (*Cooper pairs*).

Bohm-Aharanov effect. Another experiment that shows the phase dependency from the magnetic flux is shown in Fig. 3.10. Behind both slits of this

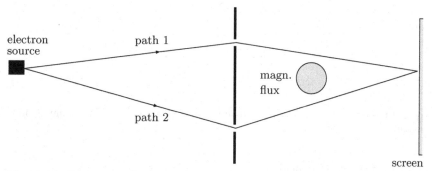

Fig. 3.10. Experimental setup for measuring the relative phase of electron waves that move along two different paths enclosing a magnetic flux.

double-slit experiment, there is an electric coil inducing a magnetic flux. The interference pattern observed on the screen is caused by the superposition of the wave functions following paths 1 and 2:

$$\psi' = \psi_1' + \psi_2' = \psi_1 \exp\left(\frac{ie}{\hbar c}\int_1 d\boldsymbol{x}'\,\boldsymbol{A}\right) + \psi_2 \exp\left(\frac{ie}{\hbar c}\int_2 d\boldsymbol{x}'\,\boldsymbol{A}\right)$$

$$= \left[\psi_1 \exp\left(\frac{ie}{\hbar c}\Phi_{\mathrm{m}}\right) + \psi_2\right] \exp\left(\frac{ie}{\hbar c}\int_2 d\boldsymbol{x}'\,\boldsymbol{A}\right) \ .$$

Therefore, when the current flows through the coil, the relative phase of the two wave functions differs from that of the no-current case by a factor $\exp\left(\frac{ie}{\hbar c}\Phi_{\mathrm{m}}\right)$. This effect was first predicted by Bohm and Aharanov and has subsequently been verified experimentally.

3.6.3 Stern-Gerlach Experiment

At the end of this section we return to the electron spin operator that we introduced, somewhat ad-hoc, in Subsection 3.4.3. To better motivate its introduction, we present an experiment first performed by Stern and Gerlach.

Consider a homogeneous beam of hydrogen atoms propagating through a strongly inhomogeneous magnetic field $\boldsymbol{B} = B(z)\boldsymbol{e}_z$ in the z-direction (Fig. 3.11).[15] If an atom in the beam has a magnetic moment \boldsymbol{M}, its potential

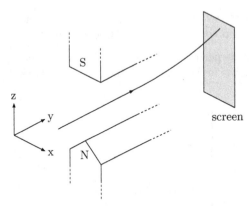

Fig. 3.11. Experimental setup to prove the existence of an intrinsic angular momentum (spin) of electrons.

energy in the magnetic field is

$$V(z) = -\boldsymbol{MB} = -M_z B(z) \ .$$

Thus, classically, a force in the z-direction applies to the atom,

$$F_z = -\frac{\partial V}{\partial z} = M_z \frac{\partial B}{\partial z} \ ,$$

so that the atom is deflected from its original path. Since M_z can take on any real value in some interval around zero, one would expect to see the ray smeared out in a fan-like manner. However, from the quantum mechanical viewpoint, following Theorem 3.21, the operator \boldsymbol{M}_z has discrete eigenvalues:

$$\frac{\hbar e}{2m_e c} m \ , \quad m = -l, \ldots, l \ .$$

So if the atoms are all in the same state, we would expect the jet to split into $2l + 1$ equidistant jets. In particular, for atoms in the ground state $(n = 1, l = 0)$, no splitting is expected at all. But what is found experimentally is that the jet splits into two. Obviously, this implies that on top of the magnetic moment resulting from the movement of the electrons around the proton, there has to be an additional dipole moment that does not vanish, even when the orbital angular momentum is null. The angular momentum operator corresponding to this effect is the already-mentioned electron spin \boldsymbol{S} whose quantum numbers $s = 1/2$ and $m_s = \pm 1/2$ follow from the splitting

[15] This experiment was originally carried out with silver atoms, but the following argument also holds for the simpler hydrogen atoms.

of the hydrogen jet into two. The size of the electron's magnetic moment can also be deduced from this experiment and is given by

$$M^{(e)} = \frac{eg_e}{2m_ec}S \ , \ g_e \approx 2 \ , \tag{3.65}$$

where g_e is the *gyromagnetic ratio* of the electron. Similar experiments can be carried out to verify that the proton has an electron-like spin S with $s = 1/2$, too, accompanied by a much smaller magnetic dipole moment (e=electron charge)

$$M^{(p)} = \frac{eg_p}{2m_pc}S \ , \ g_p \approx 5.56 \ .$$

The interaction between $M^{(e)}$ and $M^{(p)}$ is responsible for the *hyperfine splitting* of the hydrogen atom, which we will discuss in Subsection 3.7.5.

Summary

- The Hamilton operator for an electron in an external electromagnetic field is obtained by the usual operator replacements, starting from the classical equation of motion of the electron. One of the terms in the Hamilton operator can be interpreted as the **magnetic dipole moment** of the electron that is caused by its orbital angular momentum.

- Multiplication of the electron wave function with a suitable phase factor (**gauge transformation**) makes the Schrödinger equation invariant under gauge transformations of the electromagnetic field.

- The phase factor has experimentally verifiable implications as exemplified by the **quantization of the magnetic flux** and the **Bohm-Aharanov effect**.

- The **Stern-Gerlach experiment** shows that the electron and proton have intrinsic angular momenta (**spins**) confined to two values. Each of these spins is linked to a magnetic dipole moment. That of the proton is smaller than that of the electron by a factor of $m_p/m_e \approx 2000$.

Applications

46. Continuity equation. Show that for a system satisfying the Schrödinger equation

$$i\hbar\frac{d}{dt}\psi(\boldsymbol{x},t) = \boldsymbol{H}\psi(\boldsymbol{x},t) \ , \ \boldsymbol{H} = \frac{1}{2m}\left(\boldsymbol{P} - \frac{e}{c}\boldsymbol{A}(\boldsymbol{x},t)\right)^2 + e\phi(\boldsymbol{x},t) \ , \tag{3.66}$$

the continuity equation

$$\frac{d}{dt}(\psi^*\psi) + \boldsymbol{\nabla}\boldsymbol{j} = 0$$

holds, with the current density \boldsymbol{j}

$$\boldsymbol{j} = \frac{\hbar}{2im}\left(\psi^*\boldsymbol{\nabla}\psi - \psi\boldsymbol{\nabla}\psi^* - \frac{2ie}{\hbar c}\boldsymbol{A}\psi^*\psi\right) .$$

Solution. Multiplying (3.66) by ψ^* and the adjoint equation by ψ, followed by subtraction of the two, we find

$$i\hbar\frac{d}{dt}(\psi^*\psi) + \frac{\hbar^2}{2m}[\psi^*\boldsymbol{\nabla}^2\psi - \psi\boldsymbol{\nabla}^2\psi^*$$
$$-\frac{2ie}{\hbar c}(\psi^*\boldsymbol{A}\boldsymbol{\nabla}\psi + \psi\boldsymbol{A}\boldsymbol{\nabla}\psi^* + \psi^*\psi\boldsymbol{\nabla}\boldsymbol{A})] = 0$$

$$\Longrightarrow i\hbar\frac{d}{dt}(\psi^*\psi) + \frac{\hbar^2}{2m}\boldsymbol{\nabla}\left(\psi^*\boldsymbol{\nabla}\psi - \psi\boldsymbol{\nabla}\psi^* - \frac{2ie}{\hbar c}\boldsymbol{A}\psi^*\psi\right) = 0$$

$$\Longrightarrow \frac{d}{dt}(\psi^*\psi) + \boldsymbol{\nabla}\boldsymbol{j} = 0 , \; \boldsymbol{j} = \frac{\hbar}{2im}\left(\psi^*\boldsymbol{\nabla}\psi - \psi\boldsymbol{\nabla}\psi^* - \frac{2ie}{\hbar c}\boldsymbol{A}\psi^*\psi\right) .$$

Note that no gauge condition was used in this derivation.

47. Electron in a constant magnetic field. Suppose an electron of mass m_e and charge e moves in a constant magnetic field $\boldsymbol{B} = B\boldsymbol{e}_z$. What are the stationary solutions to the corresponding Schrödinger equation?

Solution. According to Theorem 3.21 with \boldsymbol{A} given by

$$\boldsymbol{A} = \frac{B}{2}\begin{pmatrix} -y \\ x \\ 0 \end{pmatrix} , \; \boldsymbol{\nabla}\boldsymbol{A} = 0 ,$$

the Hamilton operator reads

$$\boldsymbol{H} = -\frac{\hbar^2}{2m_e}\boldsymbol{\nabla}^2 - \frac{eB}{2m_ec}\boldsymbol{L}_z + \frac{e^2B^2}{8m_ec^2}(x^2 + y^2) .$$

The form of the "potential" $\frac{e^2B^2}{8m_ec^2}(x^2 + y^2)$ suggests the use of cylindrical coordinates to separate the variables:

$$x = r\cos\varphi , \; y = r\sin\varphi .$$

The differential operators \boldsymbol{L}_z and $\boldsymbol{\nabla}^2$ are then given by

$$\boldsymbol{L}_z = -i\hbar\frac{\partial}{\partial\varphi} , \; \boldsymbol{\nabla}^2 = \frac{\partial^2}{\partial z^2} + \frac{\partial^2}{\partial r^2} + \frac{1}{r}\frac{\partial}{\partial r} + \frac{1}{r^2}\frac{\partial^2}{\partial\varphi^2} ,$$

and the Hamilton operator is

$$\boldsymbol{H} = -\frac{\hbar^2}{2m_e}\left(\frac{\partial^2}{\partial z^2} + \frac{\partial^2}{\partial r^2} + \frac{1}{r}\frac{\partial}{\partial r} + \frac{1}{r^2}\frac{\partial^2}{\partial\varphi^2}\right) + \frac{i\hbar eB}{2m_ec}\frac{\partial}{\partial\varphi} + \frac{e^2B^2}{8mc^2}r^2 .$$

Since $[\boldsymbol{H}, \boldsymbol{L}_z] = 0$, it is useful to write the eigenstates of \boldsymbol{H} as

$$\Psi(r,\varphi,z) = u(r)v(\varphi)w(z) \ , \ v(\varphi) = e^{im\varphi} \ , \ w(z) = e^{ikz} \ ,$$

where $v(\varphi)$ is an eigenfunction of \mathbf{L}_z with eigenvalue $\hbar m$. With this, the stationary Schrödinger equation becomes a differential equation in u:

$$u''(r) + \frac{u'(r)}{r} + \left(\frac{2m_e E}{\hbar^2} + \frac{meB}{\hbar c} - \frac{e^2 B^2}{4\hbar^2 c^2}r^2 - k^2 - \frac{m^2}{r^2} \right) u(r) = 0. \ (3.67)$$

Introducing the new variables

$$x = \sqrt{-\frac{eB}{2\hbar c}}r \ , \ \Lambda = \frac{4m_e c}{\hbar e B}\left(\frac{\hbar^2 k^2}{2m_e} - E \right) - 2m \ ,$$

we find

$$u''(x) + \frac{1}{x}u'(x) + \left(\Lambda - x^2 - \frac{m^2}{x^2} \right) u(x) = 0 \ .$$

We see that u has the following asymptotic behavior for large and small x:

$$x \to \infty : \ u''(x) - x^2 u(x) = 0 \implies u(x) \sim e^{-x^2/2}$$

$$x \to 0 : \ u''(x) + \frac{1}{x}u'(x) - \frac{m^2}{x^2}u(x) = 0 \implies u(x) \sim x^{|m|} \ .$$

Therefore, in order to solve (3.67), we make the following ansatz:

$$u(x) = G(x)e^{-x^2/2}x^{|m|} \ .$$

This yields the differential equation

$$G''(x) + \left(\frac{2|m| + 1}{x} - 2x \right) G'(x) + (\Lambda - 2 - 2|m|)G(x) = 0 \ .$$

Substituting $y = x^2$, we finally find

$$yG''(y) + (|m| + 1 - y)G'(y) + \frac{\Lambda - 2 - 2|m|}{4}G(y) = 0 \ ,$$

which is of the same form as (3.45) with the replacements

$$l \longrightarrow \frac{|m| - 1}{2} \ , \ \lambda \longrightarrow \frac{\Lambda}{4} \ .$$

Comparison with (3.46) directly leads to the energy eigenvalues of our problem:

$$\frac{\Lambda}{4} - \frac{|m| + 1}{2} = n' = 0, 1, 2, \ldots$$

$$\iff E_{n'} = \frac{\hbar^2 k^2}{2m_e} - \frac{\hbar e B}{2m_e c}(2n' + |m| + 1 + m) \ .$$

By comparison with (3.47), the corresponding unnormalized eigenfunctions are given by

$$G(y) = L_{n'}^{|m|}(y) \ .$$

3.7 Perturbation Theory and Real Hydrogen Atom

Most quantum mechanical problems are not solvable exactly in analytic form. However, given certain conditions, one can make use of approximation schemes that come very close to an exact solution. In this section we discuss such a method, namely *time-independent perturbation theory*. It is used particularly for the case of bound states and time-independent Hamilton operators.

The section begins with a general discussion of time-independent perturbation theory for nondegenerate and degenerate cases. We then consider two realistic applications, the *Stark* and the *anomalous Zeeman effect*. Furthermore, we will see how the hydrogen atom, naively discussed in Subsection 3.5.5, can be treated in a more realistic fashion leading to *fine structure* and *hyperfine structure splitting*.

3.7.1 Time-Independent Perturbation Theory

Consider the time-independent Schrödinger equation

$$(\boldsymbol{H}^{(0)} + \boldsymbol{H}')\,|n\rangle = E_n\,|n\rangle \quad . \tag{3.68}$$

Suppose a complete set of nondegenerate and orthonormal eigenvectors $|n^{(0)}\rangle$ of $\boldsymbol{H}^{(0)}$, together with their eigenvalues $E_n^{(0)}$, have already been found:

$$\boldsymbol{H}^{(0)}\left|n^{(0)}\right\rangle = E_n^{(0)}\left|n^{(0)}\right\rangle \quad .$$

We assume \boldsymbol{H}' is "small" compared to $\boldsymbol{H}^{(0)}$, where we shall define later what we mean by this. Conventionally, $\boldsymbol{H}^{(0)}$ is called the *unperturbed Hamilton operator* and \boldsymbol{H}' the *perturbation operator* or simply *perturbation*. To solve (3.68) we introduce a fictitious perturbation parameter λ that will be removed or set to unity later on. This allows us to rewrite (3.68) as

$$(\boldsymbol{H}^{(0)} + \lambda \boldsymbol{H}')\,|n\rangle = E_n\,|n\rangle \quad ,$$

where the smallness of \boldsymbol{H}' is meant to be reflected in the parameter λ. Now it seems reasonable to assume that it is possible for E_n and $|n\rangle$ to be expanded as a series in powers of λ such that

$$E_n = E_n^{(0)} + \lambda E_n^{(1)} + \lambda^2 E_n^{(2)} + \ldots \tag{3.69}$$

$$|n\rangle = \left|n^{(0)}\right\rangle + \lambda\left|n^{(1)}\right\rangle + \lambda^2\left|n^{(2)}\right\rangle + \ldots \quad ,$$

where the perturbed quantities converge smoothly to the unperturbed values for $\lambda \to 0$:

$$\lambda \to 0 \Longrightarrow |n\rangle \to \left|n^{(0)}\right\rangle \quad , \quad E_n \to E_n^{(0)} \quad .$$

Since the basis states $|n^{(0)}\rangle$ of the unperturbed problem form a complete set, the kets $|n^{(r)}\rangle$, $r > 0$, can be expanded in this basis,

$$\left| n^{(r)} \right\rangle = \sum_m C_m^{(r)} \left| m^{(0)} \right\rangle \;,$$

and we find

$$|n\rangle = \left| n^{(0)} \right\rangle + \lambda \sum_m C_m^{(1)} \left| m^{(0)} \right\rangle + \lambda^2 \sum_m C_m^{(2)} \left| m^{(0)} \right\rangle + \ldots$$

or, rescaling $C_m^{(r)}$ suitably,

$$|n\rangle = \left| n^{(0)} \right\rangle + \lambda \sum_{m \neq n} C_m^{(1)} \left| m^{(0)} \right\rangle + \lambda^2 \sum_{m \neq n} C_m^{(2)} \left| m^{(0)} \right\rangle + \ldots \;. \quad (3.70)$$

Inserting (3.69) and (3.70) into the original equation (3.68), and ordering the terms in powers of λ, one finds the following equations in lowest orders:

0th order:

$$\boldsymbol{H}^{(0)} \left| n^{(0)} \right\rangle = E_n^{(0)} \left| n^{(0)} \right\rangle \;.$$

1st order:

$$\boldsymbol{H}^{(0)} \sum_{m \neq n} C_m^{(1)} \left| m^{(0)} \right\rangle + \boldsymbol{H}' \left| n^{(0)} \right\rangle = E_n^{(0)} \sum_{m \neq n} C_m^{(1)} \left| m^{(0)} \right\rangle$$
$$+ E_n^{(1)} \left| n^{(0)} \right\rangle \;. \quad (3.71)$$

2nd order:

$$\boldsymbol{H}^{(0)} \sum_{m \neq n} C_m^{(2)} \left| m^{(0)} \right\rangle + \boldsymbol{H}' \sum_{m \neq n} C_m^{(1)} \left| m^{(0)} \right\rangle = E_n^{(0)} \sum_{m \neq n} C_m^{(2)} \left| m^{(0)} \right\rangle$$
$$+ E_n^{(1)} \sum_{m \neq n} C_m^{(1)} \left| m^{(0)} \right\rangle$$
$$+ E_n^{(2)} \left| n^{(0)} \right\rangle \;. \quad (3.72)$$

Obviously, the lowest order naturally returns the equation for the unperturbed case. Multiplying (3.71) from the left by $\left\langle n^{(0)} \right|$, we obtain

$$E_n^{(1)} = \left\langle n^{(0)} \left| \boldsymbol{H}' \right| n^{(0)} \right\rangle \;.$$

Thus, to first order, the energy correction to the unperturbed energy $E_n^{(0)}$ is given by the expectation value of the perturbation operator \boldsymbol{H}' in the corresponding unperturbed state $\left| n^{(0)} \right\rangle$. The expansion coefficients $C_m^{(1)}$ are obtained by multiplying (3.71) from the left by $\left\langle k^{(0)} \right| \neq \left\langle n^{(0)} \right|$:

$$\sum_{m \neq n} C_m^{(1)} E_m^{(0)} \delta_{km} + \left\langle k^{(0)} \left| \boldsymbol{H}' \right| n^{(0)} \right\rangle = E_n^{(0)} \sum_{m \neq n} C_m^{(1)} \delta_{km}$$

$$\Longrightarrow C_m^{(1)} = \frac{\left\langle m^{(0)} \left| \boldsymbol{H}' \right| n^{(0)} \right\rangle}{E_n^{(0)} - E_m^{(0)}} \;.$$

Therefore, the unnormalized states of the perturbed problem to first order are given by (setting $\lambda = 1$)

$$|n\rangle = \left|n^{(0)}\right\rangle + \sum_{m \neq n} \frac{\left\langle m^{(0)}\right| \boldsymbol{H}' \left|n^{(0)}\right\rangle}{E_n^{(0)} - E_m^{(0)}} \left|m^{(0)}\right\rangle \; . \tag{3.73}$$

To obtain the second-order relations, one proceeds in just the same way. Here, we give only the second-order energy correction that is obtained from (3.72) by multiplication with $\left\langle n^{(0)}\right|$:

$$E_n^{(2)} = \sum_{m \neq n} \frac{\left|\left\langle m^{(0)}\right| \boldsymbol{H}' \left|n^{(0)}\right\rangle\right|^2}{E_n^{(0)} - E_m^{(0)}} \; .$$

With the help of (3.73), we now return to the issue of the "smallness" of \boldsymbol{H}' and check the validity of the perturbation expansion. For $\left|n^{(1)}\right\rangle$ to be small relative to $\left|n^{(0)}\right\rangle$, a necessary condition is given by the inequality

$$\left| \frac{\left\langle m^{(0)}\right| \boldsymbol{H}' \left|n^{(0)}\right\rangle}{E_n^{(0)} - E_m^{(0)}} \right| \ll 1 \; ,$$

which depends on three factors, namely

- the absolute size of the the perturbation \boldsymbol{H}',
- the matrix elements of \boldsymbol{H}' for unperturbed states,
- the energy differences between unperturbed states.

For degenerate states, $E_n^{(0)} = E_m^{(0)}$, we see that it is problematic to unconditionally satisfy the above constraint. It can only be satisfied if, in addition to the denominator, the numerator given by the nondiagonal matrix elements of \boldsymbol{H}' also vanishes. More precisely: the eigenvectors of $\boldsymbol{H}^{(0)}$ to the same energy eigenvalue have to simultaneously form an eigenbasis of \boldsymbol{H}'. But this can always be achieved as for these vectors, $\boldsymbol{H}^{(0)}$ is not just diagonal but even proportional to the unit matrix. Since \boldsymbol{H}' commutes with the unit matrix, \boldsymbol{H}' can be diagonalized in the degenerate subspace without destroying the diagonality of $\boldsymbol{H}^{(0)}$. Once again: even if the totality of eigenvectors of $\boldsymbol{H}^{(0)}$ spans an infinitely dimensional space, the simultaneous diagonalization of $\boldsymbol{H}^{(0)}$ and \boldsymbol{H}' generally affects only a small finite-dimensional subspace, which is the one spanned by the eigenkets of $\boldsymbol{H}^{(0)}$ corresponding to the same eigenvalue.

Theorem 3.22: Time-independent perturbation theory

Consider the time-independent Schrödinger equation

$$\left(\boldsymbol{H}^{(0)} + \boldsymbol{H}'\right)|n\rangle = E_n |n\rangle \; ,$$

with a small perturbation operator \boldsymbol{H}'. Suppose that for the unperturbed problem

▷

$$H^{(0)} \left| n^{(0)} \right\rangle = E_n^{(0)} \left| n^{(0)} \right\rangle ,$$

a complete set of nondegenerate normalized solution kets and their energies are known. The correction energies to first and second order are then respectively given by

$$E_n^{(1)} = \left\langle n^{(0)} \right| H' \left| n^{(0)} \right\rangle , \ E_n^{(2)} = \sum_{m \neq n} \frac{|\langle m^{(0)} | H' | n^{(0)} \rangle|^2}{E_n^{(0)} - E_m^{(0)}} . \quad (3.74)$$

In case of a degenerate energy eigenvalue, a basis has to be chosen for the corresponding degenerate subspace that diagonalizes H'.

Note that problems in perturbation theory can often be simplified by taking into account symmetries. Suppose, for example, that

$$[\boldsymbol{\Omega}, H'] = 0 , \ \boldsymbol{\Omega} |\alpha, \omega\rangle = \omega |\alpha, \omega\rangle .$$

It then follows that

$$0 = \langle \alpha_1 \omega_1 | \boldsymbol{\Omega} H' - H' \boldsymbol{\Omega} | \alpha_2, \omega_2 \rangle = (\omega_1 - \omega_2) \langle \alpha_1, \omega_1 | H' | \alpha_2, \omega_2 \rangle ,$$

and this gives the *selection rule*

$$\omega_1 \neq \omega_2 \implies \langle \alpha_1, \omega_1 | H' | \alpha_2, \omega_2 \rangle = 0 . \quad (3.75)$$

Furthermore, if we can find an operator $\boldsymbol{\Omega}$ for which

$$H' \left| n^{(0)} \right\rangle = [\boldsymbol{\Omega}, H^{(0)}] \left| n^{(0)} \right\rangle ,$$

the sum in (3.74) for $E_n^{(2)}$ can be reduced to calculating just three matrix elements. Using the completeness of the unperturbed eigenkets, we have

$$
\begin{aligned}
E_n^{(2)} &= \sum_{m \neq n} \frac{\left\langle n^{(0)} \right| H' \left| m^{(0)} \right\rangle \left\langle m^{(0)} \right| \boldsymbol{\Omega} H^{(0)} - H^{(0)} \boldsymbol{\Omega} \left| n^{(0)} \right\rangle}{E_n^{(0)} - E_m^{(0)}} \\
&= \sum_{m \neq n} \left\langle n^{(0)} \right| H' \left| m^{(0)} \right\rangle \left\langle m^{(0)} \right| \boldsymbol{\Omega} \left| n^{(0)} \right\rangle \\
&= \left\langle n^{(0)} \right| H' \boldsymbol{\Omega} \left| n^{(0)} \right\rangle - \left\langle n^{(0)} \right| H' \left| n^{(0)} \right\rangle \left\langle n^{(0)} \right| \boldsymbol{\Omega} \left| n^{(0)} \right\rangle . \quad (3.76)
\end{aligned}
$$

3.7.2 Stark Effect

As a first application of time-independent perturbation theory, we consider again the naive hydrogen atom, but now in a constant electric field of size ϵ and applied in the z-direction (*Stark effect*). The Hamilton operator in coordinate representation is given by

$$H = H^{(0)} + H' , \ H^{(0)} = -\frac{\hbar^2}{2\mu} \boldsymbol{\nabla}^2 - \frac{Ze^2}{r} , \ H' = -e\epsilon z .$$

The eigenfunctions of $H^{(0)}$ are

$$\Psi_{n,l,m}(r,\theta,\varphi) = g_{n,l}(r)Y_{l,m}(\theta,\varphi) \; .$$

The energy shift of the nondegenerate ground state ($n = 1$) to first order vanishes since $\Psi_{1,0,0}$ is of definite parity, so that $|\Psi_{1,0,0}|^2$ is an even function, whereas z is odd:

$$E^{(1)}_{1,0,0} = e\epsilon \, \langle 100| \, z \, |100\rangle = e\epsilon \int dr\, r^2 d\Omega |\Psi_{1,0,0}(r)|^2 z = 0 \; .$$

Thus, the hydrogen atom in its ground state possesses no permanent dipole moment. However, for the energy shift to second order we expect a nonzero contribution, since the external electric field deforms the surrounding electron sphere, thus inducing a dipole moment that interacts with that field:

$$E^{(2)}_{1,0,0} = e^2 \epsilon^2 \sum_{n>1,l,m} \frac{|\langle nlm| \, z \, |100\rangle|^2}{E^{(0)}_1 - E^{(0)}_n} \; .$$

Now, we can show that, in coordinate representation, the operator

$$\Omega = \frac{\mu r_0 e\epsilon}{Z\hbar^2}\left(\frac{r^2}{2} + \frac{r_0 r}{Z}\right)\cos\theta$$

suffices

$$H'\,|100\rangle = [\Omega, H^{(0)}]\,|100\rangle \; .$$

Using this and (3.76), we obtain

$$E^{(2)}_{1,0,0} = \langle 100|\, H'\Omega\, |100\rangle - 0$$

$$= -\frac{\mu r_0 e^2 \epsilon^2}{Z\hbar^2}\frac{1}{\pi}\left(\frac{Z}{r_0}\right)^3 \int dr\, r^2 d\Omega e^{-2Zr/r_0}\left(\frac{r^3}{2} + \frac{r_0 r^2}{Z}\right)\cos^2\theta$$

$$= -\frac{9}{4}\frac{\epsilon^2 r_0^3}{Z^4} \; .$$

Next, we consider the Stark effect for the degenerate $n=2$ level of the hydrogen atom. It has the following four states with the same energy:

$$\Psi_{2,0,0} = 2\left(\frac{Z}{2r_0}\right)^{3/2}\left(1 - \frac{Zr}{2r_0}\right)e^{-Zr/2r_0}Y_{0,0}$$

$$\Psi_{2,1,0} = \frac{1}{\sqrt{3}}\left(\frac{Z}{2r_0}\right)^{3/2}\frac{Zr}{r_0}e^{-Zr/2r_0}Y_{1,0}$$

$$\Psi_{2,1,1} = \frac{1}{\sqrt{3}}\left(\frac{Z}{2r_0}\right)^{3/2}\frac{Zr}{r_0}e^{-Zr/2r_0}Y_{1,1}$$

$$\Psi_{2,1,-1} = \frac{1}{\sqrt{3}}\left(\frac{Z}{2r_0}\right)^{3/2}\frac{Zr}{r_0}e^{-Zr/2r_0}Y_{1,-1} \; .$$

First, we need to use these states to create basis vectors

$$\alpha_1 \Psi_{2,0,0} + \alpha_2 \Psi_{2,1,0} + \alpha_3 \Psi_{2,1,1} + \alpha_4 \Psi_{2,1,-1} \ ,$$

which diagonalize H'. Of the total of 16 matrix elements $\langle 2lm|\, H'\, |2l'm' \rangle$, the diagonal components do not contribute, since all four states are of definite parity. Also, H' commutes with L_z so that, due to (3.75), all matrix elements with $m \neq m'$ also vanish. The eigenvalue problem that remains to be solved is

$$\begin{pmatrix} 0 & \Delta & 0 & 0 \\ \Delta & 0 & 0 & 0 \\ 0 & 0 & 0 & 0 \\ 0 & 0 & 0 & 0 \end{pmatrix} \begin{pmatrix} \alpha_1 \\ \alpha_2 \\ \alpha_3 \\ \alpha_4 \end{pmatrix} = E^{(1)} \begin{pmatrix} \alpha_1 \\ \alpha_2 \\ \alpha_3 \\ \alpha_4 \end{pmatrix} \ ,$$

with

$$\Delta = -e\epsilon \, \langle 200|\, z\, |210 \rangle = \frac{3e\epsilon r_0}{Z} \ .$$

Its solutions are

$$E^{(1)} = \pm\Delta : \alpha = \frac{1}{\sqrt{2}} \begin{pmatrix} 1 \\ \pm 1 \\ 0 \\ 0 \end{pmatrix} \ , \quad E^{(1)} = 0 : \alpha = \begin{pmatrix} 0 \\ 0 \\ 1 \\ 0 \end{pmatrix}, \begin{pmatrix} 0 \\ 0 \\ 0 \\ 1 \end{pmatrix} \ .$$

Thus, the unperturbed $n=2$ states that are stable when perturbed with H', together with their energy shifts to first order, are given by

$$\frac{1}{\sqrt{2}} \left(\Psi_{2,0,0} + \Psi_{2,1,0} \right) \ , \quad E^{(1)} = \frac{3e\epsilon r_0}{Z}$$

$$\frac{1}{\sqrt{2}} \left(\Psi_{2,0,0} - \Psi_{2,1,0} \right) \ , \quad E^{(1)} = -\frac{3e\epsilon r_0}{Z}$$

$$\Psi_{2,1,1} \ , \quad E^{(1)} = 0$$

$$\Psi_{2,1,-1} \ , \quad E^{(1)} = 0 \ .$$

3.7.3 Fine Structure Splitting

We shall now attempt a more realistic description of hydrogen-like atoms. Our starting point is the unperturbed Hamilton operator

$$H^{(0)} = \frac{P^2}{2\mu} - \frac{Ze^2}{r} \ ,$$

where the relative momentum $P = P_R$ in the center of mass system ($P_S = 0$) is related to the electron and nucleus momenta in the following way:

$$\frac{P^2}{2\mu} = \frac{P_e^2}{2m_e} + \frac{P_k^2}{2m_k} \ .$$

To start, we shall take into account relativistic kinematic effects to lowest order by replacing $P_e^2/2m_e + P_k^2/2m_k$ with

$$\sqrt{P_e^2 c^2 + m_e^2 c^4} + \frac{P_k^2}{2m_k} \approx m_e c^2 + \frac{P_e^2}{2m_e} + \frac{P_k^2}{2m_k} - \frac{P_e^4}{8m_e^3 c^2}$$

$$\approx m_e c^2 + \frac{P^2}{2\mu} - \frac{P^4}{8\mu^3 c^2} .$$

The first term on the right-hand side is an irrelevant constant; the second one yields the known nonrelativistic expression. The third term leads to the relativistic correction

$$H_T = -\frac{P_e^4}{8\mu^3 c^2} ,$$

which we treat as a small perturbation to $H^{(0)}$. Since the operator H_T is rotationally invariant, it has diagonal form in the (nlm)-basis of the unperturbed problem. In other words: the (nlm)-basis is stable under this perturbation, and we need no longer worry about the degeneracy of the unperturbed energy levels. The energy shift to first order is therefore simply given by

$$E_T^{(1)} = -\frac{1}{8\mu^3 c^2} \langle nlm| P^4 |nlm \rangle .$$

Using

$$P^4 = 4\mu^2 \left(\frac{P^2}{2\mu}\right)^2 = 4\mu^2 \left(H^{(0)} + \frac{Ze^2}{r}\right)^2$$

and (3.50), we find

$$E_T^{(1)} = -\frac{1}{2\mu c^2} \left(E_n^2 + 2E_n Ze^2 \left\langle \frac{1}{r} \right\rangle_{nlm} + Z^2 e^4 \left\langle \frac{1}{r^2} \right\rangle_{nlm} \right)$$

$$= \frac{Z^4 \alpha_e^4 \mu c^2}{2} \left(\frac{3}{4n^4} - \frac{1}{n^3 \left(l + \frac{1}{2}\right)} \right) .$$

Another effect is due to the spin of the electron, which we have neglected so far. Classically, its origin can be understood as follows: suppose an electron seen in the rest system of the nucleus is moving with velocity v. Then the nucleus has the velocity $-v$ in the rest system of the electron and, according to Biot-Savart's law, Theorem 2.16, induces a magnetic field

$$B = \frac{Ze}{c} \frac{v \times x}{|x|^3} = -\frac{Ze}{m_e c |x|^3} x \times p .$$

This field interacts with the magnetic moment $M^{(e)}$ of the electron, resulting in the energy contribution

$$H_{SB} = -M^{(e)} B .$$

From our quantum mechanical point of view, this and (3.65) yield the perturbation operator for the *spin-orbit interaction*:

$$H_{SB} = -\frac{e}{m_e c} SB = \frac{Ze^2}{m_e^2 c^2 r^3} LS , \quad r = |x| .$$

However, this is not the correct expression. It must be reduced by a factor of 2 due to relativistic effects, as well as the fact that the electron does not move in a straight line around the nucleus (*Thomas precession*). Neglecting the difference between m_e and μ, the correct perturbation operator is therefore given by

$$H_{SB} = \frac{Ze^2}{2\mu^2 c^2 r^3} \boldsymbol{LS} \ .$$

For the calculation of the energy levels, we have to diagonalize \boldsymbol{H}_{SB} in the subspace of the kets corresponding to the unperturbed energies $E_n^{(0)}$. This subspace has dimension $2(2l + 1)$, where the factor of 2 takes into account the two spin orientations of the electron. The problem can be simplified by noting that

$$\boldsymbol{J} = \boldsymbol{L} + \boldsymbol{S} \ \Longrightarrow \ \boldsymbol{J}^2 = \boldsymbol{L}^2 + \boldsymbol{S}^2 + 2\boldsymbol{LS}$$
$$\Longrightarrow \ \boldsymbol{H}_{SB} = \frac{Ze^2}{4\mu^2 c^2 r^3} \left[\boldsymbol{J}^2 - \boldsymbol{L}^2 - \boldsymbol{S}^2 \right] \ .$$

Now we can use the eigenstates of the total angular momentum $\left| l, \frac{1}{2}, J, M \right\rangle$ that we found in Subsection 3.4.5 and for which \boldsymbol{H}_{SB} is already diagonal. According to (3.34), the two possible total angular momenta $J = l \pm 1/2$, $l > 0$ correspond to energies:

$$
\begin{aligned}
E_{SB}^{(1)} &= \left\langle n, l, \frac{1}{2}, l \pm \frac{1}{2}, M \middle| \boldsymbol{H}_{SB} \middle| n, l, \frac{1}{2}, l \pm \frac{1}{2}, M \right\rangle \\
&= \left\langle n, l, \frac{1}{2}, l \pm \frac{1}{2}, M \middle| \frac{1}{r^3} \middle| n, l, \frac{1}{2}, l \pm \frac{1}{2}, M \right\rangle \\
&\quad \times \frac{Ze^2 \hbar^2}{4\mu^2 c^2} \left[\left(l \pm \frac{1}{2} \right) \left(l \pm \frac{1}{2} + 1 \right) - l(l+1) - \frac{3}{4} \right] \\
&= \left\{ \frac{l + \frac{1}{2} \pm M}{2l + 1} \left\langle n, l, M - \frac{1}{2}; + \middle| \frac{1}{r^3} \middle| n, l, M - \frac{1}{2}; + \right\rangle \right. \\
&\quad \left. + \frac{l + \frac{1}{2} \mp M}{2l + 1} \left\langle n, l, M + \frac{1}{2}; - \middle| \frac{1}{r^3} \middle| n, l, M + \frac{1}{2}; - \right\rangle \right\} \\
&\quad \times \frac{Ze^2 \hbar^2}{4\mu^2 c^2} \left[\left(l \pm \frac{1}{2} \right) \left(l \pm \frac{1}{2} + 1 \right) - l(l+1) - \frac{3}{4} \right] \\
&= \frac{Ze^2 \hbar^2}{4\mu^2 c^2} \left\langle \frac{1}{r^3} \right\rangle_{nl} \left\{ \begin{matrix} l \\ -l - 1 \end{matrix} \right\} \ .
\end{aligned}
$$

Due to the orthogonality of the spherical harmonics and the electron's spin basis states, the matrix elements

$$\left\langle n, l, M \pm \frac{1}{2}; \mp \middle| \frac{1}{r^3} \middle| n, l, M \mp \frac{1}{2}; \pm \right\rangle$$

vanish. Finally, using (3.50), we find

$$E_{SB}^{(1)} = \frac{Z^4 \alpha_e^4 \mu c^2}{4} \frac{\begin{Bmatrix} l \\ -l-1 \end{Bmatrix}}{n^3 l \left(l + \frac{1}{2}\right)(l+1)} . \tag{3.77}$$

Combining the two energy shifts $E_T^{(1)}$ and $E_{SB}^{(1)}$, we obtain the *fine structure shift* or *fine structure splitting*

$$E_{FS}^{(1)} = E_T^{(1)} + E_{SB}^{(1)} = \frac{Z^4 \alpha_e^4 \mu c^2}{2n^3}\left(\frac{3}{4n} - \frac{1}{J + \frac{1}{2}}\right), \quad J = l \pm \frac{1}{2}. \tag{3.78}$$

Note that this derivation was made for $l > 0$. If $l = 0$, then $\left\langle \frac{1}{r^3} \right\rangle_{nl}$ diverges and **LS** vanishes. However, if we insert $l = 0$ in (3.78), we obtain a finite result. In the context of a full relativistic treatment of this problem (Dirac theory), it can be shown that this is indeed the correct result for the energy shift for $l = 0$.

The energy splittings induced by $E_{SB}^{(1)}$ and $E_{FS}^{(1)}$ are depicted in Fig. 3.12. Interestingly, we see from (3.78) that the corrections conspire such that the

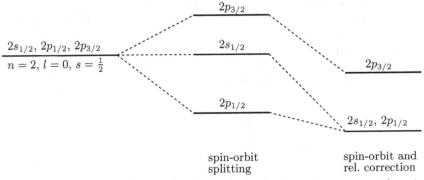

Fig. 3.12. Splittings of the $n=2,l=0$-hydrogen levels, taking into account the spin-orbit interaction as well as relativistic corrections.

states $2s_{1/2}$ and $2p_{1/2}$ overlay exactly. However, in 1947 Lamb and Retherford devised a high-precision experiment that showed there is a small deviation between the states $2s_{1/2}$ and $2p_{1/2}$. This effect, known as the *Lamb shift*, can only be understood in the context of quantum electrodynamics, which explains the small shift as due to the interaction of the electron with its own radiation field.

3.7.4 Anomalous Zeeman Effect

We wish to study the effect of the spin-orbit coupling for hydrogen-like atoms in a constant magnetic field $\mathbf{B} = B\mathbf{e}_z$ in the z-direction. To this end, we consider the unperturbed Hamilton operator

$$H^{(0)} = \frac{\boldsymbol{P}^2}{2\mu} - \frac{Ze^2}{r} + \frac{Ze^2}{2m_e^2 c^2 r^3} \boldsymbol{LS}$$

and

$$H_{\mathrm{AZ}} = -\frac{e}{2m_e c}(\boldsymbol{L} + 2\boldsymbol{S})\boldsymbol{B} = -\frac{eB}{2m_e c}(L_z + 2S_z) = -\frac{eB}{2m_e c}(\boldsymbol{J}_z + \boldsymbol{S}_z)$$

as a small perturbation. The latter takes into account the interaction of the external magnetic field with the magnetic dipole moment of the electron due to the orbital angular momentum, as well as the magnetic dipole moment of the electron spin. The form of $H^{(0)}$ forces us again to use the total angular momentum eigenstates from (3.34) to calculate the energy shifts. With these, we find

$$
\begin{aligned}
E_{\mathrm{AZ}}^{(1)} &= \left\langle n, l, \frac{1}{2}, l \pm \frac{1}{2}, M \left| H_{\mathrm{AZ}} \right| n, l, \frac{1}{2}, l \pm \frac{1}{2}, M \right\rangle \\
&= -\frac{eB}{2m_e c}\left\{ \hbar M + \left\langle n, l, \frac{1}{2}, l \pm \frac{1}{2}, M \left| S_z \right| n, l, \frac{1}{2}, l \pm \frac{1}{2}, M \right\rangle \right\} \\
&= -\frac{eB}{2m_e c}\left\{ \hbar M + \frac{l + \frac{1}{2} \pm M}{2l+1} \left\langle n, l, M - \frac{1}{2}; + \left| S_z \right| n, l, M - \frac{1}{2}; + \right\rangle \right. \\
&\quad \left. + \frac{l + \frac{1}{2} \mp M}{2l+1} \left\langle n, l, M + \frac{1}{2}; - \left| S_z \right| n, l, M + \frac{1}{2}; - \right\rangle \right\} \\
\Longrightarrow E_{\mathrm{AZ}}^{(1)} &= -\frac{eB\hbar M}{2m_e c}\left(1 \pm \frac{1}{2l+1} \right) , \quad J = l \pm \frac{1}{2} .
\end{aligned}
$$

Thus, states with fixed $J = l \pm 1/2$ are split into $2j$ equidistant lines separated by $\frac{eB\hbar}{2m_e c}\frac{2l+2}{2l+1}$, resp. $\frac{eB\hbar}{2m_e c}\frac{2l}{2l+1}$ (see Fig. 3.13). When these shifts were

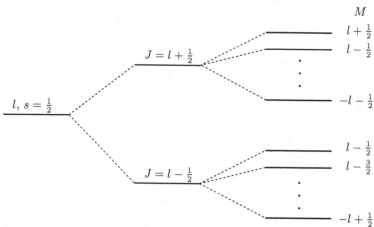

Fig. 3.13. Splitting of hydrogen levels belonging to the total angular momentum quantum numbers $J = l \pm 1/2$ in the presence of an external magnetic field (anomalous Zeeman effect).

discovered, the electron spin had not yet been introduced so that they were given the name *anomalous Zeeman effect*.

3.7.5 Hyperfine Splitting

For states with orbital angular momentum $l = 0$, (3.78) tells us that there is no fine structure splitting for hydrogen-like atoms. However, precision experiments show that even this state is split into two levels. This can be explained by the magnetic dipole moment (and its associated spin) of the nucleus, which we have so far neglected. The magnetic dipole moment operator is

$$M^{(\mathrm{k})} = \frac{Z e g_{\mathrm{k}}}{2 m_{\mathrm{k}} c} I \,, \quad I = \text{nucleus spin} \,,$$

and it interacts with the electron's magnetic dipole moment

$$M^{(\mathrm{e})} = \frac{e}{m_{\mathrm{e}} c} S^{(\mathrm{e})} \,.$$

Considering only $l=0$-states of the hydrogen atom ($Z = 1$, $g_{\mathrm{k}} = g_{\mathrm{p}}$, $I = S^{(\mathrm{p})}$), it can be shown that the perturbation operator due to the magnetic moments of electron and proton is given by[16]

$$H_{\mathrm{HF}} = \frac{4 g_{\mathrm{p}} m_{\mathrm{e}}^2 c^2 \alpha_{\mathrm{e}}^4}{3 m_{\mathrm{p}} n^3 \hbar^2} S^{(\mathrm{e})} S^{(\mathrm{p})} \,.$$

Since the spins are completely decoupled from the other degrees of freedom, we only need to find a spin basis that diagonalizes H_{HF}, in order to calculate the energy shifts. To this end, we proceed in analogy to the LS-coupling and write

$$S = S^{(\mathrm{e})} + S^{(\mathrm{p})} \implies S^2 = S^{(\mathrm{e})2} + S^{(\mathrm{p})2} + 2 S^{(\mathrm{e})} S^{(\mathrm{p})}$$

$$\implies H_{\mathrm{HF}} = \frac{4 g_{\mathrm{p}} m_{\mathrm{e}}^2 c^2 \alpha_{\mathrm{e}}^4}{6 m_{\mathrm{p}} n^3 \hbar^2} \left[S^2 - S^{(\mathrm{e})2} - S^{(\mathrm{p})2} \right] \,.$$

Now we can use the total spin basis from Subsection 3.4.5, for which H_{HF} is diagonal. For the three *triplet states* with total spin $S = 1$, we find

$$E_{\mathrm{HF}}^{(1)} = \frac{g_{\mathrm{p}} m_{\mathrm{e}}^2 c^2 \alpha_{\mathrm{e}}^4}{3 m_{\mathrm{p}} n^3}$$

and for the *singlet state* with $S = 0$,

$$E_{\mathrm{HF}}^{(1)} = -\frac{g_{\mathrm{p}} m_{\mathrm{e}}^2 c^2 \alpha_{\mathrm{e}}^4}{m_{\mathrm{p}} n^3} \,.$$

The energy difference between the triplet and singlet states for $n = 1$, $l = 0$ is

[16] For $l>0$-states the spin-spin coupling is negligible compared to the spin-orbit coupling, due to the much larger mass of the nucleus.

$$\Delta E_{\mathrm{HF}}^{(1)} = \frac{4g_{\mathrm{p}} m_{\mathrm{e}}^2 c^2 \alpha_{\mathrm{e}}^4}{3m_{\mathrm{p}}} = 5.857 \cdot 10^{-12}\,\mathrm{MeV}\ .$$

The radiation frequency corresponding to this transition is

$$\nu = \frac{\Delta E_{\mathrm{HF}}^{(1)}}{2\pi\hbar} \approx 1420\,\mathrm{MHz}\ ,$$

which lies in the microwave range ($\lambda \approx 21\,\mathrm{cm}$). This frequency plays an important role in radio astronomy as it allows predictions about the densities of atomic hydrogen gas in distant galaxies.

Summary

- Many stationary problems that are not solvable analytically can be tackled using **time-independent perturbation theory** to find approximate solutions. The corrections to the unperturbed states and energies are obtained from matrix elements of the perturbation operator between unperturbed states. If an unperturbed energy eigenvalue is degenerate, the perturbation operator has to be diagonalized in the subspace spanned by the eigenstates corresponding to that particular energy.

- Realistic examples of time-independent perturbation theory are the **Stark effect**, the **fine structure splitting**, and the **hyperfine structure splitting**.

- The fine structure splitting of hydrogen-like atoms is due to two effects: (i) a relativistic kinematic correction (\boldsymbol{P}^4-term) and (ii) the interaction of the intrinsic dipole moment of the electron with the magnetic field caused by the motion of the electron. The latter is best described using total angular momentum eigenstates.

- The hyperfine structure is due to the interaction of the magnetic moments of the nucleus and the electron spin. For $l=0$-states it is best described using the total spin eigenbasis.

Applications

48. Naive helium atom. Consider a helium atom consisting of two protons ($Z = 2$) and two electrons. To a very good approximation, the nucleus is infinitely heavy compared to the surrounding electrons. In the rest system of the nucleus we then have a two-particle problem whose Hamilton operator is given by

$$\boldsymbol{H} = \boldsymbol{H}_1 + \boldsymbol{H}_2 + V\ ,\ \ \boldsymbol{H}_i = \frac{\boldsymbol{P}_i^2}{2m_{\mathrm{e}}} - \frac{2e^2}{|\boldsymbol{x}_i|}\ ,\ \ V(\boldsymbol{x}_1, \boldsymbol{x}_2) = \frac{e^2}{|\boldsymbol{x}_1 - \boldsymbol{x}_2|}\ ,$$

where relativistic effects as well as effects due to the spins are neglected. V describes the repulsive interaction between the two electrons. Using this ansatz, calculate the ground-state energy of the system to first-order perturbation theory, where V is treated as a small perturbation.

Solution. The stationary Schrödinger equation of the unperturbed problem (0th order) is

$$(\boldsymbol{H}_1 + \boldsymbol{H}_2)\Psi(\boldsymbol{x}_1, \boldsymbol{x}_2) = E^{(0)}\Psi(\boldsymbol{x}_1, \boldsymbol{x}_2) \ .$$

To solve this, we use the product ansatz

$$\Psi(\boldsymbol{x}_1, \boldsymbol{x}_2) = \Psi_1(\boldsymbol{x}_1)\Psi_2(\boldsymbol{x}_2) \ .$$

This yields two decoupled equations for hydrogen-like atoms,

$$\boldsymbol{H}_i\Psi_i(\boldsymbol{x}_i) = E_i^{(0)}\Psi_i(\boldsymbol{x}_i) \ , \ E^{(0)} = E_1^{(0)} + E_2^{(0)} \ ,$$

whose solutions are given by

$$\Psi_i(\boldsymbol{x}) = \Psi_{n_i,l_i,m_i}(\boldsymbol{x}) = g_{n_i,l_i}(r)Y_{l_i,m_i}(\theta, \varphi) \ , \ E_i^{(0)} = -\frac{2m_ec^2\alpha_e^2}{n_i^2} \ .$$

For the nondegenerate ground-state energy of the helium atom ($n_1 = n_2 = 1$) to 0th order, we obtain

$$E_{1,1}^{(0)} = -4m_ec^2\alpha_e^2 = -108.8 \, \text{eV} \ .$$

For comparison: the experimental value is $-78.98 \, \text{eV}$. To find the first-order correction, we need to evaluate the integral

$$\begin{aligned}
E_{1,1}^{(1)} &= \int d^3x_1 \int d^3x_2 |\Psi_{1,0,0}(\boldsymbol{x}_1)|^2 |\Psi_{1,0,0}(\boldsymbol{x}_2)|^2 \frac{e^2}{|\boldsymbol{x}_1 - \boldsymbol{x}_2|} \\
&= \frac{e^2}{\pi^2}\left(\frac{2}{r_0}\right)^6 \int dr_1 r_1^2 e^{-4r_1/r_0} \int dr_2 r_2^2 e^{-4r_2/r_0} \\
&\quad \times \int d\Omega_1 \int d\Omega_2 \frac{1}{|\boldsymbol{x}_1 - \boldsymbol{x}_2|} \ .
\end{aligned}$$

Since the expression

$$\int \frac{d\Omega_2}{|\boldsymbol{x}_1 - \boldsymbol{x}_2|}$$

depends only on $|\boldsymbol{x}_1|$, we only need to calculate it for a single vector, for example, $\boldsymbol{x}_1 = r_1\boldsymbol{e}_z$. Introducing spherical coordinates for \boldsymbol{x}_2,

$$\boldsymbol{x}_2 = r_2 \begin{pmatrix} \cos\varphi\cos\theta \\ \sin\varphi\cos\theta \\ \sin\theta \end{pmatrix} \ ,$$

we find

$$\int \frac{d\Omega_2}{|\boldsymbol{x}_1 - \boldsymbol{x}_2|} = \int\limits_0^{2\pi} d\varphi \int\limits_{-1}^{1} d\cos\theta (r_1^2 + r_2^2 - 2r_1 r_2 \cos\theta)^{-1/2}$$

$$= -\frac{2\pi}{r_1 r_2}\left(\sqrt{r_1^2 + r_2^2 - 2r_1 r_2} - \sqrt{r_1^2 + r_2^2 + 2r_1 r_2}\right)$$

$$= \frac{2\pi}{r_1 r_2}(r_1 + r_2 - |r_1 - r_2|)$$

$$\Longrightarrow E_{1,1}^{(1)} = 8e^2 \left(\frac{2}{r_0}\right)^6 \int\limits_0^\infty dr_1 r_1 e^{-4r/r_0}$$

$$\times \left[2\int\limits_0^{r_1} dr_2 r_2^2 e^{-4r_2/r_0} + 2r_1 \int\limits_{r_1}^\infty dr_2 r_2 e^{-4r_2/r_0} \right]$$

$$= \frac{5}{4}m_e c^2 \alpha_e^2 = 34\,\text{eV}\ .$$

In total, we obtain a ground-state energy of the helium atom to first order of

$$E_{1,1} \approx E_{1,1}^{(0)} + E_{1,1}^{(1)} = -74.8\,\text{eV}\ .$$

3.8 Atomic Transitions

In this section we discuss transitions between atomic, hydrogen-like energy levels, accompanied by absorption and emission of radiation. In this context, we are particularly interested in the interaction between atoms and electromagnetic fields, where the fields are either external (*induced transitions*) or created spontaneously (*spontaneous transitions*). Whereas in Section 3.7 we discussed atomic effects due to the presence of constant electromagnetic fields (Stark effect, Zeeman effect), we now look at oscillating, time-dependent fields whose response is described by *time-dependent perturbation theory*.

The section starts with a general discussion of time-dependent perturbations. We derive *Fermi's golden rule*, which gives the transition probability between two atomic levels in the presence of a periodic perturbation. The discussion is then extended to the case of spontaneous emissions. These, however, can only fully be treated using quantum electrodynamics, where they are explained as quantum fluctuations of the electromagnetic field around its macroscopic zero mean. We shall see that certain atomic transition probabilities vanish in the dipole approximation, and this leads us to formulate *dipole selection rules*. Finally, we discuss as an example the *intensity ratio* of the two transitions $2p_{3/2} \to 1s_{1/2}$ and $2p_{1/2} \to 1s_{1/2}$ as well as the *transition rate* of $2p_{1/2} \to 1s_{1/2}$.

3.8.1 Time-Dependent Perturbation Theory

Our starting point is the full Schrödinger equation

$$i\hbar \frac{d}{dt}|\psi(t)\rangle = H(t)|\psi(t)\rangle \ , \ H(t) = H^{(0)} + \lambda H'(t) \ .$$

Here, $H^{(0)}$ denotes the time-independent unperturbed part, whereas $H'(t)$ is a time-dependent perturbation. As in Subsection 3.7.1, we use a fictitious perturbation parameter λ, which will be removed later on. For the unperturbed problem we assume that a complete basis in terms of eigenfunctions has already been found:

$$\left.\begin{aligned} i\hbar \frac{d}{dt}\left|n^{(0)}(t)\right\rangle &= H^{(0)}\left|n^{(0)}(t)\right\rangle \\ \left|n^{(0)}(t)\right\rangle &= e^{-i\omega_n t}\left|n^{(0)}\right\rangle \ , \ \omega_n = \frac{E_n^{(0)}}{\hbar} \ . \end{aligned}\right\} \tag{3.79}$$

We now pose the following problem: *suppose the system is in an unperturbed eigenstate $\left|i^{(0)}\right\rangle$ at time $t = 0$. What is the probability of finding this system at a later time t in the unperturbed state $\left|f^{(0)}\right\rangle$?* To answer this, we start by expanding $|\psi(t)\rangle$ in the unperturbed states,

$$|\psi(t)\rangle = \sum_n c_{ni}(t)e^{-i\omega_n t}\left|n^{(0)}\right\rangle \ ,$$

where the expansion coefficients c_{ni} satisfy the initial condition

$$c_{ni}(0) = \delta_{ni} \ .$$

Inserting this expansion into (3.79), we find

$$i\hbar \sum_n \left[\dot{c}_{ni} - i\omega_n c_{ni}\right]e^{-i\omega_n t}\left|n^{(0)}\right\rangle = \sum_n \left[E_n^{(0)} + \lambda H'(t)\right]c_{ni}e^{-i\omega_n t}\left|n^{(0)}\right\rangle$$

$$\Longrightarrow i\hbar \sum_n \dot{c}_{ni}e^{-i\omega_n t}\left|n^{(0)}\right\rangle = \sum_n \lambda H' c_{ni}e^{-i\omega_n t}\left|n^{(0)}\right\rangle \ .$$

Multiplying the last relation with $\left\langle f^{(0)}\right| e^{i\omega_f t}$, we obtain the following differential equations:

$$i\hbar \dot{c}_{fi}(t) = \sum_n \left\langle f^{(0)}\right|\lambda H'(t)\left|n^{(0)}\right\rangle e^{i\omega_{fn} t}c_{ni}(t)\acute{} \ , \ \omega_{fn} = \omega_f - \omega_n \ . \tag{3.80}$$

This system can be solved iteratively at different orders of λ by successively substituting lower-order solutions into the right-hand side. To zeroth order, we ignore the right-hand side of (3.80), as the matrix element $\left\langle f^{(0)}\right|\lambda H'\left|n^{(0)}\right\rangle$ is itself of first order. As expected, we then have

$$\dot{c}_{fi}(t) = 0 \Longrightarrow c_{fi}^{(0)}(t) = c_{fi}(0) = \delta_{fi} \ .$$

If no external perturbation is present, the system remains in the state $\left| i^{(0)} \right\rangle$ for all time. Inserting $c_{ni}^{(0)}$ on the right-hand side of (3.80) (and with $\lambda = 1$) yields to first order

$$\dot{c}_{fi}(t) = -\frac{\mathrm{i}}{\hbar} \left\langle f^{(0)} \middle| \boldsymbol{H}'(t) \middle| i^{(0)} \right\rangle \mathrm{e}^{\mathrm{i}\omega_{fi}t}$$

$$\implies c_{fi}^{(1)}(t) = \delta_{fi} - \frac{\mathrm{i}}{\hbar} \int_0^t \mathrm{d}t' \left\langle f^{(0)} \middle| \boldsymbol{H}'(t') \middle| i^{(0)} \right\rangle \mathrm{e}^{\mathrm{i}\omega_{fi}t'} .$$

Higher orders are obtained in a similar fashion. All in all, we find

$$c_{fi}(t) = \delta_{fi}$$

$$-\frac{\mathrm{i}}{\hbar} \int_0^t \mathrm{d}t' \left\langle f^{(0)} \middle| \boldsymbol{H}'(t') \middle| i^{(0)} \right\rangle \mathrm{e}^{\mathrm{i}\omega_{fi}t'}$$

$$+\left(\frac{-\mathrm{i}}{\hbar}\right)^2 \int_0^t \mathrm{d}t' \int_0^{t'} \mathrm{d}t'' \sum_n \left\langle f^{(0)} \middle| \boldsymbol{H}'(t') \middle| n^{(0)} \right\rangle$$

$$\times \left\langle n^{(0)} \middle| \boldsymbol{H}'(t'') \middle| i^{(0)} \right\rangle \mathrm{e}^{\mathrm{i}\omega_{fn}t'} \mathrm{e}^{\mathrm{i}\omega_{ni}t''}$$

$$+\dots . \tag{3.81}$$

Note that this perturbation expansion only makes sense if, for $f \neq i$, we have $\left| c_{fi}^{(1)}(t) \right| \ll 1$. Otherwise, the calculation becomes inconsistent due to $c_{fi}^{(0)}(t) = \delta_{fi}$. From the expansion coefficients or *transition amplitudes* $c_{fi}(t)$ we obtain the transition probabilities for the transition $\left| i^{(0)} \right\rangle \overset{t}{\longrightarrow} \left| f^{(0)} \right\rangle$ as

$$W_{fi}(t) = |c_{fi}(t)|^2 .$$

Interpretation of the expansion terms. The terms contributing to the transition amplitudes (3.81) allow us to interpret the interaction between the system and the perturbation in a simple way. To this end, we use the interaction picture rather than the Schrödinger picture (see Subsection 3.2.5). The relationship between these two is

$$\left| \psi_{\mathrm{S}}(t) \right\rangle = \boldsymbol{U} \left| \psi_{\mathrm{I,S}}(t_0) \right\rangle = \boldsymbol{U}^{(0)} \left| \psi_{\mathrm{I}}(t) \right\rangle$$

$$\left| \psi_{\mathrm{I}}(t) \right\rangle = \boldsymbol{U}' \left| \psi_{\mathrm{I,S}}(t_0) \right\rangle$$

$$\mathrm{i}\hbar \dot{\boldsymbol{U}} = \left[\boldsymbol{H}_{\mathrm{S}}^{(0)} + \lambda \boldsymbol{H}_{\mathrm{S}}'(t) \right] \boldsymbol{U} , \quad \boldsymbol{U} = \boldsymbol{U}^{(0)} \boldsymbol{U}'$$

$$\mathrm{i}\hbar \dot{\boldsymbol{U}}^{(0)} = \boldsymbol{H}_{\mathrm{S}}^{(0)} \boldsymbol{U}^{(0)}$$

$$\mathrm{i}\hbar \dot{\boldsymbol{U}}' = \lambda \boldsymbol{H}_{\mathrm{I}}'(t) \boldsymbol{U}' .$$

The interaction picture has the advantage that the whole temporal evolution of the system is determined solely by the perturbation term $\boldsymbol{H}_{\mathrm{I}}'$. Integration of the last equation yields

$$U'(t, t_0) = I - \frac{i}{\hbar} \int_{t_0}^{t} dt' \lambda H_I'(t') U'(t', t_0) \; ,$$

where I denotes the unit operator. The iterative solution of this equation to different orders of λ gives ($\lambda = 1$)

$$U'(t, t_0) = I \qquad\qquad\qquad\qquad\qquad \text{(0th order)}$$

$$-\frac{i}{\hbar} \int_{t_0}^{t} H_I'(t') dt' \qquad\qquad\qquad \text{(1st order)}$$

$$+\left(\frac{-i}{\hbar}\right)^2 \int_{t_0}^{t} dt' \int_{t_0}^{t'} dt'' H_I'(t') H_I'(t'') \qquad \text{(2nd order)}$$

$$+ \dots \; .$$

Multiplying this from the left with $U^{(0)}(t, t_0)$ and writing H_I' in terms of H_S', we obtain the temporal evolution of the system in the Schrödinger picture at different orders as

$$|\psi_S(t)\rangle = U(t, t_0) |\psi_S(t_0)\rangle \; ,$$

with

$$U(t, t_0) = U^{(0)}(t, t_0)$$

$$-\frac{i}{\hbar} \int_{t_0}^{t} dt' U^{(0)}(t, t') H_S'(t') U^{(0)}(t', t_0)$$

$$+\left(\frac{-i}{\hbar}\right)^2 \int_{t_0}^{t} dt' \int_{t_0}^{t'} dt'' U^{(0)}(t, t') H_S'(t') U^{(0)}(t', t'')$$

$$\times H_S'(t'') U^{(0)}(t'', t_0)$$

$$+ \dots \; . \qquad\qquad\qquad\qquad\qquad\qquad (3.82)$$

The first term on the right-hand side represents the 0th order and describes the unperturbed propagation of the system. Reading all subsequent terms from right to left, the first-order term, for example, states the following: the system propagates unperturbed from t_0 to t'. Then it interacts with the perturbation, followed again by an unperturbed propagation up to time t. The integral over t' expresses a sum over all possible intermediary times at which the interaction with the perturbation can occur. Accordingly, the pth-order term contains p interaction contacts of the system at times $t \geq t'' \geq \dots \geq t^{(p)}$, all of which are integrated up. In between these times the system propagates unperturbed. Figure 3.14 gives a pictorial representation of this interpretation.

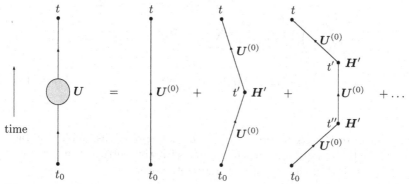

Fig. 3.14. Graphical representation of terms contributing to the perturbation expansion (3.82).

Note that the transition amplitude $c_{fi}(t)$ is the projection of the state $U(t, t_0) \left| i_{\mathrm{S}}^{(0)} \right\rangle$ onto the unperturbed state $\mathrm{e}^{-\mathrm{i}\omega_f t} \left| f_{\mathrm{S}}^{(0)} \right\rangle$ at time t,

$$c_{fi}(t) = \left\langle f^{(0)} \middle| \mathrm{e}^{\mathrm{i}\omega_f t} U(t, t_0) \middle| i^{(0)} \right\rangle ,$$

and we easily see that this expression reduces to (3.81) for $t_0 = 0$.

Theorem 3.23: Time-dependent perturbation theory

Consider the time-dependent Schrödinger equation

$$\mathrm{i}\hbar\frac{\mathrm{d}}{\mathrm{d}t}\left|\psi(t)\right\rangle = \left[H^{(0)} + H'(t)\right]\left|\psi(t)\right\rangle ,$$

with H' as a small, time-dependent perturbation. Assume that for the unperturbed system

$$\mathrm{i}\hbar\frac{\mathrm{d}}{\mathrm{d}t}\left|n^{(0)}(t)\right\rangle = H^{(0)}\left|n^{(0)}(t)\right\rangle , \quad \left|n^{(0)}(t)\right\rangle = \mathrm{e}^{-\mathrm{i}\omega_n t}\left|n^{(0)}\right\rangle, \omega_n = \frac{E_n^{(0)}}{\hbar}$$

a complete set of eigenkets is known. The probability amplitude $c_{fi}(t)$ of the transition $\left|i^{(0)}\right\rangle \xrightarrow{t} \left|f^{(0)}\right\rangle$ is then given by

$$c_{fi}(t) = \delta_{fi} \qquad\qquad\qquad \text{(0th order)}$$

$$-\frac{\mathrm{i}}{\hbar}\int_0^t \mathrm{d}t' \left\langle f^{(0)}\middle| H'(t')\middle| i^{(0)}\right\rangle \mathrm{e}^{\mathrm{i}\omega_{fi}t'} \quad \text{(1st order)}$$

\triangleright

$$+ \left(\frac{-i}{\hbar} \right)^2 \int_0^t dt' \int_0^{t'} dt'' \sum_n \left\langle f^{(0)} \middle| H'(t') \middle| n^{(0)} \right\rangle$$

$$\times \left\langle n^{(0)} \middle| H'(t'') \middle| i^{(0)} \right\rangle e^{i\omega_{fn}t'} e^{i\omega_{ni}t''} \qquad \text{(2nd order)}$$

$$+ \dots .$$

The expansion terms of order p can be interpreted as p-fold interaction contacts of the unperturbed system with the perturbation at times $t' \geq t'' \geq \dots \geq t^{(p)}$, all of which are integrated over.

Periodic perturbations and Fermi's golden rule. Other than adiabatic and instantaneous perturbations, we are mainly interested in periodic perturbations of the form[17]

$$H'(t) = \mathcal{H}' e^{\pm i\omega t} .$$

These occur, for example, during the emission and absorption within atoms. In this case, using Theorem 3.23, we obtain the transition amplitude to first order ($f \neq i$) as

$$c_{fi}(t) = -\frac{i}{\hbar} \int_0^t dt' \left\langle f^{(0)} \middle| \mathcal{H}' \middle| i^{(0)} \right\rangle e^{i(\omega_{fi} \pm \omega)t}$$

$$= -\frac{i}{\hbar} \left\langle f^{(0)} \middle| \mathcal{H}' \middle| i^{(0)} \right\rangle \frac{e^{i(\omega_{fi} \pm \omega)t} - 1}{i(\omega_{fi} \pm \omega)}$$

and the corresponding transition probability as

$$W_{fi}(t) = \frac{1}{\hbar^2} \left| \left\langle f^{(0)} \middle| \mathcal{H}' \middle| i^{(0)} \right\rangle \right|^2 \frac{4 \sin^2 \left[(\omega_{fi} \pm \omega) \frac{t}{2} \right]}{(\omega_{fi} \pm \omega)^2} .$$

In Fig. 3.15 the expression $\sin^2 \left(\frac{\Delta t}{2} \right) / \Delta^2$ is plotted as a function of Δ. It has a strongly damped oscillatory behavior around a distinct maximum at $\Delta = 0$. For large t this behavior can be related to the δ-function since

$$\int_{-\infty}^{\infty} d\Delta f(\Delta) \frac{4}{\Delta^2} \sin^2 \left(\frac{\Delta t}{2} \right) \overset{t \to \infty}{\approx} 2f(0)t \int_{-\infty}^{\infty} dy \frac{\sin^2 y}{y^2} = 2\pi t f(0)$$

$$\implies \lim_{t \to \infty} \frac{4}{\Delta^2} \sin^2 \left(\frac{\Delta t}{2} \right) = 2\pi t \delta(\Delta) .$$

Thus, in the limit of large times, we find for the transition amplitude

[17] Note that H' is not Hermitean in this form. However, this is of no consequence for what follows and simplifies the mathematical treatment.

Fig. 3.15. $\sin^2\left(\frac{\Delta t}{2}\right)/\Delta^2$ as a function of Δ.

$$\lim_{t\to\infty} W_{fi}(t) = \frac{2\pi t}{\hbar^2}\left|\left\langle f^{(0)}\left|\mathcal{H}'\right|i^{(0)}\right\rangle\right|^2 \delta(\omega_{fi}\pm\omega)\ .$$

Theorem 3.24: Fermi's golden rule

In the presence of a periodic perturbation of the form

$$H'(t) = \mathcal{H}'e^{\pm i\omega t}\ ,$$

the *transition rate* P_{fi}, $f\neq i$ (transition probability per time), for large times and to first order is given by

$$P_{fi} = \lim_{t\to\infty}\frac{W_{fi}(t)}{t} = \frac{2\pi}{\hbar^2}\left|\left\langle f^{(0)}\left|\mathcal{H}'\right|i^{(0)}\right\rangle\right|^2 \delta(\omega_{fi}\pm\omega)$$

$$= \frac{2\pi}{\hbar}\left|\left\langle f^{(0)}\left|\mathcal{H}'\right|i^{(0)}\right\rangle\right|^2 \delta\left(E_f^{(0)} - E_i^{(0)}\pm\hbar\omega\right)\ .$$

Clearly, in this theorem, the δ-function expresses an energy conservation rule: induced emission or absorption within an atom is only possible if the radiated light has exactly the frequency corresponding to the energy difference of the final and initial states. For induced absorption this difference is positive (\Rightarrow negative sign); for induced emission it is negative (\Rightarrow positive sign). When we need to apply Fermi's golden rule, it turns out that there is always an integral over the δ-function, so that in any case we have a well-defined mathematical expression.

3.8.2 Spontaneous Emission and Phase Space of Photons

In the framework of quantum mechanics there is no explanation for the spontaneous emission within atoms, as an atom will remain in its initial state $|i^{(0)}\rangle$ for all times if no perturbation is present. Within quantum electrodynamics the external electromagnetic field itself is also quantized. It is then possible to explain spontaneous emission by allowing for electromagnetic quantum fluctuations around the macroscopic zero mean of the field. In quantum electrodynamics, photons are regarded as the field quanta of the electromagnetic field and thus also have a particle-like character. They possess the energy $E = \hbar\omega$, the momentum $\boldsymbol{p} = \hbar\boldsymbol{k}$, and due to the dispersion relation $\omega = c|\boldsymbol{k}|$, a vanishing rest mass m_0 in the sense of special relativity (see Theorem 1.38). Photons also have spin $s = 1$, a fact that is coupled to the polarization $\boldsymbol{\epsilon}(\lambda)$ of light. However, due to $\boldsymbol{\epsilon}\boldsymbol{k} = 0$, there exist only two independent possible polarization vectors and therefore only two (instead of three) possible spin states (helicity $\lambda = \pm 1$). In the absence of external fields we have a zero-photon, or better: the vacuum state. Spontaneous emission must then mean that a zero-photon state is transformed into a one-photon state. In other words, the perturbation operator must contain a photon-creation operator.

In experiments, the emitted photons with momenta $\hbar\boldsymbol{k}$ are measured by a detector that always has a limited resolution, so that all photons within a certain momentum interval $[\hbar\boldsymbol{k} : \hbar(\boldsymbol{k} + \Delta\boldsymbol{k})]$ are actually counted. Therefore, the transition rate has to be summed over this interval in momentum space:

$$R_{fi} = \sum_{\Delta\boldsymbol{k}} P_{fi} \, . \tag{3.83}$$

To determine the number of photon states within this region, imagine that the photons are confined to a box of volume[18] $V = L^3$. To each photon with energy $E = \hbar\omega$, momentum $\boldsymbol{p} = \hbar\boldsymbol{k}$, and helicity λ, we associate a plane wave of the form

$$\boldsymbol{A}(\boldsymbol{x}, t) = A_0 \boldsymbol{\epsilon}(\lambda) \left(e^{-i(\boldsymbol{k}\boldsymbol{x} - \omega t)} + e^{i(\boldsymbol{k}\boldsymbol{x} - \omega t)} \right) ,$$

where only the first term is relevant for photon emission and only the second for photon absorption. The normalization constant A_0 can be determined from the constraint that the mean energy $\langle E \rangle$ of the wave \boldsymbol{A},

$$\langle E \rangle = \int_V \mathrm{d}^3 x \overline{\epsilon_{\mathrm{em}}} \, , \quad \overline{\epsilon_{\mathrm{em}}} = \frac{1}{T} \int_0^T \mathrm{d}t \frac{|\boldsymbol{E}|^2 + |\boldsymbol{B}|^2}{8\pi} \, , \quad T = \frac{2\pi}{\omega}$$

(see Theorem 2.4 and the definition on page 123) is just that of a single photon, i.e., $\hbar\omega$. It then follows that

[18] This setup circumvents possible problems with the normalization of wave functions of free particles (photons). Later on, we will take the limit $V \to \infty$.

$$A_0 = \left(\frac{2\pi c^2 \hbar}{\omega V}\right)^{1/2} .$$

Furthermore, we impose periodic boundary conditions at the edges of the box:

$$\boldsymbol{A}(x + L, y, z, t) = \boldsymbol{A}(x, y, z, t) , \qquad \text{and so on.}$$

This automatically leads to quantized wave numbers:

$$k_i = \frac{2\pi}{L} n_i , \quad n_i = 0, \pm 1 \ldots , \quad \Delta \boldsymbol{k} = \Delta k_x \Delta k_y \Delta k_z = \frac{(2\pi)^3}{V} \Delta n_x \Delta n_y \Delta n_z .$$

Now the sum in (3.83) is evaluated in the limit $V \to \infty$ as

$$R_{fi} = \sum_{\Delta \boldsymbol{k}} P_{fi} = \int \mathrm{d}^3 n P_{fi} = \frac{V}{(2\pi)^3} \int \mathrm{d}^3 k P_{fi} .$$

If we restrict ourselves to those components in Theorem 3.21 that are linear in \boldsymbol{A} (dropping the interaction of the photon-\boldsymbol{B} field with the intrinsic dipole moment of the outer electron: $-\boldsymbol{M}^{(e)}\boldsymbol{B} = -\frac{e}{m_e c}\boldsymbol{S}\boldsymbol{B}$), the perturbation operator for spontaneous emission of a photon is given by

$$\boldsymbol{H}'(t) = \mathcal{H}' \mathrm{e}^{\mathrm{i}\omega t} , \quad \mathcal{H}' = -\frac{e A_0}{m_e c} \mathrm{e}^{-\mathrm{i}\boldsymbol{k}\boldsymbol{x}} \boldsymbol{\epsilon}(\lambda) \boldsymbol{P} .$$

Using Fermi's golden rule, we find for the transition rate

$$
\begin{aligned}
R_{fi} &= \frac{V}{(2\pi\hbar)^2} \int \mathrm{d}|\boldsymbol{k}||\boldsymbol{k}|^2 \int \mathrm{d}\Omega_k \left| \left\langle f^{(0)} \middle| \mathcal{H}' \middle| i^{(0)} \right\rangle \right|^2 \delta(\omega_{fi} + \omega) \\
&= \frac{V}{(2\pi\hbar)^2} \frac{1}{c^3} \int \mathrm{d}\omega \omega^2 \int \mathrm{d}\Omega_k \left| \left\langle f^{(0)} \middle| \mathcal{H}' \middle| i^{(0)} \right\rangle \right|^2 \delta(\omega_{fi} + \omega) \\
&= \frac{V}{(2\pi\hbar)^2} \frac{\omega^2}{c^3} \int \mathrm{d}\Omega_k \left| \left\langle f^{(0)} \middle| \mathcal{H}' \middle| i^{(0)} \right\rangle \right|^2 \bigg|_{\omega = -\omega_{fi}} ,
\end{aligned}
$$

where $\mathrm{d}\Omega_k$ is the solid angle element in momentum space.

Theorem 3.25: Transition rate for spontaneous photon emission

$$R_{fi} = \frac{\alpha_e \omega}{2\pi m_e^2 c^2} \int \mathrm{d}\Omega_k |M_{fi}|^2 \bigg|_{\omega = -\omega_{fi}} , \quad M_{fi} = \left\langle f^{(0)} \middle| \mathrm{e}^{-\mathrm{i}\boldsymbol{k}\boldsymbol{x}} \boldsymbol{\epsilon}(\lambda) \boldsymbol{P} \middle| i^{(0)} \right\rangle .$$

If the experiment does not distinguish between the polarization states of the photon, the summation must be extended to include these.

3.8.3 Selection Rules in the Dipole Approximation

We now look at transitions of hydrogen-like atoms and, in particular, at the matrix element

$$M_{fi} = \left\langle f^{(0)} \middle| \mathrm{e}^{-\mathrm{i}\boldsymbol{kx}} \boldsymbol{\epsilon P} \middle| i^{(0)} \right\rangle .$$

We wish to determine which transitions are allowed, i.e., for which transitions the matrix element yields a nonvanishing result. To this end, we first note that within the whole area of extension of the electronic wave function, the exponent of the photon wave fulfills[19]

$$|\boldsymbol{k}| = \frac{\omega}{c} = \frac{E}{\hbar c} \approx \frac{m_{\mathrm{e}} c Z^2 \alpha_{\mathrm{e}}^2}{2\hbar} \qquad \text{(see (3.48))}$$

$$|\boldsymbol{x}| \lesssim \frac{\hbar}{m_{\mathrm{e}} c Z \alpha_{\mathrm{e}}} \qquad \text{(see (3.49))}$$

$$\Longrightarrow \boldsymbol{kx} \leq |\boldsymbol{k}||\boldsymbol{x}| \lesssim \frac{\alpha_{\mathrm{e}} Z}{2} .$$

Thus, for $\alpha_{\mathrm{e}} Z \ll 1$, the exponential can be expanded as

$$\mathrm{e}^{-\mathrm{i}\boldsymbol{kx}} = 1 - \mathrm{i}\boldsymbol{kx} + \dots .$$

In the following, we restrict our analysis to the dipole approximation in which only the first term in the expansion is kept, so that

$$M_{fi} \approx \left\langle f^{(0)} \middle| \boldsymbol{\epsilon P} \middle| i^{(0)} \right\rangle .$$

For the unperturbed Hamilton operator we choose

$$\boldsymbol{H}^{(0)} = \frac{\boldsymbol{P}^2}{2m_{\mathrm{e}}} + a(r)\boldsymbol{LS} + V(r) , \quad a(r) = \frac{Ze^2}{2m_{\mathrm{e}}^2 c^2 r^3} , \quad V(r) = -\frac{Ze^2}{r} ,$$

so that the unperturbed eigenstates are given by the $\left| n, l, \frac{1}{2}, J, M \right\rangle$-basis. For the calculation of M_{fi} it is advantageous to choose the following representation of \boldsymbol{P}:

$$\boldsymbol{P} = \frac{\mathrm{i}m_{\mathrm{e}}}{\hbar} \left[\frac{\boldsymbol{P}^2}{2m_{\mathrm{e}}} + V(r), \boldsymbol{x} \right] = \frac{\mathrm{i}m_{\mathrm{e}}}{\hbar} \left[\boldsymbol{H}^{(0)} - a(r)\boldsymbol{LS}, \boldsymbol{x} \right] .$$

With this we find

$$
\begin{aligned}
M_{fi} &= \frac{\mathrm{i}m_{\mathrm{e}}}{\hbar} \left\langle n', l', \frac{1}{2}, J', M' \middle| \boldsymbol{\epsilon} \left[\boldsymbol{H}^{(0)} - a(r)\boldsymbol{LS}, \boldsymbol{x} \right] \middle| n, l, \frac{1}{2}, J, M \right\rangle \\
&= \frac{\mathrm{i}m_{\mathrm{e}}}{\hbar} \left\{ (E_{n'} - E_n) \left\langle n', l', \frac{1}{2}, J', M' \middle| \boldsymbol{\epsilon x} \middle| n, l, \frac{1}{2}, J, M \right\rangle \right. \\
&\quad -\frac{1}{2} [j'(j'+1) - l'(l'+1) - j(j+1) + l(l+1)] \\
&\quad \left. \times \left\langle n', l', \frac{1}{2}, J', M' \middle| a(r)\boldsymbol{\epsilon x} \middle| n, l, \frac{1}{2}, J, M \right\rangle \right\} . \qquad (3.84)
\end{aligned}
$$

Strictly speaking, E_n has two parts: first, the energy of the naive hydrogen atom (3.48), and second, the correction energy of the spin-orbit coupling

[19] From now on we won't distinguish between the reduced mass μ and the electron mass m_{e}.

(3.77). However, for the case at hand, the latter can be ignored. The expression (3.84) can be further simplified by noting that

$$x = -\sqrt{\frac{8\pi}{3}}\frac{r}{2}\left[Y_{1,1}(\theta,\varphi) - Y_{1,-1}(\theta,\varphi)\right]$$

$$y = \sqrt{\frac{8\pi}{3}}\frac{ir}{2}\left[Y_{1,1}(\theta,\varphi) + Y_{1,-1}(\theta,\varphi)\right]$$

$$z = \sqrt{\frac{4\pi}{3}}rY_{1,0}(\theta,\varphi)$$

$$\Longrightarrow \boldsymbol{\epsilon x} = r\sqrt{\frac{4\pi}{3}}\left[\underbrace{\frac{-\epsilon_1 + i\epsilon_2}{\sqrt{2}}}_{e_1}Y_{1,1} + \underbrace{\frac{\epsilon_1 + i\epsilon_2}{\sqrt{2}}}_{e_{-1}}Y_{1,-1} + \underbrace{\epsilon_3}_{e_0}Y_{1,0}\right].$$

The transition matrix element now becomes

$$M_{fi} = \sqrt{\frac{4\pi}{3}}\frac{im_e}{\hbar}\sum_{q=-1}^{1}\left\langle l',\frac{1}{2},J',M'\left|e_qY_{1,q}\right|l,\frac{1}{2},J,M\right\rangle$$

$$\times\left\{(E_{n'} - E_n)\int_0^\infty \mathrm{d}r r^3 g_{n',l'}^*(r)g_{n,l}(r)\right.$$

$$-\frac{1}{2}\left[j'(j'+1) - l'(l'+1) - j(j+1) + l(l+1)\right]$$

$$\times\left.\int_0^\infty \mathrm{d}r r^3 a(r)g_{n',l'}^*(r)g_{n,l}(r)\right\}. \tag{3.85}$$

Using (3.34), the angular part of (3.85) evaluates to

$$\sum_{q=-1}^{1}\left\langle l',\frac{1}{2},J',M'\left|\epsilon_qY_{1,q}\right|l,\frac{1}{2},J,M\right\rangle$$

$$= \sum_{q=-1}^{1}\left\{c_1(l')c_1(l)\left\langle l',M'-\frac{1}{2};+\left|\epsilon_qY_{1,q}\right|l,M-\frac{1}{2};+\right\rangle\right.$$

$$\left. + c_2(l')c_2(l)\left\langle l',M'+\frac{1}{2};-\left|\epsilon_qY_{1,q}\right|l,M+\frac{1}{2};-\right\rangle\right\}$$

$$= \sum_{q=-1}^{1}\left\{c_1(l')c_1(l)\int \mathrm{d}\Omega Y_{l',M'-\frac{1}{2}}^*\epsilon_qY_{1,q}Y_{l,M-\frac{1}{2}}\right.$$

$$\left. + c_2(l')c_2(l)\int \mathrm{d}\Omega Y_{l',M'+\frac{1}{2}}^*\epsilon_qY_{1,q}Y_{l,M+\frac{1}{2}}\right\}, \tag{3.86}$$

with

$$c_1(l) = \pm\sqrt{\frac{l+\frac{1}{2}\pm M}{2l+1}} \; , \quad c_2(l) = \sqrt{\frac{l+\frac{1}{2}\mp M}{2l+1}} \quad \text{for} \quad J = l\pm\frac{1}{2}. \quad (3.87)$$

We can draw the following conclusions from this:

- Since the spherical harmonics in (3.86) have a definite parity, $Y_{l,m}(-e) = (-1)^l Y_{l,m}(e)$, we have

$$(-1)^{l'}(-1)^1(-1)^l = 1 \Longleftrightarrow l' + l + 1 = \text{even} \; .$$

This means the parity of the atomic state has to change when undergoing a transition.

- Using the addition theorem for spherical harmonics,

$$Y_{l_1,m_1}Y_{l_2,m_2} = \sum_{L=|l_1-l_2|}^{l_1+l_2} \langle l_1, m_1; l_2, m_2 | L, m_1+m_2 \rangle Y_{L,m_1+m_2} \; ,$$

we find

$$\int d\Omega Y^*_{l',M'\pm\frac{1}{2}} Y_{1,q} Y_{l,M\pm\frac{1}{2}}$$

$$= \sum_{L=|l-1|}^{l+1} \left\langle 1,q; l, M\pm\frac{1}{2} \middle| L, q+M\pm\frac{1}{2} \right\rangle \delta_{l'L}\delta_{M',M+q}$$

$$\Longrightarrow l' = |l-1|, l, l+1 \; , \quad M' = M + q \; .$$

Due to the first conclusion, $l' = l$ is excluded, so that in total $\Delta l = \pm 1$ and $\Delta M = 0, \pm 1$ remain as the only possibilities.

Theorem 3.26: Selection rules for atomic transitions in the dipole approximation

For atomic transitions of hydrogen-like atoms, the following transition rules hold in the dipole approximation:

$$\Delta l = \pm 1 \; , \quad \Delta M = 0, \pm 1 \; .$$

3.8.4 Intensity Rules

In this subsection we calculate the ratio of transition rates (intensity ratio) of the two allowed transitions $2p_{3/2} \to 1s_{1/2}$ and $2p_{1/2} \to 1s_{1/2}$ for hydrogen-like atoms. Ignoring the second term in (3.85), which is due to the spin-orbit coupling, the integration over the radial part yields the same result for both transitions so that these terms cancel one another. As we will see shortly, the same holds for the integration over the phase space in Theorem 3.25

if we assume that the experiment does not distinguish between the total angular momentum orientations of the initial and final states, i.e., that all possible orientations are measured. With all this, the intensity ratio of the two transitions is then solely determined by the ratio of the angular parts:

$$
\frac{R(2p_{3/2} \to 1s_{1/2})}{R(2p_{1/2} \to 1s_{1/2})} = \frac{2 \displaystyle\sum_{\substack{M' = -\frac{3}{2}, \dots, \frac{3}{2} \\ M = -\frac{1}{2}, \frac{1}{2} \\ q = -1, 0, 1}} \left| \langle 1, \tfrac{1}{2}, \tfrac{3}{2}, M' | \epsilon_q Y_{1,q} | 0, \tfrac{1}{2}, \tfrac{1}{2}, M \rangle \right|^2}{2 \displaystyle\sum_{\substack{M' = -\frac{1}{2}, \frac{1}{2} \\ M = -\frac{1}{2}, \frac{1}{2} \\ q = -1, 0, 1}} \left| \langle 1, \tfrac{1}{2}, \tfrac{1}{2}, M' | \epsilon_q Y_{1,q} | 0, \tfrac{1}{2}, \tfrac{1}{2}, M \rangle \right|^2} .
$$

Here the sums run over all possible total angular momenta of the initial and final states as well as over q. The factor of 2 in the numerator and denominator reflects the two possible polarization states of the photon. If we use (3.86) and (3.87), the matrix elements in the numerator are

$$
\left\langle 1, \frac{1}{2}, \frac{3}{2}, M' \middle| Y_{1,q} \middle| 0, \frac{1}{2}, \frac{1}{2}, M \right\rangle
$$

$$
= \frac{1}{\sqrt{4\pi}} \left\{ \sqrt{\frac{\frac{3}{2} + M'}{3}} \sqrt{\frac{1}{2} + M} \, \delta_{M,\frac{1}{2}} \delta_{M' - \frac{1}{2}, q} \right.
$$

$$
\left. + \sqrt{\frac{\frac{3}{2} - M'}{3}} \sqrt{\frac{1}{2} - M} \, \delta_{M,-\frac{1}{2}} \delta_{M' + \frac{1}{2}, q} \right\}
$$

$$
= \sqrt{\frac{2+q}{12\pi}} \, \delta_{M,\frac{1}{2}} \delta_{M', q+\frac{1}{2}} + \sqrt{\frac{2-q}{12\pi}} \, \delta_{M,-\frac{1}{2}} \delta_{M', q-\frac{1}{2}}
$$

and those of the denominator are

$$
\left\langle 1, \frac{1}{2}, \frac{1}{2}, M' \middle| Y_{1,q} \middle| 0, \frac{1}{2}, \frac{1}{2}, M \right\rangle
$$

$$
= \frac{1}{\sqrt{4\pi}} \left\{ -\sqrt{\frac{\frac{3}{2} - M'}{3}} \sqrt{\frac{1}{2} + M} \, \delta_{M,\frac{1}{2}} \delta_{M' - \frac{1}{2}, q} \right.
$$

$$
\left. + \sqrt{\frac{\frac{3}{2} + M'}{3}} \sqrt{\frac{1}{2} - M} \, \delta_{M,-\frac{1}{2}} \delta_{M' + \frac{1}{2}, q} \right\}
$$

$$
= -\sqrt{\frac{1-q}{12\pi}} \, \delta_{M,\frac{1}{2}} \delta_{M', q+\frac{1}{2}} + \sqrt{\frac{1+q}{12\pi}} \, \delta_{M,-\frac{1}{2}} \delta_{M', q-\frac{1}{2}} .
$$

The ratio is therefore given by

$$\frac{R(2p_{3/2} \rightarrow 1s_{1/2})}{R(2p_{1/2} \rightarrow 1s_{1/2})}$$

$$= \frac{2 \sum_{\substack{M' = -\frac{3}{2}, \ldots, \frac{3}{2} \\ M = -\frac{1}{2}, \frac{1}{2} \\ q = -1, 0, 1}} |\epsilon_q|^2 \left(\frac{2+q}{12\pi} \delta_{M', q+\frac{1}{2}} \delta_{M, \frac{1}{2}} + \frac{2-q}{12\pi} \delta_{M, -\frac{1}{2}} \delta_{M', q-\frac{1}{2}} \right)}{2 \sum_{\substack{M' = -\frac{1}{2}, \ldots, \frac{1}{2} \\ M = -\frac{1}{2}, \frac{1}{2} \\ q = -1, 0, 1}} |\epsilon_q|^2 \left(\frac{1-q}{12\pi} \delta_{M', q+\frac{1}{2}} \delta_{M, \frac{1}{2}} + \frac{1+q}{12\pi} \delta_{M, -\frac{1}{2}} \delta_{M', q-\frac{1}{2}} \right)}$$

$$= \frac{2/(3\pi)}{2/(6\pi)} , \tag{3.88}$$

where we have used the fact, that $|\epsilon_{-1}|^2 + |\epsilon_0|^2 + |\epsilon_{+1}|^2 = 1$. As we see, the ϵ_q-dependencies vanish due to the summation over M and M', so that the phase space integrations in the numerator and denominator are indeed the same and could be dropped right from the beginning. The final result is

$$\frac{R(2p_{3/2} \rightarrow 1s_{1/2})}{R(2p_{1/2} \rightarrow 1s_{1/2})} = 2 .$$

As the $2p_{3/2}$-state allows more possible total angular momentum values than the $2p_{1/2}$-state, the intensity of the transition $2p_{3/2} \rightarrow 1s_{1/2}$ is twice that of the transition $2p_{1/2} \rightarrow 1s_{1/2}$, as long as the contribution of the LS-coupling is ignored.

3.8.5 $2p_{3/2} \rightarrow 1s_{1/2}$-Transition

At the end of this section we calculate the transition rate for $2p_{3/2} \rightarrow 1s_{1/2}$, again ignoring the LS-coupling as in the previous subsection. The radial part of the transition matrix element (3.85) is

$$\left| \int_0^\infty dr\, r^3 g_{2,1}^*(r) g_{1,0}(r) \right|^2 = \frac{2^{15}}{3^9} \frac{r_0^2}{Z^2} .$$

The angular part has already been evaluated and is just the numerator of (3.88). However, this result needs to be multiplied by a factor of $1/3$ because, according to Theorem 3.26, to a given $M = \pm 1$ there are three (and not four) possible M'-values, which all statistically contribute one third to the total number of excited $2p_{3/2}$-atoms:

$$\frac{2}{3} \sum_{\substack{M' = -\frac{3}{2}, \ldots, \frac{3}{2} \\ M = -\frac{1}{2}, \frac{1}{2} \\ q = -1, 0, 1}} \left| \left\langle 1, \frac{1}{2}, \frac{3}{2}, M' \middle| \epsilon_q Y_{1,q} \middle| 0, \frac{1}{2}, \frac{1}{2}, M \right\rangle \right|^2 = \frac{2}{9\pi} .$$

Thus, the transition matrix element reads

$$|M_{fi}|^2 = \frac{2^{18}}{3^{12}} \frac{r_0^2 m_e^2 \omega_{fi}^2}{Z^2} .$$

Inserting this expression into Theorem 3.25 and noting that the radiation frequency is

$$\omega_{fi} = \frac{E_1 - E_2}{\hbar} = -\frac{m_e c^2 Z^2 \alpha_e^2}{2\hbar} \left(1 - \frac{1}{4}\right) = -\frac{3 m_e c^2 Z^2 \alpha_e^2}{8\hbar} ,$$

we finally get

$$R(2p_{3/2} \to 1s_{1/2}) = -\frac{\alpha_e \omega_{fi}}{2\pi m_e^2 c^2} 4\pi \frac{2^{18}}{3^{12}} \frac{r_0^2 m_e^2 \omega_{fi}^2}{Z^2} , \quad r_0 = \frac{\hbar}{m_e c \alpha_e}$$

$$= \frac{2^{10}}{3^9} \frac{\alpha_e^5 m_e c^2 Z^4}{\hbar} \approx 0.8 \cdot 10^9 Z^4 \, \mathrm{s}^{-1} .$$

Summary

- Using **time-dependent perturbation theory** we can solve quantum mechanical problems containing a small time-dependent perturbation. In particular, we can calculate **transition rates** between unperturbed atomic states in the presence of periodic perturbation fields (**Fermi's golden rule**).

- In contrast to **induced transitions** for which the perturbations are externally applied fields, **spontaneous transitions** are created by quantum fluctuations of electromagnetic fields. This effect can only be properly explained within quantum electrodynamics. To determine the corresponding transition rates, one has to integrate the individual rates over the **phase space of photons**.

- In the **dipole approximation** the transition rates and their matrix elements can be calculated in a straightforward fashion. One obtains a set of **dipole selection rules** as a necessary criterion for a nonvanishing transition probability.

- Examples for the calculation of transition rates are the intensity ratio of the transitions $2p_{3/2} \to 1s_{1/2}$ and $2p_{1/2} \to 1s_{1/2}$ as well as the transition rate of $2p_{3/2} \to 1s_{1/2}$.

Applications

49. Lightelectric effect. Consider a hydrogen-like atom exposed to electromagnetic radiation

$$\boldsymbol{A}(\boldsymbol{x}, t) = A_0 \boldsymbol{\epsilon} e^{\mathrm{i}(\boldsymbol{k}\boldsymbol{x} - \omega t)} .$$

If the energy $\hbar\omega$ of this radiation is larger than the binding energy of the outer electron in the atom, the atom is *ionized*, and the electron moves freely with kinetic energy

$$\frac{\boldsymbol{p}_f^2}{2m_e} = \hbar\omega + E_i^{(0)} = \hbar(\omega + \omega_i) \ .$$

Calculate the transition rate for the freed electron to be found in the angle element $d\Omega$ in direction \boldsymbol{p}_f if, prior to being ionized, the atom is in its ground state. Disregard relativistic effects and effects due to the spins ("naive hydrogen atom").

Solution. The perturbation operator has the form

$$\boldsymbol{H}'(t) = \mathcal{H}' e^{-i\omega t} \ , \quad \mathcal{H}' = -\frac{eA_0}{m_ec} e^{i\boldsymbol{k}\boldsymbol{x}} \boldsymbol{\epsilon P} \ .$$

The electron wave functions of the initial and final states are

$$\Psi_i^{(0)}(\boldsymbol{x}) = \Psi_{1,0,0}(\boldsymbol{x}) = \frac{2}{\sqrt{4\pi}} \left(\frac{Z}{r_0}\right)^{3/2} e^{-Zr/r_0}$$

$$\Psi_f^{(0)}(\boldsymbol{x}) = \frac{1}{\sqrt{V}} e^{i\boldsymbol{p}_f\boldsymbol{x}/\hbar} \ ,$$

where $\Psi_f^{(0)}$ has been normalized to unity within a box of volume $V = L^3$. In the dipole approximation and using Fermi's golden rule, the transition rate P_{fi} is given by

$$P_{fi} = \frac{2\pi}{\hbar} \left(\frac{eA_0}{m_ec}\right)^2 |M_{fi}|^2 \delta\left(\frac{\boldsymbol{p}_f^2}{2m_e} - \hbar(\omega_i + \omega)\right) \ ,$$

with

$$M_{fi} = \left\langle \Psi_f^{(0)} \left| \boldsymbol{\epsilon P} \right| \Psi_i^{(0)} \right\rangle = \left\langle \Psi_i^{(0)} \left| \boldsymbol{\epsilon P} \right| \Psi_f^{(0)} \right\rangle^*$$

$$= N\boldsymbol{\epsilon p}_f \int d^3x e^{-Zr/r_0} e^{-i\boldsymbol{p}_f\boldsymbol{x}/\hbar}$$

and

$$N = \frac{1}{\sqrt{\pi V}} \left(\frac{Z}{r_0}\right)^{3/2} \ .$$

Choosing the z-axis along \boldsymbol{p}_f, we further have

$$M_{fi} = N\boldsymbol{\epsilon p}_f \int_0^\infty dr r^2 \int_0^{2\pi} d\varphi \int_{-1}^1 d\cos\theta e^{-Zr/r_0} e^{-ip_f r\cos\theta/\hbar} \tag{3.89}$$

$$= 2\pi i\hbar N\boldsymbol{\epsilon} \frac{\boldsymbol{p}_f}{p_f} \int_0^\infty dr r \left[e^{-r(Z/r_0+ip_f/\hbar)} - e^{-r(Z/r_0-ip_f/\hbar)}\right]$$

$$= 2\pi i\hbar N\epsilon \frac{\boldsymbol{p}_f}{p_f} \left[\frac{1}{\left(\frac{Z}{r_0} + \frac{i}{\hbar}p_f\right)^2} - \frac{1}{\left(\frac{Z}{r_0} - \frac{i}{\hbar}p_f\right)^2} \right]$$

$$= \frac{8\sqrt{\frac{\pi}{V}}\left(\frac{r_0}{Z}\right)^{3/2}\epsilon\boldsymbol{p}_f}{\left[1 + \left(\frac{p_f r_0}{\hbar Z}\right)^2\right]^2}$$

$$\Longrightarrow P_{fi} = \frac{128\pi^2 r_0^3 e^2 A_0^2}{\hbar V m_e^2 c^2 Z^3} \frac{|\epsilon\boldsymbol{p}_f|^2}{\left[1 + \left(\frac{p_f r_0}{\hbar Z}\right)^2\right]^4} \delta\left(\frac{p_f^2}{2m_e} - \hbar(\omega_i + \omega)\right) .$$

Since in experiments the resolution of any detector is limited, we need to integrate over all electron states within a certain range around the \boldsymbol{p}_f-state. To this end, we proceed as with the photons in Subsection 3.8.2 and impose periodic boundary conditions for the box. This results in quantized momenta:

$$\boldsymbol{p}_f' = \frac{2\pi\hbar}{L}\boldsymbol{n} , \quad n_x, n_y, n_z = 0, \pm 1, \dots .$$

The transition rate is now given by summation over all possible electron states and can be approximated, as with the photons previously, by an integral:

$$R_{fi}\mathrm{d}\Omega = \mathrm{d}\Omega \frac{V}{(2\pi\hbar)^3} \int \mathrm{d}p_f' p_f'^2 P_{f'i} .$$

If we use

$$\delta\left(\frac{p_f'^2}{2m_e} - \hbar(\omega_i + \omega)\right) = \delta\left(\frac{p_f'^2}{2m_e} - \frac{p_f^2}{2m_e}\right) = \frac{m_e}{p_f'}\delta(p_f' - p_f) ,$$

it follows that

$$R_{fi}\mathrm{d}\Omega = \frac{16 r_0^3 e^2 A_0^2 p_f |\epsilon\boldsymbol{p}_f|^2}{\pi\hbar^4 m_e c^2 Z^3 \left[1 + \left(\frac{p_f r_0}{\hbar Z}\right)^2\right]^4} \mathrm{d}\Omega .$$

Apparently, this rate depends on the amplitude of the absorbed photon field, the angle between the photon polarization ϵ and the electron momentum \boldsymbol{p}_f, and finally on the absolute value p_f (or the photon frequency ω). But it is independent of the direction of the incoming radiation. This is a consequence of working in the dipole approximation: $e^{i\boldsymbol{k}\boldsymbol{x}} \approx 1$. Note, however, that the corresponding results in the nonapproximated case can be easily obtained by the replacement

$$\boldsymbol{p}_f \longrightarrow \boldsymbol{p}_f - \hbar\boldsymbol{k}$$

from (3.89) onward. Integrating R_{fi} over all angles, we obtain the *total ionization rate*. For convenience, we put the z-axis in direction of ϵ and find

$$R_{fi}^{\text{tot}} = \frac{16 r_0^3 e^2 A_0^2 p_f^3}{\pi \hbar^4 m_e c^2 Z^3 \left[1 + \left(\frac{p_f r_0}{\hbar Z}\right)^2\right]^4} \int\limits_0^{2\pi} d\varphi \int\limits_{-1}^{1} d\cos\theta \cos^2\theta$$

$$= \frac{64 r_0^3 e^2 A_0^2 p_f^3}{3 \hbar^4 m_e c^2 Z^3 \left[1 + \left(\frac{p_f r_0}{\hbar Z}\right)^2\right]^4} \, .$$

3.9 *N*-Particle Systems

So far, we have been considering one- and three-dimensional systems with few degrees of freedom. In this section we study the implications of quantum mechanics for many-particle systems. We shall be interested particularly in the different attributes of *distinguishable* and *identical* particles. As we will see, the quantum mechanical treatment of systems consisting of identical particles yields new and surprising types of behavior that have no analogy in classical mechanics. On our way to such an analysis we shall first consider a system of distinguishable particles and review some of their properties with respect to the interpretation of quantum mechanical measurements.

3.9.1 Distinguishable Particles

Consider a three-dimensional system of N spinless particles that we take to be distinguishable, so that they differ in at least one of their intrinsic properties, such as mass or charge. In classical mechanics these particles are described by their coordinate and momentum vectors $(\boldsymbol{x}_1, \boldsymbol{p}_1)$, ...,$(\boldsymbol{x}_N, \boldsymbol{p}_N)$. We arrive at the corresponding quantum mechanical description by the operator replacements (postulate III)

$$\boldsymbol{x}_i \longrightarrow \boldsymbol{X}_i \, , \quad \boldsymbol{p}_i \longrightarrow \boldsymbol{P}_i \, ,$$

where the coordinate and momentum operators obey the canonical commutation relations

$$[\boldsymbol{X}_{ik}, \boldsymbol{P}_{jl}] = i\hbar\delta_{ij}\delta_{kl} \, , \quad [\boldsymbol{X}_{ik}, \boldsymbol{X}_{jl}] = [\boldsymbol{P}_{ik}, \boldsymbol{P}_{jl}] = 0 \, .$$

In some cases, such as the harmonic oscillator, Subsection 3.3.5, it is possible to deduce all the physics from these commutation relations. But, most often, one uses a particular basis given by the simultaneous eigenkets

$$|\omega_1\rangle \otimes \cdots \otimes |\omega_N\rangle = |\omega_1, \ldots, \omega_N\rangle$$

of commuting observables $\boldsymbol{\Omega}_i(\boldsymbol{X}_i, \boldsymbol{P}_i), i = 1, \ldots, N$ (Ω-basis), which spans the N-particle Hilbert space:

$$\mathcal{H} = \mathcal{H}_1 \otimes \cdots \otimes \mathcal{H}_N \, .$$

If the system is described by the state vector $|\psi\rangle$, the probability of an experimental measurement yielding the first particle in state ω_1, the second particle in state ω_2, and so on (discrete, nondegenerate case) is

$$W(\omega_1,\ldots,\omega_N) = |\langle\omega_1,\ldots,\omega_N|\psi\rangle|^2 \; ,$$

provided that $|\psi\rangle$ is normalized to unity:

$$1 = \langle\psi|\psi\rangle = \sum_{\omega_1,\ldots,\omega_N} W(\omega_1,\ldots,\omega_N) \; .$$

Accordingly, the probability of finding the first particle in state ω_1, the second in state ω_2, and the remaining particles in an arbitrary state is

$$W(\omega_1,\omega_2,\text{rest arbitrary}) = \sum_{\omega_3,\ldots,\omega_N} W(\omega_1,\ldots,\omega_N) \; .$$

Choosing the continuous coordinate representation, the corresponding relations are

$$|\boldsymbol{x}_1\rangle \otimes \cdots \otimes |\boldsymbol{x}_N\rangle = |\boldsymbol{x}_1,\ldots,\boldsymbol{x}_N\rangle$$

$$\psi(\boldsymbol{x}_1,\ldots,\boldsymbol{x}_N) = \langle\boldsymbol{x}_1,\ldots,\boldsymbol{x}_N|\psi\rangle$$

$$W(\boldsymbol{x}_1,\ldots,\boldsymbol{x}_N) = |\psi(\boldsymbol{x}_1,\ldots,\boldsymbol{x}_N)|^2$$

$$1 = \langle\psi|\psi\rangle = \int \mathrm{d}^3x_1 \cdots \mathrm{d}^3x_N W(\boldsymbol{x}_1,\ldots,\boldsymbol{x}_N)$$

$$W(\boldsymbol{x}_1,\boldsymbol{x}_2,\text{rest arbitrary}) = \int \mathrm{d}^3x_3 \cdots \mathrm{d}^3x_N W(\boldsymbol{x}_1,\ldots,\boldsymbol{x}_N) \; .$$

Here, $W(\boldsymbol{x}_1,\boldsymbol{x}_2,\text{rest arbitrary})$ is the probability density of finding the first particle in volume element $[\boldsymbol{x}_1 : \boldsymbol{x}_1 + \mathrm{d}^3x]$, the second in volume element $[\boldsymbol{x}_2 : \boldsymbol{x}_2 + \mathrm{d}^3x]$, and the remaining particles anywhere. The temporal evolution of the coordinate wave function ψ is given by the Schrödinger equation

$$\mathrm{i}\hbar\frac{\mathrm{d}}{\mathrm{d}t}\psi(\boldsymbol{x}_1,\ldots,\boldsymbol{x}_N,t) = H\left(\boldsymbol{X}_i \to \boldsymbol{x}_i, \boldsymbol{P}_i \to \frac{\hbar}{\mathrm{i}}\boldsymbol{\nabla}_i\right)\psi(\boldsymbol{x}_1,\ldots,\boldsymbol{x}_N,t) \; .$$

3.9.2 Identical Particles and the Pauli Principle

In classical mechanics it is always possible, in principle, to distinguish between two particles, even if they have the same intrinsic properties, by following their distinct trajectories – but, clearly, without interfering with them. This implies that two configurations (sets of many particles) that differ only by the exchange of two identical particles are physically different in the classical sense. In quantum mechanics there is no such way of distinguishing two particles, since they do not have a well-defined trajectory, but are instead

described by a probabilistic wave function.[20] Therefore, two configurations obtained by the exchange of two identical particles must be regarded quantum mechanically as physically equivalent and be described by the same state vector.

To understand this better, consider a system of two identical particles in coordinate representation. Since two states differing only by a phase α are physically equivalent, the above constraint for the exchange $x_1 \leftrightarrow x_2$ implies that

$$\psi(x_1, x_2) \xrightarrow{x_1 \leftrightarrow x_2} \alpha\psi(x_2, x_1) \xrightarrow{x_1 \leftrightarrow x_2} \alpha^2\psi(x_1, x_2) \ .$$

Since a twofold application of this operator has to return us to the original wave function, we see that $\alpha = \pm 1$. Thus, the wave function can be either symmetric ($\alpha = +1$) or antisymmetric ($\alpha = -1$) under exchange of two particles. If the identical particles have a spin, this degree of freedom also needs to be taken into account, so that

$$\psi(x_1, m_1, x_2, m_2) = \pm 1\psi(x_2, m_2, x_1, m_1) \ .$$

Whether we need to choose the symmetric or the antisymmetric version depends on the type of particles under consideration. More precisely, it depends on the particle spin. Within quantum field theory, it can be shown that identical particles with integer spin quantum number[21] s (*bosons*) are described by symmetric wave functions, whereas particles with half-integer spin (*fermions*) have antisymmetric wave functions.[22] The generalization to N particles is unproblematic, and we obtain

Theorem 3.27: Symmetrization rule

The states of a system consisting of N particles are necessarily symmetric (\to bosons) or antisymmetric (\to fermions) under exchange of two particles. In other words: if P is a permutation of $1, \ldots, N$,

$$P = \begin{pmatrix} 1 & \ldots & N \\ P_1 & \ldots & P_N \end{pmatrix} ,$$

we have for fermions (half-integer spin)

$$\psi(x_1, m_1, \ldots, x_N, m_N) = \epsilon(P)\psi(x_{P_1}, m_{P_1}, \ldots, x_{P_N}, m_{P_N})$$

and for bosons (integer spin) \triangleright

[20] The only special case in which identical particles can be distinguished quantum mechanically is given when their respective areas of nonvanishing probabilities are totally disjunct. Think, for example, of two electrons cased in two different boxes, or one of them being on earth and the other on the moon.

[21] Recall that s fulfills the eigenvalue equation $S^2\psi = \hbar^2 s(s+1)\psi$.

[22] Another way of putting this is: fermions obey *Fermi-Dirac statistics*, bosons obey *Bose-Einstein statistics*.

$$\psi(\boldsymbol{x}_1, m_1, \ldots, \boldsymbol{x}_N, m_N) = \psi(\boldsymbol{x}_{P_1}, m_{P_1}, \ldots, \boldsymbol{x}_{P_N}, m_{P_N}) \, ,$$

where $\epsilon(P) = +1$ for even permutations and $\epsilon(P) = -1$ for odd permutations.

The symmetrization rule has a far-reaching consequence: suppose a measurement of an observable Ω on a fermionic two-particle system yields the values ω_1 and ω_2. Immediately after the measurement, its antisymmetric state vector is then given by

$$|\psi\rangle = \frac{1}{\sqrt{2}} (|\omega_1, \omega_2\rangle - |\omega_2, \omega_1\rangle) \, .$$

Now setting $\omega_1 = \omega_2$, we have $|\psi\rangle = 0$. This yields

Theorem 3.28: Pauli's exclusion principle

Two identical fermions cannot be in the same quantum state.

This principle has many interesting implications in statistical mechanics, in our understanding of the structure and chemical properties of atoms, and in many other topics.

Bosonic and fermionic Hilbert spaces. We return again to the system of two identical particles. If we choose the Ω-basis for its representation and assume that Ω has a discrete, nondegenerate spectrum, the two-particle Hilbert space $\mathcal{H}_1 \otimes \mathcal{H}_2$ consists of all vectors of the form $|\omega_1, \omega_2\rangle$. For each pair of vectors $|\omega_1, \omega_2\rangle$ and $|\omega_2, \omega_1\rangle$, there exists exactly one bosonic vector

$$|\omega_1, \omega_2, \mathrm{S}\rangle = \frac{1}{\sqrt{2}} (|\omega_1, \omega_2\rangle + |\omega_2, \omega_1\rangle)$$

and one orthogonal fermionic vector

$$|\omega_1, \omega_2, \mathrm{A}\rangle = \frac{1}{\sqrt{2}} (|\omega_1, \omega_2\rangle - |\omega_2, \omega_1\rangle) \, ,$$

and vice versa. If $\omega_1 = \omega_2$, the vector $|\omega_1, \omega_1\rangle$ is already symmetric and therefore bosonic; due to the Pauli principle, there exists no corresponding fermionic state. Therefore, the two-particle Hilbert space consists of the sum of a symmetric (S) and an antisymmetric (A) Hilbert space

$$\mathcal{H}_1 \otimes \mathcal{H}_2 = \mathcal{H}^{(\mathrm{S})} \oplus \mathcal{H}^{(\mathrm{A})} \, ,$$

where the dimension of $\mathcal{H}^{(\mathrm{S})}$ is slightly larger than half the dimension of $\mathcal{H}_1 \otimes \mathcal{H}_2$. If a bosonic (fermionic) two-particle system is given by the ket $|\psi_\mathrm{S}\rangle$ ($|\psi_\mathrm{A}\rangle$), we interpret

$$
\begin{aligned}
W_\mathrm{S}(\omega_1, \omega_2) &= |\langle \omega_1, \omega_2, \mathrm{S} | \psi_\mathrm{S}\rangle|^2 \\
&= \frac{1}{2} |\langle \omega_1, \omega_2 | \psi_\mathrm{S}\rangle + \langle \omega_2, \omega_1 | \psi_\mathrm{S}\rangle|^2 \\
&= 2 |\langle \omega_1, \omega_2 | \psi_\mathrm{S}\rangle|^2
\end{aligned}
$$

and

$$W_{\mathrm{A}}(\omega_1, \omega_2) = |\langle \omega_1, \omega_2, \mathrm{A}| \psi_{\mathrm{A}} \rangle|^2$$

$$= \frac{1}{2} |\langle \omega_1, \omega_2| \psi_{\mathrm{A}} \rangle - \langle \omega_2, \omega_1| \psi_{\mathrm{A}} \rangle|^2$$

$$= 2 |\langle \omega_1, \omega_2| \psi_{\mathrm{A}} \rangle|^2$$

as the probability of finding one of the two particles in the state ω_1 and the other in the state ω_2 when a measurement is performed.[23] The normalization of $|\psi_{\mathrm{S,A}}\rangle$ is determined by

$$1 = \langle \psi_{\mathrm{S,A}}| \psi_{\mathrm{S,A}} \rangle = \sum_{\mathrm{diff.}} W_{\mathrm{S,A}}(\omega_1, \omega_2)$$

$$= 2 \sum_{\mathrm{diff.}} |\langle \omega_1, \omega_2| \psi_{\mathrm{S,A}} \rangle|^2 \ ,$$

where we only sum physically different states. In the continuous coordinate basis, the corresponding relations are

$$|\boldsymbol{x}_1, \boldsymbol{x}_2, \mathrm{S}, \mathrm{A}\rangle = \frac{1}{\sqrt{2}} \left(|\boldsymbol{x}_1, \boldsymbol{x}_2\rangle \pm |\boldsymbol{x}_2, \boldsymbol{x}_1\rangle \right)$$

$$W_{\mathrm{S,A}}(\boldsymbol{x}_1, \boldsymbol{x}_2) = |\langle \boldsymbol{x}_1, \boldsymbol{x}_2, \mathrm{S}, \mathrm{A}| \psi_{\mathrm{S,A}} \rangle|^2 = 2 |\langle \boldsymbol{x}_1, \boldsymbol{x}_2| \psi_{\mathrm{S,A}} \rangle|^2$$

$$1 = \frac{1}{2} \int \mathrm{d}^3 x_1 \mathrm{d}^3 x_2 W_{\mathrm{S,A}}(\boldsymbol{x}_1, \boldsymbol{x}_2) = \int \mathrm{d}^3 x_1 \mathrm{d}^3 x_2 \, |\langle \boldsymbol{x}_1, \boldsymbol{x}_2| \psi_{\mathrm{S,A}} \rangle|^2 \ .$$

Here the factor of $1/2$ takes into account the double counting of equivalent states. (States with $\boldsymbol{x}_1 = \boldsymbol{x}_2$ for which the factor of $1/2$ does not apply only contribute an infinitesimal amount to the integration in the $\boldsymbol{x}_1 \boldsymbol{x}_2$-hyperspace.)

3.9.3 Fermionic Pressure

The meaning of the Pauli principle can best be demonstrated using an N-particle system of free electrons enclosed in a box of dimension L. The corresponding stationary Schrödinger equation

$$\sum_{i=1}^{N} H_i \Psi = E\Psi \ , \ H_i = -\frac{\hbar^2}{2m_{\mathrm{e}}} \boldsymbol{\nabla}_i^2$$

can be solved using the ansatz

$$\Psi = \prod_{i=1}^{N} \Psi_{\boldsymbol{k}_i}(\boldsymbol{x}_i, m_i) \ , \ E = \sum_{i=1}^{N} E_i \ ,$$

with

[23] Note that $\langle a, b| \psi_{\mathrm{S}} \rangle = \langle b, a| \psi_{\mathrm{S}} \rangle$ and $\langle a, b| \psi_{\mathrm{A}} \rangle = - \langle b, a| \psi_{\mathrm{A}} \rangle$.

$$\Psi_{\boldsymbol{k}_i}(\boldsymbol{x}, m) = \sin(k_{ix}x)\sin(k_{iy}y)\sin(k_{iz}z)\chi_i(m) .$$

$\chi(m)$ denotes the two-component electron spinor. Since the particles are confined, the wave function vanishes at the walls of the box, i.e., the wave vectors are quantized as

$$\boldsymbol{k}_i = \frac{\pi}{L}\boldsymbol{n}_i , \quad \boldsymbol{n}_i = (n_{ix}, n_{iy}, n_{iz}) , \quad n_{ix}, n_{iy}, n_{iz} = 1, 2, \dots .$$

Since electrons are fermions, the total wave function still has to be made antisymmetric. This can be achieved using the *Slater determinant* (this is true whenever the wave function factorizes).[24]

$$\Psi(\boldsymbol{x}_1, m_1, \dots, \boldsymbol{x}_N, m_N) = \frac{1}{\sqrt{N!}} \begin{vmatrix} \Psi_{\boldsymbol{k}_1}(\boldsymbol{x}_1, m_1) & \dots & \Psi_{\boldsymbol{k}_1}(\boldsymbol{x}_N, m_N) \\ \Psi_{\boldsymbol{k}_2}(\boldsymbol{x}_1, m_1) & \dots & \Psi_{\boldsymbol{k}_2}(\boldsymbol{x}_N, m_N) \\ \vdots & \vdots & \vdots \\ \Psi_{\boldsymbol{k}_N}(\boldsymbol{x}_1, m_1) & \dots & \Psi_{\boldsymbol{k}_N}(\boldsymbol{x}_N, m_N) \end{vmatrix} .$$

As we can see, the exchange of two particles corresponds to the exchange of two columns, which results in an additional sign to the determinant. Due to the form of the determinant, the antisymmetric wave function vanishes if two particles have the same spin direction ($m_i = m_j$) and come close to one another ($\boldsymbol{x}_i \approx \boldsymbol{x}_j$). This means that the probability density for finding both particles close together is small. In other words: the symmetrization rule effectively works like a repulsive force between the particles. Additionally, the wave function also vanishes if two particles are in the same state (Pauli principle): $(\boldsymbol{k}_i, m_i) = (\boldsymbol{k}_j, m_j)$. The state of smallest energy is therefore not simply given by the case of all N particles having the smallest possible wave vector $|\boldsymbol{k}_i| = \pi/L$. Rather, each wave vector \boldsymbol{k}_i can only be "populated" with two electrons: one electron with spin up ($m = 1/2$) and another with spin down ($m = -1/2$). Thus, the ground-state energy is given by summing all lowest particle energies,

$$E = 2\frac{\hbar^2}{2m_e}\left(\frac{\pi}{L}\right)^2 \sum_{|n|\leq n_F} n^2 ,$$

where n_F denotes an a priori unknown maximal value. If N is sufficiently large, a good approximation is given by demanding that all triplets (n_x, n_y, n_z) lie within the positive octant of a sphere of radius n_F. The number of these triplets is then (due to the double occupancy)

$$\frac{N}{2} = \frac{1}{8} \int_{|n|\leq n_F} d^3n = \frac{1}{8}\frac{4\pi}{3}n_F^3 \implies n_F = \left(\frac{3N}{\pi}\right)^{1/3} .$$

For the ground-state energy of the total system, it follows that

[24] The Slater determinant can also be used to symmetrize bosonic wave functions. In this case, all signs within the determinant's expansion have to be chosen positive.

$$E = 2\frac{\hbar^2}{2m_e}\left(\frac{\pi}{L}\right)^2\frac{1}{8}\int\limits_{|n|\leq n_F} d^3nn^2$$

$$= 2\frac{\hbar^2}{2m_e}\left(\frac{\pi}{L}\right)^2\frac{1}{8}4\pi\int\limits_0^{n_F} dnn^4$$

$$= \frac{\pi^3\hbar^2}{10m_eL^2}\left(\frac{3N}{\pi}\right)^{5/3}.$$

Note that the energy grows more than linearly with the particle number N. Therefore, the energy per particle E/N itself increases with the particle number and decreases with the volume L^3 of the box confining them.

Summary

- Contrary to classical mechanics, where particles can always be distinguished by their different trajectories, quantum mechanical systems can consist of **distinguishable** and **identical** particles. The latter are physically equivalent in all manner and are therefore described by the same state vector.

- The **symmetrization rule** states that the wave function of identical **bosons** (integer spin) or **fermions** (half-integer spin) is symmetric or antisymmetric under the exchange of two particles.

- This law yields **Pauli's exclusion principle**, which states that no two fermions can be in the same quantum state.

Applications

50. Population of different particle configurations. Consider a system of three particles, each of which can be in one of the three states $|a\rangle$, $|b\rangle$, and $|c\rangle$. Show that the total possible number of different system configurations is given by

a. 27 for nondistinguishable particles,

b. 10 for identical bosons,

c. 1 for identical fermions.

Solution.

To a. For indistinguishable particles the most general state vector is $|\omega_1, \omega_2, \omega_3\rangle$, where all three indices can take on the values a, b, or c. Each of these cases represents a different physical state. In total, this gives $3\cdot3\cdot3 = 27$ different configurations.

To b. The most general state vector that is symmetric under exchange of any two particles is

$$|\omega_1\omega_2\omega_3\rangle + |\omega_1\omega_3\omega_2\rangle + |\omega_2\omega_1\omega_3\rangle + |\omega_2\omega_3\omega_1\rangle + |\omega_3\omega_1\omega_2\rangle + |\omega_3\omega_2\omega_1\rangle \ ,$$

where, again, ω_i can take on any of the three possible values. If all three indices are different, we have 1 configuration. If two equal indices are present, there are 6 configurations, and if all three indices are the same, there are 3 different cases. In total, we have $1+6+3 = 10$ distinguishable configurations.

To c. For fermions the state vector must be antisymmetric under exchange of any two particles. The most general state is

$$|\omega_1\omega_2\omega_3\rangle - |\omega_1\omega_3\omega_2\rangle - |\omega_2\omega_1\omega_3\rangle + |\omega_2\omega_3\omega_1\rangle + |\omega_3\omega_1\omega_2\rangle - |\omega_3\omega_2\omega_1\rangle \ .$$

According to Pauli's principle, all three indices must be different. Since the exchange of two indices just results in an irrelevant change of sign, we only have a single configuration.

51. Identical two-particle system. Two identical one-dimensional particles of mass m are confined to the region $0 \le x \le L$ within a box. A measurement of the energy of the system yields the values

$$\text{(a)} \ E = \frac{\hbar^2\pi^2}{mL^2} \qquad , \qquad \text{(b)} \ E = \frac{5\hbar^2\pi^2}{mL^2} \ .$$

What are the system's wave functions if the particles are identical spin-1/2 fermions or identical spin-0 bosons? (Assume that the spin does not affect the energy measurement.)

Solution. The normalized solution for spin-1/2 fermions is

$$\Psi(x_1, m_1, x_2, m_2) = \Psi_{k_1}(x_1, m_1)\Psi_{k_2}(x_2, m_2) \ ,$$

with

$$\Psi_{k_i}(x, m) = \sqrt{\frac{2}{L}} \sin(k_i x)\chi_i(m)$$

and for spin-0 bosons

$$\Psi(x_1, x_2) = \Psi_{k_1}(x_1)\Psi_{k_2}(x_2) \ , \ \Psi_{k_i}(x) = \sqrt{\frac{2}{L}} \sin(k_i x) \ ,$$

where the wave vectors k_i are quantized as

$$k_i = k_i(n) = \frac{n\pi}{L} \ , \ n = 1, 2, \ldots \ .$$

The total energy of the two-particle system is

$$E = \frac{\hbar^2}{2m} \left[k_1(n_1)^2 + k_2(n_2)^2 \right] = \frac{\hbar^2\pi^2}{2mL^2}(n_1^2 + n_2^2) \ .$$

In case a, both particles are in the ground state ($n_1 = n_2 = 1$). Thus, the identical fermions must be distinguishable by their spins due to the Pauli principle, and the total antisymmetric wave function reads

$$\Psi^{(A)}(x_1, +, x_2, -) = \frac{1}{\sqrt{2}} \begin{vmatrix} \Psi_{k_1(1)}(x_1, +) & \Psi_{k_1(1)}(x_2, -) \\ \Psi_{k_2(1)}(x_1, +) & \Psi_{k_2(1)}(x_2, -) \end{vmatrix}$$

$$= -\frac{1}{\sqrt{2}} \begin{vmatrix} \Psi_{k_1(1)}(x_1, -) & \Psi_{k_1(1)}(x_2, +) \\ \Psi_{k_2(1)}(x_1, -) & \Psi_{k_2(1)}(x_2, +) \end{vmatrix}$$

$$= \frac{\sqrt{2}}{L} \sin(\pi x_1/L) \sin(\pi x_2/L)$$

$$\times [\chi_1(+)\chi_2(-) - \chi_1(-)\chi_2(+)] \ .$$

For identical bosons the symmetric wave function is

$$\Psi^{(S)}(x_1, x_2) = \Psi_{k_1(1)}(x_1)\Psi_{k_1(1)}(x_2) = \frac{2}{L} \sin(\pi x_1/L) \sin(\pi x_2/L) \ .$$

In case b, we have either $n_1 = 1, n_2 = 2$ or $n_1 = 2, n_2 = 1$. For identical fermions this results in the following four configurations and corresponding wave functions:

(i)+(ii) Both spins are parallel:

$$\Psi^{(A)}(x_1, \pm, x_2, \pm) = \frac{1}{\sqrt{2}} \begin{vmatrix} \Psi_{k_1(1)}(x_1, \pm) & \Psi_{k_1(1)}(x_2, \pm) \\ \Psi_{k_2(2)}(x_1, \pm) & \Psi_{k_2(2)}(x_2, \pm) \end{vmatrix}$$

$$= -\frac{1}{\sqrt{2}} \begin{vmatrix} \Psi_{k_1(2)}(x_1, \pm) & \Psi_{k_1(2)}(x_2, \pm) \\ \Psi_{k_2(1)}(x_1, \pm) & \Psi_{k_2(1)}(x_2, \pm) \end{vmatrix}$$

$$= \frac{\sqrt{2}}{L} [\sin(2\pi x_1/L) \sin(\pi x_2/L)$$

$$- \sin(\pi x_1/L) \sin(2\pi x_2/L)] \chi_1(\pm)\chi_2(\pm) \ .$$

(iii) The particle with $n = 1$ has spin up, that with $n = 2$ spin down:

$$\Psi^{(A)}(x_1, +, x_2, -) = \frac{1}{\sqrt{2}} \begin{vmatrix} \Psi_{k_1(1)}(x_1, +) & \Psi_{k_1(1)}(x_2, -) \\ \Psi_{k_2(2)}(x_1, +) & \Psi_{k_2(2)}(x_2, -) \end{vmatrix}$$

$$= -\frac{1}{\sqrt{2}} \begin{vmatrix} \Psi_{k_1(1)}(x_1, -) & \Psi_{k_1(1)}(x_2, +) \\ \Psi_{k_2(2)}(x_1, -) & \Psi_{k_2(2)}(x_2, +) \end{vmatrix}$$

$$= \frac{\sqrt{2}}{L} [\sin(\pi x_1/L) \sin(2\pi x_2/L)\chi_1(+)\chi_2(-)$$

$$- \sin(2\pi x_1/L) \sin(\pi x_2/L)\chi_1(-)\chi_2(+)] \ .$$

(iv) The particle with $n = 1$ has spin down, that with $n = 2$ spin up:

$$\Psi^{(A)}(x_1, +, x_2, -) = \frac{1}{\sqrt{2}} \begin{vmatrix} \Psi_{k_1(2)}(x_1, +) & \Psi_{k_1(2)}(x_2, -) \\ \Psi_{k_2(1)}(x_1, +) & \Psi_{k_2(1)}(x_2, -) \end{vmatrix}$$

$$= -\frac{1}{\sqrt{2}} \begin{vmatrix} \Psi_{k_1(2)}(x_1, -) & \Psi_{k_1(2)}(x_2, +) \\ \Psi_{k_2(1)}(x_1, -) & \Psi_{k_2(1)}(x_2, +) \end{vmatrix}$$

$$= \frac{\sqrt{2}}{L} [\sin(2\pi x_1/L) \sin(\pi x_2/L)\chi_1(+)\chi_2(-)$$

$$- \sin(\pi x_1/L) \sin(2\pi x_2/L)\chi_1(-)\chi_2(+)] \ .$$

For identical bosons we have

$$\Psi^{(S)}(x_1, x_2) = \frac{\sqrt{2}}{L} \left[\sin(\pi x_1/L) \sin(2\pi x_2/L) + \sin(2\pi x_1/L) \sin(\pi x_2/L) \right] .$$

3.10 Scattering Theory

One of the most successful methods of understanding the structure of particles and their interactions is given by the study of scattering processes. Just as in the classical theory, one is particularly interested in cross sections. In the quantum case, these are closely related to the asymptotic behavior of the stationary solutions of the Schrödinger equation.

We begin our discussion with quantum mechanical scattering on a fixed center or potential, where the corresponding calculation of the differential cross section is traced back to the *scattering amplitude* of the scattered particles. We then discuss the method of *scattering phases*, which lends itself to the case of centrally symmetric interactions of the projectile and the scattering potential. To this end, we use a *partial wave decomposition* of the scattering amplitude, which splits the process into components of different angular momentum modes. Subsequently, we consider the more general case of particle-particle scattering using the center of mass and the laboratory reference frame. Here we will again see the differences between the handling of identical and distinguishable particles.

In the following, we restrict ourselves to time-independent potentials $V(x)$, which fall off faster than $1/|x|$ when approaching infinity, so that the incoming and outgoing particles can both be taken as asymptotically free.

3.10.1 Scattering Amplitude and Cross Section

The problem of quantum mechanical scattering of particles on a fixed target (scattering center) is the same as in Subsections 1.5.4 and 1.5.5 for the case of classical particles. Figure 3.16 shows a graphical representation of the setup. A beam of particles with mean momentum $\langle P \rangle = \hbar k e_z$ and moving in the positive z-direction onto a fixed local target at $x = 0$ is deflected (scattered). Just as in classical mechanics, we wish to determine the number of scattered particles at a large distance to the scatterer within a solid angle element $d\Omega$ as measured by a detector. The relevant quantity is the differential cross section $d\sigma/d\Omega$ defined as

$$\frac{d\sigma}{d\Omega} d\Omega = \frac{(\text{number of scattered particles toward } d\Omega)/s}{(\text{number of incoming particles})/s/m^2} .$$

For us to calculate $d\sigma/d\Omega$, each particle would actually have to be described by a wave packet propagating in time. And in the asymptotic region $|x| \to \infty$, $t \to \pm\infty$, the incoming and scattered part would have to be isolated

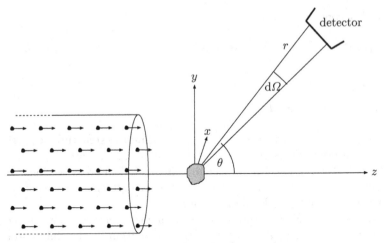

Fig. 3.16. Scattering of particles on a fixed scattering center.

to obtain the final differential cross section. However, in Subsection 3.3.3 we already showed that the characteristic scattering quantities can all be obtained from the static case (Theorem 3.15), and this result carries over to the three-dimensional problem; the more we narrow the incoming wave packets in momentum space, the wider the corresponding wave packets become in coordinate space. Eventually, they turn into the solutions of the stationary Schrödinger equation with sharp momentum,

$$\left(\nabla^2 + k^2\right)\Psi_k(\boldsymbol{x}) = \frac{2m}{\hbar^2}V(\boldsymbol{x})\Psi_k(\boldsymbol{x}) \ , \quad \boldsymbol{k} = \begin{pmatrix} 0 \\ 0 \\ k \end{pmatrix} \ , \quad k^2 = \frac{2mE}{\hbar^2} \ . \qquad (3.90)$$

In this limit, the incoming and outgoing waves coexist, so that the actual scattering process is no longer bounded in time but extends over the whole time axis. As we will show shortly, in the limit $|\boldsymbol{x}| = r \to \infty$, the eigenfunctions Ψ_k can be split into two parts,

$$\Psi_k \overset{r \to \infty}{\longrightarrow} \Psi_{\text{in}} + \Psi_{\text{sc}} \ , \quad \Psi_{\text{in}}(\boldsymbol{x}) = e^{ikz} \ , \quad \Psi_{\text{sc}}(\boldsymbol{x}) = f(\theta, \varphi)\frac{e^{ikr}}{r} \ , \qquad (3.91)$$

where Ψ_{in} denotes the incident wave, which is a solution of the free Schrödinger equation, and Ψ_{sc} is the scattered wave moving away from the center of the scattering potential. f is the *scattering amplitude*, which contains all the information of the scattering process. If $\boldsymbol{j}_{\text{in}}(k, \boldsymbol{x})$ and $\boldsymbol{j}_{\text{sc}}(k, \boldsymbol{x})$ denote the probability densities corresponding to Ψ_{in} and Ψ_{sc}, defined by the three-dimensional equivalent of (3.17),

$$\boldsymbol{j} = \frac{\hbar}{2im}\left(\Psi^*\nabla\Psi - \Psi\nabla\Psi^*\right) \ ,$$

then

$$r^2 d\Omega j_{\text{sc}}(k, \boldsymbol{x}) \boldsymbol{n}_{\text{sc}} \ , \ \ \boldsymbol{n}_{\text{sc}} = \begin{pmatrix} \cos\varphi \sin\theta \\ \sin\varphi \sin\theta \\ \cos\theta \end{pmatrix}$$

is the probability per time to find a particle in the solid angle element $d\Omega$. Correspondingly, the incoming particle current density in the z-direction is given by

$$\boldsymbol{j}_{\text{in}}(k, \boldsymbol{x}) \boldsymbol{n}_{\text{in}} \ , \ \ \boldsymbol{n}_{\text{in}} = \begin{pmatrix} 0 \\ 0 \\ 1 \end{pmatrix} \ .$$

With this notation, the differential scattering amplitude becomes

$$\frac{d\sigma}{d\Omega} = \lim_{|\boldsymbol{x}| \to \infty} \frac{r^2 \boldsymbol{j}_{\text{sc}}(k, \boldsymbol{x}) \boldsymbol{n}_{\text{sc}}}{\boldsymbol{j}_{\text{in}}(k, \boldsymbol{x}) \boldsymbol{n}_z} = |f(\theta, \varphi)|^2 \ ,$$

where we have used

$$\boldsymbol{j}_{\text{in}} \boldsymbol{n}_{\text{in}} = \frac{\hbar}{2mi} \left(\Psi_{\text{in}}^* \frac{\partial}{\partial z} \Psi_{\text{in}} - \Psi_{\text{in}} \frac{\partial}{\partial z} \Psi_{\text{in}}^* \right) = \frac{\hbar k}{m}$$

$$\begin{aligned} \boldsymbol{j}_{\text{sc}} \boldsymbol{n}_{\text{sc}} &= \frac{\hbar}{2mi} \left(\Psi_{\text{sc}}^* \boldsymbol{n}_{\text{sc}} \boldsymbol{\nabla} \Psi_{\text{sc}} - \Psi_{\text{sc}} \boldsymbol{n}_{\text{sc}} \boldsymbol{\nabla} \Psi_{\text{sc}}^* \right) \\ &= \frac{\hbar}{2mi} |f(\theta, \varphi)|^2 \left(\frac{\text{e}^{-ikr}}{r} \frac{\partial}{\partial r} \frac{\text{e}^{ikr}}{r} - \frac{\text{e}^{ikr}}{r} \frac{\partial}{\partial r} \frac{\text{e}^{-ikr}}{r} \right) \\ &= \frac{\hbar k}{mr^2} |f(\theta, \varphi)|^2 \ . \end{aligned}$$

We must still prove the validity of (3.91). Using the method of Green functions, we write (3.90) as an integral equation,

$$\Psi_k(\boldsymbol{x}) = \Psi_{\text{in}}(\boldsymbol{x}) + \int d^3 x' G(\boldsymbol{x} - \boldsymbol{x}') V(\boldsymbol{x}') \Psi_k(\boldsymbol{x}') \ , \tag{3.92}$$

where $G(\boldsymbol{x} - \boldsymbol{x}')$ is the Green function satisfying

$$\left(\boldsymbol{\nabla}^2 + k^2 \right) G(\boldsymbol{x} - \boldsymbol{x}') = \frac{2m}{\hbar^2} \delta(\boldsymbol{x} - \boldsymbol{x}') \ . \tag{3.93}$$

Its physical solution is given by

$$G(\boldsymbol{x} - \boldsymbol{x}') = -\frac{2m}{4\pi\hbar^2} \frac{\text{e}^{ik|\boldsymbol{x} - \boldsymbol{x}'|}}{|\boldsymbol{x} - \boldsymbol{x}'|} \ ,$$

and we find

$$\Psi_k(\boldsymbol{x}) = \text{e}^{ikz} - \frac{2m}{4\pi\hbar^2} \int d^3 x' \frac{\text{e}^{ik|\boldsymbol{x} - \boldsymbol{x}'|}}{|\boldsymbol{x} - \boldsymbol{x}'|} V(\boldsymbol{x}') \Psi_k(\boldsymbol{x}') \ . \tag{3.94}$$

In general, the effective reach of the scattering potential is restricted to a small area r_0, whereas the particles are detected at a large distance to the scattering center,

$$|\boldsymbol{x}'| \leq r_0 \ll |\boldsymbol{x}| \to \infty \ ,$$

so that we can expand $\frac{e^{ik|\boldsymbol{x}-\boldsymbol{x}'|}}{|\boldsymbol{x}-\boldsymbol{x}'|}$ just as in (2.61) and (2.62) in Subsection 2.4.3:

$$\frac{e^{ik|\boldsymbol{x}-\boldsymbol{x}'|}}{|\boldsymbol{x}-\boldsymbol{x}'|} \xrightarrow{|\boldsymbol{x}|\to\infty} \frac{e^{ik|\boldsymbol{x}|}e^{-ik\boldsymbol{x}\boldsymbol{x}'/|\boldsymbol{x}|}}{|\boldsymbol{x}|}\left(1+\frac{\boldsymbol{x}\boldsymbol{x}'}{\boldsymbol{x}^2}\right) \approx \frac{e^{ikr}}{r}e^{-ik\boldsymbol{n}_{\mathrm{sc}}\boldsymbol{x}'} \ . \tag{3.95}$$

Here, $r = |\boldsymbol{x}|$, and $\boldsymbol{n}_{\mathrm{sc}} = \boldsymbol{x}/|\boldsymbol{x}|$ denotes the unit vector in direction of the scattered particles. Inserting (3.95) into (3.94) completes the proof:

$$\Psi_k(\boldsymbol{x}) \xrightarrow{r\to\infty} e^{ikz} + f(\theta,\varphi)\frac{e^{ikr}}{r} \ , \tag{3.96}$$

with[25]

$$f(\theta,\varphi) = -\frac{m}{2\pi\hbar^2}\int d^3x' e^{-ik\boldsymbol{n}_{\mathrm{sc}}\boldsymbol{x}'}V(\boldsymbol{x}')\Psi_k(\boldsymbol{x}') \ .$$

To be able to calculate the scattering amplitude, we need to know what the wave function Ψ_k is. This can be found by iteratively solving the integral equation (3.92) to different orders in the potential V:

$$\Psi_k(\boldsymbol{x}) = \Psi_{\mathrm{in}}(\boldsymbol{x}) \qquad\qquad\qquad\qquad\text{(0th order)}$$

$$+ \int d^3x' G(\boldsymbol{x}-\boldsymbol{x}')V(\boldsymbol{x}')\Psi_{\mathrm{in}}(\boldsymbol{x}') \qquad\text{(1st order)}$$

$$+ \int d^3x' \int d^3x''$$

$$\times G(\boldsymbol{x}-\boldsymbol{x}')V(\boldsymbol{x}')G(\boldsymbol{x}'-\boldsymbol{x}'')V(\boldsymbol{x}'')\Psi_{\mathrm{in}}(\boldsymbol{x}'') \quad\text{(2nd order)}$$

$$+ \dots \ .$$

This expansion is called the *Born series*. Working only to lowest order defines the *Born approximation* and leads to the scattering amplitude

$$f(\theta,\varphi) = -\frac{m}{2\pi\hbar^2}\int d^3x' e^{i\boldsymbol{\Delta}\boldsymbol{x}'}V(\boldsymbol{x}') \ , \quad \boldsymbol{\Delta} = k(\boldsymbol{n}_{\mathrm{in}}-\boldsymbol{n}_{\mathrm{sc}}) \ .$$

Theorem 3.29: Scattering amplitude and differential cross section

Consider the scattering of particles moving along the z-axis with average momentum $\langle\boldsymbol{P}\rangle = \hbar k\boldsymbol{e}_z$ toward a scattering potential $V(\boldsymbol{x})$. If the effective reach r_0 of the potential is small compared to the distance r of the scattering center to the detector, $r_0 \ll r$, the asymptotic solution of the time-independent Schrödinger equation can be written as

\triangleright

[25] Note that the second solution to the differential equation (3.93),

$$G(\boldsymbol{x}-\boldsymbol{x}') = -\frac{2m}{4\pi\hbar^2}\frac{e^{-ik|\boldsymbol{x}-\boldsymbol{x}'|}}{|\boldsymbol{x}-\boldsymbol{x}'|} \ ,$$

does not apply to the case at hand, since it would lead to the unphysical situation of an incoming spherical wave in (3.96).

$$\Psi_k(x) \overset{r \to \infty}{\longrightarrow} \underbrace{e^{ikz}}_{\Psi_{\text{in}}} + \underbrace{f(\theta, \varphi) \frac{e^{ikr}}{r}}_{\Psi_{\text{sc}}} \ ,$$

where the scattering amplitude f is given by

$$f(\theta, \varphi) = -\frac{m}{2\pi\hbar^2} \int d^3x' e^{-ikn_{\text{sc}}x'} V(x') \Psi_k(x') \ . \tag{3.97}$$

The differential cross section is

$$\frac{d\sigma}{d\Omega} = |f(\theta, \varphi)|^2 \ .$$

In the Born approximation, (3.97) is reduced to

$$f(\theta, \varphi) = f(\boldsymbol{\Delta}) = -\frac{2m}{4\pi\hbar^2} \int d^3x' e^{i\boldsymbol{\Delta}x'} V(x') \ , \quad \boldsymbol{\Delta} = k(n_{\text{in}} - n_{\text{sc}}) \ .$$

Ignoring a constant factor, this is just the Fourier transform of the potential V as a function of the transferred momentum $\hbar\boldsymbol{\Delta}$. For centrally symmetrical potentials, $V(x) = V(|x|)$, the scattering amplitude is independent of φ, so that $f = f(\theta) = f(|\boldsymbol{\Delta}|)$.

Coulomb scattering. As an example of this theorem we calculate the differential cross section in Born approximation for the scattering of particles with mass m and charge $Z_1 e$ on a Coulomb potential of charge $Z_2 e$. For reasons that will become clear in a minute, we first consider the more general case of a *Yukawa potential*:

$$V(r) = g\frac{e^{-\beta r}}{r} \ .$$

Due to its rotational symmetry, it is sufficient to calculate $f(|\boldsymbol{\Delta}|)$ for $\boldsymbol{\Delta} = e_z$. Using spherical coordinates,

$$x' = r\cos\varphi\sin\theta \ , \quad y' = r\sin\varphi\sin\theta \ , \quad z' = r\cos\theta \ ,$$

we find

$$f(|\boldsymbol{\Delta}|) = -\frac{2m}{4\pi\hbar^2} 2\pi g \int_0^\infty dr\, r^2 \frac{e^{-\beta r}}{r} \int_{-1}^1 d\cos\theta\, e^{i\Delta r\cos\theta}$$

$$= -\frac{mg}{\hbar^2} \int_0^\infty dr\, r e^{-\beta r} \frac{1}{i\Delta r} \left(e^{i\Delta r} - e^{-i\Delta r} \right) \tag{3.98}$$

$$= \frac{img}{\hbar^2 \Delta} \left(\frac{1}{\beta - i\Delta} - \frac{1}{\beta + i\Delta} \right)$$

$$= -\frac{2mg}{\hbar^2} \frac{1}{\beta^2 + \Delta^2} \ ,$$

where

$$\Delta^2 = k^2(n_{\text{in}}^2 + n_{\text{sc}}^2 - 2n_{\text{in}}n_{\text{sc}}) = 2k^2(1 - \cos\theta) = 4k^2\sin^2\frac{\theta}{2}\ .$$

Setting

$$g = Z_1 Z_2 e^2\ ,\quad \beta = 0\ ,$$

we find for the differential cross section of Coulomb scattering the well-known Rutherford scattering amplitude [see (1.63)]

$$\frac{d\sigma}{d\Omega} = \left(\frac{2mZ_1Z_2e^2}{4\hbar^2k^2\sin^2\frac{\theta}{2}}\right)^2 = \left(\frac{Z_1Z_2e^2}{4E\sin^2\frac{\theta}{2}}\right)^2\ . \tag{3.99}$$

Now it is clear why we started with the Yukawa potential: we need the *screening factor* β, so that the r integration in (3.98) converges. Strangely, (3.99) does not just hold in Born approximation but is, in fact, true exactly.

3.10.2 Phase Shift Analysis for Centrally Symmetric Potentials

For centrally symmetric potentials the angular momentum is a conserved quantity, and it is advantageous to expand the scattering amplitude $f = f(\theta)$ in the Legendre polynomials

$$P_l(\cos\theta) = \sqrt{\frac{4\pi}{2l+1}}Y_{l,0}(\theta)\ ,\quad Y_{l,0}(\theta) = Y_{l,0}(\theta,\varphi)\ .$$

This allows us to study the scattering process for each l individually. Using Theorem 3.20, we find for the asymptotic form of the wave function Ψ_k

$$\Psi_k(\boldsymbol{x}) \xrightarrow{r\to\infty} e^{ikz} + f(\theta)\frac{e^{ikr}}{r} = \sum_{l=0}^{\infty}\left[(2l+1)i^l j_l(kr) + a_l\frac{e^{ikr}}{r}\right]P_l(\cos\theta)\ .$$

Since j_l behaves asymptotically as

$$j_l(kr) \xrightarrow{r\to\infty} \frac{\sin(kr - l\pi/2)}{kr}\ ,$$

we can rewrite this equation in such a way that incoming and outgoing waves are separated:

$$\Psi_k(\boldsymbol{x}) \xrightarrow{r\to\infty} \frac{e^{ikr}}{r}\sum_l\left[\frac{(2l+1)i^l}{2ik}e^{-il\pi/2} + a_l\right]P_l(\cos\theta)$$

$$-\frac{e^{-ikr}}{r}\sum_l\frac{(2l+1)i^l}{2ik}e^{il\pi/2}P_l(\cos\theta)\ . \tag{3.100}$$

On the other hand, Ψ_k can generally be expanded as

$$\Psi_k(\boldsymbol{x}) = \sum_{l=0}^{\infty}A_l g_l(r)Y_{l,0}(\theta) = \sum_{l=0}^{\infty}A_l\sqrt{\frac{2l+1}{4\pi}}g_l(r)P_l(\cos\theta)\ .$$

At infinity, the radial part of this reduces asymptotically to the regular solution $j_l(kr)$ for free particles, up to an additional phase shift δ_l called *scattering phase*, which contains all the information about the scattering process in the angular momentum l-sector:

$$g_l(r) \xrightarrow{r \to \infty} \frac{\sin(kr - l\pi/2 + \delta_l)}{kr} .$$

Thus, we have

$$\Psi_k(x) \xrightarrow{r \to \infty} \frac{e^{ikr}}{r} \sum_l A_l \sqrt{\frac{2l+1}{4\pi}} \frac{e^{i\delta_l}}{2ik} e^{-il\pi/2} P_l(\cos\theta)$$

$$- \frac{e^{-ikr}}{r} \sum_l A_l \sqrt{\frac{2l+1}{4\pi}} \frac{e^{-i\delta_l}}{2ik} e^{il\pi/2} P_l(\cos\theta) . \qquad (3.101)$$

Comparing the coefficients in (3.100) and (3.101), we see that

$$A_l = \sqrt{4\pi(2l+1)} i^l e^{i\delta_l} , \quad a_l = \frac{2l+1}{k} e^{i\delta_l} \sin\delta_l .$$

Finally, this gives us the *partial wave expansion* of the scattering amplitude:

$$f(\theta) = \frac{1}{k} \sum_l (2l+1) e^{i\delta_l} \sin\delta_l P_l(\cos\theta) .$$

From this result, we obtain an interesting relationship between the total cross section and the scattering amplitude, known as *optical theorem*:

$$\sigma = \int d\Omega |f(\theta)|^2 = \frac{1}{k^2} \int d\Omega \left| \sum_l \sqrt{4\pi(2l+1)} e^{i\delta_l} \sin\delta_l Y_{l,0}(\theta) \right|^2$$

$$= \frac{1}{k^2} \sum_l 4\pi(2l+1) \sin^2\delta_l = \frac{4\pi}{k} \text{Im} f(\theta = 0) .$$

According to this, the total cross section is equal to the imaginary part of the scattering amplitude in the forward direction.

To calculate the scattering phases for a given scattering potential, we consider the radial Schrödinger equation (3.39) with and without the potential term,

$$u_l''(r) - \frac{l(l+1)}{r^2} u_l(r) + k^2 u_l(r) = \frac{2m}{\hbar^2} V(r) u_l(r)$$

$$v_l''(r) - \frac{l(l+1)}{r^2} v_l(r) + k^2 v_l(r) = 0 ,$$

where we denote the free wave functions by v_l, and those for the scattering case by u_l. Multiplying the first equation by v_l, the second by u_l, and subtracting both, we find

$$\frac{d}{dr} [u_l'(r) v_l(r) - v_l'(r) u_l(r)] = \frac{2m}{\hbar^2} V(r) u_l(r) v_l(r)$$

or

$$[u_l'(r)v_l(r) - v_l'(r)u_l(r)]_0^\infty = \frac{2m}{\hbar^2} \int\limits_0^\infty dr V(r)u_l(r)v_l(r) \ . \tag{3.102}$$

Due to the asymptotic behavior of u_l and v_l,[26]

$$r \to 0: \quad u_l(r), v_l(r) \sim r^{l+1}$$

$$r \to \infty: \quad \begin{cases} u_l(r) = A_l(\delta_l)\dfrac{\sin(kr - l\pi/2 + \delta_l)}{k} \\[2mm] v_l(r) = A_l(\delta_l = 0)\dfrac{\sin(kr - l\pi/2)}{k} \ , \end{cases}$$

it follows from (3.102) that

$$e^{i\delta_l} \sin \delta_l = -\frac{1}{4\pi(2l+1)i^{2l}}\frac{2mk}{\hbar^2} \int\limits_0^\infty dr V(r)u_l(r)v_l(r)$$

and with $v_l(r) = A_l(\delta_l = 0)rj_l(kr)$

$$e^{i\delta_l} \sin \delta_l = -\frac{1}{\sqrt{4\pi(2l+1)}i^l}\frac{2mk}{\hbar^2} \int\limits_0^\infty dr V(r)u_l(r)rj_l(kr) \ . \tag{3.103}$$

If $V(r)$ is sufficiently small, u_l differs only marginally from the solution v_l of the free radial Schrödinger equation, and the scattering phase δ_l is close to zero. In this case, u_l can be replaced by v_l in (3.103), which, in the Born approximation (i.e., to 0th order), yields

$$e^{i\delta_l} \sin \delta_l = -\frac{2mk}{\hbar^2} \int\limits_0^\infty dr V(r)r^2 j_l^2(kr) \ .$$

**Theorem 3.30: Partial wave decomposition
of the scattering amplitude and optical theorem**

Consider a centrally symmetric potential $V(r)$ with $\lim\limits_{r \to 0} r^2 V(r) = 0$ and $\lim\limits_{r \to \infty} rV(r) = 0$. The scattering amplitude f can be expanded in Legendre polynomials,

$$f(\theta) = \frac{1}{k}\sum_l (2l+1)e^{i\delta_l}\sin \delta_l P_l(\cos \theta) \ ,$$

where δ_l denote the scattering phases. They contain the complete information of the scattering process in the corresponding l-sectors and obey the relation

\triangleright

[26] We are assuming that $\lim\limits_{r \to 0} r^2 V(r) = 0$ and, as before, $\lim\limits_{r \to \infty} rV(r) = 0$.

$$e^{i\delta_l}\sin\delta_l = -\frac{1}{\sqrt{4\pi(2l+1)}i^l}\frac{2mk}{\hbar^2}\int_0^\infty dr V(r)u_l(r)r j_l(kr) \ . \tag{3.104}$$

Here, u_l are the solutions to the radial Schrödinger equation with a potential V. In the Born approximation these solutions are replaced by those of the free case, such that (3.104) becomes

$$e^{i\delta_l}\sin\delta_l = -\frac{2mk}{\hbar^2}\int_0^\infty dr V(r)r^2 j_l^2(kr) \ .$$

The total cross section obeys the optical theorem:

$$\sigma = \int d\Omega |f(\theta)|^2 = \frac{1}{k^2}\sum_l 4\pi(2l+1)\sin^2\delta_l = \frac{4\pi}{k}\mathrm{Im}f(\theta=0) \ .$$

If the scattering potential is restricted to a finite range r_0, $V(r > r_0) = 0$, the infinite range of integration in (3.104) can be replaced by a finite one. Furthermore, taking into account the asymptotic behavior of the spherical Bessel functions $j_l(kr)$ in the limit $k \to 0$,

$$j_l(\rho) \xrightarrow{\rho \to 0} \frac{\rho^l}{(2l+1)!!} \ ,$$

(3.104) becomes

$$e^{i\delta_l}\sin\delta_l \xrightarrow{k \to 0} -\frac{1}{\sqrt{4\pi(2l+1)}i^l}\frac{2mk^{l+1}}{\hbar^2(2l+1)!!}\int_0^{r_0} dr V(r)u_l(r)r^{l+1} \ .$$

We see that for small energies the lowest partial waves $l = 0, 1, \ldots$ are dominant. The differential cross section for this case can be written as

$$\frac{d\sigma}{d\Omega} \xrightarrow{k \to 0} \frac{1}{k^2}\left[\sin^2\delta_0 + 6\sin\delta_0\sin\delta_1\cos(\delta_0 - \delta_1)\cos\theta \right.$$
$$\left. + 9\sin^2\delta_1\cos^2\theta + \ldots\right] \ .$$

To lowest order it is isotropic, i.e., independent of the scattering angle θ. In the Born approximation, the scattering phases behave like

$$\sin\delta_l \overset{k \to 0}{\approx} \delta_l \sim -k^{2l+1} \ . \tag{3.105}$$

It turns out that this limiting type of behavior holds for a fairly large class of potentials and is also not restricted to the Born approximation.

3.10.3 Resonance Scattering

The term *resonance scattering* is used for those cases where the differential cross section at a particular energy, called *resonance energy*, is dominated by a certain partial wave $l = L$:

$$\frac{d\sigma}{d\Omega} \approx \frac{d\sigma_L}{d\Omega} = \frac{2L+1}{k^2}|T_L|^2 P_L^2(\cos\theta) \ , \quad T_L = e^{i\delta_L}\sin\delta_L = \frac{1}{2i}\left(e^{2i\delta_L}-1\right) \ .$$

Clearly, this implies that the corresponding *partial wave amplitude* T_L has a large local maximum at this energy. In fact, we speak of a resonance only if the resonant scattering phase δ_L runs through a half-integer multiple of π:

$$\delta_L = \left(n+\frac{1}{2}\right)\pi \ , \quad n = 0,1,\dots \ .$$

Figure 3.17 shows a typical resonance situation. As the energy increases, the

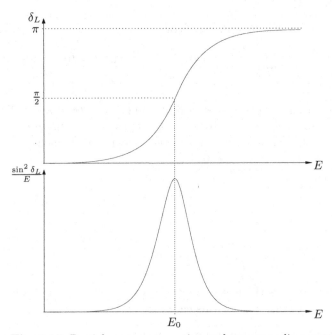

Fig. 3.17. Partial wave cross section and corresponding scattering phase.

scattering phase δ_L quickly grows from 0 to π [more generally: from $n\pi$ to $(n+1)\pi$] while passing through $\pi/2$ at the resonance energy E_0. At this energy the cross section is maximal. To understand how the differential cross section depends on the energy in the vicinity of the resonance energy E_0, we expand δ_L around E_0:

$$\delta_L(E) \approx \delta_L(E_0) + (E-E_0)\delta_L'(E_0) \ .$$

In so doing, we have to ensure that the relations $\left|e^{2i\delta_L(E)}\right| = 1$ and $e^{2i\delta_L(E_0)} = -1$ remain unaffected. This can be achieved by

$$e^{2i\delta_L(E)} \approx \frac{e^{i\delta_L(E_0)}e^{i(E-E_0)\delta_L'(E_0)}}{e^{-i\delta_L(E_0)}e^{-i(E-E_0)\delta_L'(E_0)}} \approx -\frac{1+i(E-E_0)\delta_L'(E_0)}{1-i(E-E_0)\delta_L'(E_0)} \ .$$

It then follows that

$$T_L(E) \approx -\frac{\frac{\Gamma}{2}}{E - E_0 + \frac{i\Gamma}{2}} \, , \quad \frac{\Gamma}{2} = \frac{1}{\delta'_L(E_0)} \, ,$$

and we obtain

Theorem 3.31: Breit-Wigner equation

In the vicinity of a resonance energy E_0, the cross section is dominated by the cross section contribution σ_L of the resonant partial wave. It is given by

$$\sigma_L(E) = \frac{4\pi(2l+1)}{k^2} \sin^2 \delta_l(E) = \frac{4\pi(2l+1)}{k^2} \frac{\left(\frac{\Gamma}{2}\right)^2}{(E-E_0)^2 + \left(\frac{\Gamma}{2}\right)^2} \, ,$$

with the *resonance width* Γ:

$$\frac{\Gamma}{2} = \frac{1}{\delta'_L(E_0)} \, .$$

Generally, scattering resonance occurs when the effective potential in the radial Schrödinger equation (scattering potential plus centrifugal barrier) is strongly attractive at small distances and repulsive at large distances. Consider, for example, the deep spherically symmetric potential well

$$V(r) = \left\{ \begin{array}{ll} -V_0 & \text{for } r < a \\ 0 & \text{for } r \geq a \end{array} \right\} \, , \quad V_0 > 0 \, .$$

The corresponding effective potential V_{eff} for $l > 0$ is shown in Fig. 3.18. Ignoring tunnel processes, a particle with energy $0 < E_0 < V_{\text{max}}$ can form a bound state within the attractive region. But at some stage, since tunneling is of course present, the particle will escape to $r \to \infty$. Similarly, a free particle coming from infinity can tunnel through the centrifugal barrier and form a *meta-stable state* in the attractive region. With growing l, the centrifugal term starts to dominate, so that the tunneling probability and with it the resonance width become smaller. Accordingly, the lifetime T of the meta-stable state becomes larger. Generally, due to Heisenberg's uncertainty principle, we have

$$T \sim \frac{\hbar}{\Gamma} \, .$$

For $l = 0$ no repulsive barrier is present, and if $V = V_{\text{eff}}$ is purely attractive, there can only be true bound states with negative energy and infinite lifetime. Here, states that come most closely to resonances are those with energies near to zero (see Application 53).

3.10.4 Particle-Particle Scattering

In a typical scattering experiment one collides a target of some particle type 2 with a mono-energetic beam consisting of particles of type 1. One then counts

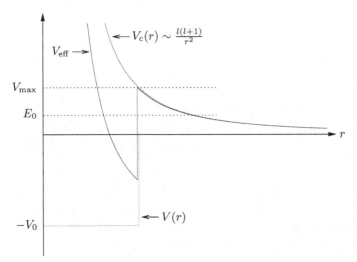

Fig. 3.18. Effective potential (*solid line*), composed of a deep potential well (*dotted line*) and the centrifugal barrier (*dashed line*).

the number of particles of one of these types, say type 1, which are scattered in a particular solid angle element. In the following, we shall assume that the interaction potential between the two types of particles depends only on their relative distance. In this case, we have a two-particle problem with a Hamilton operator of the form

$$H = \frac{P_1^2}{2m_1} + \frac{P_2^2}{2m_2} + V(x_1 - x_2) \ .$$

In the absence of external forces, the center of mass motion is that of a free particle and can be separated by the use of center of mass and relative coordinates. The remaining relative motion is determined by

$$\left(-\frac{\hbar^2}{2\mu}\boldsymbol{\nabla}^2 + V(\boldsymbol{x})\right)\Psi(\boldsymbol{x}) = E\Psi(\boldsymbol{x}) \ , \quad \mu = \frac{m_1 m_2}{m_1 + m_2} \ , \quad \boldsymbol{x} = \boldsymbol{x}_1 - \boldsymbol{x}_2 \ ,$$

with the reduced mass μ and the distance vector \boldsymbol{x} of both particles. In the center of mass system, where the center of mass is at rest, one obtains the scattering amplitude from the asymptotic behavior of the wave function of relative motion,

$$\Psi(\boldsymbol{x}) = \mathrm{e}^{\mathrm{i}kz} + f(\theta, \varphi)\frac{\mathrm{e}^{\mathrm{i}kr}}{r} \ ,$$

where θ and φ are the scattering angles in that frame. To calculate the scattering amplitude, the differential cross section, and the partial wave expansion, we can carry over the results from Subsections 3.10.1 and 3.10.2, in particular Theorems 3.29 and 3.30, by simply substituting m by μ and $\mathrm{d}\Omega$ by the solid angle element $\mathrm{d}\Omega^*$ of the center of mass frame. For the center of mass

momenta of the particles before $(\boldsymbol{p}_1^{I*}, \boldsymbol{p}_2^{I*})$ and after $(\boldsymbol{p}_1^{F*}, \boldsymbol{p}_2^{F*})$ scattering, we have the situation as shown in Fig. 3.19a: Particles 1 and 2 move toward one

Fig. 3.19. Particle-particle scattering in the center of mass system (**a**) and in the laboratory system (**b**).

another with equal and opposite initial momenta $\boldsymbol{p}_1^{I*} = -\boldsymbol{p}_2^{I*}$, are scattered, and carry on with final momenta $\boldsymbol{p}_1^{F*} = -\boldsymbol{p}_2^{F*}$.

To move from the center of mass to the laboratory frame (Fig. 3.19b), where the second particle is at rest long before the scattering ($\boldsymbol{p}_2^{I} = \boldsymbol{0}$), we need to move to the left with velocity $\boldsymbol{p}_2^{I*}/m_2$. In this system each center of mass momentum has a component in the positive z-direction, just as in the classical case. The transformation is the same as described in Subsection 1.5.5 with the replacements $\chi \to \theta$, $\theta_1 \to \theta_L$. Using Theorem 1.34, we therefore have

Theorem 3.32: Particle-particle scattering

For the particle-particle scattering in the center of mass system, the results from Theorems 3.29 and 3.30 are valid with the replacements $m \to \mu$ and $d\Omega \to d\Omega^*$. The relation between the differential cross sections in the center of mass reference frame, $d\sigma/d\Omega^*$, and in the laboratory system, $d\sigma/d\Omega_L$, is given by

$$\frac{d\sigma}{d\Omega_L} = \frac{d\sigma}{d\Omega^*} \frac{\left[\left(\frac{m_1}{m_2}\right)^2 + 1 + 2\frac{m_1}{m_2}\cos\theta(\theta_L)\right]^{3/2}}{\frac{m_1}{m_2}\cos\theta(\theta_L) + 1},$$

with

$$\cos\theta_L = \frac{\frac{m_1}{m_2} + \cos\theta}{\sqrt{\left(\frac{m_1}{m_2}\right)^2 + 1 + 2\frac{m_1}{m_2}\cos\theta}}, \quad \varphi_L = \varphi$$

and

$$d\Omega^* = \sin\theta d\theta d\varphi = \text{solid angle element in the center of mass system,}$$

$$d\Omega_L = \sin\theta_L d\theta_L d\varphi = \begin{array}{l}\text{solid angle element in the laboratory system,} \\ \text{in which the projectiles are scattered.}\end{array}$$

Scattering of distinguishable particles. For particle-particle scattering of distinguishable particles, the differential cross section in the center of mass system for particle 1 to be scattered in the direction of (θ, φ) and particle 2 in the direction of $(\pi - \theta, \varphi + \pi)$ is

$$\frac{d\sigma}{d\Omega^*} = |f(\theta, \varphi)|^2 \ .$$

Conversely,

$$\frac{d\sigma}{d\Omega^*} = |f(\pi - \theta, \varphi + \pi)|^2$$

is the differential cross section for particle 1 to be scattered in the direction of $(\pi - \theta, \varphi + \pi)$ and particle 2 in the direction of (θ, φ). If the detector measures all particles without differentiating between particles of type 1 and 2, we need to add the partial cross sections to obtain the full differential cross section

$$\frac{d\sigma}{d\Omega^*} = |f(\theta, \varphi)|^2 + |f(\pi - \theta, \varphi + \pi)|^2 \tag{3.106}$$

for one of the particles to be scattered in the direction of (θ, φ).

Scattering of identical spin-0 bosons. For the scattering of identical particles, we must take into account the symmetrization rule. In particular, for zero-spin bosons, for example, π mesons, the total wave function and its asymptotic solution must be symmetric:

$$\Psi(\boldsymbol{x}) \overset{r \to \infty}{\longrightarrow} e^{ikz} + e^{-ikz} + [f(\theta, \varphi) + f(\pi - \theta, \varphi + \pi)] \frac{e^{ikr}}{r} \ .$$

Calculating the current densities of the incoming and outgoing particles by using this symmetrized form, we find the cross section to be

$$\frac{d\sigma}{d\Omega^*} = |f(\theta, \varphi) + f(\pi - \theta, \varphi + \pi)|^2 \ . \tag{3.107}$$

Note the difference between (3.106) and (3.107): in (3.106), the individual cross sections are added, whereas in (3.107) the scattering amplitudes are summed.

Scattering of spin-1/2 fermions. For identical spin-1/2 fermions, for example, electrons, the total wave function must be antisymmetric. As long as there are no explicitly spin-dependent terms in the Hamilton operator, the wave function is composed of symmetrized and antisymmetrized spatial and spin wave functions:

$$\Psi^{(\pm)}(\boldsymbol{x}, m_1, m_2) = \left\{ e^{ikz} \pm e^{-ikz} + [f(\theta, \varphi) \pm f(\pi - \theta, \varphi + \pi)] \frac{e^{ikr}}{r} \right\}$$

$$\times \chi^{\binom{s}{A}}(m_1, m_2) \ , \quad m_1, m_2 = \pm \frac{1}{2} \ .$$

Here,

$$\chi^{(S)}(+,+) = \chi_1(+)\chi_2(+)$$

$$\chi^{(S)}(+,-) = \frac{1}{\sqrt{2}}\left(\chi_1(+)\chi_2(-) + \chi_1(-)\chi_2(+)\right) = \chi^{(S)}(-,+)$$

$$\chi^{(S)}(-,-) = \chi_1(-)\chi_2(-)$$

$$\chi^{(A)}(+,-) = \frac{1}{\sqrt{2}}\left(\chi_1(+)\chi_2(-) - \chi_1(-)\chi_2(+)\right) = -\chi^{(A)}(-,+)$$

denote the spinors corresponding to total spin $S = 1$ (triplet) and $S = 0$ (singlet). For the differential cross section this implies

$$\frac{d\sigma}{d\Omega^*} = \begin{cases} |f(\theta,\varphi) + f(\pi - \theta, \varphi + \pi)|^2 & \text{triplet state} \\ |f(\theta,\varphi) - f(\pi - \theta, \varphi + \pi)|^2 & \text{singlet state .} \end{cases}$$

If both particle beams are totally unpolarized, so that each of the four possible spin configurations has the same weight, the total cross section is given by the arithmetic mean of the cross sections of the three triplet states and of the one singlet state:

$$\frac{d\sigma}{d\Omega^*} = \frac{3}{4}|f(\theta,\varphi) + f(\pi - \theta, \varphi + \pi)|^2 + \frac{1}{4}|f(\theta,\varphi) - f(\pi - \theta, \varphi + \pi)|^2 .$$

Summary

- To describe quantum mechanical scattering processes, it is sufficient to consider the static solutions of the corresponding Schrödinger equation. The asymptotic wave function of particles scattered on a fixed target is composed of an incoming plane wave and a spherical wave moving away from the scattering center. The latter contains the **scattering amplitude**, which encompasses all information about the scattering process.

- The differential cross section is given by the probability current densities (in beam direction) of the incoming and the scattered parts and is equal to the absolute square value of the scattering amplitude.

- The calculation of the differential cross section for Coulomb scattering in the **Born approximation**, and also exactly, gives the familiar classical Rutherford scattering formula.

- For centrally symmetric scattering potentials the scattering process can be studied for each angular momentum l-sector separately by decomposing the scattering amplitude into Legendre polynomials (**partial wave expansion**). This expansion comprises **scattering phases**, which represent the phase shifts of the asymptotic wave functions relative to the solutions for free particles. The partial wave expansion leads to the **optical theorem**, according to which the total cross section is equal to the imaginary part of the scattering amplitude in the forward direction.

▷

- **Resonance scattering** occurs if the differential cross section is dominated by a certain partial wave at some particular energy. The **resonance width** is obtained from the **Breit-Wigner formula**. In general, scattering resonances result in meta-stable states with positive energy, which are formed in the attractive part of the effective potential screened by the centrifugal barrier.

- As in classical mechanics, the particle-particle scattering can be traced back to an effective one-particle scattering by separating the center of mass movement. The relations between the differential cross sections in the center of mass and in the laboratory system are the same as for classical scattering.

- If, in the case of distinguishable particles, the detector registers both types of particles, the differential cross sections for the scattering processes particle $1 \to (\theta, \varphi)$ and particle $2 \to (\pi - \theta, \varphi + \pi)$ and vice versa have to be added. In the case of identical particles, the corresponding scattering amplitudes must be summed or subtracted.

Applications

52. Scattering on a hard sphere. Consider the scattering on a hard sphere given by the potential

$$V(r) = \begin{cases} 0 \text{ for } r \le a \\ \infty \text{ for } r > a \ . \end{cases}$$

Find the behavior of the various scattering phases for small energies. What is the differential and the total cross section for the angular momentum $l=0$-sector?

Solution. The physical solution of the radial Schrödinger equation (3.38) for the outer area is

$$g_l(r) = A_l j_l(kr) + B_l n_l(kr) \ .$$

On the surface of the sphere, the wave function must vanish for reasons of continuity, so that

$$g_l(a) = 0 \implies \frac{B_l}{A_l} = -\frac{j_l(ka)}{n_l(ka)} \ .$$

The usual constraint that g_l reduces to the regular solution for free particles up to the scattering phase δ_l,

$$g_l(r) \stackrel{r \to \infty}{\longrightarrow} \frac{\sin(kr - l\pi/2 + \delta_l)}{kr} \ ,$$

yields the desired relation for δ_l:

$$A_l \sin(kr - l\pi/2) + B_l \cos(kr - l\pi/2) = C_l \sin(kr - l\pi/2 + \delta_l)$$
$$= C_l[\sin(kr - l\pi/2)\cos\delta_l$$
$$+ \cos(kr - l\pi/2)\sin\delta_l]$$

$$\implies \tan\delta_l = -\frac{j_l(ka)}{n_l(ka)} .$$

For small energies, we find

$$j_l(ka) \overset{k\to 0}{\approx} \frac{(ka)^l}{(2l+1)!!} , \quad n_l(ka) \overset{k\to 0}{\approx} \frac{(2l-1)!!}{(ka)^{l+1}} ,$$

and this leads to the characteristic behavior

$$\delta_l \overset{k\to 0}{\approx} \tan\delta_l \approx -\frac{(2l-1)!!}{(2l+1)!!}(ka)^{2l+1}$$

already mentioned in (3.105). It is consistent with our expectation that for small energies, higher l-sectors can be neglected. Using Theorem 3.30, we find for $l = 0$

$$\delta_0 = \arctan\left(-\frac{\sin ka}{\cos ka}\right) = -ka \implies \begin{cases} f_0(\theta) = -\dfrac{1}{k}e^{-ika}\sin ka \\[2mm] \dfrac{d\sigma_0}{d\Omega} = \dfrac{\sin^2 ka}{k^2} \\[2mm] \sigma_0 = \dfrac{4\pi}{k^2}\sin^2 ka . \end{cases}$$

53. Scattering on a spherically symmetric potential well. Consider the scattering of s-waves at a spherically symmetric potential well of the form

$$V(r) = \left\{ \begin{array}{ll} -V_0 & \text{for } r < a \\ 0 & \text{for } r \geq a \end{array} \right\} , \quad V_0 > 0 .$$

Show that for small particle energies, the scattering phase δ_0 is resonant for wave numbers

$$k_1 = \frac{\left(n + \frac{1}{2}\right)\pi}{a} , \quad n = 0, 1, \dots$$

within the inner area of the well. What are the corresponding resonance widths?

Solution. As shown in Subsection 3.5.4, the solutions of the radial Schrödinger equation for a spherically symmetric potential well and unbound states $(E > 0)$ are

$$u_l(r) = \left\{ \begin{array}{ll} A_l r j_l(k_1 r) , \ k_1 = \sqrt{\dfrac{2m(E + V_0)}{\hbar^2}} & \text{for } r < a \\[4mm] B_l[r j_l(k_2 r)\cos\delta_l + r n_l(k_2 r)\sin\delta_l] , \ k_2 = \sqrt{\dfrac{2mE}{\hbar^2}} & \text{for } r \geq a . \end{array} \right.$$

For large distances, this takes on the asymptotic form

$$u_l(r) \overset{r\to\infty}{\sim} \frac{\sin(k_2 r - l\pi/2)}{k_2} \cos\delta_l + \frac{\cos(k_2 r - l\pi/2)}{k_2} \sin\delta_l$$

$$= \frac{\sin(k_2 r + l\pi/2 + \delta_l)}{k_2},$$

so that δ_l can directly be identified with the scattering phases. For small particle energies, the scattering takes place mainly in s-states ($l = 0$), for which we already know the corresponding relation between energy and scattering phase [see (3.43)]:

$$\tan(k_2 a + \delta_0) = \frac{k_2}{k_1} \tan k_1 a$$

$$\Longrightarrow \delta_0 = \arctan\left(\frac{k_2}{k_1} \tan k_1 a\right) - k_2 a \overset{k_2\to 0}{\approx} \arctan\left(\frac{k_2}{k_1} \tan k_1 a\right).$$

From this, we see that the resonance energies are at

$$k_{1,n} = \frac{\left(n + \frac{1}{2}\right)\pi}{a} \quad \text{or} \quad E_n = \frac{\hbar^2 \left(n + \frac{1}{2}\right)^2 \pi^2}{2ma^2} - V_0 .$$

To determine the corresponding widths, we need the derivative of δ_0 at E_n:

$$\delta_0'(E_n) = \left.\frac{\left(\frac{K_2}{K_1}\right)' \tan(k_1 a) + \frac{K_2}{K_1} k_1' a \left(1 - \tan^2 k_1 a\right)}{1 - \left(\frac{K_2}{K_1}\right)^2 \tan^2 k_1 a}\right|_{k_{1,2}=k_{1,2}(E_n)}$$

$$= \frac{a k_1'(E_n) k_1(E_n)}{k_2(E_n)}$$

$$= \frac{ma}{\hbar^2 k_2(E_n)}$$

$$\Longrightarrow \frac{\Gamma_n}{2} = \frac{1}{\delta_0'(E_n)} = \frac{\hbar^2 k_2(E_n)}{ma} .$$

For $V_0 = \hbar^2\pi^2/(8ma^2)$, the lowest resonant s-state has a resonance energy of $E_0 = 0$. This *null-energy state* can be interpreted as a *pseudo meta-stable bound state* in a purely attractive potential and is closest to a true scattering resonance. If the potential well is lowered further, a second null-state occurs at $V_0 = 9\hbar^2\pi^2/(8ma^2)$, and so on. The succeeding states with larger resonance energy are *virtual states*, which have little to do with true scattering resonances.

4. Statistical Physics and Thermodynamics

The fundament for the theoretical description of macroscopic systems (gases, liquids, and solids) as we know it today was laid in the 18th century. Empirical laws found by Davy, Mayer, Joule, and many more since then are, still today, the basis of thermodynamics as first formulated in a consistent fashion by Lord Kelvin in 1850. Around this time, scientists became more and more convinced that matter has substructure in form of atoms and molecules, and they began with microscopic investigations of macroscopic systems. The development of quantum mechanics at the start of the 20th century finally delivered the adequate formalism for a realistic description of such systems.

Today, microscopic interactions between individual particles are understood, in principle, their description being given by classical and quantum mechanics. Nevertheless, without additional statistical assumptions, it is not possible to apply these laws on macroscopic systems. The reasons are manifold:

- To specify the state of a system, it would be necessary to determine on the order of 10^{23} degrees of freedom. This is not only unpractical but also theoretically impossible due to Heisenberg's uncertainty principle.

- Even if the initial state of such a large number of individual particles and all the equations of motion were known, it would still take today's computers an unacceptably long time to solve for the evolution of all particles.

- For large numbers of particles, qualitatively new phenomena arise, which necessitate a deeper understanding of the interactions than that provided purely by classical or quantum mechanics. An example is the abrupt condensation of a gas into a liquid state with qualitatively new properties.

From our everyday practical experience, we already know that it is not necessary to have a detailed microscopic understanding of the individual particles and their motion in time if all we aspire to is the macroscopic description of the system. If we consider, for example, a gas in a fixed volume, we know that in the absence of externally induced changes, the temperature and pressure of the gas will remain unchanged, although microscopically we expect the system to change continuously. This observation constitutes the starting point of statistical physics. Here, the picture of a single macroscopic system

evolving in time is replaced by another, where a very large number of ficti-
tious, macroscopically identical but microscopically different states coexist,
which are weighted with their respective occupation probabilities.

In this chapter we discuss statistical physics and thermodynamics. We
stress that we consider only *systems in equilibrium* that are characterized by
time-independent macroscopic properties. The time evolution of states from
nonequilibrium to equilibrium states is the topic of *kinetic theory* and will not
be touched upon. Normally, we will be looking at systems consisting of a large
number of particles ($N \approx 10^{23}$) contained within a large volume ($V \approx 10^{23}$
molecular volumes). We therefore often consider the *thermodynamic limit*,
which is defined by the following limiting process:

$$N \to \infty \,, \ V \to \infty \,, \ \frac{N}{V} = \text{const} \,.$$

The first section of this chapter addresses the fundamental ideas of statistical
physics. In particular, the probabilistic concept of a statistical ensemble is
introduced, and its temporal evolution with a view to the classification of
equilibrium systems is discussed.

The following two sections, 4.2 and 4.3, are concerned with three partic-
ular types of ensembles: the microcanonical, canonical, and grand canonical
ensembles. All of them describe equilibrium systems but with different, exter-
nally given boundary conditions. Within the context of the microcanonical
ensemble, we introduce the concept of entropy, which is of fundamental im-
portance in statistical physics as well as in thermodynamics. An important
result of these sections will be that all the ensembles yield equivalent descrip-
tions of macroscopic systems in the thermodynamic limit.

Next to the statistical approach another possible way of describing macro-
scopic systems and defining entropy is provided by information theory, which
describes a system from the point of view of its information content. This
will be discussed in Section 4.4. Entropy, in this context, is a measure of our
lack of knowledge of a system, and we will see that this is equivalent to the
statistical definition.

Section 4.5 deals with the phenomenological theory of thermodynamics.
Starting from the three thermodynamic laws, we discuss equilibrium and
stability conditions of open systems by introducing suitable thermodynamic
potentials as well as the description of state changes with the help of thermic
coefficients. Furthermore, we look at heat machines particularly with respect
to their general realizability.

In Section 4.6 we discuss classical Maxwell-Boltzmann statistics. Starting
from either the classical Hamilton function or the quantum mechanical Hamil-
ton operator, this type of statistics describes the behavior of many-particle
systems without taking into account the quantum nature of individual parti-
cles, i.e., their bosonic or fermionic character. We consider the "proper classi-
cal" limit, discuss the virial and equipartition theorems for "proper classical"

systems, and finish with a discussion of the harmonic oscillator system and ideal spin systems in different ensembles.

The last section of this chapter is reserved for quantum statistics, which yields the complete quantum mechanical description of statistical systems while also taking into account the quantum nature of particles. We compare Fermi-Dirac, Bose-Einstein, and Maxwell-Boltzmann statistics and derive the state equations of ideal Fermi and Bose gases. This will lead us to some interesting new effects that do not occur in Maxwell-Boltzmann statistics.

It should be emphasized once again that this chapter deals solely with equilibrium systems. For the treatment of nonequilibrium states (phase transitions, kinetic energy, etc.), we refer to the literature list given in the appendix.

Annotation. For convenience, in Chapter 4, we will generally use the notation

$$\left(\frac{\partial f}{\partial E}\right)_{V,N}$$

for the partial differentiation of a function f. This means that f – as a function of the variables E, V, and N – is differentiated with respect to E while V and N are held fixed. If the variables E, V, and N are transformed into new variables, say T, P, and μ, then we also use the convenient and unmathematical notation $f(T, P, \mu)$, although $f(T, P, \mu)$ describes a different functional relationship than $f(E, V, N)$.

4.1 Foundations of Statistical Physics

Statistical physics attempts to describe macroscopic systems by combining microscopically valid physical laws of interactions with a statistical viewpoint. This allows one to make predictions about the macroscopic behavior of such systems. Thus, statistical physics can be regarded as a fundamental theory that provides a deeper level of explanation for the purely phenomenologically motivated laws of thermodynamics.

In this section we discuss the basic ideas underlying all of statistical physics. The most important conceptual step consists of replacing the temporal evolution of a single microstate by a statistical description in which all possible microstates exist simultaneously and are weighted with their respective probabilities of occurrence. This *ensemble* propagates as a whole through the phase space and is described classically by a density and quantum mechanically by a *density operator*. Furthermore, we show which constraints these densities must obey, either classically or quantum mechanically, for them to describe equilibrium systems that are characterized by time-independent macroscopic properties.

4.1.1 States, Phase Space, Ensembles, and Probabilities

Within statistical physics, two types of states are used to describe physical systems, *microstates* and *macrostates*. A microstate encompasses detailed information about the movement of all the system's individual particles. Generally, the states of these particles are determined by the laws of quantum mechanics, and the overall system state can be described by a quantum state vector $|\psi(n_1, \ldots, n_f)\rangle$, where n_1, \ldots, n_f represents a system-specific set of f quantum numbers. This description can be called complete in that the temporal evolution of a state vector $|\psi\rangle$ known at a particular time is uniquely determined through the quantum mechanical equations of motion. If we enumerate all possible quantum states of the system in a suitable way ($r = 1, 2, 3, \ldots$), a specific microstate is uniquely identified by the index r:

$$\text{Quantum mechanical microstate: } r = (n_1, \ldots, n_f) \ . \tag{4.1}$$

Although quantum mechanics generally provides the adequate frame for the specification of a system, classical physics is often a good and useful approximation giving important insights into the dynamics of the system. Herein, the microstate of an N-particle system is given by the $3N$ generalized coordinates and $3N$ generalized momenta of all individual particles at any one time. The temporal evolution of the system is then also uniquely defined by the classical equations of motion.

$$\text{Classical microstate: } r = (q_1, \ldots, q_{3N}, p_1, \ldots, p_{3N}) \ . \tag{4.2}$$

A very important concept for the description of classical systems is that of the phase space, which is spanned by the $6N$ generalized coordinates and momenta. It implies that assigning the $6N$ particle values is equivalent to the specification of a single point in the phase space. Now, contrary to (4.1), there are continuous quantities on the right-hand side of (4.2). But for a statistical treatment of a system, it is necessary that the states r are countable. Therefore, it is reasonable to discretize the phase space into cells, such that each cell contains exactly one microstate. The size of the cells must be determined by comparison with quantum mechanics. Studying simple quantum systems (for example, the harmonic oscillator; see Application 54) shows that a quantum mechanical state in phase space corresponds to a phase space volume of size $(2\pi\hbar)^{3N} = h^{3N}$. We therefore think of phase space as subdivided into cells with

$$\text{cell size} = h^{3N} \ .$$

Numbering these cells ($r = 1, 2, \ldots$), we can now, as in quantum mechanics, define a microstate exactly by simply specifying its discrete index r.

The second way of describing the state of a system is provided by the macrostate. It is fully specified by the relevant macroscopic quantities, such as pressure and temperature. In principle, the macrostate of the system should

be uniquely determined by the microstate. However, as stated in the intro-
duction to this section, it is neither possible nor meaningful to attempt to
describe and to trace the microstate of a system containing something like
10^{23} particles. For the same reason, it is not possible to follow the process
of measuring within a microcanonical calculation since this implies averaging
over a suitable proportion of the temporal evolution of the microstate or of
the corresponding phase space trajectory.

Around 1900, Gibbs developed what is now known as *ensemble theory*. It
replaces the picture of a single temporally evolving microstate by a picture
in which all microstates a system can possibly evolve through exist all at
once at some fixed time, and constitute the initial situation for many sim-
ilar macrosystems. These fictitious systems constitute a *statistical ensemble*
whose elements propagate independently of one another and do not interact
among themselves. Classically, this corresponds to the coexistence of all ac-
cessible phase space points assembled into one phase space trajectory. The
ensemble concept is the basis of statistical physics. It is founded on the fol-
lowing two assumptions.

The first assumption is the *equivalence of time and ensemble average of a
macroscopic system in statistical equilibrium*. Here, "statistical equilibrium"
is taken to be a state in which the measurement of macroscopic quantities such
as pressure and temperature is time-independent. To see what this means,
consider a gas enclosed in a fixed volume. If this system is heated from the
outside, we intuitively expect the system not to be in an equilibrium state
immediately after it has been heated. However, after some system charac-
teristic time, the *relaxation time*, the system will have returned to a state
of equilibrium at its original temperature. In phase space terminology, the
equivalence of time and ensemble average implies that the phase space trajec-
tory of a system contains all phase space points that are in accordance with
the system-specific physics or, put differently, that each possible microstate
will have been reached after sufficiently long time (*ergodic hypothesis*).[1] In
contrast, the term "statistical equilibrium" implies that the phase space tra-
jectory (or phase space density) is invariant with time; see Fig. 4.1.

Having familiarized ourselves with the ensemble concept, it should now be
plausible that *none of the states (systems) within an ensemble is favored over
the others*. This is the second fundamental postulate of statistical physics.

Theorem 4.1: Fundamental postulates of statistical physics

1. postulate: in an isolated system in equilibrium, time and ensemble
average are equivalent.

\triangleright

[1] It is known that nonergodic systems exist; we shall not be studying them here.

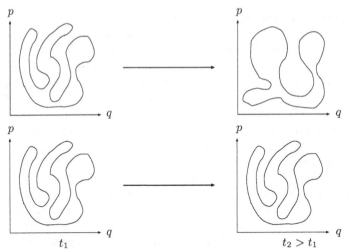

Fig. 4.1. Temporal evolution of the phase space trajectory of a system in disequilibrium (*above*) and in equilibrium (*below*).

> **2. postulate:** the probability of finding a randomly picked element of the ensemble in a particular microstate is proportional to the overall number of ensemble elements in exactly this microstate. Put differently, all microstates are equally probable. This is called the *postulate of a priori equal probabilities*.

We emphasize that no deeper theoretical explanation is available for these postulates. They are justified a posteriori by checking for consistency between experiment and predictions following from the postulates. The two postulates express the probabilistic nature of statistical physics: the detailed structure of particular microstates are not of interest, but only the number of different possible microstates within the statistical ensemble.

Generally, for most problems, some properties of a physical system are given. For example, the total energy may be held fixed. In this case, the system can only be in microstates consistent with this total energy (accessible microstates), and the statistical ensemble consists just of all macrosystems with this energy. Thus, the macrostate of a system is uniquely determined by the relative probabilities P_r of all accessible microstates:

Macrostate: $\{P_r\} = (P_1, P_2, \ldots)$.

The above two hypotheses can therefore be summarized as follows: *macroscopic quantities of a system in equilibrium are determined by the ensemble averages of all accessible microstates, where each microstate is weighted with its relative probability.*

Theorem 4.2: Ensemble average

Let P_r be the probability of finding an equilibrium system in a microstate for which the macroscopic quantity A has the stationary value A_r. The ensemble average \overline{A} of A is then

$$\overline{A} = \sum_r P_r A_r \ , \tag{4.3}$$

where the probabilities P_r are normalized as

$$\sum_r P_r = 1 \ , \ 0 \le P_r \le 1 \ .$$

The sum in (4.3) runs over all accessible microstates, where the term "accessible" depends on the choice of A. It also requires a very large number M of equivalent systems, of which M_r are in the microstate r:

$$P_r = \lim_{M \text{ sufficiently large}} \frac{M_r}{M} \ .$$

Furthermore, we see from (4.3) that only probability distributions satisfying $\mathrm{d}P_r/\mathrm{d}t = 0$ lead to time-independent ensemble averages.

4.1.2 Classical Statistical Physics: Probability Density

The goal of statistical physics is to determine the number of ensemble members in different microstates and therefore in different regions of phase space. If we assume that the number M of elements in the ensemble is very large, we can describe the system by a *probability density* ρ that determines how the M elements are distributed in phase space. If

$$\mathrm{d}\Gamma = \mathrm{d}q_1 \cdots \mathrm{d}q_{3N} \mathrm{d}p_1 \cdots \mathrm{d}p_{3N}$$

is a volume element of this phase space, it contains exactly

$$\mathrm{d}M = \frac{1}{h^{3N}} \rho(q, p, t) \mathrm{d}\Gamma$$

elements. The probability that a randomly picked point in phase space falls within the volume element $\mathrm{d}\Gamma$ is given by

$$\frac{\mathrm{d}M}{M} = \frac{1}{h^{3N}} \frac{\rho \mathrm{d}\Gamma}{M} \ .$$

The total number of ensemble elements is, of course,

$$M = \int \mathrm{d}M = \frac{1}{h^{3N}} \int \mathrm{d}\Gamma \rho(q, p, t) \ .$$

In what follows we will always assume that the phase space density is normalized to the total number M of ensemble elements. The ensemble average of a classical observable $A(q, p)$ is then given by

$$\overline{A} = \frac{1}{h^{3N}} \int d\Gamma \rho(q,p,t) A(q,p) \ . \tag{4.4}$$

If we assume further that A is not explicitly time-dependent, as is the case of isolated systems, (4.4) shows again that only densities with $\partial \rho / \partial t = 0$ yield stationary ensemble averages.

4.1.3 Quantum Statistics: Density Operator

In quantum mechanics, observables are represented by Hermitean operators \boldsymbol{A}, and measurements of \boldsymbol{A} lead to expectation values

$$\langle \boldsymbol{A} \rangle = \langle \psi | \, \boldsymbol{A} \, | \psi \rangle \ . \tag{4.5}$$

A quantum mechanical state is prepared by measuring a complete set of commuting observables. These *pure states* are described by Hilbert vectors in a complex Hilbert space. However, generally, pure states are an idealization. Particularly for large systems, it is impossible to perform a complete preparation. The more general case is that of a *mixed state*, for which the set of measured observables is not complete. In this case the state is no longer describable by a Hilbert vector. It is then useful to consider an ensemble of pure states $| \psi_r \rangle$ and their relative probabilities P_r. According to (4.3) and (4.5), the quantum mechanical ensemble average is[2]

$$\overline{A} = \sum_r P_r \langle \psi_r | \, \boldsymbol{A} \, | \psi_r \rangle \ , $$

where the two averaging procedures in this equation are fundamentally different. The quantum mechanical average leads to the well-known interference phenomena within quantum theory. By contrast, the ensemble average is an average over incoherent states $| \psi_r \rangle$, so that no interference effects can arise (see Application 55). Now, if we suppose that the states $| \psi_r \rangle$ are normalized (but not necessarily orthogonal), then \overline{A} can be expanded in a complete orthonormal basis $| u_i \rangle$ with $\sum_i | u_i \rangle \langle u_i | = 1$ as follows:

$$\begin{aligned}
\overline{A} &= \sum_{r,i,j} \langle \psi_r | u_j \rangle \langle u_j | \, \boldsymbol{A} \, | u_i \rangle \langle u_i | \psi_r \rangle \, P_r \\
&= \sum_{i,j} \left\{ \sum_r P_r \langle u_i | \psi_r \rangle \langle \psi_r | u_j \rangle \right\} \langle u_j | \, \boldsymbol{A} \, | u_i \rangle \\
&= \sum_{i,j} \rho_{ij} A_{ji} = \mathrm{tr}(\boldsymbol{\rho A}) \ .
\end{aligned}$$

The quantity

[2] We denote by \overline{A} the combined average of quantum mechanical expectation value and ensemble averaging.

$$\rho = \sum_r |\psi_r\rangle\, P_r \,\langle\psi_r|$$

is called the *density operator* and

$$A_{ji} = \langle u_j|\, A\, |u_i\rangle \ , \quad \rho_{ij} = \sum_r P_r \,\langle u_i|\,\psi_r\rangle\,\langle\psi_r|\,u_j\rangle = \langle u_i|\,\rho\,|u_j\rangle$$

are the matrix elements of A and ρ in the u-basis. The following properties of ρ can be noted:

- ρ is Hermitean.

- If $W(\alpha, \psi_r)$ and $W(\alpha, \rho)$ are the probabilities of finding the eigenvalue α when performing a measurement of an observable A on a system in a pure state $|\psi_r\rangle$ or in a mixed state characterized by ρ, it follows that [compare to (3.5)]

$$W(\alpha, \rho) = \sum_r P_r W(\alpha, \psi_r) = \sum_r P_r \,\langle\psi_r|\,\mathcal{P}_\alpha\,|\psi_r\rangle = \mathrm{tr}(\rho\mathcal{P}_\alpha) \ ,$$

 where \mathcal{P}_α is the projection operator onto the eigenspace of A with eigenvalue α.

- ρ is normalized:

$$\mathrm{tr}(\rho) = \sum_r P_r = 1 \ .$$

- ρ is positive definite since for any $|v\rangle$, we have

$$\langle v|\,\rho\,|v\rangle = \sum_r \langle v|\,\psi_r\rangle\, P_r \,\langle\psi_r|\,v\rangle = \sum_r P_r |\langle v|\,\psi_r\rangle|^2 \geq 0 \ .$$

Therefore, ρ is a positive definite Hermitean operator whose complete discrete eigenspectrum lies between 0 and 1.

Theorem 4.3: Density operator

A statistical quantum mechanical state is completely described by the Hermitean density operator

$$\rho = \sum_r |\psi_r\rangle\, P_r \,\langle\psi_r| \ , \quad \mathrm{tr}(\rho) = 1 \ .$$

The sum is to be taken over all pure states $|\psi_r\rangle$. For a measurement of the observable A, the probability of finding the eigenvalue α is given by

$$W(\alpha, \rho) = \mathrm{tr}(\rho\mathcal{P}_\alpha) \ .$$

The expectation value of A is the ensemble average

$$\overline{A} = \mathrm{tr}(\rho A) \ .$$

The above formalism can also be used in the case of pure states. If it is known with certainty that the system is in a pure state $|\psi\rangle$, the operator ρ reduces to the projection operator

$$\rho = \boldsymbol{P}_\psi = |\psi\rangle \langle \psi| \ ,$$

and we obtain

$$\rho^2 = \rho \Longrightarrow \mathrm{tr}\left(\rho^2\right) = 1 \ .$$

If, on the other hand, $\rho^2 = \rho$, we find

$$\rho^2 = \sum_{r,m} |\psi_r\rangle \, P_r \langle \psi_r| \psi_m\rangle \, P_m \langle \psi_m| = \sum_r P_r^2 |\psi_r\rangle \langle \psi_r|$$

$$= \sum_r P_r |\psi_r\rangle \langle \psi_r| = \rho \ .$$

Since the probabilities P_r are normalized, the condition $P_r^2 = P_r$ for all r can only be satisfied if one of the P_r is identical 1 and all others vanish. Therefore, $\rho^2 = \rho$ or $\mathrm{tr}\left(\rho^2\right) = 1$ is a necessary and sufficient criterion for ρ to describe a pure state.

Quantum mechanical measurement of a statistical system. In Subsection 3.2.3 we discussed the process of a measurement on pure states. Performing a measurement of an observable \boldsymbol{A} on a system in a pure and normalized state $|\psi\rangle$ will collapse the system with probability $\langle \psi| \boldsymbol{P}_\alpha |\psi\rangle$ into the normalized state $\dfrac{\boldsymbol{P}_\alpha|\psi\rangle}{\sqrt{\langle \psi| \boldsymbol{P}_\alpha|\psi\rangle}}$. For the mixed case, the density operator after the measurement is therefore given by

$$\rho_r' = \sum_\alpha \frac{\boldsymbol{P}_\alpha |\psi\rangle}{\sqrt{\langle \psi| \boldsymbol{P}_\alpha |\psi\rangle}} \, \langle \psi| \boldsymbol{P}_\alpha |\psi\rangle \, \frac{\langle \psi| \boldsymbol{P}_\alpha}{\sqrt{\langle \psi| \boldsymbol{P}_\alpha |\psi\rangle}} \ ,$$

with probability P_r. Thus, the complete statistical operator is

$$\rho' = \sum_r P_r \rho_r' = \sum_\alpha \boldsymbol{P}_\alpha \rho \boldsymbol{P}_\alpha \ .$$

Theorem 4.4: Quantum mechanical measurement and density operator

Let ρ be the density operator of a mixed state prior to a measurement of an observable \boldsymbol{A}. Then, immediately after the measurement, the system is described by the density operator

$$\rho' = \sum_\alpha \boldsymbol{P}_\alpha \rho \boldsymbol{P}_\alpha \ .$$

The sum is to be taken over all eigenvalues α of \boldsymbol{A}.

4.1.4 Temporal Evolution of an Ensemble

We have already seen, classically and quantum mechanically, that we need $\partial\rho/\partial t = 0$ to hold for the ensemble average to be time-independent, as we expect it to be for systems in equilibrium. In this subsection we investigate more generally the temporal behavior of the classical probability density and of the quantum mechanical density operator. For these we derive equations of motion that will help us to specify more precisely the condition of stationarity.

Classical picture: Liouville equation. In the ensemble interpretation a system is represented by a statistical ensemble. The members of this ensemble are identical imaginary macroscopic systems. To each of these corresponds a phase space point that propagates independently in time through phase space. Now consider a fixed volume element $d\Gamma$ of this phase space. The number of elements it contains will generally vary with time since the coordinates and momenta of the ensemble members also vary in accordance with Hamilton's equations

$$\frac{\partial H}{\partial p_i} = \dot{q}_i , \quad \frac{\partial H}{\partial q_i} = -\dot{p}_i . \tag{4.6}$$

However, since points in phase space cannot be created or annihilated,[3] we can write down a continuity equation, which states that the rate of change of the density within the volume element $d\Gamma$ is proportional to the flux of points through the surface $d\omega$ of this volume element (see Fig. 4.2):

$$\frac{\partial}{\partial t} \int_{\Gamma} d\Gamma \rho = - \oint_{\omega} d\omega \rho(\boldsymbol{vn}) .$$

Here, \boldsymbol{v} denotes the velocity of the phase space points, and \boldsymbol{n} is a normal vector pointing outward from the surface element $d\omega$. Using (4.6) and the

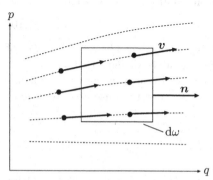

Fig. 4.2. Temporal flux of phase space points through a phase space volume element.

[3] During their movement, two different points in phase space can never overlap since Hamilton's equations have unique solutions.

generalized Gauss theorem, it follows that

$$\frac{\partial \rho}{\partial t} + \boldsymbol{\nabla}(\rho \boldsymbol{v}) = 0 \; , \tag{4.7}$$

with

$$
\begin{aligned}
\boldsymbol{\nabla}(\rho \boldsymbol{v}) &= \sum_i \left[\frac{\partial}{\partial q_i}(\rho \dot{q}_i) + \frac{\partial}{\partial p_i}(\rho \dot{p}_i) \right] \\
&= \sum_i \left[\frac{\partial \rho}{\partial q_i} \frac{\partial H}{\partial p_i} - \frac{\partial \rho}{\partial p_i} \frac{\partial H}{\partial q_i} + \rho \left(\frac{\partial^2 H}{\partial q_i \partial p_i} - \frac{\partial^2 H}{\partial p_i \partial q_i} \right) \right] \\
&= \{\rho, H\} \; .
\end{aligned}
\tag{4.8}
$$

For the total temporal derivative of ρ, we find[4]

$$
\begin{aligned}
\frac{\mathrm{d}\rho}{\mathrm{d}t} &= \frac{\partial \rho}{\partial t} + \sum_i \left[\frac{\partial \rho}{\partial q_i} \frac{\partial H}{\partial p_i} - \frac{\partial \rho}{\partial p_i} \frac{\partial H}{\partial q_i} \right] \\
&= \frac{\partial \rho}{\partial t} + \{\rho, H\} \; .
\end{aligned}
\tag{4.9}
$$

Combining (4.7), (4.8), and (4.9), we finally obtain

Theorem 4.5: Liouville's equation

The total temporal derivative of the phase space density vanishes:

$$\frac{\mathrm{d}\rho}{\mathrm{d}t} = \frac{\partial \rho}{\partial t} + \{\rho, H\} = 0 \; .$$

This theorem implies that for an observer travelling on a phase space element of an ensemble, the phase space density is time-invariant. In other words: the ensemble moves through phase space like an incompressible fluid. For a stationary ensemble, $\partial \rho / \partial t = 0$, Liouville's theorem implies the condition

$$\{\rho, H\} = 0 \; . \tag{4.10}$$

Thus, a stationary ensemble is a constant of motion. The simplest way of satisfying this condition is to choose a phase space density that is constant in some subspace Ω of Γ and zero elsewhere. In such an ensemble all microstates are uniformly distributed over Ω, so that the ensemble average of an observable A is given by

$$\overline{A} = \frac{1}{\Omega h^{3N}} \int_{\Omega} A(p, q) \mathrm{d}\Gamma \; , \quad \Omega = \frac{1}{h^{3N}} \int_{\Omega} \mathrm{d}\Gamma \; .$$

A slightly less restrictive possibility to fulfill the condition (4.10) is a phase space density that explicitly depends on q and p via a constant of motion $h(q, p)$:

[4] See Theorem 1.25 and its discussion.

$$\rho(q, p) = \rho[h(q, p)] \ .$$

In this case, we have

$$\frac{\partial \rho}{\partial q_i} = \frac{\partial \rho}{\partial h} \frac{\partial h}{\partial q_i} = 0 \ , \quad \frac{\partial \rho}{\partial p_i} = \frac{\partial \rho}{\partial h} \frac{\partial h}{\partial p_i} = 0 \Longrightarrow \{\rho, H\} = 0 \ .$$

From mechanics, we know that there are seven constants of motion, not counting the center of mass motion. These are energy, momentum, and angular momentum (see Theorem 1.11). For very large systems, we can assume that it is always possible to find a coordinate transformation that results in zero momentum and angular momentum. In the following, we shall therefore restrict ourselves to stationary ensembles that are functions of the generalized coordinates via Hamilton's function $H(q, p) = E$, only.

Quantum mechanical picture: Von Neumann equation. In order to find the quantum mechanical analogue of Liouville's equation, we work in the Schrödinger representation (see Subsection 3.2.4) and, with the help of

$$\boldsymbol{H} \, |\psi\rangle = \mathrm{i}\hbar \frac{\mathrm{d}}{\mathrm{d}t} \, |\psi\rangle \ , \quad \langle\psi| \, \boldsymbol{H} = -\mathrm{i}\hbar \frac{\mathrm{d}}{\mathrm{d}t} \, \langle\psi|$$

and $\mathrm{d}P_r/\mathrm{d}t = 0$, calculate as follows:

$$\begin{aligned}
\mathrm{i}\hbar \frac{\mathrm{d}\boldsymbol{\rho}}{\mathrm{d}t} &= \mathrm{i}\hbar \frac{\mathrm{d}}{\mathrm{d}t} \sum_r |\psi_r\rangle \, P_r \, \langle\psi_r| \\
&= \sum_r (\boldsymbol{H} \, |\psi_r\rangle \, P_r \, \langle\psi_r| - |\psi_r\rangle \, P_r \, \langle\psi_r| \, \boldsymbol{H}) = [\boldsymbol{H}, \boldsymbol{\rho}] \ .
\end{aligned}$$

Theorem 4.6: Von Neumann equation

The temporal evolution of the density operator is given by

$$\frac{\mathrm{d}\boldsymbol{\rho}}{\mathrm{d}t} = -\frac{\mathrm{i}}{\hbar} [\boldsymbol{H}, \boldsymbol{\rho}] \ .$$

The solution of this generalized Schrödinger equation is

$$\boldsymbol{\rho}(t) = \mathrm{e}^{-\mathrm{i}Ht/\hbar} \boldsymbol{\rho}(0) \mathrm{e}^{\mathrm{i}Ht/\hbar} \ .$$

In analogy to the classical case, the density matrix is stationary if

$$\mathrm{i}\hbar \frac{\mathrm{d}\boldsymbol{\rho}}{\mathrm{d}t} = [\boldsymbol{H}, \boldsymbol{\rho}] = \boldsymbol{0} \ . \tag{4.11}$$

From this, the time dependency of operators follows as

$$\mathrm{i}\hbar \frac{\mathrm{d}}{\mathrm{d}t} \overline{\boldsymbol{A}} = \mathrm{i}\hbar \frac{\mathrm{d}}{\mathrm{d}t} \mathrm{tr}(\boldsymbol{\rho}\boldsymbol{A}) = \mathrm{tr} \left([\boldsymbol{H}, \boldsymbol{\rho}] \boldsymbol{A} + \mathrm{i}\hbar\boldsymbol{\rho} \frac{\partial \boldsymbol{A}}{\partial t} \right) = \mathrm{i}\hbar \overline{\frac{\partial \boldsymbol{A}}{\partial t}} \ .$$

As expected, given stationarity, measurements of not explicitly time-dependent observables yield time-independent results. In the following, we will

honor the condition of stationarity (4.11) by considering density operators of the type

$$\rho = \rho(\boldsymbol{H}, \boldsymbol{x}) \ , \tag{4.12}$$

where \boldsymbol{x} is a set of time-independent observables (e.g., particle number operator \boldsymbol{N}) that commute with \boldsymbol{H}.

From our above considerations, we can already obtain some quite insightful statements about the general form of the density operator and the classical probability density: suppose we have two subsystems with independent distributions ρ_1 and ρ_2. For the combined system, we must have

$$\rho_{12} = \rho_1 \rho_2 \Longrightarrow \ln \rho_{12} = \ln \rho_1 + \ln \rho_2 \ . \tag{4.13}$$

Thus, the logarithm of the density operator is a linear function of the additive conserved quantities. Next, suppose that the particle number N and the energy E of both systems are given and fixed. This implies that the density operators of the subsystems are constant, and their microstates are uniformly distributed (*microcanonical ensemble*). If the subsystems can exchange energy between one another (*canonical ensemble*), then, due to (4.13), we must have

$$\rho_{1,2} \sim \mathrm{e}^{-\beta_{1,2} \boldsymbol{H}} \ .$$

If both energy and particles can be exchanged (*grand canonical ensemble*), then (4.13) implies that

$$\rho_{1,2} \sim \mathrm{e}^{-\beta_{1,2} \boldsymbol{H} + \alpha_{1,2} \boldsymbol{N}} \ .$$

The signs of $\alpha_{1,2}$ and $\beta_{1,2}$ are arbitrary. Overall, we see that the general form of the density matrices can be inferred directly from the stationarity condition (4.12) and the assumption of (4.13). In the following two sections we will derive the three above ensembles in a different manner.

Special role of energy. It is apparently very easy to calculate the density operator in the energy eigenbasis $\{|\psi_r\rangle\}$ since, for this case, the density matrix is diagonal if, as assumed, ρ and \boldsymbol{H} commute:

$$[\boldsymbol{H}, \rho] = 0 \Longrightarrow \rho_{mn} = \langle \psi_n | \rho(\boldsymbol{H}) | \psi_m \rangle = \rho(E_n) \delta_{nm} \ .$$

We shall practically always be working in this basis. However, it should be clear that all results are basis-independent.

Summary

- By using the concept of a **statistical ensemble**, statistical physics bypasses the practically impossible description of **microstates** of a macroscopic system. The statistical ensemble consists of a large number of identical (fictitious) macrostates based on different microstates. Therefore, the original problem is reduced to determining the number of possible microstates that yield a particular macrostate.

▷

- The macrostate of a system is defined by a set of relative **occupation probabilities** of the different microstates of the ensemble. This is the **ensemble interpretation** of statistical physics.

- The **time average** and **ensemble average** of a system in **statistical equilibrium (stationary system)** are identical.

- Statistical ensembles are described using a **probability density** (classically) or a **density operator** (quantum mechanically).

- For stationary systems, the Poisson bracket of the probability density and the Hamilton function (classically) or the commutator of the density operator and the Hamilton operator (quantum mechanically) vanishes.

Applications

54. Classical phase space of the harmonic oscillator. Determine the phase space trajectory of a classical one-dimensional harmonic oscillator, as well as the phase space volume enclosed by $E - \delta E$ and E. How large is this volume if δE is the distance of two successive energy levels of the quantum mechanical oscillator?

Solution. The classical Hamilton function of this problem is given by

$$H(q,p) = \frac{k}{2}q^2 + \frac{p^2}{2m} \ .$$

The corresponding Hamilton equations are solved by

$$q(t) = A\cos(\omega t + \varphi) \ , \quad p(t) = -m\omega A\sin(\omega t + \varphi) \ , \quad \omega = \sqrt{\frac{k}{m}} \ .$$

Using

$$H(q,p) = E = \frac{m\omega^2 A^2}{2} = \text{const} \ ,$$

we can rewrite this into a phase space trajectory:

$$(q,p) = \left(\sqrt{\frac{2E}{m\omega^2}}\cos(\omega t + \varphi), -\sqrt{2mE}\sin(\omega t + \varphi) \right)$$

$$\Longrightarrow \frac{q^2}{(2E/m\omega^2)} + \frac{p^2}{2mE} = 1 \ .$$

This expression corresponds to an ellipse with an area of $2\pi E/\omega$. Note that after one period $T = 2\pi/\omega$, each point of the ellipse has been traversed, so that the ergodic hypothesis is exactly fulfilled. Now we assume that the energy

of the oscillator is somewhat blurred and confined to the area $[E - \delta E : E]$. The phase space trajectory is then confined to the volume

$$\int\limits_{E-\delta E \leq H \leq E} \mathrm{d}q\mathrm{d}p = \int\limits_{E-\delta E \leq H \leq E} \mathrm{d}\Gamma = \frac{2\pi\delta E}{\omega} \ .$$

Using the quantum mechanical energy eigenvalues

$$E_n = \hbar\omega \left(n + \frac{1}{2} \right) \ ,$$

the number of eigenstates within the allowed energy interval and for large energies is practically identical to $\delta E / \hbar\omega$. In this case, the phase space volume of an energy eigenstate is

$$\nu = \frac{(2\pi\delta E/\omega)}{(\delta E/\hbar\omega)} = 2\pi\hbar = h \ .$$

If we have a system of N harmonic oscillators, this becomes $\nu = h^N$. When counting the microstates within a phase space volume element, we therefore need to divide this volume by ν.

55. Postulate of a priori random phases. The quantum mechanical expectation value of an observable A in a basis $\{ | \psi_r \rangle \}$ is formally obtained from

$$\overline{A} = \sum_r P_r \langle \psi_r | A | \psi_r \rangle \ , \tag{4.14}$$

where P_r is the occupation probability of the pure state $| \psi_r \rangle$. Now consider a system held at constant energy. Then, in (4.14), only those states will contribute that are consistent with this energy. If the kets $| \psi_r \rangle$ are the energy eigenstates of the system, then, following the second statistical postulate, the probabilities P_r are all equal: $P_r = P = \text{const}$. Using this, derive the *postulate of a priori random phases* by transforming from the energy eigenbasis to another basis (of the same Hilbert space).

Solution. We assume that the energy eigenbasis $\{ | \psi_r \rangle \}$ can be expanded in a basis $\{ | n \rangle \}$ as

$$| \psi_r \rangle = \sum_n | n \rangle \, a_{nr} \ , \ a_{nr} = \langle n | \psi_r \rangle \ .$$

For the average value of A in a mixed state, we then have

$$\overline{A} = \sum_{r,n,m} P \langle \psi_r | n \rangle \langle n | A | m \rangle \langle m | \psi_r \rangle$$

$$= \sum_{r,n} P a_{nr}^* a_{nr} \langle n | A | n \rangle + \sum_{r,n,m} P a_{nr}^* a_{mr} \langle n | A | m \rangle (1 - \delta_{nm}) \ .$$

For $n \neq m$, comparison with (4.14) leads to the condition

$$\sum_r a_{nr}^* a_{mr} = 0 \iff \sum_r |a_{nr}||a_{mr}|e^{i(\phi_{mr}-\phi_{nr})} = 0 \ ,$$

which can only be satisfied if it is postulated that the phases ϕ are randomly distributed over the ensemble:

$$\overline{a_{nr}^* a_{mr}} = 0 \ , \ n \neq m \ .$$

This postulate of random phases implies that the states $|\psi_r\rangle$ must be incoherent superpositions, so that correlations between the individual ensemble members are excluded. For mixed states, this condition must be added to the postulate of a priori equal probabilities.

4.2 Ensemble Theory I: Microcanonical Ensemble and Entropy

In the preceding section we established that a statistical ensemble in equilibrium is described by a probability density or a density operator that depends only on conserved quantities, such as the energy or the total number of particles. Three such ensembles commonly used in statistical physics are

- *Microcanonical ensemble:* it describes an isolated system in equilibrium at constant energy E.

- *Canonical ensemble:* the system may exchange energy with its environment, while its temperature is held fixed.

- *Grand canonical ensemble:* temperature and *chemical potential* are held fixed, and the system may exchange energy and particles with its environment.

In this section we consider the microcanonical ensemble; canonical and grand canonical ensemble are discussed in the next section. We also introduce *entropy*, which is one of the most important concepts of statistical physics. It is closely related to the *microcanonical partition function*.[5] A discussion of the *principle of maximal entropy* for equilibrium states follows. This principle will finally lead us to the formulation of equilibrium conditions by introducing the concepts of *temperature* and *generalized forces*.

4.2.1 Microcanonical Ensemble

The microcanonical ensemble describes a closed system at constant energy E. The ensemble elements consist of a very large number of microstates, all of which have energy E and which are consistent with a possibly given set of external parameters, such as volume V or total number of particles N. We

[5] Sometimes the partition function is also called "sum-over-states", reflecting its German origin "Zustandssumme".

denote the complete set of parameters by $x = (V, N, \ldots)$. Now, the second statistical postulate states that all microstates are equally likely to occur and therefore the probability $P_r(E, x)$ of finding the macrostate in a particular microstate r is constant and independent of r:

$$P_r(E, x) = \begin{cases} \dfrac{1}{\Omega(E, x)} & \text{for all accessible microstates with } E_r = E \\ 0 & \text{else .} \end{cases}$$

Ω is the *microcanonical partition function*. It equals the total number of accessible microstates ("sum-over-states"), i.e.,

$$\Omega = \Omega(E, x) = \sum_{r:E_r(x)=E} 1 . \tag{4.15}$$

Since the energy E can only ever be measured with a finite precision δE (for theoretical and practical reasons), it is reasonable to rewrite (4.15) as

$$\Omega(E, x) = \sum_{r:E-\delta E \leq E_r(x) \leq E} 1 .$$

However, for this to make sense, we still have to show that, in the thermo-dynamical limit, the uncertainty δE can be chosen such that the partition function Ω is independent thereof. We define

Definition: Microcanonical ensemble

The microcanonical ensemble determines the equilibrium state of a closed system at given energy E and possibly given external parameters x:

$$P_r(E, x) = \frac{1}{\Omega(E, x)} , \quad \sum_r P_r(E, x) = 1$$

$$\Omega(E, x) = \begin{cases} \displaystyle\sum_{r:E-\delta E \leq E_r(x) \leq E} 1 & \text{(quantum mechanically)} \\[2mm] \dfrac{1}{h^{3N} N!} \displaystyle\int_{E-\delta E \leq H(q,p,x) \leq E} d\Gamma & \text{(classically) .} \end{cases}$$

To allow the inclusion of identical particles in the above, we have attached an additional factor $1/N!$ to the classical partition function. From the quantum mechanical viewpoint, this factor is necessary since arbitrary permutations of identical (and therefore indistinguishable) particles do not yield new ensemble members.[6]

In the quantum mechanical density operator formalism, the microcanonical probabilities P_r correspond to the degenerate energy eigenstates $|\psi_r\rangle$

[6] If more than one particle type is present, the factor $1/N!$ must be replaced by $1/\left(\prod_i N_i!\right)$ with $\sum_i N_i = N$. In Application 57 we show that omission of the factor $1/N!$ results in contradictions within thermodynamical relations (Gibbs paradox).

of H with energy E. According to Theorem 4.3, the microcanonical density operator is therefore given by

$$\rho = \frac{1}{\Omega} \sum_r |\psi_r\rangle \langle \psi_r| = \begin{cases} \dfrac{1}{\Omega} & \text{for all accessible microstates} \\ 0 & \text{else} \end{cases}$$

and is diagonal in any representation.

Density of microcanonical states. Without taking into account any specifics, we shall derive a rough estimate of how the microcanonical partition function depends on energy. Let us define, to this end, the following quantities:

> **Definition: Phase space volume ω and phase space density g**
>
> $$\omega(E) = \sum_{r: E_r \leq E} 1 \;\;\overset{\text{classically}}{\longrightarrow}\;\; \frac{1}{h^{3N} N!} \int_{H(q,p)\leq E} d\Gamma$$
>
> $$g(E) = \frac{\partial \omega(E)}{\partial E} = \lim_{\delta E \to 0} \frac{\Omega(E)}{\delta E} . \tag{4.16}$$

According to this, the phase space volume determines the number of microstates with energy less than or equal to E, whereas the phase space density gives the number of states per unit energy interval.

If the energy E of the system is not too small, it is plausible that the energy is distributed evenly over the f degrees of freedom of the system, so that each degree of freedom corresponds to the same amount of energy $\epsilon = E/f$. Furthermore, we can assume that the phase space volume $\omega_1(\epsilon)$ of any degree of freedom, i.e., the possible values of a degree of freedom contributing the energy amount ϵ or less to the total system, is approximately proportional to ϵ:

$$\omega_1(\epsilon) \sim \epsilon^\alpha , \quad \alpha \approx 1 .$$

For the total phase space volume, this gives

$$\omega(E) \sim [\omega_1(\epsilon)]^f = \epsilon^f , \quad \epsilon = \frac{E}{f} .$$

Therefore, the number of states in the energy interval $[E - \delta E : E]$ reads

$$\Omega(E) = \omega(E) - \omega(E - \delta E) = \frac{\partial \omega}{\partial E} \delta E \sim f \omega_1^{f-1} \frac{\partial \omega_1}{\partial E} \delta E . \tag{4.17}$$

For macroscopic systems, f and with it also the exponent of this equation is of the order of 10^{23}. This means that Ω grows extremely fast with the energy of the system. We shall see that this is in fact a general property of partition functions. Taking the logarithm of (4.17), it follows that

$$\ln \Omega \approx (f-1)\ln \omega_1 + \ln \left(f \frac{\partial \omega_1}{\partial E} \delta E \right) .$$

The two terms on the right-hand side are of the order of f and of $\ln f$, so that, for large f, the second term is negligible compared to the first. Therefore, we find

$$\ln \Omega(E) \approx \ln \omega(E) \approx \ln g(E) \approx f \ln \omega_1(\epsilon) \approx \mathcal{O}(f) .$$

To summarize

**Theorem 4.7: Energy dependency
of the microcanonical partition function**

For macroscopic systems (large number of degrees of freedom, f) the quantities

$$\ln \Omega(E) , \ \ln \omega(E) , \ \ln g(E)$$

are equivalent. Approximately, we have

$$\Omega(E) \sim E^f .$$

This theorem shows that for macroscopic systems, the partition function Ω is practically independent of the energy uncertainty δE. Put differently: for any interval $[E - \delta E : E]$, practically all microstates are very close to E, even on the scale set by δE (see Application 56).

4.2.2 Principle of Maximal Entropy

Entropy is a concept of central importance to statistical physics. It is related in the following manner to the microcanonical partition function:

Definition: Entropy S

The statistical entropy of a system in equilibrium is defined by the *Boltzmann equation*

$$S(E, x) = k \ln \Omega(E, x) , \tag{4.18}$$

where the constant of proportionality

$$k = 1.38054 \cdot 10^{-23} \frac{J}{K} \quad (K = \text{Kelvin})$$

is called the *Boltzmann constant*.

We now show that the entropy of a closed system in equilibrium is maximal. Let x be an extensive[7] macroscopic quantity that can have different values

[7] Quantities are called *extensive* if they are proportional to the number of particles N. Quantities independent of N are called *intensive*.

independently of the energy E of the system. According to the second statistical postulate, all microstates $\Omega(E,x) = \exp[S(E,x)/k]$ are equally likely. The probability $W(x)$ of finding the system in a state that is in accordance with x is therefore proportional to the number $\Omega(E,x)$ of microstates with this value:

$$W(x) \sim \exp\left(\frac{S(E,x)}{k}\right) .$$

We expand $\ln W(x)$ around an arbitrary value \tilde{x}:

$$\ln W(x) = \ln W(\tilde{x}) + \frac{1}{k}\frac{\partial S}{\partial x}\bigg|_{x=\tilde{x}}(x-\tilde{x}) + \frac{1}{2k}\frac{\partial^2 S}{\partial x^2}\bigg|_{x=\tilde{x}}(x-\tilde{x})^2 + \ldots . \quad (4.19)$$

Assuming that \tilde{x} yields the maximum of $W(x)$,

$$\frac{\partial S}{\partial x}\bigg|_{x=\tilde{x}} = 0 ,$$

we can write $W(x)$ as a normalized Gaussian distribution

$$W(x) = \frac{1}{\sqrt{2\pi}\Delta x}\exp\left(-\frac{(x-\tilde{x})^2}{2(\Delta x)^2}\right) , \quad \int_{-\infty}^{\infty} W(x)\mathrm{d}x = 1 ,$$

where

$$\Delta x = \sqrt{-\frac{k}{\frac{\partial^2 S}{\partial x^2}\big|_{x=\tilde{x}}}}$$

is the variance (1σ-deviation) of the distribution. Since for large N the variance is relatively small, we are justified in dropping higher-order terms in the expansion (4.19). It also means that practically all microstates lie close to the maximum. Thus, for the expectation value \bar{x}, it follows that $\bar{x} = \tilde{x}$.

Theorem 4.8: Law of maximal entropy

Let x be an extensive macroscopic quantity independent of the energy E. For a closed system in equilibrium, its expectation value is determined by the maximum condition

$$S(E,x) = \text{maximal} \iff \frac{\partial S}{\partial x}\bigg|_{x=\bar{x}} = 0 .$$

Entropy is defined only for equilibrium states. Therefore, the question naturally arises how it is possible for S to become maximal without ever not being in a state of equilibrium. The following argument allows for the definition of the entropy of macroscopic states away from equilibrium. Divide the system into subsystems that are each in a *local equilibrium* state, and assume that the relaxation times of the subsystems are small with respect to the observation time. It is then possible to evaluate the subsystem entropies so that, due to the additiveness of entropy, the total entropy can be defined.

4.2.3 Equilibrium Conditions and Generalized Forces

Using Theorem 4.8, we can now find equilibrium conditions for systems that are in contact with one another and can therefore exchange, for example, energy, volume, or particles. To this end, we consider quite generally an isolated system divided into two subsystems 1 and 2 by a partitioning wall. Due to the isolation, the total energy of the system is constant,

$$E_1 + E_2 = E = \text{const} ,$$

so that we can work in the formalism of the microcanonical ensemble. Furthermore, we allow the systems 1 and 2 to exchange one external parameter x, whose total value is also conserved:

$$x = x_1 + x_2 = \text{const} .$$

In this context, we note in advance that one distinguishes two types of energy exchange:

- *Heat exchange* or *thermic exchange*: energy exchange at constant external parameters.

- *Mechanical energy exchange* at thermic isolation, solely caused by an exchange of external parameters.

According to this, most generally, energy exchange is a combined exchange of heat and mechanical energy.

We now wish to answer the question: *which value $\overline{x_1}$ does system 1 have in equilibrium?* Following Theorem 4.8, the corresponding maximum condition is

$$S(x_1) = S_1(E_1, x_1) + S_2(E - E_1, x - x_1)$$

$$
\begin{aligned}
0 = \mathrm{d}S &= \frac{\partial S_1}{\partial x_1}\mathrm{d}x_1 + \frac{\partial S_2}{\partial x_2}\mathrm{d}x_2 + \frac{\partial S_1}{\partial E_1}\mathrm{d}E_1 + \frac{\partial S_2}{\partial E_2}\mathrm{d}E_2 \\
&= \frac{\partial S_1}{\partial x_1}\mathrm{d}x_1 - \frac{\partial S_2}{\partial x_2}\mathrm{d}x_1 + \frac{\partial S_1}{\partial E_1}\mathrm{d}E_1 - \frac{\partial S_2}{\partial E_2}\mathrm{d}E_1 ,
\end{aligned}
\tag{4.20}
$$

where the derivatives are taken at $\overline{x_1}$ and $\overline{x_2} = x - \overline{x_1}$, respectively. If we initially restrict ourselves to the case that the dividing wall allows only thermic contact, i.e., $\mathrm{d}x_1 = 0$, (4.20) reduces to

$$\mathrm{d}S = \left(\frac{\partial S_1}{\partial E_1} - \frac{\partial S_2}{\partial E_2} \right) \mathrm{d}E_1 = 0 .$$

We define

Definition: Temperature T

The temperature T of an equilibrium system with energy E is defined as[8]

$$\frac{1}{T(E,x)} = \left(\frac{\partial S}{\partial E} \right)_x = \frac{\partial S(E,x)}{\partial E} = k\frac{\partial \ln \Omega(E,x)}{\partial E} .$$
$$\tag{4.21}$$

For the total system in thermic equilibrium, we therefore have

$$T_1 = T_2 . \qquad (4.22)$$

Now, if we also allow both systems to exchange the parameter x (exchange of heat and mechanical energy), (4.20) can be simplified by the following definition:

Definition: Generalized force X

The generalized force X is the conjugate force to the external parameter x:

$$X = T \left(\frac{\partial S}{\partial x} \right)_E = T \frac{\partial S(E,x)}{\partial x} = kT \frac{\partial \ln \Omega(E,x)}{\partial x} . \qquad (4.23)$$

With this definition, (4.20) becomes

$$0 = dS = \left(\frac{X_1}{T_1} - \frac{X_2}{T_2} \right) dx_1 + \left(\frac{1}{T_1} - \frac{1}{T_2} \right) dE_1 . \qquad (4.24)$$

To interpret this equation correctly, one has to bear in mind that the variation dE_1 is generally not independent of dx_1. This is due to the two possible exchanges of energy, introduced above. We will show later that for *reversible* exchange processes, the following relation holds:

$$dE = -X dx \qquad \left(\text{generally: } dE = - \sum_i X_i dx_i \right) .$$

Using this result, it follows from (4.24) that

$$\left(\frac{X_1}{T_1} - \frac{X_2}{T_2} - \frac{X_1}{T_1} + \frac{X_1}{T_2} \right) dx_1 = 0 \implies X_1 = X_2 .$$

Since both temperatures are equal [see (4.22)], we finally arrive at

Theorem 4.9: Equilibrium conditions of a closed two-part system

An isolated system built from two subsystems 1 and 2 is in equilibrium with respect to energy exchange and the exchange of an external extensive parameter x if the temperatures and the generalized forces of both subsystems obey the following conditions:

$$T_1 = T_2 \qquad \text{(equilibrium at thermic exchange)}$$

$$X_1 = X_2 \qquad \text{(equilibrium at x-exchange) .}$$

[8] Note the bracket notation, which is explained on page 381 and which will be repeatedly used in the following. Furthermore, the quantity kT corresponds to the energy per degree of freedom, since, following Theorem 4.7, we have

$$\frac{1}{kT} = \frac{\partial \ln \Omega}{\partial E} = f \frac{\partial \ln E}{\partial E} = \frac{f}{E} .$$

In addition to temperature T, we shall frequently need two more intensive quantities, namely

$$P = T \left(\frac{\partial S}{\partial V} \right)_{E,N} \qquad (pressure)$$

$$\mu = -T \left(\frac{\partial S}{\partial N} \right)_{E,V} \qquad (chemical\ potential) \ .$$

These are the conjugate forces to the extensive parameters volume V and number of particles N.

To illustrate the meaning of Theorem 4.9, consider a gas enclosed in a glass box with a movable piston. Suppose the gas is heated for a while and, simultaneously, the piston is pushed down, compressing the gas into a smaller volume. Once the process is stopped and the gas is left to its own devices, it will attempt to regain equilibrium. According to Theorem 4.9, this implies an exchange of heat via the glass walls as well as an exchange of volume via the piston, until the temperatures and pressures of the gas and its external environment are equal. It is therefore correct to view temperature differences and pressure differences as the driving forces for heat exchange and volume exchange. This justifies the term "generalized forces".

Statistical physics and thermodynamics. At the end of this section we wish to emphasize the fundamental nature of the definition of entropy (4.18). It provides the connection between the microscopic view of statistical physics and the macroscopic view of thermodynamics. Knowing the microcanonical partition function of an equilibrium system is equivalent to knowing the macroscopic state variable entropy. And knowing either of these allows the determination of all other macroscopic thermodynamical state variables, such as temperature, pressure, chemical potential, and so on (see Section 4.5). Therefore, we can write down the following simple scheme in order to derive the thermodynamical relations from a given Hamilton operator or Hamilton function:

$$\boldsymbol{H}(x) \longrightarrow E_r(x) \longrightarrow \Omega(E,x) \longrightarrow S(E,x) \longrightarrow \left\{ \begin{array}{l} \text{thermodyn.} \\ \text{relations} \end{array} \right\} \ . \quad (4.25)$$

Summary

- The **microcanonical ensemble** describes an equilibrium system at given energy E. It is composed of many similar systems, all of which are consistent with this energy and possibly other given external parameters within an uncertainty tolerance δE. All members of the ensemble are equiprobable.

- The corresponding **microcanonical partition function** grows extremely fast with E, so that it is independent of the exact size of the tolerance δE.

\triangleright

- The relation between statistical physics and thermodynamics is given by the **Boltzmann equation**, which relates the microcanonical partition function with the macroscopic quantity **entropy**.

- For a closed system in equilibrium, the entropy is maximal (**principle of maximal entropy**).

- This principle yields **equilibrium conditions** for interacting systems (**thermic** and/or **mechanical exchange**).

Applications

56. Ideal gas I: Phase space volume, microcanonical partition function, and state equation. Consider an *ideal gas* consisting of N indistinguishable noninteracting particles of mass m in a volume $L^3 = V$.

a. By comparing the classical and quantum mechanical phase space volumes $\omega(E)$, show that the size of the classical phase space cell is h^{3N}.

b. Using the phase space density, compute the microcanonical partition function $\Omega(E)$ and verify that it is independent of the size of the energy uncertainty δE.

c. Determine the *caloric state equation* $E = E(T, V, N)$ and the *thermic state equation* $P = P(T, V, N)$ of the ideal gas.

Solution.

To a. The classical Hamilton function of the system is given by

$$H(q,p) = \sum_{i=1}^{N} \frac{p_i^2}{2m} = \sum_{i=1}^{3N} \frac{p_i^2}{2m} = E = \text{const}.$$

This yields the classical phase space volume

$$\omega(E) = \frac{1}{\nu^{3N} N!} \int\limits_{\sum_i p_i^2 \leq 2mE} d\Gamma = \frac{V^N}{\nu^{3N} N!} \int\limits_{\sum_i p_i^2 \leq 2mE} dp, \qquad (4.26)$$

where ν^{3n} is the size of the phase space cell that we wish to determine. To find the quantum mechanical anologon, we use the fact that the N particles are confined to a volume $L^3 = V$, so that each of the $3N$ Cartesian momenta are restricted to the quantized values

$$p_i = \frac{\pi \hbar}{L} n_i, \ n_i = 0, 1, \ldots.$$

The energy eigenvalues are therefore

$$E_n = \sum_{i=1}^{3N} \frac{\pi^2 \hbar^2}{2mL^2} n_i^2 \ .$$

Next, we assume that the energy states are so close that the sum $\omega(E) = \sum_{n:E_n \leq E} 1$ can be replaced by an integral. This is true if the particle momenta are much larger than the smallest possible quantum of momentum. For the quantum mechanical phase space volume, this implies

$$\omega(E) = \frac{1}{2^{3N}} \int_{E_n \leq E} dn \ ,$$

where the factor $1/2^{3N}$ also allows negative values of n. Replacing dn_i by dp_i, we find

$$\omega(E) = \frac{V^N}{(2\pi\hbar)^{3N}} \int_{\sum_i p_i^2 \leq 2mE} dp \ . \tag{4.27}$$

Taking into account the indistinguishability of the particles by attaching a factor $1/N!$ and then comparing this relation with the classical result (4.26), we see that

$$\nu = h \ .$$

In Application 54, we already obtained the same result studying the one-dimensional harmonic oscillator.

To b. To find the microcanonical partition function, we need to evaluate the integral in (4.27). It represents the volume K of a $3N$-dimensional sphere with radius $R = \sqrt{2mE}$ and is given by[9]

$$K = \frac{\pi^{3N/2}}{\left(\frac{3N}{2}\right)!} R^{3N} \ .$$

Thus, we find

$$\omega(E,V,N) = \frac{1}{N!} \left(\frac{V}{h^3}\right)^N K = \frac{1}{N!} \left(\frac{V}{h^3}\right)^N \frac{\pi^{3N/2}}{\left(\frac{3N}{2}\right)!} (2mE)^{3N/2} \ ,$$

where the correction factor $1/N!$ is taken into account. Using Stirling's relation

$$\ln N! \overset{N \gg 1}{\approx} N(\ln N - 1) \ , \tag{4.28}$$

we further have

[9] Strictly speaking, this result is correct only for even values of N. If N is odd, we can simply add or subtract a single particle; for large N, this cannot be of importance.

$$\ln \omega(E, V, N) = N \left\{ \ln \left[\frac{V}{N} \left(\frac{4\pi m}{3h^2} \frac{E}{N} \right)^{3/2} \right] + \frac{5}{2} \right\}. \qquad (4.29)$$

With this equation, we can verify that the microcanonical partition function Ω is independent of δE. According to (4.16),

$$\Omega(E, V, N) \approx \delta E g(E) = \delta E \frac{\partial \omega(E, V, N)}{\partial E} = \frac{3N}{2} \frac{\delta E}{E} \omega(E, V, N)$$

$$\Longrightarrow \ln \Omega \approx \ln \omega + \ln \left(\frac{3N}{2} \right) + \ln \left(\frac{\delta E}{E} \right).$$

Now, we can safely assume that δE is small compared to the energy E; typically, we have $\delta E / E = \mathcal{O}(1/\sqrt{N})$. Therefore, the last two $\mathcal{O}(\ln N)$-terms can be neglected compared to $\ln \omega = \mathcal{O}(N)$. This shows that the energy width δE is of no physical consequence. For very large N, the number of microstates grows so fast with growing energy that the major contribution to the partition function always comes from states close to the hypersurface of energy E. We can therefore sum over all states between 0 and E without the additional states making a noticeable contribution.

To c. From the microcanonical partition function, one obtains the caloric and thermic equations of state as

$$\frac{1}{T} = \left(\frac{\partial S}{\partial E} \right)_{V,N} = k \left(\frac{\partial \ln \Omega}{\partial E} \right)_{V,N} = \frac{3Nk}{2E} \Longrightarrow E = \frac{3}{2} NkT \qquad (4.30)$$

$$\frac{P}{T} = \left(\frac{\partial S}{\partial V} \right)_{E,N} = k \left(\frac{\partial \ln \Omega}{\partial V} \right)_{E,N} = \frac{Nk}{V} \Longrightarrow P = \frac{NkT}{V}. \qquad (4.31)$$

Inserting the first equation into (4.29), we find the *Sackur-Tetrode equation* for the microcanonical partition function of the ideal gas:

$$\ln \omega(T, V, N) = \ln \Omega(T, V, N) = N \left\{ \ln \left[\frac{V}{N} \left(\frac{2\pi mkT}{h^2} \right)^{3/2} \right] + \frac{5}{2} \right\} \qquad (4.32)$$

or

$$\ln \Omega(T, V, N) = N \left(\ln V - \ln N + \frac{3}{2} \ln T + \sigma \right), \qquad (4.33)$$

with

$$\sigma = \frac{3}{2} \ln \left(\frac{2\pi mk}{h^2} \right) + \frac{5}{2}.$$

From (4.33), we see that for very small temperatures, the entropy diverges to infinity. However, this is in contradiction to the third law of thermodynamics (see Subsection 4.5.1). Replacing the sum by an integral (which we did in our calculation of the phase space volume) is therefore not valid at zero temperature, where the lowest state with $p = 0$ becomes very dominant. The correct way of approaching this issue is provided by quantum statistics (Section 4.7), which will resolve this contradiction.

57. Ideal gas II: Gibbs paradox. Consider an ideal gas of N identical atoms in a volume V. Inserting a partition wall divides the gas into two volumes V_1 and V_2. If this is done in a reversible fashion, we must have

$$S = S_1 + S_2 , \tag{4.34}$$

where S is the entropy of the total volume before and S_1 and S_2 are the entropies of the two subvolumes after the wall has been inserted. Show that (4.34) is satisfied only if the Gibbs factor $1/N!$ is used in the microcanonical partition function.

Solution. To start, we note that the particle density is constant throughout:

$$\rho = \frac{N}{V} = \frac{N_1}{V_1} = \frac{N_2}{V_2} = \text{const} .$$

Using (4.33), in which $1/N!$ is taken into account, we find

$$S_i = N_i k \left(\ln V_i - \ln \rho - \ln V_i + \frac{3}{2} \ln T + \sigma \right)$$

$$= N_i k \left(-\ln \rho + \frac{3}{2} \ln T + \sigma \right)$$

$$\Longrightarrow S_1 + S_2 = N k \left(-\ln \rho + \frac{3}{2} \ln T + \sigma \right)$$

$$= N k \left(\ln V - \ln N + \frac{3}{2} \ln T + \sigma \right) = S .$$

To make the corresponding calculation without $1/N!$, we need an equation analogous to (4.33) with $1/N!$ missing. This results in

$$S_i = N_i k \left(\ln V + \frac{3}{2} \ln T + \sigma - 1 \right)$$

and

$$S_1 + S_2 = N k \left(\ln V + \frac{3}{2} \ln T + \sigma - 1 \right) \neq S .$$

This contradiction was Gibbs's reason for introducing the factor $1/N!$ long before it was justified by quantum statistics.

4.3 Ensemble Theory II: Canonical and Grand Canonical Ensembles

In the previous section we saw how the macroscopic state of a system in equilibrium with constant energy is related to the microcanonical ensemble. Using this as our starting point for this section, we consider two more types of ensembles, called *canonical* and *grand canonical ensemble*. We will also show that all three ensembles yield consistent and equivalent descriptions of systems in thermodynamical equilibrium.

4.3.1 Canonical Ensemble

Generally, it is almost impossible to control the energy of a physical system since it is hard to keep it totally isolated. A much more practical concept is the less stringent requirement of constant temperature. This quantity is not only measured easily (e.g., with the help of a thermometer), but can also be controlled very precisely by bringing the system into touch with a heat bath.

In statistical physics, systems with a given temperature are described by the *canonical ensemble*, whose occupation probabilities can be deduced by the following reasoning: consider a closed system consisting of two subsystems 1 and 2 in thermic contact. Suppose the second system is a lot larger than the first, so that the temperature of the composite system is practically controlled by system 2 (see Fig. 4.3). For this to hold, it is necessary that the energy E

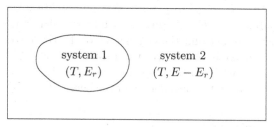

Fig. 4.3. Physical system in contact with a much larger heat bath that controls the temperature of both systems.

of the combined system is much larger than the possible energy levels E_r of the small system 1:

$$E_r \ll E . \tag{4.35}$$

In other words, system 2 has to be macroscopically large, whereas system 1 is not subject to this condition. Since, by design, the total energy of the system is constant,

$$E_r + E_2 = E ,$$

we can describe the combined system with the help of the microcanonical ensemble. Of the total of $\Omega(E, x)$ states, there are $\Omega_2(E - E_r, x)$ states for which system 1 is in a particular microstate with energy E_r. Since, due to the second statistical postulate, all microstates are equally probable, the probability $P_r(T, x)$ for system 1 to be in the microstate r at temperature T is proportional to $\Omega_2(E - E_r, x)$:

$$P_r(T, x) \sim \Omega_2(E - E_r, x) .$$

Now, we expand the logarithm of $\Omega_2(E - E_r, x)$ around E, expecting a fast convergence due to (4.35):

$$\ln \Omega_2(E - E_r, x) = \ln \Omega_2(E, x) + \frac{\partial \ln \Omega_2}{\partial E}(E - E_r) + \dots$$

$$\approx \text{const} - \frac{E_r}{kT} .$$

From this, we find the (normalized) probability distribution

$$P_r(T, x) = \frac{1}{Z(T, x)} e^{-\beta E_r(x)} , \quad \beta = \frac{1}{kT} ,$$

where

$$Z(T, x) = \sum_r e^{-\beta E_r(x)}$$

denotes the *canonical partition function*.

Probability distribution of the energy. $P_r(T, x)$ is the probability for a system with given temperature T to be in a particular microstate r. In order to determine the probability $W(E_r)$ to find an arbitrary microstate with energy E_r, we have to take into account the degeneracy of the Hamilton operator of the system. For a given temperature, the probability $P_r(T, x)$ depends only on the energy. Thus, we can use the phase space density $g(E_r)$ from (4.16), which actually determines the degeneracy of the energy level E_r, to rewrite $W(E_r)$ as

$$W(E_r) = g(E_r)P_r(T, x) = \frac{g(E_r)}{Z} e^{-\beta E_r}$$

$$Z(T, x) = \sum_r g(E_r)e^{-\beta E_r} .$$

In case of densely distributed energy levels, these equations become

$$W(E)\mathrm{d}E = \frac{g(E)}{Z(T, x)} e^{-\beta E} \mathrm{d}E , \quad Z(T, x) = \int_0^\infty \mathrm{d}E g(E)e^{-\beta E} . \tag{4.36}$$

The right-hand equation of (4.36) shows that the canonical partition function Z is just the Laplace transform of the phase space density g, so that

$$g(E) = \frac{1}{2\pi\mathrm{i}} \int_{c-\mathrm{i}\infty}^{c+\mathrm{i}\infty} \mathrm{d}\beta Z(\beta, x)e^{\beta E} , \quad \mathrm{Re}(\beta) = c > 0 .$$

Here, β is a complex variable, and the integration is carried out parallel to the imaginary axis along $c > 0$.

Characteristic energy. The average energy \overline{E} of the canonical ensemble is given by

$$\overline{E(T, x)} = -\left(\frac{\partial \ln Z}{\partial \beta}\right)_x .$$

The corresponding mean-square deviation is

$$(\Delta E)^2 = \overline{E^2} - \overline{E}^2 = \left(\frac{\partial^2 \ln Z}{\partial \beta^2}\right)_x = -\left(\frac{\partial \overline{E}}{\partial \beta}\right)_x = kT^2 \left(\frac{\partial \overline{E}}{\partial T}\right)_x$$
$$= kT^2 C_x \ ,$$

where $C_x = \left(\partial \overline{E}/\partial T\right)_x$ is the *specific heat* at fixed external parameters x.[10]
Thus, we find

$$\frac{\sqrt{(\Delta E)^2}}{\overline{E}} \sim \mathcal{O}\left(\frac{1}{\sqrt{N}}\right) \ . \tag{4.37}$$

This shows that practically all systems in the canonical ensemble and in the limit $N \to \infty$ are in a state with energy \overline{E}. We conclude that the energy distribution $W(E)$ has a sharp maximum around the average value \overline{E}, which can be obtained from the following maximum condition:

$$\left.\frac{\partial W}{\partial E}\right|_{E=\overline{E}} = 0 \Longrightarrow \left.\frac{\partial}{\partial E} \ln g(E)\right|_{E=\overline{E}} = \beta \Longrightarrow \left.\frac{\partial S}{\partial E}\right|_{E=\overline{E}} = \frac{1}{T} \ .$$

In the last relation, we have used the identity $S = k \ln g$.[11]

In the thermodynamic limit, we see that the microcanonical and canonical ensembles describe the same physical situation. In both descriptions, the energy has a sharp value, although in the canonical ensemble only the system's temperature is held fixed. However, for small systems with relatively few degrees of freedom, the two ensembles represent very different situations. In this case, the mean-square deviation (4.37) is no longer negligible, so that, within the canonical ensemble, a physical system with fixed temperature undergoes very large energy fluctuations.

Free energy. Using

$$\frac{\partial^2 \ln g(\overline{E})}{\partial \overline{E}^2} = \frac{1}{k} \left(\frac{\partial^2 S}{\partial \overline{E}^2}\right)_x = \left(\frac{\partial}{\partial E}\right)_x \left(\frac{1}{kT}\right) = \left(\frac{\partial E}{\partial T}\right)_x^{-1} \left(\frac{\partial}{\partial T}\right)_x \left(\frac{1}{kT}\right)$$
$$= -\frac{1}{kT^2 C_x} \ ,$$

the expansion of $\ln[ZW(E)]$ around the maximum \overline{E} gives

$$\ln[ZW(E)] = -\beta E + \ln g(E) = -\beta \overline{E} + \frac{S}{k} - \frac{(E-\overline{E})^2}{2kT^2 C_x} \ ,$$

so that

$$W(E) = \frac{1}{Z} e^{-\beta(\overline{E} - TS)} \exp\left(-\frac{(E-\overline{E})^2}{2kT^2 C_x}\right) \ .$$

[10] In the canonical ensemble, the external parameters are most often taken to be $x = (V, N)$. In this case, $C_x = C_V$ is the specific heat at constant volume and constant number of particles.

[11] Remember that S is the entropy of the small system 1 (rather than that of the combined system 1+2).

Applying the normalization condition $\int dE W(E) = 1$, we find for the canonical partition function

$$-kT \ln Z(T,x) = (\overline{E} - TS) - \frac{1}{2}kT \ln \left(2\pi kT^2 C_x\right) \ .$$

The last term on the right-hand side of this equation is of the order $\mathcal{O}(\ln N)$ and can be neglected compared to the other terms of order $\mathcal{O}(N)$ if $N \to \infty$. This yields

$$-kT \ln Z(T,x) = (\overline{E} - TS) \ .$$

Theorem 4.10: Canonical ensemble

The canonical ensemble describes the equilibrium state of a system at fixed temperature T:

$$P_r(T,x) = \frac{1}{Z(T,x)}e^{-\beta E_r(x)} \ , \ Z(T,x) = \sum_r e^{-\beta E_r(x)} \ , \ \beta = \frac{1}{kT} \ .$$

All average values of state variables can be calculated from this. For example, the thermodynamic energy is given by

$$\overline{E(T,x)} = \sum_r P_r(T,x)E_r(x) = -\left(\frac{\partial \ln Z}{\partial \beta}\right)_x \ .$$

In the thermodynamic limit, the energy fluctuations in the canonical ensemble are negligibly small. The connection to thermodynamics is provided by

$$-kT \ln Z(T,x) = F(T,x) = \overline{E} - TS \ ,$$

where F is the *free energy*. The term $e^{-\beta E}$ is called the *Boltzmann factor*.

Interpreting the states $|\psi_r\rangle$ from Theorem 4.3 as the energy eigenstates of the Hamilton operator \boldsymbol{H} with energy E_r, the canonical density operator can be written as

$$\rho = \frac{1}{Z}\sum_r |\psi_r\rangle\, e^{-\beta E_r} \,\langle\psi_r| = \frac{e^{-\beta \boldsymbol{H}}}{Z}\sum_r |\psi_r\rangle\,\langle\psi_r| = \frac{e^{-\beta \boldsymbol{H}}}{\mathrm{tr}\left(e^{-\beta \boldsymbol{H}}\right)} \ , \quad (4.38)$$

where we have used

$$Z = \sum_r e^{-\beta E_r} = \sum_r \langle\psi_r|\,e^{-\beta \boldsymbol{H}}\,|\psi_r\rangle = \mathrm{tr}\left(e^{-\beta \boldsymbol{H}}\right)$$

in the denominator. The corresponding density matrix is diagonal in the energy eigenbasis:

$$\rho_{nm} = \langle\psi_n|\,\rho\,|\psi_m\rangle = \frac{e^{-\beta E_n}}{\sum\limits_n e^{-\beta E_n}}\delta_{nm} \ . \quad (4.39)$$

4.3.2 Grand Canonical Ensemble

The canonical ensemble was introduced to describe physical situations where the rather restrictive condition of constant energy, used in the microcanonical formalism, is replaced by the experimentally easier controllable condition of constant temperature. We now relax the conditions even further by allowing both energy and particle exchange, while the corresponding intensive quantities temperature and chemical potential are held fixed. This case will prove to be relevant particularly for processes in quantum field theory and in chemistry, where particles can be generated and destroyed. The statistical description of this scenario is provided by the *grand canonical ensemble*. Its occupation probabilities can best be determined by using the same argument as applied to the canonical ensemble and by introducing a controlling heat bath and particle reservoir.

Consider, therefore, a small system 1 in contact with a much larger system 2 with which it can exchange energy and particles (Fig. 4.4). Thus, the large

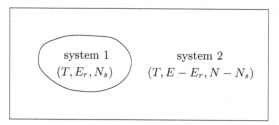

Fig. 4.4. Physical system in contact with a much larger heat and particle bath that controls temperature and chemical potential of both systems.

system controls the temperature T and the chemical potential μ of the small system. Although energy and particles can be exchanged, the total values must remain constant,

$$E_r + E_2 = E = \text{const} , \quad N_s + N_2 = N = \text{const} ,$$

and we assume that

$$E_r \ll E , \quad N_s \ll N . \tag{4.40}$$

The question we wish to answer is: *what is the probability $P_{r,s}(T, \mu, x)$ for the small system 1 in equilibrium, at given temperature T and given chemical potential μ, to be in the microstate (r, N_s)?*[12] As before, our starting point is the second statistical postulate, which states that all microstates of the isolated combined system are equally likely. Of these $\Omega(E, N)$ possible states, there are $\Omega_2(E - E_r, N - N_s)$ states for which system 1 is in a particular microstate with (E_r, N_s). Thus, the probability $P_{r,s}(T, \mu, x)$ is given by

[12] Note that in r we collect all quantum numbers, while holding fixed the number of particles N_s.

$$P_{r,s}(T,\mu,x) \sim \Omega_2(E - E_r, N - N_s) \ .$$

If we use (4.40), the expansion of $\ln P_{r,s}$ around the point (E, N) yields

$$\ln \Omega_2(E - E_r, N - N_s) = \ln \Omega_2(E, N) + \frac{\partial \ln \Omega_2}{\partial E}(E - E_r)$$

$$+ \frac{\partial \ln \Omega_2}{\partial N}(N - N_s) + \dots$$

$$\approx \text{const} - \frac{\partial \ln \Omega_2}{\partial E} E_r - \frac{\partial \ln \Omega_2}{\partial N} N_s \ .$$

With

$$\beta = \frac{\partial \Omega_2}{\partial E} \ , \quad -\beta\mu = \frac{\partial \ln \Omega_2}{\partial N} \ ,$$

we finally obtain the (normalized) probability distribution

$$P_{r,s}(T,\mu,x) = \frac{1}{Y(T,\mu,x)} e^{-\beta(E_r(N_s) - \mu N_s)}$$

$$= \frac{1}{Y(T,\mu,x)} z^{N_s} e^{-\beta E_r(N_s)} \ ,$$

where

$$Y(T,\mu,x) = \sum_{r,s} e^{-\beta(E_r(N_s) - \mu N_s)}$$

is the *grand canonical partition function* and

$$z = e^{\beta\mu}$$

is the *fugacity*.

Characteristic energy and particle number. The average energy \overline{E} and average particle number \overline{N} are calculated in now-customary fashion:

$$\overline{E}(T,\mu,x) = -\left(\frac{\partial \ln Y}{\partial \beta}\right)_{z,x} \ , \quad \overline{N}(T,\mu,x) = \frac{1}{\beta}\left(\frac{\partial \ln Y}{\partial \mu}\right)_{T,x} \ .$$

The relative deviation of the particle number is

$$\frac{(\Delta N)^2}{\overline{N}^2} = \frac{\overline{N^2} - \overline{N}^2}{\overline{N}^2} = \frac{1}{\overline{N}^2 \beta^2}\left(\frac{\partial^2 \ln Y}{\partial \mu^2}\right)_{T,x} = \frac{kT}{\overline{N}^2}\left(\frac{\partial \overline{N}}{\partial \mu}\right)_{T,x} \ . \quad (4.41)$$

To simplify this equation further, we shall assume that x represents the volume V (this is the most common case) and make use of the thermodynamic relation (see Subsection 4.5.2)

$$d\mu = \frac{V}{\overline{N}} dP - \frac{S}{\overline{N}} dT \ .$$

With this, we find

$$\left(\frac{\partial \mu}{\partial v}\right)_T = v\left(\frac{\partial P}{\partial v}\right)_T \ , \quad v = \frac{V}{\overline{N}} \implies -\frac{\overline{N}^2}{V}\left(\frac{\partial \mu}{\partial \overline{N}}\right)_{T,V} = V\left(\frac{\partial P}{\partial V}\right)_{T,\overline{N}} \ ,$$

so that (4.41) can be written as

$$\frac{(\Delta N)^2}{\overline{N}^2} = -\frac{kT}{V^2}\kappa_T \ , \quad \kappa_T = \left(\frac{\partial V}{\partial P}\right)_{T,\overline{N}} ,$$

where κ_T is the *isothermic compressibility*. We see that the particle density fluctuates as $\mathcal{O}\left(1/\sqrt{N}\right)$, which is negligibly small for large particle numbers. A similar calculation for the energy fluctuation ($x = V$) yields

$$(\Delta E)^2 = -\left(\frac{\partial \overline{E}}{\partial \beta}\right)_{V,\mu} = kT^2\left(\frac{\partial \overline{E}}{\partial T}\right)_{V,\mu}$$

$$= kT^2\left[\left(\frac{\partial \overline{E}}{\partial T}\right)_{V,N} + \left(\frac{\partial \overline{E}}{\partial \overline{N}}\right)_{T,V}\left(\frac{\partial \overline{N}}{\partial T}\right)_{V,\mu}\right] . \tag{4.42}$$

Using one of the *Maxwell relations* [see the last row of (4.59) in Subsection 4.5.2]

$$\left(\frac{\partial \overline{N}}{\partial T}\right)_{V,\mu} = -\frac{\partial^2 J}{\partial T \partial \mu} = -\frac{\partial^2 J}{\partial \mu \partial T} = \left(\frac{\partial S}{\partial \mu}\right)_{T,V} ,$$

we can simplify (4.42) as follows:

$$(\Delta E)^2 = kT^2 C_V + kT^2\left(\frac{\partial \overline{E}}{\partial N}\right)_{T,V}\left(\frac{\partial S}{\partial \overline{E}}\right)_{T,V}\left(\frac{\partial \overline{E}}{\partial \mu}\right)_{T,V}$$

$$= (\Delta E)^2_{\text{kan}} + kT\left(\frac{\partial \overline{E}}{\partial \overline{N}}\right)^2_{T,V}\left(\frac{\partial \overline{N}}{\partial \mu}\right)_{T,V}$$

$$= (\Delta E)^2_{\text{kan}} + \left(\frac{\partial \overline{E}}{\partial \overline{N}}\right)^2_{T,V}(\Delta N)^2 .$$

We see that the energy fluctuations in the grand canonical ensemble are identical to those in the canonical ensemble with an extra term stemming from the particle fluctuations. We wish to point out that there are situations where the energy and particle fluctuations can deviate from the $\mathcal{O}\left(1/\sqrt{N}\right)$ behavior. This happens close to phase transitions, where the compressibility κ_T can grow exponentially. κ_T is then of the order of $\mathcal{O}(N)$.

 In the thermodynamical limit, practically all members of the grand canonical ensemble have the same energy $\overline{E} = E$ and the same number of particles $\overline{N} = N$. With regard to the results of the previous subsection, we can therefore conclude that microcanonical, canonical, and grand canonical ensembles yield equivalent descriptions of macroscopic systems in equilibrium.

Grand canonical potential. The grand canonical partition function can also be written as

$$Y(T, z, x) = \sum_{N=1}^{\infty} z^N Z(T, N, x) , \tag{4.43}$$

where $Z(T, N, x)$ is the canonical partition function of the N-particle state. The upper limit of summation can be shifted to infinity since only systems with particle number N_s much smaller than the total number of particles of the combined system particle+reservoir contribute significantly to the sum. Now, from (4.43), we see that the probability $W_{\overline{E}}(N)$ of finding the system with energy \overline{E} and particle number N is given by

$$W_{\overline{E}}(N) = \frac{1}{Y} z^N Z(T, N, x) = \frac{1}{Y} e^{-\beta(\overline{E} - TS - \mu N)} \ .$$

Since this distribution has a sharp maximum around the average value \overline{N} in the thermodynamical limit, the grand canonical partition function must be equivalent to the canonical totality with \overline{N} particles. It therefore follows that

$$Y(T, z, x) = z^{\overline{N}} Z(T, \overline{N}, x) \implies -kT \ln Y = \overline{E} - TS - \mu \overline{N} \ .$$

Theorem 4.11: Grand canonical ensemble

The grand canonical ensemble describes the equilibrium state of a system with given temperature T and chemical potential μ:

$$P_{r,s}(T, \mu, x) = \frac{1}{Y(T, \mu, x)} e^{-\beta(E_r(N_s) - \mu N_s)}$$

$$Y(T, \mu, x) = \sum_{r,s} e^{-\beta(E_r(N_s) - \mu N_s)}$$

$$= \sum_{N=1}^{\infty} z^N Z(T, N, x) \ , \ z = e^{\beta \mu} \ .$$

All expectation values can be obtained from the grand canonical partition function. In particular, the thermodynamic energy and average particle number are

$$\overline{E}(T, \mu, x) = \sum_{r,s} P_{r,s}(T, \mu, x) E_r(N_s, x) = -\left(\frac{\partial \ln Y}{\partial \beta} \right)_{z,x}$$

$$\overline{N}(T, \mu, x) = \sum_{r,s} P_{r,s}(T, \mu, x) N_s = \frac{1}{\beta} \left(\frac{\partial \ln Y}{\partial \mu} \right)_{T,x} \ .$$

In the thermodynamical limit the energy and particle fluctuations of the grand canonical ensemble are vanishingly small (away from phase transitions). The connection to thermodynamics is given by

$$-kT \ln Y(T\mu, x) = J(T, \mu, x) = \overline{E} - TS - \mu \overline{N} \ ,$$

where J is the *grand canonical potential*.

In the energy basis, the relations for the density operator corresponding to (4.38) and (4.39) are

$$\rho = \frac{e^{-\beta(\mathbf{H}-\mu\mathbf{N})}}{\operatorname{tr}\left(e^{-\beta(\mathbf{H}-\mu\mathbf{N})}\right)} \ , \quad \rho_{nm} = \frac{e^{-\beta(E_n-\mu N)}}{\sum\limits_{n} e^{-\beta(E_n-\mu N)}}\delta_{nm} \ .$$

4.3.3 Comparison of the Ensembles

Following the results of this and the previous sections, we see a simple principle emerge that relates the microcanonical, canonical, and grand canonical ensembles: they are all connected via Laplace transformations. The canonical partition function Z is given by the sum over the "microcanonical partition functions" $g(E)$ [phase space density, degeneracy function; see (4.16)], each weighted with a Boltzmann factor $e^{-\beta E}$:

$$Z(T, N, x) = \sum_{E} g(E)e^{-\beta E(N,x)} \ .$$

Unlike in the microcanonical ensemble, the energy is not fixed. Instead, the temperature T is constant and this determines the average energy \overline{E}. The grand canonical partition function Y is obtained by summing all canonical partition functions Z at given temperatures T, volume V, and particle number N, each weighted with a factor $e^{\beta\mu N}$:

$$Y(T\mu, x) = \sum_{N} e^{\beta\mu N} Z(T, N, x) \ . \tag{4.44}$$

Here, the system can exchange energy and particles with its environment. The temperature T and the chemical potential μ are constant and determine the average energy \overline{E} and average particle number \overline{N}. For systems of noninteracting particles, the canonical partition function factorizes, so that for indistinguishable particles, we find

$$Z(T, N, x) = \frac{1}{N!}Z^{N}(T, 1, x) \ . \tag{4.45}$$

Inserting this into (4.44) yields

$$Y(T, \mu, x) = \sum_{N} \frac{1}{N!} \left[e^{\beta\mu} Z(T, 1, x)\right]^{N} = \exp\left[e^{\beta\mu} Z(T, 1, x)\right] \ .$$

In principle, all of these ensembles can be equally well used to describe the thermodynamic properties of a system. Which one is chosen is often a matter of ease. As we have seen, the fluctuations of the energy and the particle number are negligible in the limit $N \to \infty$, so that expectation values of observables are sharp and therefore all three ensembles give equivalent descriptions. Clearly, the equivalence of the ensembles no longer holds for microscopically small systems with few degrees of freedom. For example, in the canonical ensemble, the probability distribution of a single particle is proportional to the Boltzmann factor and thus very broad:

$$\frac{\sqrt{(\Delta E)^2}}{\overline{E}} = \mathcal{O}(1) \ .$$

The canonical and grand canonical partition functions are each related to a thermodynamic potential (free energy and grand canonical potential) just as the microcanonical partition function is related to entropy. In Section 4.5, we will show that each of these potentials contains the complete thermodynamic information of a system in equilibrium. We can therefore expand our schematic diagram (4.25) from the end of the previous section for the determination of thermodynamic relations as follows:

$$H(N,x) \rightarrow E_r(N,x) \rightarrow \left\{ \begin{array}{l} \Omega(E,N,x) \rightarrow S(E,N,x) \\ Z(T,N,x) \rightarrow F(T,N,x) \\ Y(T,\mu,x) \rightarrow J(T,\mu,x) \end{array} \right\} \rightarrow \left\{ \begin{array}{l} \text{thermody-} \\ \text{namical} \\ \text{relations} \end{array} \right\}.$$

How these calculations are done in detail is the subject of Section 4.5.

Summary

- The **canonical ensemble** describes an equilibrium system at given temperature T. It is composed of many similar (but not necessarily macroscopic) systems with different energies E_r, each weighted by the **Boltzmann factor** $e^{-E_r/kT}$.

- The **grand canonical ensemble** describes an equilibrium system at given temperature T and given **chemical potential** μ. The ensemble members are a large number of similar (but not necessarily macroscopic) systems with different energies E_r and different particle numbers N_s, each weighted with the factor $e^{-(E_r-\mu N_s)/kT}$.

- In the thermodynamic limit, the energy and particle number fluctuations of the canonical and grand canonical ensembles are negligibly small. Therefore, microcanonical, canonical, and grand canonical ensembles yield equivalent descriptions of macroscopic systems.

- The canonical partition function is related to the **free energy**, whereas the grand canonical partition function is related to the **grand canonical potential**.

- Microcanonical, canonical, and grand canonical ensembles are related via Laplace transformations.

Applications

58. Ideal gas III: Canonical and grand canonical ensembles. Using the ideal gas, show that in the thermodynamic limit (where $\overline{N} = N$ and $\overline{E} = E$), microcanonical, canonical, and grand canonical ensembles are equivalent. Proceed by

- calculating the canonical and grand canonical partition functions and the corresponding potentials,

- determining the corresponding entropies using the relations [see (4.59) in Subsection 4.5.2]

$$S(T,V,N) = -\left(\frac{\partial F}{\partial T}\right)_{V,N} \;,\; S(T,V,\mu) = -\left(\frac{\partial J}{\partial T}\right)_{\mu,V} \tag{4.46}$$

and comparing them to the microcanonical entropy (4.32).

Solution. The canonical partition function is

$$Z(T,V,N) = \sum_r \exp\left(-\sum_{i=1}^{3N} \frac{p_i^2}{2mkT}\right)\;,$$

where

$$r = (x_1,\ldots,x_{3N},p_1,\ldots,p_{3N})$$

specifies a classical microstate of the ideal gas. We can replace the sum over microstates by an integral over the phase space,

$$\sum_r \longrightarrow \frac{1}{N!h^{3N}} \int d\Gamma\;,$$

where the indistinguishability of the particles is taken into account by the Gibbs factor $1/N!$. It follows that

$$
\begin{aligned}
Z(T,V,N) &= \frac{1}{N!h^{3N}} \int d\Gamma \exp\left(-\sum_{i=1}^{3N} \frac{p_i^2}{2mkT}\right) \\
&= \frac{V^N}{N!h^{3N}} \left[\int_{-\infty}^{\infty} dp \exp\left(-\frac{p^2}{2mkT}\right)\right]^{3N} \\
&= \frac{V^N}{N!}\left(\frac{2\pi mkT}{h^2}\right)^{3N/2} = \frac{Z_1^N}{N!}\;,
\end{aligned}
\tag{4.47}
$$

where

$$Z_1 = Z(T,V,1) = V\left(\frac{2\pi mkT}{h^2}\right)^{3/2}$$

is the one-particle partition function. With the help of Stirling's formula (4.28), the free energy adds up to

$$F(T,V,N) = -NkT\ln\left[\frac{V}{N}\left(\frac{2\pi mkT}{h^2}\right)^{3/2}\right] - NkT\;.$$

Using the first equation from (4.46), we find

$$S(T,V,N) = Nk \ln \left[\frac{V}{N} \left(\frac{2\pi mkT}{h^2} \right)^{3/2} \right] + \frac{5}{2} Nk .$$

This relation is identical to the microcanonical entropy following from (4.32). For the calculation of the grand canonical partition function and its potential, we use the result (4.47) and proceed as follows:

$$Y(T,V,\mu) = \sum_N e^{\frac{N\mu}{kT}} Z(T,V,N) = \sum_N \frac{1}{N!} \left(Z_1 e^{\mu/kT} \right)^N = \exp \left(Z_1 e^{\mu/kT} \right)$$

$$= \exp \left[V \left(\frac{2\pi mkT}{h^2} \right)^{3/2} e^{\mu/kT} \right] \tag{4.48}$$

$$\Longrightarrow J(T,V,\mu) = -kTV \left(\frac{2\pi mkT}{h^2} \right)^{3/2} e^{\mu/kT} .$$

Using the second equation from (4.46), this results in

$$S(T,V,\mu) = kV \left(\frac{2\pi mkT}{h^2} \right)^{3/2} e^{\mu/kT} \left(\frac{5}{2} - \frac{\mu}{kT} \right) .$$

To compare this expression with (4.32), we need to eliminate μ. For this, we calculate

$$N = kT \left(\frac{\partial \ln Y}{\partial \mu} \right)_T = V \left(\frac{2\pi mkT}{h^2} \right)^{3/2} e^{\mu/kT}$$

$$\Longrightarrow \frac{\mu}{kT} = \ln \left[\frac{N}{V} \left(\frac{2\pi mkT}{h^2} \right)^{-3/2} \right]$$

$$\Longrightarrow S(T,V,N) = Nk \left\{ \ln \left[\frac{V}{N} \left(\frac{2\pi mkT}{h^2} \right)^{3/2} \right] + \frac{5}{2} \right\} .$$

This equation is also in agreement with (4.32).

59. Maxwell's velocity distribution. Calculate the velocity distribution for an atom of an ideal gas confined to a volume V.

Solution. From the viewpoint of the atom of interest, the remaining atoms of the gas can be seen as a heat bath that keeps its temperature at constant value. We shall therefore use the canonical ensemble. Following (4.47), the one-particle partition function is

$$Z(T,V,1) = Z_1 = \frac{V}{h^3} \int d^3p \exp \left(-\frac{p^2}{2mkT} \right) = V \left(\frac{2\pi mkT}{h^2} \right)^{3/2} .$$

The average momentum is calculated as

$$\bar{p} = \frac{V}{h^3 Z_1} \int d^3p \, p \exp \left(-\frac{p^2}{2mkT} \right) .$$

This shows that

$$W(\boldsymbol{p})\mathrm{d}^3 p = \frac{V}{h^3 Z_1} \exp\left(-\frac{\boldsymbol{p}^2}{2mkT}\right) \mathrm{d}^3 p$$

is the probability of finding the atom with a momentum in the interval $[\boldsymbol{p} : \boldsymbol{p} + \mathrm{d}^3 p]$ and that

$$W(p)\mathrm{d}p = \frac{V}{h^3 Z_1} \exp\left(-\frac{p^2}{2mkT}\right) 4\pi p^2 \mathrm{d}p$$

is the probability that the atom has an absolute momentum in the interval $[p : p + \mathrm{d}p]$. Setting $p = mv$, the normalized Maxwell velocity distribution for a single atom is found to be

$$W(v) = 4\pi \left(\frac{m}{2\pi kT}\right)^{3/2} v^2 \exp\left(-\frac{mv^2}{2kT}\right) .$$

4.4 Entropy and Information Theory

In this section we discuss information theory as developed by Shannon. The goal of this theory is to make predictions based on incomplete information. This is clearly similar to what we have been trying to do so far in the framework of statistical physics. Thus, it will come as no surprise that the methods of information theory, if correctly interpreted, can be used in statistical physics.

Having shown that Shannon entropy is equivalent to statistical entropy, we return to the three ensembles discussed in the previous sections and show how these ensembles arise naturally as solutions to a variational problem of the Shannon entropy.

4.4.1 Information Theory and Shannon Entropy

The macrostate of a system is defined via a set of probabilities $\{P_1, P_2, \ldots\} = \{P\}$ of microstates. The allocation of a set of probabilities to a set of occurrences (microstates) can be said to represent information.

Derivation of the Shannon entropy. Shannon showed quite generally that a measure of information or of lack of information is given by an *entropy function* $S(\{P\})$, which is defined by the following conditions:

- $S(\{P\})$ is a continuous-differentiable and unique function of the (normalized) probabilities $\{P\}$.

- For N equiprobable states, $P_i = 1/N$, the entropy is a monotonously growing function of N. For this specific case, we introduce the notation

$$I(N) = S\left(\left\{P_i = \frac{1}{N}\right\}\right) .$$

- The uncertainty associated with a set of probabilities remains unchanged if the states are grouped into subsets. Suppose such a partitioning is as follows:

$$w_1 = \sum_{i=1}^{n_1} P_i \ , \ w_2 = \sum_{i=n_1+1}^{n_2} P_i \ , \ \dots \ , \ w_k = \sum_{i=n_{k-1}+1}^{n_k=N} P_i \ .$$

Then, we have

$$S(\{P\}) = S(\{w\}) + \sum_{j=1}^{k} w_j S\left(\frac{\{P\}}{w_j}\right) \ .$$

The second term necessarily contains a factor w_j as this is the probability of an event actually being part of the group w_j.

To find $S(\{P\})$, we start by looking at the case of identical probabilities,

$$P_1 = \frac{1}{nm} \ , \ \dots \ , \ P_{n\cdot m} = \frac{1}{nm} \ , \ n, m \in \mathbb{N},$$

which are grouped into m equally sized subsets:

$$w_1 = \sum_{i=1}^{n} P_i = \frac{1}{m} \ , \ \dots \ , \ w_m = \sum_{i=(m-1)n+1}^{nm} P_i = \frac{1}{m} \ .$$

The last of Shannon's conditions implies that

$$I(nm) = I(m) + \sum_{j=1}^{m} \frac{1}{m} I(n) = I(m) + I(n) \ .$$

Since $S(\{P\})$ is continuous, we can differentiate with respect to n. Letting $p = nm$, we find

$$m\frac{\mathrm{d}}{\mathrm{d}p} I(p) = \frac{\mathrm{d}}{\mathrm{d}n} I(n) \ .$$

Multiplication of both sides by n yields

$$p\frac{\mathrm{d}}{\mathrm{d}p} I(p) = n\frac{\mathrm{d}}{\mathrm{d}n} I(n) = \text{const} \ .$$

This term has to be a constant since we can change p without variation of n. So we finally arrive at

$$I(n) = k \ln n \ , \quad k = \text{const} \ . \tag{4.49}$$

Next, consider the case of identical probabilities that are grouped into subsets of different sizes:

$$P_i = \frac{1}{n} \ , \ w_j = \frac{\alpha_j}{n} \ , \ \sum_{j} \alpha_j = n \ , \ n, \alpha_j \in \mathbb{N} \ .$$

Using (4.49) and Shannon's third condition, we see that

$$I(n) = S(\{w\}) + \sum_j \left(\frac{\alpha_j}{n}\right) I(\alpha_j)$$

$$\Longrightarrow S(\{w\}) = \sum_j \left(\frac{\alpha_j}{n}\right) [I(n) - I(\alpha_j)] = -k \sum_j \left(\frac{\alpha_j}{n}\right) [\ln \alpha_j - \ln n]$$

$$= -k \sum_j \left(\frac{\alpha_j}{n}\right) \ln \left(\frac{\alpha_j}{n}\right) = -k \sum_j w_j \ln w_j \ .$$

Substituting $w_i \to P_i$, we can therefore define:

Definition: Shannon entropy

The Shannon entropy of a system with relative probabilities $\{P_1, P_2, \ldots\}$
is defined as

$$S = -k \sum_i P_i \ln P_i \ , \ k > 0 \ , \ \sum_i P_i = 1 \ .$$

The quantity $- \ln P_i$ is called *surprise*.

Note that $P_i = 0$ implies that $P_i \ln P_i$ is identical to zero. This means that
impossible events do not contribute to the entropy. The constant k in the
above definition is generally set to 1 or to $1/\ln 2$ (information theory). It is
now easy to see that Shannon entropy and statistical entropy are equivalent
if k is identified with the Boltzmann constant: if all P_i are set to the constant
microcanonical probability $1/\Omega$, the entropy becomes

$$S = k \sum_i \frac{1}{\Omega} \ln \Omega = k \ln \Omega \ . \tag{4.50}$$

Properties of the Shannon entropy. The Shannon entropy has the fol-
lowing properties:

- The entropy is nonnegative since k is a positive constant.

- If any one of the probabilities is one, $P_i = 1$, implying that all other prob-
abilities are identically zero, then S is identical to zero. This corresponds
to an experiment with a unique outcome.

- For a set of equiprobable states, $P_i = P$, the entropy S is maximal since
with $\sum_i dP_i = 0$, we have

$$dS = -k \sum_i (\ln P_i + 1) dP_i = -k \ln P \sum_i dP_i - k \sum_i dP_i = 0 \ .$$

- For independent events, the entropy S is an additive quantity:

$$S_{12} = -k \sum_{i,j} P_i P_j \ln(P_i P_j)$$

$$= -k \sum_{i,j} (P_i P_j \ln P_i + P_i P_j \ln P_j)$$

$$= S_1 \sum_j P_j + S_2 \sum_i P_i = S_1 + S_2 \ .$$

- If $\{P\}$ corresponds to a mixed state with the density matrix

$$\rho = \sum_i |i\rangle\, P_i \,\langle i| \ ,$$

the Shannon entropy is given by

$$S(\rho) = -k \sum_i P_i \ln P_i = -k \mathrm{tr}\,(\rho \ln \rho) \ .$$

We now prove that Shannon entropy is maximal for an equilibrium distribution. Consider the density operator of a given equilibrium distribution $\{P\}$,

$$\rho = \sum_n |n\rangle\, P_n \,\langle n| \ ,$$

and, additionally, another arbitrary distribution $\{P'\}$ with the corresponding density matrix

$$\rho' = \sum_{n'} |n'\rangle\, P'_{n'} \,\langle n'| \ ,$$

where it is assumed that both basis systems $\{|n\rangle\}$ and $\{|n'\rangle\}$ span the same Hilbert space. Next, we introduce a function known as *Boltzmann's \mathcal{H}-function*,

$$\mathcal{H} = \mathrm{tr}\,[\rho'(\ln \rho - \ln \rho')] \ ,$$

into which we insert the identity operators $\sum_{n'} |n'\rangle\,\langle n'|$ and $\sum_n |n\rangle\,\langle n|$:

$$\mathcal{H} = \sum_{n'} P'_{n'} \left[\langle n'|\ln \rho |n'\rangle - \ln P'_{n'}\langle n'|n'\rangle \right]$$

$$= \sum_{n,n'} P'_{n'} \left[\langle n'|\ln \rho |n\rangle\langle n|n'\rangle - \ln P'_{n'}\langle n'|n\rangle\langle n|n'\rangle \right]$$

$$= \sum_{n,n'} P'_{n'} |\langle n|n'\rangle|^2 (\ln P_n - \ln P'_{n'})$$

$$= \sum_{n,n'} P'_{n'} |\langle n|n'\rangle|^2 \ln\left(\frac{P_n}{P'_{n'}} \right) \ . \tag{4.51}$$

Using $\ln x \le x - 1$ yields

$$\mathcal{H} \le \sum_{n,n'} |\langle n|n'\rangle|^2 (P_n - P'_{n'}) = \mathrm{tr}(\rho - \rho') = 0 \ . \tag{4.52}$$

Inserting the microcanonical probabilities $P_n = 1/\Omega$ into the equilibrium distribution ρ and using (4.50), we find that

$$\mathcal{H} = -\ln \Omega \sum_{n'} P'_{n'} \sum_{n} |\langle n| n'\rangle|^2 - \sum_{n'} P'_{n'} \ln P'_{n'} \sum_{n} |\langle n| n'\rangle|^2$$

$$= -\frac{S}{k} + \frac{S'}{k} \leq 0$$

$$\Longrightarrow S \geq S' .$$

Thus, we have found[13]

Theorem 4.12: Shannon's entropy law

The Shannon entropy of a closed system in equilibrium is maximal:

$$S = \text{maximal} .$$

This theorem provides the connection between information theory and statistical physics. In the following subsection, we will see how the ensembles are obtained by maximizing Shannon's entropy, taking into account certain boundary conditions.

4.4.2 Variation of the Entropy

Prior to reconsidering the microcanonical, canonical, and grand canonical ensembles, we consider more generally an equilibrium system that is characterized by the following averages:

$$\sum_{i} P_i A_i^{(1)} = \overline{A^{(1)}} , \quad \sum_{i} P_i A_i^{(2)} = \overline{A^{(2)}} , \quad \dots .$$

To find the corresponding set of probabilities $\{P\}$, we follow Theorem 4.12 and determine the stationary points of the Shannon entropy with the above boundary conditions. To this end, we use the method of Lagrange parameters, which yields the following variation condition:

$$\delta F(P_1, P_2, \dots) = 0 ,$$

with

$$F(P_1, P_2, \dots) = -k \sum_{i} P_i \ln P_i - \beta_1 \sum_{i} P_i A_i^{(1)} - \beta_2 \sum_{i} P_i A_i^{(2)} + \dots .$$

This results in

$$-k(\ln P_i + 1) - \beta_1 A_i^{(1)} - \beta_2 A_i^{(2)} - \dots = 0$$

$$\Longrightarrow P_i = \frac{1}{\Phi} \exp\left(-\sum_{j} \beta_j A_i^{(j)}\right) , \quad \Phi = \sum_{i} \exp\left(-\sum_{j} \beta_j A_i^{(j)}\right) ,$$

[13] In Application 61 we will use the extremum condition (4.52) for the canonical and grand canonical distributions to find similar minimum principles for the free energy and the grand canonical potential.

where β_j have been rescaled in the last step. Note that the extremal points are indeed maxima since

$$\delta^2 F(P_1, P_2, \ldots) = -k \sum_i \frac{(\delta P_i)^2}{P_i} < 0 .$$

The Lagrange parameters β_j must be chosen such that the above boundary conditions are satisfied. Differentiation of the general partition function Φ gives

$$\frac{\partial \ln \Phi}{\partial \beta_m} = -\frac{1}{\Phi} \sum_i A_i^{(m)} \exp\left(-\sum_j \beta_j A_i^{(j)}\right) = -\overline{A^{(m)}} \tag{4.53}$$

$$\frac{\partial^2 \ln \Phi}{\partial \beta_m^2} = \overline{A^{(m)2}} - \overline{A^{(m)}}^2 = \left(\Delta A^{(m)}\right)^2 .$$

Since the order of differentiation is interchangeable, it follows from (4.53) that

$$\frac{\partial \overline{A^{(m)}}}{\partial \beta_n} = -\frac{\partial^2 \ln \Phi}{\partial \beta_n \partial \beta_m} = -\frac{\partial^2 \ln \Phi}{\partial \beta_m \partial \beta_n} = \frac{\partial \overline{A^{(n)}}}{\partial \beta_m} .$$

These equations are known as *Maxwell's integrability conditions*, and we shall be using them extensively in the following. For the entropy, we find

$$S = -k \sum_i P_i \ln P_i$$

$$= -\frac{k}{\Phi} \sum_i \exp\left(-\sum_j \beta_j A_i^{(j)}\right)\left(-\ln \Phi - \sum_j \beta_j A_i^{(j)}\right)$$

$$= k \ln \Phi + k \sum_j \beta_j \overline{A^{(j)}} . \tag{4.54}$$

Using

$$\frac{\partial \ln \Phi}{\partial \overline{A^{(m)}}} = \frac{1}{\Phi} \sum_i \exp\left(-\sum_j \beta_j A_i^{(j)}\right)\left(-\sum_j A_i^{(j)} \frac{\partial \beta_j}{\partial \overline{A^{(m)}}}\right)$$

$$= -\sum_j \overline{A^{(j)}} \frac{\partial \beta_j}{\partial \overline{A^{(m)}}}$$

and (4.54) finally leads to

$$\frac{\partial S}{\partial \overline{A^{(m)}}} = k\left(\frac{\partial \ln \Phi}{\partial \overline{A^{(m)}}} + \beta_m + \sum_j \overline{A^{(j)}} \frac{\partial \beta_j}{\partial \overline{A^{(m)}}}\right) = k\beta_m .$$

Theorem 4.13: Principle of maximal entropy and partition function

The principle of maximal entropy with general boundary conditions of the type

$$\sum_i P_i A_i^{(1)} = \overline{A^{(1)}} \,,\quad \sum_i P_i A_i^{(2)} = \overline{A^{(2)}} \,,\quad \dots$$

yields the (normalized) probability distribution

$$P_i = \frac{1}{\Phi} \exp\left(-\sum_j \beta_j A_i^{(j)}\right) \,,\quad \sum_i P_i = 1 \,,$$

with the partition function

$$\Phi = \sum_i \exp\left(-\sum_j \beta_j A_i^{(j)}\right)$$

and entropy

$$S = k \ln \Phi + k \sum_j \beta_j \overline{A^{(j)}} \,.$$

Furthermore, we have the relations

$$\overline{A^{(m)}} = -\frac{\partial \ln \Phi}{\partial \beta_m} \,,\quad \left(\Delta A^{(m)}\right)^2 = \frac{\partial^2 \ln \Phi}{\partial \beta_m^2} \,,\quad \frac{\partial S}{\partial \overline{A^{(m)}}} = k\beta_m$$

as well as Maxwell's integrability conditions

$$\frac{\partial \overline{A^{(m)}}}{\partial \beta_n} = \frac{\partial \overline{A^{(n)}}}{\partial \beta_m} \,.$$

With the help of this theorem, we can now easily deduce the probabilities for any ensemble, given its respective boundary conditions:

Generalized grand canonical ensemble. Let us first consider the *generalized grand canonical ensemble* that is subject to the conditions

$$\sum_i P_i E_i = \overline{E} \,,\quad \sum_i P_i N_i = \overline{N} \,,\quad \sum_i P_i A_i^{(j)} = \overline{A^{(j)}} \,,\quad j = 1, 2, \dots \,.$$

The corresponding probabilities are

$$P_i = \frac{1}{\Phi} \exp\left(-\beta_E E_i - \beta_N N_i - \sum_j \beta_j A_i^{(j)}\right) \,,$$

with the generalized grand canonical partition function

$$\Phi = \sum_i \exp\left(-\beta_E E_i - \beta_N N_i - \sum_j \beta_j A_i^{(j)}\right) \; .$$

The entropy is

$$S = k \ln \Phi + k\beta_E \overline{E} + k\beta_N \overline{N} + k \sum_j \beta_j \overline{A^{(j)}} \; .$$

Its derivatives with respect to the averages are given by

$$\frac{\partial S}{\partial \overline{E}} = k\beta_E \; , \quad \frac{\partial S}{\partial \overline{N}} = k\beta_N \; , \quad \frac{\partial S}{\partial \overline{A^{(j)}}} = k\beta_j \; .$$

Replacing the Lagrange parameters β_E, β_N, and β_j by the quantities T, μ, and β_j', respectively, such that

$$\beta_E = \frac{1}{kT} \; , \quad \beta_N = -\frac{\mu}{kT} \; , \quad \beta_j = \frac{\beta_j'}{kT} \; ,$$

we can write the grand canonical potential as

$$-kT \ln \Phi = \overline{E} - TS - \mu\overline{N} + \sum_j \beta_j' \overline{A^{(j)}} \; .$$

Now it is straightforward to deduce the corresponding relations for the grand canonical, canonical, and microcanonical ensembles by successively dropping the redundant boundary conditions from the equations of the generalized grand canonical ensemble.

Grand canonical ensemble.

Constraints: $\sum_i P_i E_i = \overline{E} \; , \quad \sum_i P_i N_i = \overline{N}$

$$\Longrightarrow \begin{cases} P_i = \dfrac{1}{Y}\mathrm{e}^{-\beta_E E_i - \beta_N N_i} \; , \quad Y = \sum_i \mathrm{e}^{-\beta_E E_i - \beta_N N_i} \\[2mm] S = k \ln Y + k\beta_E \overline{E} + k\beta_N \overline{N} \\[2mm] -kT \ln Y = \overline{E} - TS - \mu\overline{N} \; . \end{cases}$$

Canonical ensemble.

Constraints: $\sum_i P_i E_i = \overline{E}$

$$\Longrightarrow \begin{cases} P_i = \dfrac{1}{Z}\mathrm{e}^{-\beta_E E_i} \; , \quad Z = \sum_i \mathrm{e}^{-\beta_E E_i} \\[2mm] S = k \ln Z + k\beta_E \overline{E} \\[2mm] -kT \ln Z = \overline{E} - TS \; . \end{cases}$$

Microcanonical ensemble.

No constraints in form of averages

$$\implies \begin{cases} P_i = \dfrac{1}{\Omega} \ , \ \Omega = \sum_i 1 \\ S = k \ln \Omega \ . \end{cases}$$

Obviously, the information theoretical approach leads to the same ensemble equations as the statistical one (Sections 4.2 and 4.3). This shows the equivalence of the Shannon and the statistical entropy. Many textbooks prefer to use Shannon's ansatz rather than the statistical approach due to its formal convenience and transparency. We have decided to present both approaches in order to highlight their fundamental relationship.

Gibbs's fundamental form. To end this section, we consider the energy exchange of an equilibrium system with its environment where we work in the canonical ensemble. The exchange of energy can be engineered either by a change of the relative probabilities P_i or by a change of the energy values E_i, which in turn is caused by a (slow) change of the external parameters x_j. Due to

$$\overline{E} = \sum_i P_i E_i \ ,$$

we have

$$\begin{aligned} \mathrm{d}\overline{E} &= \sum_i (\mathrm{d}P_i E_i + P_i \mathrm{d}E_i) = \sum_i \mathrm{d}P_i E_i + \sum_i P_i \sum_j \frac{\partial E_i}{\partial x_j} \mathrm{d}x_j \\ &= \sum_i \mathrm{d}P_i E_i - \sum_j X_j \mathrm{d}x_j \ , \end{aligned} \qquad (4.55)$$

where the quantities

$$X_j = -\sum_i P_i \frac{\partial E_i}{\partial x_j} = -\frac{\partial \overline{E}}{\partial x_j} = \left(\frac{\partial S}{\partial \overline{E}}\right)^{-1} \frac{\partial S}{\partial x_j} = T \frac{\partial S}{\partial x_j}$$

are the conjugate generalized forces to x_j [compare to (4.21) and (4.23)]. Using the canonical probabilities $P_i = \mathrm{e}^{-\beta E_i}/Z$, the change in entropy is given by ($\sum_i \mathrm{d}P_i = 0$)

$$\begin{aligned} \mathrm{d}S &= -k\mathrm{d}\left(\sum_i P_i \ln P_i\right) = -k \sum_i \mathrm{d}P_i(\ln P_i + 1) = -k \sum_i \mathrm{d}P_i \ln P_i \\ &= k\beta \sum_i \mathrm{d}P_i E_i \ . \end{aligned}$$

With this, (4.55) finally turns into

$$\mathrm{d}\overline{E} = \frac{1}{k\beta}\mathrm{d}S - \sum_j X_j \mathrm{d}x_j = T\mathrm{d}S - \sum_j X_j \mathrm{d}x_j \ . \qquad (4.56)$$

This is called *Gibbs's fundamental form*. We shall derive it in another way in the next section. It is fundamental to thermodynamics in that all thermodynamical relations can be obtained from it.

Summary

- **Shannon's entropy** is the central concept of **information theory**. It is a measure of the incomplete information inherent to a physical system.

- Shannon's entropy is maximal for a system in equilibrium.

- The principle of maximal entropy can be used to formulate a variational principle that yields results in agreement with those of statistical physics.

- The different definitions of entropy (information theoretical, statistical, and thermodynamical) are equivalent.

Applications

60. Master equation and Boltzmann's \mathcal{H}-theorem. Consider the case of time-dependent probabilities and find an equation for their temporal evolution (*master equation*). Using this equation, calculate *Boltzmann's \mathcal{H}-function*

$$\mathcal{H} = \sum_r P_r \ln P_r = \overline{\ln P}$$

and study its time dependence.

Solution. Recall the discussion of atomic transitions in Section 3.8 and consider a small time-dependent perturbation \mathbf{H}' of the Hamilton operator $\mathbf{H}(t) = \mathbf{H}^{(0)} + \mathbf{H}'(t)$, with $\mathbf{H}' \ll \mathbf{H}^{(0)}$. The perturbation induces transition probabilities W_{rs} between states r and s of equal energy, where we have the symmetry $W_{rs} = W_{sr}$. Thus, it follows that

$$\frac{\mathrm{d}P_r}{\mathrm{d}t} = \sum_s (P_s W_{sr} - P_r W_{rs}) = \sum_s W_{rs}(P_s - P_r) \ .$$

This is the master equation. Note that it is not invariant under time reversal $t \to -t$, which implies that it describes an irreversible process. For Boltzmann's \mathcal{H}-function, we find

$$\frac{\mathrm{d}\mathcal{H}}{\mathrm{d}t} = \sum_r \left(\frac{\mathrm{d}P_r}{\mathrm{d}t} \ln P_r + \frac{\mathrm{d}P_r}{\mathrm{d}t} \right) = \sum_{r,s} W_{rs} (P_s - P_r)(\ln P_r + 1) \ .$$

Interchanging the summation indices and then adding both expressions yields

$$\frac{\mathrm{d}\mathcal{H}}{\mathrm{d}t} = -\frac{1}{2} \sum_{r,s} W_{rs}(P_r - P_s)(\ln P_r - \ln P_s) \ . \tag{4.57}$$

Now, $\ln P_r$ is a monotonously growing function of P_r, so that if $P_r > P_s$, then $\ln P_r > \ln P_s$, too, and vice versa. This implies

$$(P_r - P_s)(\ln P_r - \ln P_s) \geq 0 .$$

Since W_{rs} is always positive, each term in the sum (4.57) must also be positive. We therefore find

$$\frac{d\mathcal{H}}{dt} \leq 0 .$$

This is *Boltzmann's \mathcal{H}-theorem* [compare to (4.52)]. Since \mathcal{H} and entropy are related via $S = -k\mathcal{H}$, this theorem also implies that the entropy only ever increases. The equality holds only if $P_r = P_s$ for all states for which transitions are possible, and this is equivalent to all possible states satisfying $P_r = $ const. Obviously, this is true for an equilibrium system and corresponds to the second postulate of statistical physics. Thus, we have shown once again that entropy is maximal in equilibrium.

61. Extremum conditions in the canonical and grand canonical ensembles. Using Boltzmann's \mathcal{H}-function, derive the extremum conditions of the free energy $F = \overline{E} - TS$ and of the grand canonical potential $J = \overline{E} - TS - \mu\overline{N}$ for an equilibrium system in contact with a heat bath (former case) and with a heat and particle bath.

Solution. Inserting the canonical distribution $P_n = e^{-\beta E_n}/Z$ into (4.51), we find

$$k\mathcal{H} = -k \sum_{n,n'} P'_{n'} |\langle n| n'\rangle|^2 (\ln Z + \beta E_n) + S'$$

$$= -k\ln Z - k\beta \operatorname{tr}(\boldsymbol{\rho}' \boldsymbol{H}) + S' = -k\ln Z - k\beta\overline{E'} + S' \leq 0 .$$

This yields

$$-kT\ln Z = F = \overline{E} - TS \leq \overline{E'} - TS' .$$

Thus, the free energy of an equilibrium system with given temperature is minimal. Repeating the derivation for the grand canonical distribution by inserting the probabilities $P_n = e^{-\beta(E_n - \mu N_n)}/Y$ into (4.51), we find

$$k\mathcal{H} = -k \sum_{n,n'} P'_{n'} |\langle n| n'\rangle|^2 (\ln Y + \beta E_n - \beta\mu N_n) + S'$$

$$= -k\ln Y - k\beta \operatorname{tr}(\boldsymbol{\rho}' \boldsymbol{H}) + k\beta\mu \operatorname{tr}(\boldsymbol{\rho}' \boldsymbol{N}) + S'$$

$$= -k\ln Y - k\beta\overline{E'} + \beta\mu\overline{N'} + S' \leq 0 .$$

It follows that

$$-kT\ln Y = J = \overline{E} - TS - \mu\overline{N} \leq \overline{E'} - TS' - \mu\overline{N'} .$$

This means: at given temperature and chemical potential, the grand canonical potential of an equilibrium system is minimal. In the framework of thermodynamics, Section 4.5, we will derive these extremum conditions in a different way.

4.5 Thermodynamics

Thermodynamics is concerned with the macroscopic description of systems in equilibrium. The theory was formulated toward the middle of the 19th century, at a time when the microscopic structure of matter had not yet been fully exposed and when a statistical approach in form of ensemble theories was not yet available. Thermodynamics is therefore a purely phenomenological theory, whose fundamental laws are based on experimental experience and are axiomatically postulated. In particular, the existence of an entropy function S and its extremal property is postulated. Its exact form is only verifiable indirectly through experiment. We have mentioned before that statistical physics and thermodynamics are closely connected: statistical physics provides the microscopic explanation for the macroscopically derived concepts and definitions of thermodynamics. This connection is manifest in the Boltzmann equation $S = k \ln \Omega$.

This section deals with the theory of thermodynamics. To start, we introduce and discuss the three *laws of thermodynamics*. Next, we look at *thermodynamic potentials*, which describe macroscopic systems by various sets of independent state variables. *Thermic coefficients* are introduced to describe changes of states, and relations are derived between them. Equilibrium conditions for open systems are found from the thermodynamic potentials in the form of extremum conditions. They are related to stability conditions of the thermic coefficients. Finally, we discuss *heat machines* and use the first two laws of thermodynamics to find criteria for their realizability.

Before we start, we give a short summary of key terms in thermodynamics, some of which may have already been used in the context of statistical physics in prior sections.

- Any macroscopic system is a thermodynamic system. Its thermodynamic state is described by a set of *thermodynamic state variables*. These are macroscopic measurable quantities such as temperature T, pressure P, etc., which possess a well-defined value.

- A system is in *thermodynamic equilibrium* if its thermodynamic state is time-independent.

- An *equation of state* is a functional relation between the thermodynamic state variables of the equilibrium system. In particular, we call

$$P = P(T, V, N) \qquad \text{thermic state equation}$$
$$E = E(T, V, N) \qquad \text{caloric state equation .}$$

- A change of state of a system is said to be *quasi-static* if the external parameters change so slowly that the system moves through a set of equilibrium states. Furthermore, the change of the system is called *reversible* if a reversal of the temporal direction of change of the external parameters results in the system moving through the same states in reverse order. If

this does not hold, the system change is called *irreversible*. Any reversible system change is therefore also quasi-static, but the reverse is in general not true.

- The following types of system changes are specifically defined:

 Adiabatic: no exchange of heat: $\Delta Q = 0$

 Isentropic: no exchange of entropy: $\Delta S = 0$

 Isochore: no exchange of volume: $\Delta V = 0$

 Isothermic: no exchange of temperature: $\Delta T = 0$

 Isobaric: no exchange of pressure: $\Delta P = 0$.

- If a system is described by an intensive state variable, we know from statistical physics that the corresponding extensive quantity is specified only as an average value. However, for a macroscopic system, this average is very sharp and can, in practice, also be regarded as a state variable. We will do this in the following by using the symbols E and N instead of \overline{E} and \overline{N}. In this context, we also note that in many textbooks, the statistical average of the energy \overline{E} is denoted by *inner energy U*.

4.5.1 Laws of Thermodynamics

Thermodynamics is based on the following empirical laws:

Laws of thermodynamics

First law
For any system, the energy E is an extensive state variable that is conserved if the system is closed. If a system exchanges energy with its environment, the total differential dE is given by

$$dE = \hat{d}Q + \hat{d}W .$$

Here, $\hat{d}Q$ denotes the total amount of heat the system has absorbed, and $\hat{d}W$ is the total mechanical work done on the system.

Second law
Part 1: there exists an extensive state variable, entropy S, and an intensive quantity, the absolute temperature T, with the following property: for a nonisolated system, which absorbs the heat $\hat{d}Q$ during a quasi-static process, we have

$$dS = \frac{\hat{d}Q}{T} \qquad \text{for reversible processes}$$

$$dS > \frac{\hat{d}Q}{T} \qquad \text{for irreversible processes .}$$

\triangleright

Part 2: the entropy of an isolated system can only increase with time and is maximal at equilibrium: $\Delta S \geq 0$.

Third law (Nernst's heat law)
At absolute zero temperature $T = 0$, the entropy of an equilibrium system approaches zero: $S \xrightarrow{T \to 0} 0$.

To the first law. The first law distinguishes two different types of energy. First, mechanical work $\hat{\mathrm{d}}W$, for which examples are a change of volume or number of particles. Second, the system can absorb energy in the form of heat $\hat{\mathrm{d}}Q$ without work being done on the system.

For the infinitesimal changes $\hat{\mathrm{d}}Q$ and $\hat{\mathrm{d}}W$, we have introduced the symbol $\hat{\mathrm{d}}$, which indicates that Q and W are not state variables and do not have well-defined values in an equilibrium state. For example, consider a *cyclic process (cycle)*, in which a system returns to its original state via a particular set of intermediate states. For such a process, the total change of any state variable is identical to zero, which means for the case of energy

$$\oint \mathrm{d}E = 0 \ .$$

However, the changes of heat and work are related by (see Subsection 4.5.5)

$$\oint \hat{\mathrm{d}}Q = -\oint \hat{\mathrm{d}}W \neq 0 \ .$$

In other words: the change of heat and mechanical work depends on the type of exchange process at hand, whereas the total change of energy is completely determined by just the initial and final states. Thus, contrary to $\mathrm{d}E$, the quantities $\hat{\mathrm{d}}Q$ and $\hat{\mathrm{d}}W$ are not total differentials.

To the second law. During a quasi-static change of a system, the external parameters are changed so slowly that all intermediate states are also equilibrium states. In this case, the work done on the system is given by

$$\hat{\mathrm{d}}W = -\sum_i X_i \mathrm{d}x_i \ , \quad X_i = \text{generalized force} \ .$$

Using the first and second laws, this gives Gibbs's fundamental form [compare to (4.56)]:

Theorem 4.14: Gibbs's fundamental form

For any quasi-static change of state,

$$\mathrm{d}E = \hat{\mathrm{d}}Q + \hat{\mathrm{d}}W \leq T\mathrm{d}S - \sum_i X_i \mathrm{d}x_i \ .$$

The identity holds for reversible processes.

For reversible processes, this equation solved for $\mathrm{d}S$ is

$$dS = \frac{dE}{T} + \sum_i \frac{X_i}{T} dx_i .$$

On the other hand, taking the total differential of the entropy,

$$dS = \frac{\partial S(E, x)}{\partial E} dE + \sum_i \frac{\partial S(E, x)}{\partial x_i} dx_i ,$$

and comparing the last two equations, we find

$$\frac{1}{T} = \frac{\partial S(E, x)}{\partial E} , \quad X_i = T \frac{\partial S(E, x)}{\partial x_i} .$$

These expressions correspond exactly to the definitions of temperature (4.21) and to those of the generalized forces (4.23) of statistical physics. Thus, we see explicitly the equivalence of the statistical and the thermodynamic concepts of "entropy", "temperature" and "generalized forces". In particular, we can carry over Theorem 4.9 of Subsection 4.2.3 on the equilibrium conditions of a closed two-part system without any changes.

In thermodynamics, the temperature scale is defined by setting a particular reference temperature. One conventionally chooses the triple point of water, where all three phases of water (vapor, water, and ice) are in equilibrium. This is taken as $T_t = 273.16$ K (Kelvin). Accordingly, 1 K is the $1/273.16$th part of the temperature difference between $T = 0$ and $T = T_t$. This convention fixes the Boltzmann constant to be $k = 1.38054 \cdot 10^{-23}$ J/K.

The second part of the second law defines a direction of time, since $\Delta S \geq 0$ implies $dS/dt \geq 0$. However, we know that entropy can be explained microscopically using the laws of (quantum) mechanics, and these are theories that remain invariant under time reversal. It is still a matter of debate how the thermodynamic direction of time can be explained quantum mechanically, so we will take $\Delta S \geq 0$ as a simple experimental result. Note that even in electrodynamics and quantum mechanics we introduced, by hand, a preferred direction of time by choosing retarded rather than advanced potentials (Subsection 2.2.4) and by choosing outgoing rather than incoming spherical waves (Subsection 3.10.1).

To the third law. This law defines not only entropy differences, as the second law, but also entropy itself. Experimental support comes from measurements of the specific heat, which should also vanish at zero temperature. This has been verified for all investigated systems so far.

4.5.2 Thermodynamic Potentials

As mentioned previously, a state equation describes the functional interdependence of different state variables of a thermodynamic system. It is often the case that we wish to eliminate some quantities and keep others, for example, because they are more easily accessible in experiment. In general, each set of extensive system variables (S, x_1, x_2, \ldots) is related to a conjugate

set of intensive quantities (T, X_1, X_2, \ldots), so that the adequate variables can be chosen freely from either set, e.g., $(T, x_1, X_2, x_3, \ldots)$. We now determine the state equations for the following sets of independent state variables (the generalization to other combinations is straightforward):

$$(S, V, N) , \ (T, V, N) , \ (S, P, N) , \ (T, P, N) , \ (T, V, \mu) . \tag{4.58}$$

Our starting point is Gibbs's fundamental form (see Theorem 4.14)

$$dE = TdS + \mu dN - PdV ,$$

with entropy S, volume V, and particle number N as the only independent state variables. All other total differentials are obtained from this using Legendre transformations with the corresponding independent pairs of variables as given in (4.58).

Definition: Thermodynamic potentials

- *Energy E* (independent state variables: S, V, N):

$$dE = TdS - PdV + \mu dN$$

$$\Longrightarrow E = E(S, V, N) .$$

- *Free energy F* (independent state variables: T, V, N):

$$dF = d(E - TS) = -SdT - PdV + \mu dN$$

$$\Longrightarrow F = F(T, V, N) = E - TS .$$

- *Enthalpy H* (independent state variables: S, P, N):

$$dH = d(E + PV) = TdS + VdP + \mu dN$$

$$\Longrightarrow H = H(S, P, N) = E + PV .$$

- *Free enthalpy G* (independent state variables: T, P, N):

$$dG = d(H - TS) = -SdT + VdP + \mu dN$$

$$\Longrightarrow G = G(T, P, N) = E - TS + PV .$$

- *Grand canonical potential J* (independent state variables: T, V, μ):

$$dJ = d(F - \mu N) = -SdT - PdV - Nd\mu$$

$$\Longrightarrow J = J(T, V, \mu) = E - TS - \mu N .$$

The state variables E, F, H, G, and J are called *thermodynamic potentials* if they are given as a function of the corresponding *natural variables*.

As an example, to obtain the free enthalpy, one would have to take the corresponding functional relations $E(T, P, N)$, $S(T, P, N)$, and $V(T, P, N)$ and insert them into the definition $G = E - TS + PV$.

Thermodynamic forces. Writing down the total differentials for each of the above-mentioned potentials and comparing them with the corresponding definition equations yield the associated thermodynamical forces. They are

$$\left.\begin{aligned}
\left(\frac{\partial E}{\partial S}\right)_{V,N} &= T \quad, \quad \left(\frac{\partial E}{\partial V}\right)_{S,N} = -P \quad, \quad \left(\frac{\partial E}{\partial N}\right)_{S,V} = \mu \\[4pt]
\left(\frac{\partial F}{\partial T}\right)_{V,N} &= -S \quad, \quad \left(\frac{\partial F}{\partial V}\right)_{T,N} = -P \quad, \quad \left(\frac{\partial F}{\partial N}\right)_{T,V} = \mu \\[4pt]
\left(\frac{\partial H}{\partial S}\right)_{P,N} &= T \quad, \quad \left(\frac{\partial H}{\partial P}\right)_{S,N} = V \quad, \quad \left(\frac{\partial H}{\partial N}\right)_{S,P} = \mu \\[4pt]
\left(\frac{\partial G}{\partial T}\right)_{P,N} &= -S \quad, \quad \left(\frac{\partial G}{\partial P}\right)_{T,N} = V \quad, \quad \left(\frac{\partial G}{\partial N}\right)_{T,P} = \mu \\[4pt]
\left(\frac{\partial J}{\partial T}\right)_{V,\mu} &= -S \quad, \quad \left(\frac{\partial J}{\partial V}\right)_{T,\mu} = -P \quad, \quad \left(\frac{\partial J}{\partial \mu}\right)_{T,V} = -N \; .
\end{aligned}\right\} \quad (4.59)$$

Maxwell relations. From the interchangeability of partial differentiations of a function $f(x,y)$, i.e., $\partial^2 f/\partial x \partial y = \partial^2 f/\partial y \partial x$, we obtain a set of three relations for each row of (4.59). These are called *Maxwell relations*. Here, we give just one relation for each of the first four rows; they correspond to the case of constant particle number N:

$$\left.\begin{aligned}
\left(\frac{\partial T}{\partial V}\right)_{S,N} &= -\left(\frac{\partial P}{\partial S}\right)_{V,N} \\[4pt]
\left(\frac{\partial S}{\partial V}\right)_{T,N} &= \left(\frac{\partial P}{\partial T}\right)_{V,N} \\[4pt]
\left(\frac{\partial T}{\partial P}\right)_{S,N} &= \left(\frac{\partial V}{\partial S}\right)_{P,N} \\[4pt]
-\left(\frac{\partial S}{\partial P}\right)_{T,N} &= \left(\frac{\partial V}{\partial T}\right)_{P,N} \; .
\end{aligned}\right\} \quad (4.60)$$

Gibbs-Duhem relations. There is a particularly simple relation between the chemical potential μ and the free enthalpy G. Since free enthalpy is extensive, we must have

$$G(T,P,N) = Ng(T,P) \; ,$$

where g is some intensive quantity. From this and using (4.59), we obtain the *Gibbs-Duhem relation*

$$\mu = \left(\frac{\partial G}{\partial N}\right)_{T,P} = g(T,P) = \frac{G}{N} \Longrightarrow G = \mu N \; .$$

Inserting this, for example, into the free energy F and into the grand canonical potential J, we obtain two further Gibbs-Duhem relations as

$$F = \mu N - PV \; , \quad J = -PV \; . \tag{4.61}$$

Complete thermodynamic information. Clearly, knowledge of just one of the thermodynamic potentials as a function of its natural variables yields all other potentials (via Legendre transformations) and thus all other thermodynamic state variables. Therefore, each and any of the potentials contains the complete thermodynamic information of the system under consideration. As an example, we consider the free enthalpy $H = H(S, P, N)$ and show how to obtain the caloric state equation $E = E(T, V, N)$. Generally, one starts with the partial derivatives of the potential:

$$T = \left(\frac{\partial H}{\partial S}\right)_{P,N} = T(S, P, N) \ .$$

Solving this equation for S, we obtain $S(T, P, N)$. Furthermore, we have

$$V = \left(\frac{\partial H}{\partial P}\right)_{S,N} = V(S, P, N) = V[S(T, P, N), P, N] = V(T, P, N) \ .$$

This equation solved for P yields $P(T, V, N)$. Now, inserting $S(T, P, N)$ and $P(T, V, N)$ into $H(S, P, N)$, it follows that

$$H(S, P, N) = H\{S[T, P(T, V, N), N], P(T, V, N), N\} = H(T, V, N)$$

and, therefore, finally

$$E = H(T, V, N) - P(T, V, N)V = E(T, V, N) \ .$$

Incidentally, from $S(T, P, N)$, $P(T, V, N)$, and $E(T, V, N)$, we can also obtain the free energy as a function of its natural variables:

$$F = E(T, V, N) - TS[T, P(T, V, N), N] = F(T, V, N) \ .$$

4.5.3 State Changes and Thermic Coefficients

From an experimental point of view, one obtains information about the relations of macroscopic quantities by studying their behavior when changing other quantities. To this end, one defines thermic coefficients, which are most accessible to such experimentation:[14]

Expansion coefficient:	$\alpha = \dfrac{1}{V}\left(\dfrac{\partial V}{\partial T}\right)_P = -\dfrac{1}{V}\left(\dfrac{\partial S}{\partial P}\right)_T$
Pressure coefficient:	$\beta = \left(\dfrac{\partial P}{\partial T}\right)_V = \left(\dfrac{\partial S}{\partial V}\right)_T$
Isobaric heat capacity:	$C_P = T\left(\dfrac{\partial S}{\partial T}\right)_P = \left(\dfrac{\partial H}{\partial T}\right)_P$
Isochore heat capacity:	$C_V = T\left(\dfrac{\partial S}{\partial T}\right)_V = \left(\dfrac{\partial E}{\partial T}\right)_V \ .$

[14] In the following, we suppress the fixed particle number N.

The first two of these relations consist of second derivatives of the thermo-dynamic potentials and are contained in the Maxwell relations (4.60). The last two contain first derivatives and are obtained from the total differen-tials of dH and dE, as given above in the definitions of the thermodynamic potentials. In addition, we introduce the following compressibilities:

$$\text{Isothermic compressibility:} \quad \kappa_T = -\frac{1}{V}\left(\frac{\partial V}{\partial P}\right)_T$$

$$\text{Adiabatic compressibility:} \quad \kappa_S = -\frac{1}{V}\left(\frac{\partial V}{\partial P}\right)_S.$$

There are varying degrees of difficulty in measuring these quantities. It is therefore useful to derive some relations between them. For example, swap-ping the fixed and free variable in the pressure coefficient following the rule (A.4), we find

$$\beta = \frac{\alpha}{\kappa_T}. \tag{4.62}$$

Another relation is obtained using $S = S(T, P) = S[T, V(T, P)]$:

$$T\left(\frac{\partial S}{\partial T}\right)_P = T\left(\frac{\partial S}{\partial T}\right)_V + T\left(\frac{\partial S}{\partial V}\right)_T\left(\frac{\partial V}{\partial T}\right)_P$$

$$\Longrightarrow C_P - C_V = T\left(\frac{\partial S}{\partial V}\right)_T\left(\frac{\partial V}{\partial T}\right)_P.$$

Following rule (A.4) and the second Maxwell relation given in (4.60), we can rewrite this as

$$\left(\frac{\partial S}{\partial V}\right)_T = \left(\frac{\partial P}{\partial T}\right)_V = -\left(\frac{\partial P}{\partial V}\right)_T\left(\frac{\partial V}{\partial T}\right)_P$$

$$\Longrightarrow C_P - C_V = -T\left(\frac{\partial P}{\partial V}\right)_T\left(\frac{\partial V}{\partial T}\right)_P^2 = \frac{\alpha^2 TV}{\kappa_T}. \tag{4.63}$$

From $V = V(S, P) = V[T(S, P), P]$, we can obtain a third relation in a similar fashion:

$$\left(\frac{\partial V}{\partial P}\right)_S = \left(\frac{\partial V}{\partial T}\right)_P\left(\frac{\partial T}{\partial P}\right)_S + \left(\frac{\partial V}{\partial P}\right)_T$$

$$\Longrightarrow \kappa_T - \kappa_S = \frac{1}{V}\left(\frac{\partial V}{\partial T}\right)_P\left(\frac{\partial T}{\partial P}\right)_S.$$

If we use the third of Maxwell relations (4.60) and the chain rule (A.6), this equation becomes

$$\left(\frac{\partial T}{\partial P}\right)_S = \left(\frac{\partial V}{\partial S}\right)_P = \left(\frac{\partial V}{\partial T}\right)_P\left(\frac{\partial T}{\partial S}\right)_P$$

$$\Longrightarrow \kappa_T - \kappa_S = \frac{1}{V}\left(\frac{\partial V}{\partial T}\right)_P^2 \left(\frac{\partial T}{\partial S}\right)_P = \frac{\alpha^2 TV}{C_P} \ . \tag{4.64}$$

Finally, combining (4.63) and (4.64) gives another set of relations:

$$C_P = \frac{\alpha^2 TV}{\kappa_T - \kappa_S} \ , \quad C_V = \frac{\alpha^2 TV \kappa_S}{(\kappa_T - \kappa_S)\kappa_T} \ , \quad \frac{C_P}{C_V} = \frac{\kappa_T}{\kappa_S} \ .$$

4.5.4 Equilibrium and Stability

We shall now investigate how equilibrium conditions for nonclosed (open) systems can be formulated. We assume that such systems can exchange heat and mechanical energy in the form of volume V and particles N with their environment in a quasi-static manner. (The generalization to other forms of mechanical exchange is unproblematic.) Our starting point is Gibbs's fundamental form following from the first part of the second thermodynamical law (see Theorem 4.14):

$$T\mathrm{d}S \geq \mathrm{d}E - \mu\mathrm{d}N + P\mathrm{d}V \ . \tag{4.65}$$

Consider first the case of a closed system, so that $\mathrm{d}E = \mathrm{d}V = \mathrm{d}N = 0$. It follows that

$$(\mathrm{d}S)_{E,V,N} \geq 0 \ .$$

This means *the entropy of a closed system with constant energy, constant particle number, and constant volume never decreases and is maximal at equilibrium.* This is exactly the content of the second part of the second thermodynamical law. Thus, the second part is a necessary conclusion from the first. Next suppose the system is held externally at constant entropy and is otherwise isolated ($\mathrm{d}S = \mathrm{d}N = \mathrm{d}V = 0$). Then, we find from (4.65)

$$(\mathrm{d}E)_{S,V,N} \leq 0 \ .$$

So: *the energy of an isolated system held at constant entropy never increases and is minimal in equilibrium.* We continue with a system that is externally held at fixed temperature and is mechanically isolated, so that ($\mathrm{d}T = \mathrm{d}N = \mathrm{d}V = 0$). Using (4.65) as before, we see that

$$\begin{aligned}
T\mathrm{d}S &\geq \mathrm{d}E - \mu\mathrm{d}N + P\mathrm{d}V \\
&\geq \mathrm{d}(E - TS) + T\mathrm{d}S + S\mathrm{d}T - \mu\mathrm{d}N + P\mathrm{d}V \\
\Longrightarrow 0 &\geq \mathrm{d}F + S\mathrm{d}T - \mu\mathrm{d}N + P\mathrm{d}V \\
\Longrightarrow (\mathrm{d}F)_{T,V,N} &\leq 0 \ .
\end{aligned}$$

The free energy of a mechanically isolated system at constant temperature never increases and is minimal at equilibrium. In the same way as demonstrated, the corresponding relations for any open system with some other constant parameters can be deduced. For $\mathrm{d}S = \mathrm{d}P = \mathrm{d}N = 0$, we have

$$T\mathrm{d}S \geq \mathrm{d}E - \mu\mathrm{d}N + P\mathrm{d}V$$
$$\geq \mathrm{d}(E + PV) - V\mathrm{d}P - \mu\mathrm{d}N$$
$$\implies 0 \geq \mathrm{d}H - V\mathrm{d}P - \mu\mathrm{d}N$$
$$\implies (\mathrm{d}H)_{S,P,N} \leq 0 \ .$$

For $\mathrm{d}T = \mathrm{d}P = \mathrm{d}N = 0$, we find

$$T\mathrm{d}S \geq \mathrm{d}E - \mu\mathrm{d}N + P\mathrm{d}V$$
$$\geq \mathrm{d}(E - TS + PV) + T\mathrm{d}S + S\mathrm{d}T - V\mathrm{d}P - \mu\mathrm{d}N$$
$$\implies 0 \geq \mathrm{d}G + S\mathrm{d}T - V\mathrm{d}P - \mu\mathrm{d}N$$
$$\implies (\mathrm{d}G)_{T,P,N} \leq 0 \ ,$$

and finally for $\mathrm{d}T = \mathrm{d}\mu = \mathrm{d}V = 0$, we see that

$$T\mathrm{d}S \geq \mathrm{d}E - \mu\mathrm{d}N + P\mathrm{d}V$$
$$\geq \mathrm{d}(E - TS - \mu N) + S\mathrm{d}T + T\mathrm{d}S + N\mathrm{d}\mu + P\mathrm{d}V$$
$$\implies 0 \geq \mathrm{d}J + S\mathrm{d}T + N\mathrm{d}\mu + P\mathrm{d}V$$
$$\implies (\mathrm{d}J)_{T,V,\mu} \leq 0 \ .$$

The following simple rules emerge:

Theorem 4.15: Equilibrium conditions and thermodynamic potentials

If the natural variables of a certain thermodynamic potential are held constant, this potential never increases and has a minimum in the equilibrium state:

$$E, V, N \text{ constant} \implies (\mathrm{d}S)_{E,V,N} \geq 0 \ \text{(closed system: maximum)}$$

$$S, V, N \text{ constant} \implies (\mathrm{d}E)_{S,V,N} \leq 0$$

$$T, V, N \text{ constant} \implies (\mathrm{d}F)_{T,V,N} \leq 0$$

$$S, P, N \text{ constant} \implies (\mathrm{d}H)_{S,P,N} \leq 0$$

$$T, P, N \text{ constant} \implies (\mathrm{d}G)_{T,P,N} \leq 0$$

$$T, V, \mu \text{ constant} \implies (\mathrm{d}J)_{T,V,\mu} \leq 0 \ .$$

As an example for the minimum principle of the free energy, consider a gas in a cylinder with constant temperature, constant particle number, as well as constant volume. Inside the cylinder, we have a freely movable piston, which splits the volume V into two parts V_1 and V_2 with pressures P_1 and P_2. We seek the equilibrium position of the piston when it is released at an

arbitrary starting position. The corresponding equilibrium condition reads $(dV_2 = -dV_1)$

$$
\begin{aligned}
(dF)_{T,V,N} &= \left(\frac{\partial F}{\partial V_1}\right)_{T,N} dV_1 + \left(\frac{\partial F}{\partial V_2}\right)_{T,N} dV_2 \\
&= \left[\left(\frac{\partial F}{\partial V_1}\right)_{T,N} - \left(\frac{\partial F}{\partial V_2}\right)_{T,N}\right] dV_1 = 0 .
\end{aligned}
$$

As expected, this gives us

$$
\left(\frac{\partial F}{\partial V_1}\right)_{T,N} = \left(\frac{\partial F}{\partial V_2}\right)_{T,N} \Longleftrightarrow P_1 = P_2 .
$$

Stability. The minimum conditions of thermodynamic potentials in Theorem 4.15 are necessary but not sufficient to determine an equilibrium state. Consider, for example, the energy $E(S,V)$ and a small variation of S and V. For a stable equilibrium, we need to require not only $dE = 0$, but also

$$
d^2 E > 0 .
$$

Going through the variation, we obtain

$$
\begin{aligned}
E(S + dS, V + dV) = E(S,V) &+ \left(\frac{\partial E}{\partial S}\right)_V dS + \left(\frac{\partial E}{\partial V}\right)_S dV \\
&+ \left[\left(\frac{\partial^2 S}{\partial S^2}\right)_V (dS)^2 + 2\frac{\partial^2 S}{\partial S \partial V} dS dV \right. \\
&\left. + \left(\frac{\partial^2 E}{\partial V^2}\right)_S (dV)^2\right] + \dots \\
\Longrightarrow \left(\frac{\partial^2 E}{\partial S^2}\right)_V (dS)^2 &+ 2\frac{\partial^2 E}{\partial S \partial V} dS dV + \left(\frac{\partial^2 E}{\partial V^2}\right)_S (dV)^2 > 0 .
\end{aligned}
$$

For an arbitrary quadratic form $ax^2 + 2bxy + cy^2$ to be positive, we must have $a > 0$, $c > 0$, and $(ac - b^2) > 0$. In the present case, this implies

$$
\left(\frac{\partial^2 E}{\partial S^2}\right)_V = \left(\frac{\partial T}{\partial S}\right)_V = \frac{T}{C_V} > 0 \Longrightarrow C_V > 0 \tag{4.66}
$$

$$
\left(\frac{\partial^2 E}{\partial V^2}\right)_S = -\left(\frac{\partial P}{\partial V}\right)_S = \frac{1}{V\kappa_S} > 0 \Longrightarrow \kappa_S > 0 \tag{4.67}
$$

$$
\left(\frac{\partial^2 E}{\partial S^2}\right)_V \left(\frac{\partial^2 E}{\partial V^2}\right)_S - \left(\frac{\partial^2 E}{\partial S \partial V}\right)^2 = \frac{T}{VC_V\kappa_S} - \left(\frac{\partial T}{\partial V}\right)_S^2 > 0
$$

$$
\Longrightarrow \frac{T}{VC_V\kappa_S} > \left(\frac{\partial T}{\partial V}\right)_S^2 .
$$

In the same way, the condition $d^2 F > 0$, for example, leads to the inequality

$$\left(\frac{\partial^2 F}{\partial V^2}\right)_T = -\left(\frac{\partial P}{\partial V}\right)_T = \frac{1}{V\kappa_T} > 0 \Longrightarrow \kappa_T > 0 \;. \qquad (4.68)$$

The conditions (4.66), (4.67), and (4.68) imply that a system can be in equilibrium only if

- the temperature increases, while the system is heated at constant volume ($C_V > 0$),

- the volume decreases, while the system is compressed at constant entropy ($\kappa_S > 0$),

- the volume decreases, while the system is compressed at constant temperature ($\kappa_T > 0$).

These are specific examples of the *Le Chatelier principle*, which states that spontaneous changes of an equilibrium system induces processes that are inclined to return the system to its equilibrium state. From $C_V, \kappa_T > 0$, it also follows that $C_P > 0$ due to (4.63). And from $\kappa_T > 0$ and (4.62), one sees that α and β have the same sign; normally they are positive. An exception is water at its freezing point.

4.5.5 Heat Machines and Cyclic Processes

Historically, thermodynamics started with the study of heat machines. These are devices that transform heat into different forms of energy. Given their technological importance as well as their historical importance in the context of the industrial revolution, we shall now take a closer look at some properties of heat machines, restricting ourselves to *cyclic* machines, which, having run through a set of processes, return to their original state.

Perpetuum mobile of first and second type. Consider the closed system depicted in Fig. 4.5. It consists of a heat reservoir R at constant temperature T, a heat machine M, and a work storage S. Within one cycle, the machine withdraws a certain amount of heat ΔQ from the reservoir, transforms it into work ΔW, and deposits this into the work storage. Since the machine is now back in its initial state, we have $\Delta E_{\mathrm{M}} = 0$. According to the first law of thermodynamics, this implies

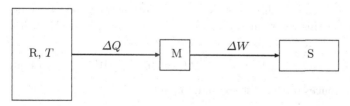

Fig. 4.5. Schematical representation of a heat machine M transforming absorbed heat ΔQ into work ΔW.

$$\Delta E_M = \Delta Q - \Delta W = 0 \implies \Delta Q = \Delta W \ .$$

This means that the machine cannot produce more work than it has taken out in the form of heat. In other words: the first law implies the impossibility of a *perpetuum mobile of the first type*. On the other hand, the machine is also not able to produce just as much work as it has absorbed in the form of heat, as this contradicts the second law of thermodynamics. To see this, we look at the changes in entropy of R, M, and S after one cycle. Since the reservoir loses an amount of heat ΔQ, we have

$$\Delta S_R \geq -\frac{\Delta Q}{T} = -\frac{\Delta W}{T} \ .$$

When the machine returns to its original state, we have

$$\Delta S_M = 0 \ .$$

Now, it is reasonable to assume that the work storage consists of very few degrees of freedom. For example, it could be a spring with just one degree of freedom. Its change in entropy will then be much smaller than that of R:

$$\Delta S_S \approx 0 \ .$$

According to the second law, the total change in entropy of the closed system as a whole must be larger than or equal to zero, such that

$$\Delta S_R + \Delta S_M + \Delta S_S \geq -\frac{\Delta W}{T} \geq 0 \implies \Delta W \leq 0 \ .$$

From this, we conclude that there are no machines whose sole action is to transform an amount of heat ΔQ into work $\Delta W = \Delta Q$ (*perpetuum mobile of the second type*). This is *Kelvin's formulation of the second law of thermodynamics*. An equivalent statement is *Clausius's formulation of the second law*: there are no machines whose sole action is to absorb the heat ΔQ from a colder heat reservoir and to deposit it to a warmer one. This claim can be easily checked by considering the balances of energy and entropy for the system $R_1 + M + R_2$.

Heat machines and efficiency. For any functional heat machine, i.e., one in accordance with the first and second laws of thermodynamics, the total amount of entropy during a single cycle must not decrease. Starting from the above-sketched system, this can be achieved by adding a further heat reservoir into which the machine deposits a part of the absorbed heat, so that the entropy of this reservoir increases by the amount required. Such a system is shown in Fig. 4.6. From the first law, we now have

$$\Delta E_M = \Delta Q_1 - \Delta Q_2 - \Delta W = 0 \implies \Delta Q_2 = \Delta Q_1 - \Delta W \ . \tag{4.69}$$

The entropy differences of the system components are

$$\Delta S_{R_1} \geq -\frac{\Delta Q_1}{T_1} \ , \quad \Delta S_{R_2} \geq \frac{\Delta Q_2}{T_2} \ , \quad \Delta S_M = 0 \ , \quad \Delta S_S \approx 0 \ .$$

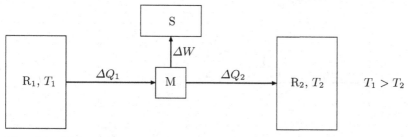

Fig. 4.6. Schematical representation of a realizable heat machine M giving out a part ΔQ_2 of the absorbed heat ΔQ_1 and transforming the rest into work ΔW.

The second law imposes the condition

$$\Delta S_{R_1} + \Delta S_{R_2} + \Delta S_M + \Delta S_S \geq \Delta Q_1 \left(\frac{1}{T_2} - \frac{1}{T_1} \right) - \frac{\Delta W}{T_2} \geq 0 \ .$$

Therefore, the maximal (positive) amount of work that M can do is bounded from above by

$$\Delta W \leq \Delta Q_1 \left(\frac{T_1 - T_2}{T_1} \right) \ , \ T_1 > T_2 \ . \tag{4.70}$$

In practice, the heat reservoir R_1 is not infinitely large, so that the heat ΔQ_1 extracted from it must be continuously replenished, for example, by burning coal or oil. Therefore, the *efficiency η* of a heat machine is usually defined as

$$\eta = \frac{\text{generated work}}{\text{invested heat}} = \frac{\Delta W}{\Delta Q_1} = \frac{\Delta Q_1 - \Delta Q_2}{\Delta Q_1} \ .$$

For realizable heat machines, we then have

$$\eta \leq \eta_{\text{ideal}} = \frac{T_1 - T_2}{T_1} \ ,$$

where the ideal efficiency η_{ideal} is possible only for reversible processes. In practice, realistic efficiencies are found to be around $\eta = 30\%$.

Of course, the process shown in Fig. 4.6 can also be reversed, so that the machine M, while performing the work ΔW, extracts heat ΔQ_2 from the reservoir R_2 and supplies heat ΔQ_1 to the reservoir R_1 (see Fig. 4.7). In this case, the signs of ΔQ_1, ΔQ_2, and ΔW are reversed, so that the absolute value equation of (4.70) turns into

$$\Delta W \geq \Delta Q_1 \left(\frac{T_1 - T_2}{T_1} \right) \ , \ T_1 > T_2$$

or, using (4.69),

$$\Delta W \geq \Delta Q_2 \left(\frac{T_1 - T_2}{T_2} \right) \ , \ T_1 > T_2 \ .$$

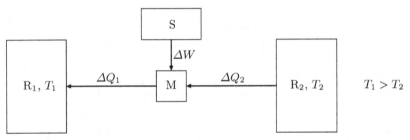

Fig. 4.7. Schematical representation of a realizable heat machine M, which, by performing the work ΔW, absorbs the heat ΔQ_2 and transforms it into the heat $\Delta Q_1 > \Delta Q_2$ (heat pump, refrigerator).

In the case of a refrigerator, the benefit is given by the heat ΔQ_2 extracted from R_2, whereas for the heat pump, it is the heat ΔQ_1 transported to R_1. For the refrigerator, we therefore define

$$\eta = \frac{\Delta Q_2}{\Delta W} \leq \eta_{\text{ideal}} = \frac{T_2}{T_1 - T_2}$$

and for the heat pump

$$\eta = \frac{\Delta Q_1}{\Delta W} \leq \eta_{\text{ideal}} = \frac{T_1}{T_1 - T_2} \, .$$

Carnot cycle. Carnot engines are *ideal heat machines*, which only ever run cyclically through reversible processes, so that their efficiency is equal to their ideal efficiency. Such a machine can be constructed (at least theoretically) as follows: let x denote an external parameter of the machine M. Then a change in x is equivalent to an amount of work done by M. Suppose the initial state of M is given by $x = x_a$ and $T = T_2$, where T_2 is the temperature of the cooler heat reservoir R_2. The machine now runs reversibly through the following four steps:

1. *Adiabatic step:* x is changed slowly, while M is kept thermically isolated until M reaches the temperature $T_1 > T_2$ ($x_a \rightarrow x_b$, $T_2 \rightarrow T_1$).

2. *Isothermic step:* M is brought into thermic contact with the warmer reservoir R_1 of temperature T_1. x is changed further, so that M absorbs the heat ΔQ_1 from R_1, while the temperature T_1 is held fixed ($x_b \rightarrow x_c$).

3. *Adiabatic step:* M is isolated thermically once again. x is changed in such a way that the temperature of M turns from T_1 back to T_2 ($x_c \rightarrow x_d$, $T_1 \rightarrow T_2$).

4. *Isothermic step:* M is brought into thermic contact with the colder reservoir R_2 of temperature T_2. The parameter x is altered until it returns to its original value x_a, while M gives up the heat ΔQ_2 to R_2 at fixed temperature T_2.

Now the cycle is closed, and the machine is back in its original state. The energy and entropy balances of the machine are as follows:

$$\Delta S_{\mathrm{M}} = 0 + \frac{\Delta Q_1}{T_1} + 0 - \frac{\Delta Q_2}{T_2} = 0 \implies \frac{\Delta Q_1}{T_1} = \frac{\Delta Q_2}{T_2}$$

$$\Delta E_{\mathrm{M}} = \Delta Q_1 - \Delta Q_2 - \Delta W = 0 \implies \Delta W = \Delta Q_1 - \Delta Q_2 \, ,$$

where

$$\Delta W = \int_{x_a}^{x_b} \mathrm{d}x X + \int_{x_b}^{x_c} \mathrm{d}x X + \int_{x_c}^{x_d} \mathrm{d}x X + \int_{x_d}^{x_a} \mathrm{d}x X = \oint \mathrm{d}x X$$

is the work done by the system. The entropy change of the combined system consisting of the reservoirs R_1 and R_2, the machine M, and the work storage S, is ($\Delta S_{\mathrm{S}} \approx 0$)

$$\Delta S = -\frac{\Delta Q_1}{T_1} + \frac{\Delta Q_2}{T_2} = 0 \, .$$

This means the Carnot cycle is indeed reversible. It can therefore be run in reverse order and thus be used for a refrigerator or a heat pump.

Consider, as an example of a Carnot process, a (not necessarily ideal) gas, which is enclosed in a cylinder with a movable piston (Fig. 4.8). In this case,

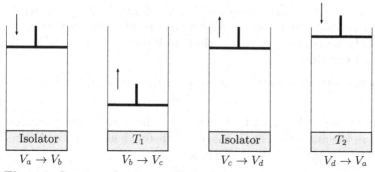

Fig. 4.8. Carnot cycle of a gas-filled cylinder with a movable piston.

the volume V is the external parameter. The four steps of the Carnot cycle can be represented in a PV-diagram as shown in Fig. 4.9. The work done by the gas during a full cycle is equal to the area enclosed by the isothermals and the adiabatics.

Summary

- **Thermodynamics** describes the behavior of macroscopic systems in equilibrium using a strictly macroscopic point of view. It is based on the three empirically motivated **laws of thermodynamics**.

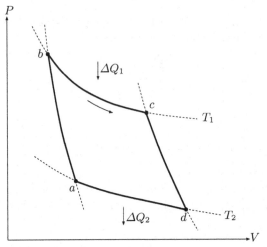

Fig. 4.9. Schematical representation of a Carnot cycle for a gas in terms of a PV-diagram. The four steps are represented by two isothermals ($\Delta T = 0$) and two adiabatics ($\Delta S = 0$). The area enclosed by these curves equals the work done within the cycle.

- The **thermodynamic potentials**, given as functions of their natural variables, are state equations with which equilibrium states of open systems can be formulated as simple minimum principles. The partial derivatives of the potentials lead to the **thermodynamic forces** and to the **Maxwell relations**.

- Each of the potentials contains the complete thermodynamic information of a system.

- State changes are described by **thermic coefficients**. Stability criteria related to the minimum conditions of the thermodynamic potentials restrict the behavior of these coefficients.

- The first two laws of thermodynamics forbid the existence of **perpetuum mobiles of the first and second type**. Realizable heat machines can only transform part of the absorbed heat into work. The rest of the heat is necessarily lost to the environment, so that the overall entropy does not decrease.

- A **Carnot cycle** describes the mechanism of heat machines in which all steps of a cycle are performed reversibly, so that they reach their ideal **efficiency**.

Applications

62. Complete thermodynamic information. Show that the state equations

$$P = P(T, V) , \quad C_V(T, V_0) \tag{4.71}$$

contain the complete thermodynamic information, i.e., that these equations yield the thermodynamic potential $E(S, V)$. Verify this statement by means of the ideal gas, for which

$$P(T, V) = \frac{TNk}{V} , \quad C_V(T, V_0) = C_V = \frac{3}{2} Nk$$

are given. N is held constant. What is the isobaric heat capacity C_P of the ideal gas?

Solution. Differentiating the second Maxwell relation in (4.60) with respect to V at constant T, we obtain

$$\left(\frac{\partial C_V}{\partial V} \right)_T = T \left(\frac{\partial^2 P}{\partial T^2} \right)_V .$$

Combining this with (4.71) leads to

$$C_V(T, V) = C_V(T, V_0) + T \int_{V_0}^{V} dV' \frac{\partial^2 P(T, V')}{\partial T^2} . \tag{4.72}$$

The same Maxwell relation also yields

$$dS = \left(\frac{\partial S}{\partial T} \right)_V dT + \left(\frac{\partial S}{\partial V} \right)_T dV = \frac{C_V}{T} dT + \left(\frac{\partial P}{\partial T} \right)_V dV$$

$$\Longrightarrow dE = TdS - PdV = C_V dT + \left[T \left(\frac{\partial P}{\partial T} \right)_V - P \right] dV .$$

Since the right-hand side of this equation is known due to (4.71) and (4.72), we obtain the functions $S(T, V)$ and $E(T, V)$ up to a constant as

$$E(T, V) = \int dT C_V + \int dV \left[T \left(\frac{\partial P}{\partial T} \right)_V - P \right] + \text{const}$$

$$S(T, V) = \int dT \frac{C_V}{T} + \int dV \left(\frac{\partial P}{\partial T} \right)_V + \text{const} .$$

Eliminating T from these equations results in $E(S, V)$ and $S(E, V)$. For the ideal gas, the last two equations yield

$$E(T, V) = E(T) = \frac{3}{2} NkT$$

$$S(T, V) = \frac{3}{2} Nk \ln T + Nk \ln V + \text{const} . \tag{4.73}$$

This gives [compare to (4.29)]

$$S(E,V) = Nk \left\{ \ln \left[V \left(\frac{2c_1 E}{3Nk} \right)^{3/2} \right] + c_2 \right\} , \quad c_1, c_2 = \text{const} .$$

The heat capacity C_P is now easily calculated:

$$S(T,P) = S[T, V(T,P)]$$
$$= Nk \left(\frac{3}{2} \ln T + \ln(TkN) - \ln P + \text{const} \right) \tag{4.74}$$

$$\implies C_P = T \left(\frac{\partial S}{\partial T} \right)_P = \frac{5}{2} Nk .$$

63. Adiabatic expansion of the ideal gas.

a. Show that for the adiabatic expansion of an ideal gas, we have

$$PV^\gamma = \text{const} , \quad \gamma = \frac{C_P}{C_V} .$$

b. Consider an ideal gas with volume V_1 and temperature T_1. Calculate its temperature T_2 if the gas is expanded adiabatically to volume V_2.

Solution.

To a. We have

$$\left(\frac{\partial V}{\partial P} \right)_S = \left(\frac{\partial V}{\partial T} \right)_S \left(\frac{\partial T}{\partial P} \right)_S . \tag{4.75}$$

From (4.73), we obtain the total differential

$$dS = \left(\frac{\partial S}{\partial T} \right)_V dT + \left(\frac{\partial S}{\partial V} \right)_T dV = \frac{C_V}{T} dT + \frac{Nk}{V} dV$$

and therefore

$$\left(\frac{\partial V}{\partial T} \right)_S = -\frac{V C_V}{NkT} . \tag{4.76}$$

On the other hand, we find another total differential from (4.74),

$$dS = \left(\frac{\partial S}{\partial T} \right)_P dT + \left(\frac{\partial S}{\partial P} \right)_T dP = \frac{C_P}{T} dT - \frac{Nk}{P} dP ,$$

so that

$$\left(\frac{\partial T}{\partial P} \right)_S = \frac{NkT}{PC_P} .$$

With (4.75), we now see that

$$\left(\frac{\partial V}{\partial P} \right)_S = -\frac{1}{\gamma} \frac{V}{P} \implies \gamma \int \frac{dV}{V} = -\int \frac{dP}{P} \implies PV^\gamma = \text{const} .$$

To b. From (4.76), it follows that

$$\int_{V_1}^{V_2} \frac{dV}{V} = -\frac{3}{2} \int_{T_1}^{T_2} \frac{dT}{T} \implies \ln\left(\frac{V_2}{V_1}\right) = -\frac{3}{2} \ln\left(\frac{T_2}{T_1}\right) \implies T_2 = T_1 \left(\frac{V_2}{V_1}\right)^{-2/3}.$$

4.6 Classical Maxwell-Boltzmann Statistics

Statistical physics encompasses two ways of describing many-particle systems: the *classical* and the *quantum statistical* approach. The classical approach is represented by *classical Maxwell-Boltzmann statistics*, whereas the quantum statistical one is described by *Fermi-Dirac* and *Bose-Einstein statistics*. These three types of statistics differ mainly by the way in which the microstates of a system are counted.

Classical Maxwell-Boltzmann statistics considers classical systems in which the particles are found on well-defined trajectories in phase space. From the quantum mechanical point of view, this is equivalent to an independent movement of the wave packets of distinguishable particles, so that the N-particle wave function is a simple tensor product of N one-particle wave functions. For the calculation of the corresponding partition functions, one can use the purely classical energies of the Hamilton function but also the energy eigenvalues of the quantum mechanical Hamilton operator; some examples are to be found in the applications of the previous sections.

By contrast, Fermi-Dirac and Bose-Einstein statistics take full account of quantum effects. In particular, the fermionic and bosonic character of identical particles is incorporated by using appropriate antisymmetric or symmetric wave functions. This, of course, has a considerable effect on the counting of microstates. We shall look at these types of statistics in the next section.

In this section we discuss two particular systems in the context of classical Maxwell-Boltzmann statistics, the N-particle oscillator and the N-particle dipole system. In both examples, we will use the corresponding classical Hamilton function H ("proper classical system") as well as the Hamilton operator \boldsymbol{H}. To start, however, we investigate the conditions under which classical Maxwell-Boltzmann statistics is valid and deduce the *equipartition theorem* for "proper classical systems".

4.6.1 Classical Limit

In classical mechanics, momenta and coordinates of an N-particle system are all simultaneously specifiable. By contrast, due to Heisenberg's uncertainty principle, this no longer holds in quantum mechanics. Therefore, the classical approximation is valid if

$$\Delta q \Delta p \gg h \ . \tag{4.77}$$

Consider, for example, an ideal gas of N identical particles in a box. For a single particle with average momentum \bar{p} and average separation \bar{r} to the other particles, the above condition (4.77) corresponds to

$$\bar{p}\bar{r} \gg h$$

or, using the average de Broglie wavelength $\bar{\lambda} = h/\bar{p}$,

$$\bar{r} \gg \bar{\lambda} \,.$$

Since $\bar{\lambda}$ can be interpreted as a measure for the quantum mechanical extent of the particles in space, the classical description is possible if the particles' wave functions do not overlap; they are then distinguishable by their positions. Now, we assume further, that each particle takes up an average volume \bar{r}^3. Then, due to (4.30), the average energy and momentum of the ideal gas are

$$\frac{\bar{p}^2}{2m} \approx \bar{E} = \frac{3}{2}kT \,, \quad \bar{p} \approx (3mkT)^{1/2} \,,$$

and the average wavelength becomes

$$\bar{\lambda} \approx \frac{h}{(3mkT)^{1/2}} \,.$$

Classical Maxwell-Boltzmann is therefore valid if

$$\bar{r} \approx \left(\frac{V}{N}\right)^{1/3} \gg \frac{h}{(3mkT)^{1/2}} \,.$$

This corresponds to the following conditions:

- N is small.

- T is large.

- m is not very small.

On the other hand,

$$\bar{r} \lesssim \bar{\lambda}$$

implies that the state of the gas in the box is given by an N-particle wave function that cannot be factorized into N one-particle wave functions. This is the case of *quantum statistics*, to which we return in the next section.

4.6.2 Virial and Equipartition Theorems

In this subsection we derive some general properties for average values of classical systems. To this end, we consider a three-dimensional N-particle system with constant energy E, which is described by a Hamilton function $H(q,p)$. Letting x_i be an arbitrary component of $\{q_1,\ldots,q_{3N},p_1,\ldots,p_{3N}\}$ and using the microcanonical ensemble, we calculate as follows:

$$\overline{x_i \frac{\partial H}{\partial x_j}} = \frac{1}{\Omega(E)} \int\limits_{E-\delta E \leq H(q,p) \leq E} d\Gamma x_i \frac{\partial H}{\partial x_j} = \frac{\delta E}{\Omega(E)} \frac{\partial}{\partial E} \int\limits_{H(q,p) \leq E} d\Gamma x_i \frac{\partial H}{\partial x_j} .$$

Due to $\partial E/\partial x_j = 0$, the integral can be transformed into

$$\int\limits_{H(q,p) \leq E} d\Gamma x_i \frac{\partial H}{\partial x_j} = \int\limits_{H(q,p) \leq E} d\Gamma x_i \frac{\partial}{\partial x_j} [H(q,p) - E]$$

$$= \int\limits_{H(q,p) \leq E} d\Gamma \frac{\partial}{\partial x_j} \{x_i [H(q,p) - E]\}$$

$$- \delta_{ij} \int\limits_{H(q,p) \leq E} d\Gamma [H(q,p) - E] . \tag{4.78}$$

The second last term of this equation contains the expression

$$x_i [H(q,p) - E]\big|_{(x_j)_1}^{(x_j)_2} ,$$

where $(x_j)_1$ and $(x_j)_2$ are the extremal values of the coordinate x_j. Since a phase space value $(q_1, \ldots, q_{3N}, p_1, \ldots, p_{3N})$ with any one arbitrary extremal coordinate has to lie on the energy-hyperplane $H(q,p) = E$, the second last term in (4.78) vanishes. Overall , we find

$$\overline{x_i \frac{\partial H}{\partial x_j}} = \frac{\delta_{ij}}{g(E)} \frac{\partial}{\partial E} \int\limits_{H(q,p) \leq E} d\Gamma [E - H(q,p)] = \frac{\delta_{ij}}{g(E)} \int\limits_{H(q,p) \leq E} d\Gamma$$

$$= \frac{\delta_{ij}}{g(E)} \omega(E) = \delta_{ij} \frac{\omega(E)}{\frac{\partial \omega(E)}{\partial E}} = \delta_{ij} \left(\frac{\partial}{\partial E} \ln \omega(E) \right)^{-1}$$

$$= k \delta_{ij} \left(\frac{\partial S}{\partial E} \right)^{-1} = \delta_{ij} kT .$$

Theorem 4.16: Virial and equipartition theorems

For a proper classical three-dimensional N-particle system with the Hamilton function $H(q,p)$, we have the following ensemble averages:

$$\overline{x_i \frac{\partial H}{\partial x_j}} = \delta_{ij} kT , \quad x_i \in \{q_1, \ldots, q_{3N}, p_1, \ldots, p_{3N}\} .$$

In particular,

$$\overline{p_i \frac{\partial H}{\partial p_i}} = \overline{p_i \dot{q}_i} = kT , \quad \overline{q_i \frac{\partial H}{\partial q_i}} = -\overline{q_i \dot{p}_i} = \overline{q_i F_i} = kT .$$

The average value of the kinetic energy is therefore

\triangleright

$$\overline{T} = \frac{1}{2}\overline{\sum_{i=1}^{3N} p_i \dot{q}_i} = \frac{1}{2}\overline{p\dot{q}} = \frac{3}{2}NkT \ ,$$

and, for the *virial* of the forces, we find

$$\overline{\sum_{i=1}^{3N} q_i F_i} = \overline{qF} = -3NkT \ .$$

The last two relations yield

$$\overline{T} = -\frac{1}{2}\overline{qF} = \frac{3}{2}NkT \qquad\qquad (\textit{virial theorem}) \ .$$

For homogeneous potentials, $V(q) = \alpha|q|^d$, $F = -\nabla_q V(q)$, the average values of T and E are

$$\overline{T} = \frac{d}{2}\overline{V} = \frac{3}{2}NkT \Longrightarrow \overline{E} = \frac{3d+6}{2d}NkT \qquad (\textit{equipartition theorem}) \ .$$

This last equation is termed "equipartition theorem" as it shows that the energy E is evenly distributed on average over the system's degrees of freedom. The reader may recall that whereas here we have deduced the virial theorem for ensemble averages, in mechanics we found a similar relation for time averages (Subsection 1.1.1, Theorem 1.6).

4.6.3 Harmonic Oscillator

We consider an oscillator system consisting of N one-dimensional distinguishable harmonic oscillators and evaluate the corresponding thermodynamical relations.

I: Harmonic oscillator in the microcanonical ensemble. As our starting point for the description in the microcanonical ensemble, we choose a quantum oscillator system with constant particle number N and constant energy

$$E = \hbar\omega\left(M + \frac{N}{2}\right) \ .$$

To find its microcanonical partition function $\Omega(E, N)$, we need to calculate the number of possibilities to distribute M quanta across N oscillators. The combinatoric result is given by

$$\Omega(E, N) = \frac{N(N+M-1)!}{M!N!} = \frac{(N+M-1)!}{M!(N-1)!} \ .$$

From this and using $M = E/(\hbar\omega) - N/2$ as well as Stirling's formula (4.28), we find the entropy of the system as

$$S(E, N) = k\left(\frac{E}{\hbar\omega} + \frac{N}{2}\right)\ln\left(\frac{E}{\hbar\omega} + \frac{N}{2}\right)$$
$$-k\left(\frac{E}{\hbar\omega} - \frac{N}{2}\right)\ln\left(\frac{E}{\hbar\omega} - \frac{N}{2}\right) - Nk\ln N \ ,$$

from which we obtain the relations

$$\frac{1}{T} = \left(\frac{\partial S}{\partial E}\right)_N = \frac{k}{\hbar\omega}\left[\ln(M + N) - \ln M\right]$$

$$\Longrightarrow M = \frac{N}{e^{\beta\hbar\omega} - 1}$$

$$\Longrightarrow E(T, N) = \hbar\omega\left(\frac{N}{e^{\beta\hbar\omega} - 1} + \frac{N}{2}\right) = \frac{N\hbar\omega}{2}\coth\left(\frac{\beta\hbar\omega}{2}\right) \tag{4.79}$$

$$\Longrightarrow S(T, N) = -Nk\ln\left(e^{\beta\hbar\omega} - 1\right) + \frac{Nk\beta\hbar\omega}{1 - e^{-\beta\hbar\omega}}$$

$$= -Nk\ln\left[2\sinh\left(\frac{\beta\hbar\omega}{2}\right)\right] + \frac{Nk\beta\hbar\omega}{2\tanh\left(\frac{\beta\hbar\omega}{2}\right)} \ . \tag{4.80}$$

Considering the classical limit $\beta\hbar\omega \to 0 \Longleftrightarrow T \to \infty$, we see that

$$S(T, N) = Nk\left[1 - \ln(\beta\hbar\omega)\right] \ , \ E(T, N) = NkT \ , \tag{4.81}$$

which is in accordance with the equipartition theorem with $d = 2$ and $N \to N/3$. In contrast, for very low temperatures ($T \to 0$), we find

$$E = \frac{N\hbar\omega}{2} \ .$$

II: Harmonic oscillator in the canonical ensemble. Next, we consider the same oscillator system in the canonical ensemble. The canonical one-particle partition function of an oscillator with energy

$$E_n = \hbar\omega\left(n + \frac{1}{2}\right)$$

is

$$Z(T, 1) = Z_1(T) = \sum_{n=0}^{\infty} e^{-\beta E_n} = e^{-\beta\hbar\omega/2}\sum_{n=0}^{\infty}\left(e^{-\beta\hbar\omega}\right)^n = \frac{e^{-\beta\hbar\omega/2}}{1 - e^{-\beta\hbar\omega}}$$

$$= \frac{1}{2\sinh\left(\frac{\beta\hbar\omega}{2}\right)} \ . \tag{4.82}$$

This gives us the following N-particle partition function as well as the free energy (distinguishable particles!):

$$Z(T, N) = Z_1(T)^N \Longrightarrow F(T, N) = NkT\ln\left[2\sinh\left(\frac{\beta\hbar\omega}{2}\right)\right] \ .$$

The equations for entropy and energy are then

$$S(T, N) = -\left(\frac{\partial F}{\partial T}\right)_N = k \ln Z - k\beta \frac{\partial \ln Z}{\partial \beta}$$

$$= -Nk \ln \left[2 \sinh\left(\frac{\beta\hbar\omega}{2}\right) \right] + \frac{Nk\beta\hbar\omega}{2 \tanh\left(\frac{\beta\hbar\omega}{2}\right)}$$

$$E(T, N) = F(T, N) + TS(T, N) = \frac{N\hbar\omega}{2 \tanh\left(\frac{\beta\hbar\omega}{2}\right)} \ ,$$

in agreement with the microcanonical results (4.79) and (4.80). For comparison, we also evaluate the corresponding relations for a classical oscillator system with energy

$$E = H(q, p) = \sum_{i=1}^{N} \left(\frac{p_i^2}{2m} + \frac{m\omega^2}{2} q_i^2 \right) \ .$$

The canonical partition function is

$$Z(T, N) = \frac{1}{h^N} \int \mathrm{d}\Gamma e^{-\beta H}$$

$$= \frac{1}{h^N} \prod_{i=1}^{N} \left[\int \mathrm{d}q_i \exp\left(-\frac{\beta m\omega^2 q_i^2}{2}\right) \int \mathrm{d}p_i \exp\left(-\frac{\beta p_i^2}{2m}\right) \right]$$

$$= Z(T, 1)^N \ ,$$

with

$$Z(T, 1) = Z_1(T) = \frac{1}{h} \int \mathrm{d}q \exp\left(-\frac{\beta m\omega^2 q^2}{2}\right) \int \mathrm{d}p \exp\left(-\frac{\beta p^2}{2m}\right) = \frac{kT}{\hbar\omega} \ .$$

The free energy follows as

$$F(T, N) = -NkT \ln\left(\frac{kT}{\hbar\omega}\right) \ .$$

Entropy and energy evaluate to [see (4.81)]

$$S(T, N) = Nk \left[\ln\left(\frac{kT}{\hbar\omega}\right) + 1 \right] \ , \quad E(T, N) = NkT \ .$$

III: Harmonic oscillator in the grand canonical ensemble. We have seen that the canonical partition function of an oscillator system consisting of N distinguishable particles factorizes in the following way:

$$Z(T, N) = \phi(T)^N \ , \quad \phi(T) = \left\{ \begin{array}{ll} \dfrac{kT}{\hbar\omega} & \text{classically} \\[2ex] \dfrac{1}{2 \sinh\left(\frac{\hbar\omega}{2kT}\right)} & \text{quantum mech.} \end{array} \right\} \quad (4.83)$$

Therefore, the grand canonical partition function is

$$Y(T,\mu) = \sum_{N=0}^{\infty} [z\phi(T)]^N = \frac{1}{1 - z\phi(T)} \ , \ z = e^{\beta\mu} \ ,$$

which converges if $z\phi(T) < 1$. From the grand canonical potential, we can obtain the thermodynamical relation

$$J(T,\mu) = kT \ln[1 - z\phi(T)] \ .$$

This gives

$$N(T,\mu) = -\left(\frac{\partial J}{\partial \mu}\right)_T = \frac{z\phi(T)}{1 - z\phi(T)} \implies z = \frac{N}{\phi(T)(N+1)}$$

$$S(T,\mu) = -\left(\frac{\partial J}{\partial T}\right)_\mu = -k\ln[1 - z\phi(T)] + \frac{zkT\phi'(T)}{1 - z\phi(T)} - Nk\ln z$$

$$E(T,\mu) = J(T,\mu) + TS(T,\mu) + NkT\ln z = \frac{zkT^2\phi'(T)}{1 - z\phi(T)} \ .$$

For large N, we have

$$z\phi(T) \approx 1 - \frac{1}{N} \ , \ 1 - z\phi(T) \approx \frac{1}{N} \ , \ \ln z = -\ln\phi(T) \ ,$$

so that we can write

$$S(T,N) = Nk\left(\frac{T\phi'(T)}{\phi(T)} + \ln\phi(T)\right) \ , \ E(T,N) = NkT^2\frac{\phi'(T)}{\phi(T)} \ .$$

If we now substitute $\phi(T)$ by (4.83), we can recover the familiar results for the classical case (4.81) as well as for the quantum mechanical one, (4.79) and (4.80).

IV: Harmonic oscillator in the canonical density operator formalism. In the energy basis $\{|n\rangle\}$ with

$$\boldsymbol{H}|n\rangle = E_n|n\rangle \ , \ E_n = \hbar\omega\left(n + \frac{1}{2}\right) \ ,$$

the canonical density matrix of the one-particle oscillator is trivial:

$$\langle n|\boldsymbol{\rho}|m\rangle = \frac{\langle n|e^{-\beta\boldsymbol{H}}|m\rangle}{Z_1} = \frac{e^{-\beta E_m}\langle n|m\rangle}{Z_1} = \frac{e^{-\beta E_n}}{Z_1}\delta_{nm} \ .$$

The corresponding one-particle partition function is

$$Z_1 = \mathrm{tr}\left(e^{-\beta\boldsymbol{H}}\right) = \sum_{n=0}^{\infty} \langle n|e^{-\beta E_n}|n\rangle = \sum_{n=0}^{\infty} e^{-\beta E_n} = \frac{1}{2\sinh\left(\frac{\beta\hbar\omega}{2}\right)} \ ,$$

which, as expected, is in agreement with (4.82). To demonstrate the strength of the quantum mechanical density matrix formalism, we show how to calculate the matrix elements of $\boldsymbol{\rho}$ in the coordinate representation. For this, we require the energy eigenfunctions of the quantum mechanical oscillator in this representation. They are given by (see Subsection 3.3.5)

$$\Psi_n(q) = \left(\frac{m\omega}{\pi\hbar}\right)^{1/4} \frac{H_n(x)}{\sqrt{2^n n!}} e^{-x^2/2} \ , \quad x = q\sqrt{\frac{m\omega}{\hbar}} \ .$$

To calculate $\langle q' | \boldsymbol{\rho} | q \rangle$, we use the integral representation of the Hermite polynomials,

$$H_n(x) = (-1)^n e^{x^2} \left(\frac{d}{dx}\right)^n e^{-x^2} = \frac{e^{x^2}}{\sqrt{\pi}} \int\limits_{-\infty}^{\infty} du (-2iu)^n e^{-u^2 + 2ixu} \ ,$$

and write

$$
\begin{aligned}
\langle q' | \boldsymbol{\rho} | q \rangle &= \sum_n \langle q' | n \rangle \, P_n \, \langle n | q \rangle \\
&= \frac{1}{Z_1} \sum_n \psi_n(q') \psi_n^*(q) e^{-\beta\hbar\omega\left(n+\frac{1}{2}\right)} \\
&= \frac{1}{Z_1} \left(\frac{m\omega}{\pi\hbar}\right)^{1/2} e^{-\frac{x^2+x'^2}{2}} \sum_n \frac{1}{2^n n!} H_n(x) H_n(x') e^{-\beta\hbar\omega\left(n+\frac{1}{2}\right)} \\
&= \frac{1}{Z_1\pi} \left(\frac{m\omega}{\pi\hbar}\right)^{1/2} e^{\frac{x^2+x'^2}{2}} \int\limits_{-\infty}^{\infty} du \int\limits_{-\infty}^{\infty} dv\, e^{-u^2 + 2ixu} e^{-v^2 + 2ix'v} \\
&\quad \times \sum_{n=0}^{\infty} \frac{(-2uv)^n}{n!} e^{-\beta\hbar\omega\left(n+\frac{1}{2}\right)} \ .
\end{aligned}
$$

The sum over n can be carried out directly and yields

$$\sum_{n=0}^{\infty} \frac{(-2uv)^n}{n!} e^{-\beta\hbar\omega\left(n+\frac{1}{2}\right)} = e^{-\beta\hbar\omega/2} e^{-2uv \exp(-\beta\hbar\omega)} \ .$$

For the matrix elements, we then find

$$
\begin{aligned}
\langle q' | \boldsymbol{\rho} | q \rangle = \frac{1}{Z_1\pi} &\left(\frac{m\omega}{\pi\hbar}\right)^{1/2} e^{\frac{x^2+x'^2}{2}} e^{-\beta\hbar\omega/2} \int\limits_{-\infty}^{\infty} du \int\limits_{-\infty}^{\infty} dv \\
&\times \exp\left[-u^2 + 2ixu - v^2 + 2ix'v - 2uv \exp(-\beta\hbar\omega)\right] \ . \quad (4.84)
\end{aligned}
$$

To further simplify this equation, we use

$$
\begin{aligned}
\int\limits_{-\infty}^{\infty} dx_1 \cdots \int\limits_{-\infty}^{\infty} dx_n &\exp\left(-\frac{1}{2} \sum_{j,k=1}^{n} a_{jk} x_j x_k + i \sum_{k=1}^{n} b_k x_k\right) \\
&= \frac{(2\pi)^{n/2}}{\sqrt{\det A}} \exp\left(-\frac{1}{2} A_{jk}^{-1} b_j b_k\right) \ ,
\end{aligned}
$$

which holds for invertible, symmetric matrices A. In our case,

$$A = 2 \begin{pmatrix} 1 & e^{-\beta\hbar\omega} \\ e^{-\beta\hbar\omega} & 1 \end{pmatrix} \ , \quad \det A = 4\left(1 - e^{-2\beta\hbar\omega}\right) \ ,$$

so that (4.84) becomes

$$\langle q' | \boldsymbol{\rho} | q \rangle = \frac{1}{Z_1} \left(\frac{m\omega}{\pi\hbar} \right)^{1/2} \frac{e^{-\beta\hbar\omega/2}}{(1 - e^{-2\beta\hbar\omega})^{1/2}}$$
$$\times \exp \left(\frac{x^2 + x'^2}{2} - \frac{x^2 + x'^2 - 2xx'e^{-\beta\hbar\omega}}{1 - e^{-2\beta\hbar\omega}} \right)$$
$$= \frac{1}{Z_1} \left(\frac{m\omega}{2\pi\hbar \sinh(\beta\hbar\omega)} \right)^{1/2}$$
$$\times \exp \left(-\frac{x^2 + x'^2}{2} \coth(\beta\hbar\omega) + \frac{xx'}{\sinh(\beta\hbar\omega)} \right)$$
$$= \frac{1}{Z_1} \left(\frac{m\omega}{2\pi\hbar \sinh(\beta\hbar\omega)} \right)^{1/2}$$
$$\times \exp \left\{ -\frac{m\omega}{4\hbar} \left[(q + q')^2 \tanh \left(\frac{\beta\hbar\omega}{2} \right) \right. \right.$$
$$\left. \left. + (q - q')^2 \coth \left(\frac{\beta\hbar\omega}{2} \right) \right] \right\} .$$

The diagonal elements of this density matrix yield the average density distribution of a quantum mechanical oscillator with temperature T:

$$\langle q | \boldsymbol{\rho} | q \rangle = \left[\frac{m\omega}{\pi\hbar} \tanh \left(\frac{\beta\hbar\omega}{2} \right) \right]^{1/2} \exp \left[-\frac{m\omega}{\hbar} \tanh \left(\frac{\beta\hbar\omega}{2} \right) q^2 \right] .$$

This is a Gaussian distribution. In the classical limit $\beta\hbar\omega \ll 1$, it simplifies to

$$\langle q | \boldsymbol{\rho} | q \rangle \approx \left(\frac{m\omega^2}{2\pi kT} \right)^{1/2} \exp \left(-\frac{m\omega^2 q^2}{2kT} \right) ,$$

which is the result we also obtain starting from the classical phase space density. On the other hand, for the pure quantum case $\beta\hbar\omega \gg 1$, we find

$$\langle q | \boldsymbol{\rho} | q \rangle \approx \left(\frac{m\omega}{\pi\hbar} \right)^{1/2} \exp \left(-\frac{m\omega q^2}{\hbar} \right) .$$

Note that this expression is identical to the probability density $|\Psi_0(q)|^2$ of an oscillator in its ground state.

4.6.4 Ideal Spin Systems and Paramagnetism

This subsection deals with systems consisting of N localized magnetic dipoles of charge e and mass m (solid crystal) in an external magnetic field. As we know from Subsection 2.5.2, each dipole is subject to a torque that tries to align it in the direction of the field. However, for $T > 0$, this does not result in total magnetization (minimal energy, all dipoles aligned) since the dipoles are also subject to thermic movements that attempt to bring the system into a

state of maximal entropy. Apparently, the two opposing limiting cases are that of vanishing thermic motion at $T \to 0$ and that of vanishing magnetization at $T \to \infty$. Quantum mechanically, the magnetic moment M is generally related to the angular momentum J via

$$M = \frac{ge}{2mc} J \qquad (g = \text{gyromagnetic ratio})$$

(compare to Section 3.6), where the possible eigenvalues j, m of J and J_z are given by[15]

$$J \,|j,m\rangle = \hbar^2 j(j+1)\,|j,m\rangle \quad , \quad j = 0, \frac{1}{2}, 1, \frac{3}{2}, \dots$$

$$J_z \,|j,m\rangle = \hbar m \,|j,m\rangle \qquad , \quad m = -j, -j+1, \dots, j-1, j \; .$$

If we apply a magnetic field $B = B e_z$ in the z-direction, the energy ϵ of each magnetic dipole is

$$\epsilon = -MB = -g\mu_{\mathrm{B}} m B \; , \quad \mu_{\mathrm{B}} = \frac{e\hbar}{2mc} \; ,$$

where μ_{B} denotes the *Bohr magneton*.

In the following, we use the microcanonical ensemble to consider a system of spin-1/2 particles for which $g = 2$ and $j = 1/2$. Such a system exhibits the interesting phenomenon of negative temperature. Subsequently, we use the canonical ensemble to discuss systems with arbitrary j. This leads us to the *law of Curie*.

I: Paramagnetism ($j = 1/2$) in the microcanonical ensemble. Consider a system of N spin-1/2 dipoles, whose magnetic moments are either parallel or antiparallel to an external magnetic field of strength B in the z-direction ($m = \pm 1/2$). Let N_+ (N_-) be the number of dipoles with energy $+\mu_{\mathrm{B}} B$ ($-\mu_{\mathrm{B}} B$). We then have

$$N = N_+ + N_- \; , \; n = N_+ - N_- \implies \begin{cases} N_+ = \dfrac{N+n}{2} \\[2mm] N_- = \dfrac{N-n}{2} \; . \end{cases}$$

With this, the constant total energy of the isolated system can be written as

$$E = n\mu_{\mathrm{B}} B \; .$$

Determining the possible number of microstates corresponding to this energy is a simple combinatoric problem: N particles can be arranged in $N!$ different ways. However, $N_+!$ ($N_-!$) exchanges of the N_+ (N_-) particles with each other does not yield new microstates. The total number of possible states is therefore given by

[15] The differentiation between the mass m and the magnetic quantum number m is left to the reader.

$$
\begin{aligned}
\ln \Omega(E, N) &= \frac{S(E, N)}{k} = \frac{N!}{N_+! N_-!} \\
&= N \ln N - N_+ \ln N_+ - N_- \ln N_- \\
&= -\left[N_+ \ln\left(\frac{N_+}{N}\right) + N_- \ln\left(\frac{N_-}{N}\right) \right] \\
&= -\left[\frac{N+n}{2} \ln\left(\frac{N+n}{2N}\right) + \frac{N-n}{2} \ln\left(\frac{N-n}{2N}\right) \right] \\
&= N \ln 2 - \frac{N+n}{2} \ln\left(1 + \frac{n}{N}\right) - \frac{N-n}{2} \ln\left(1 - \frac{n}{N}\right) ,
\end{aligned}
$$

where we have used Stirling's formula (4.28). Since the particles are localized and hence distinguishable, it is not necessary to take into account the Gibbs factor $1/N!$. It is now easy to see that the entropy is maximal for a system with equal number of dipoles pointing either way: $n = 0$. This corresponds to a system of N noninteracting particles with two possible states each, i.e., 2^N states in total.

Now, let $\Delta E = 2\mu_{\mathrm{B}} B$ denote the energy gap of the system. Then, its temperature, energy, and specific heat evaluate to

$$
\left.
\begin{aligned}
\frac{1}{T} &= \left(\frac{\partial S}{\partial E}\right)_{N,B} = \frac{1}{\mu_{\mathrm{B}} B} \left(\frac{\partial S}{\partial n}\right)_{N,B} = \frac{k}{\Delta E} \ln\left(\frac{N_-}{N_+}\right) \\
\Longrightarrow \frac{N_-}{N_+} &= \exp\left(\frac{\Delta E}{kT}\right) , \quad \frac{N_-}{N} = \frac{1}{1 + \exp\left(-\frac{\Delta E}{kT}\right)} , \quad \frac{N_+}{N} = \frac{1}{1 + \exp\left(\frac{\Delta E}{kT}\right)} \\
\Longrightarrow E(T, N, B) &= -N \mu_{\mathrm{B}} B \tanh\left(\frac{\Delta E}{2kT}\right) \\
\Longrightarrow C(T, N, B) &= \left(\frac{\partial E}{\partial T}\right)_{N,B} = N k \left(\frac{\Delta E}{2kT}\right)^2 \cosh^{-2}\left(\frac{\Delta E}{2kT}\right) .
\end{aligned}
\right\} \quad (4.85)
$$

Here, N_+/N and N_-/N are the relative probabilities of finding a randomly selected dipole with energy $+\mu_{\mathrm{B}} B$ and $-\mu_{\mathrm{B}} B$, respectively.

Let us take a closer look at the temperature. At zero temperature $T = 0$, all dipoles are in a state of minimal energy $-\mu_{\mathrm{B}} B$, and we find $S = 0$, as expected for a totally ordered system. If the system is fed with energy, the entropy grows, eventually approaching its maximum at $n = 0$ and $T = \infty$, in accordance with our expectation for maximal disorder. However, it is possible to inject even more energy into the system. The upper level will then contain more states than the lower one. The entropy has to drop again since order is now being reintroduced, and the temperature becomes negative. So far, we have only allowed positive temperatures (such systems are called *normal*) in order to prevent partition functions from becoming ill-defined. This is the case if the energy of the system can be chosen arbitrarily large, as can be seen most easily in the canonical ensemble. However, if the energy of the system is constrained by an upper bound, such a restriction is no longer

necessary, as with the case at hand. In the region $E > 0$ and $T < 0$, the system is *anomalous*, since there, the magnetization is opposed to the external field. Such a situation, in which more particles are present in the upper half of the system, is called *inversion*. Lasers, for example, can be forced into an inverted state by *pumping*. Purcell and Pound first created an inverted state of nuclear spins within crystals of type LiF. They achieved this by applying a strong magnetic field and after a long enough relaxation time rapidly switching the field direction. The spins are not able to follow the field at once, so that a state of nonequilibrium emerges, in which the energy is higher than that of the equilibrium state, which is reached a little bit later. It takes around 10^{-5} seconds for the nuclear spin system to reach internal equilibrium with negative magnetization and negative temperature. Five minutes later the system is finally back in a state of equilibrium between the spins and the lattice with positive energy. Note that it is the spins that are in a temporary state of negative temperature, whereas the lattice remains at positive temperature throughout.

Figure 4.10 shows the energy and specific heat (4.85) of our dipole system. A specific heat with such a characteristic peak is called the *Schottky effect*; it is typical for systems with an energy gap.

Fig. 4.10. Energy (*left*) and specific heat (*right*) of an ideal spin system with $j = 1/2$ [see (4.85)].

II: Paramagnetism (j arbitrary) in the canonical ensemble. For the special case of $j = 1/2$, the canonical partition function of the N-dipole system is calculated as follows:

$$Z(T, N, B) = \sum_n e^{-\beta E_n} = \left(\sum_{m = \pm \frac{1}{2}} \exp(2\beta\mu_B Bm) \right)^N = Z_1(T, B)^N \ ,$$

with

$$Z_1(T,B) = \sum_{m=\pm\frac{1}{2}} \exp(2\beta\mu_{\mathrm{B}}Bm) = 2\cosh(\beta\mu_{\mathrm{B}}B) \ .$$

From this, we can easily verify the previously calculated results within the microcanonical ensemble. Instead, let us consider the more general case of arbitrary spin j. The one-particle partition function is then given by

$$Z_1(T,B) = \sum_{m=-j}^{j} \exp(\beta g\mu_{\mathrm{B}}mB) = \sum_{m=-j}^{j} \exp\left(\frac{mx}{j}\right) \ , \quad x = \beta g\mu_{\mathrm{B}}Bj$$

$$= \frac{\exp\left(\frac{(j+1)x}{j}\right) - \exp(-x)}{\exp\left(\frac{x}{j}\right) - 1}$$

$$= \frac{\exp\left(\frac{x}{2j}\right)\left[\exp\left(\frac{(2j+1)x}{2j}\right) - \exp\left(-\frac{(2j+1)x}{2j}\right)\right]}{\exp\left(\frac{x}{2j}\right)\left[\exp\left(\frac{x}{2j}\right) - \exp\left(-\frac{x}{2j}\right)\right]}$$

$$= \frac{\sinh\left[x\left(1+\frac{1}{2j}\right)\right]}{\sinh\left(\frac{x}{2j}\right)} \ .$$

The average magnetization of the system follows as

$$M_z(T,N,B) = N\overline{g\mu_{\mathrm{B}}m} = \frac{N}{\beta}\frac{\partial}{\partial B}\ln Z_1 = Ng\mu_{\mathrm{B}}jB_j(x) \ ,$$

where

$$B_j(x) = \left(1+\frac{1}{2j}\right)\coth\left[x\left(1+\frac{1}{2j}\right)\right] - \frac{1}{2j}\coth\left(\frac{x}{2j}\right)$$

is the *Brillouin function of jth order*. Figure 4.11 shows this function for several values of j. For large x, i.e., for strong magnetic fields or low temperatures, $B_j(x)$ tends to its maximal value of 1 for all j. This corresponds to the case of maximal magnetization called *saturation*. For weak fields or high temperatures, $x \ll 1$, we can expand the Brillouin function as

$$B_j(x) = \frac{x}{3}\left(1+\frac{1}{j}\right) + \mathcal{O}(x^3) \ .$$

This gives

$$M_z \approx \frac{Ng^2\mu_{\mathrm{B}}^2Bj(j+1)}{3kT} \ ,$$

from which we obtain the magnetic susceptibility

$$\chi_{\mathrm{m}} = \frac{M_z}{B} = \frac{C_j}{T} \ , \quad C_j = \frac{Ng^2\mu_{\mathrm{B}}^2j(j+1)}{3k} \ .$$

This is *Curie's law*, which states that the magnetic susceptibility is proportional to $1/T$. The above simple model of a magnetizable solid crystal is in

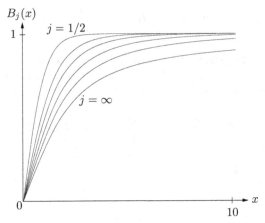

Fig. 4.11. Brillouin functions for $j = 1/2, 1, 3/2, 5/2, 5, \infty$.

rather good agreement with experimental results. In the case of only two possible magnetizations, $j = 1/2$, $g = 2$, we find

$$M_z = N\mu_B B_{1/2}(x) = N\mu_B \tanh x \ .$$

So, for $x \gg 1$, we have $M_z \approx N\mu_B$ and for $x \ll 1$: $M_z \approx N\mu_B x$.

III: Paramagnetism ($j = 1/2$) in the canonical density operator formalism. As above, we start by considering a single particle with $g = 2$ and $\boldsymbol{J} = \hbar/2 \cdot \boldsymbol{\sigma}$, in an external magnetic field $\boldsymbol{B} = B\boldsymbol{e}_z$. The corresponding Hamilton operator is

$$H = -\mu_B B\sigma_z \ .$$

The representation-independent canonical density operator is calculated using $\sigma_z^2 = \boldsymbol{I}$ and $\operatorname{tr}(\sigma_z) = 0$:

$$e^{\beta\mu_B B\sigma_z} = e^{x\sigma_z} \ , \ x = \beta\mu_B B$$

$$= \sum_{i=0}^{\infty} \frac{x^i \sigma_z^i}{i!} = \boldsymbol{I} \sum_{i=0}^{\infty} \frac{x^{2i}}{(2i)!} + \sigma_z \sum_{i=1}^{\infty} \frac{x^{2i-1}}{(2i-1)!}$$

$$= \boldsymbol{I} \cosh x + \sigma_z \sinh x$$

$$\implies \operatorname{tr}\left(e^{\beta\mu_B B\sigma_z}\right) = 2\cosh x$$

$$\implies \rho = \frac{e^{\beta\mu_B B\sigma_z}}{\operatorname{tr}\left(e^{\beta\mu_B B\sigma_z}\right)} = \frac{1}{2}(\boldsymbol{I} + \sigma_z \tanh x) \ .$$

From this, the expectation value of the spin operator follows as

$$\overline{\sigma_z} = \operatorname{tr}(\rho\sigma_z) = \tanh x \ ,$$

and the the total energy of the N-particle system is

$$E = -N\mu_B B\overline{\sigma_z} = -N\mu_B B \tanh(\beta\mu_B B) \ ,$$

which is in agreement with the first equation of (4.85). Note that we did not use a specific representation of the σ-matrices in calculating $\overline{\sigma_z}$ (see Application 64).

Summary

- **Classical Maxwell-Boltzmann statistics** does not take into account the quantum nature of particles. This approximation is valid for large temperatures, small number of particles, and not too small particle mass.

- For "proper classical systems", which are described by classical Hamilton functions, the **virial** and the **equipartition theorem** hold.

- Using the one-dimensional N-particle oscillator system, different aspects of the classical Boltzmann statistics can be studied nicely. In particular, we see that the equipartition theorem does not hold if the quantum mechanical Hamilton operator is used.

- The ideal spin system of N noninteracting dipoles is a model of paramagnetic systems. It exhibits the phenomenon of **negative temperature**, which, for example, is used in lasers, and explains **Curie's law** for the susceptibility of paramagnets.

Applications

64. Canonical density matrix of the electron in a magnetic field using particular representations. Calculate the canonical density matrix of an electron ($g = 2$, $j = 1/2$) in a magnetic field $\boldsymbol{B} = B\boldsymbol{e}_z$ in two different representations, where (a) σ_z or (b) σ_x is diagonal.

Solution.

To a. The σ-matrices in the representation, in which σ_z is diagonal, are given by

$$\sigma_x = \begin{pmatrix} 0 & 1 \\ 1 & 0 \end{pmatrix} \, , \quad \sigma_y = \begin{pmatrix} 0 & -i \\ i & 0 \end{pmatrix} \, , \quad \sigma_z = \begin{pmatrix} 1 & 0 \\ 0 & -1 \end{pmatrix} \, .$$

Therefore, the canonical density matrix is

$$\rho = \frac{1}{e^x + e^{-x}} \begin{pmatrix} e^x & 0 \\ 0 & e^{-x} \end{pmatrix} \, , \quad x = \beta \mu_B B \, ,$$

and it follows that

$$\overline{\sigma_z} = \mathrm{tr}(\rho \sigma_z) = \frac{e^x - e^{-x}}{e^x + e^{-x}} = \tanh(\beta \mu_B B) \, .$$

To b. To diagonalize σ_x, we need a similarity transformation $\sigma'_x = U\sigma_x U^{-1}$, with

$$U = \frac{1}{\sqrt{2}}\begin{pmatrix} 1 & 1 \\ -1 & 1 \end{pmatrix}, \quad U^{-1} = \frac{1}{\sqrt{2}}\begin{pmatrix} 1 & -1 \\ 1 & 1 \end{pmatrix}.$$

For σ'_x, σ'_z, and ρ', we obtain

$$\sigma'_x = U\begin{pmatrix} 0 & 1 \\ 1 & 0 \end{pmatrix}U^{-1} = \begin{pmatrix} 1 & 0 \\ 0 & -1 \end{pmatrix}$$

$$\sigma'_z = U\begin{pmatrix} 1 & 0 \\ 0 & -1 \end{pmatrix}U^{-1} = \begin{pmatrix} 0 & -1 \\ -1 & 0 \end{pmatrix}$$

$$\rho' = U\begin{pmatrix} e^x & 0 \\ 0 & e^{-x} \end{pmatrix}U^{-1} = \frac{1}{2}\begin{pmatrix} 1 & -\tanh x \\ -\tanh x & 1 \end{pmatrix}.$$

With these transformed matrices, the expectation value of σ'_z becomes

$$\overline{\sigma'_z} = \mathrm{tr}(\rho'\sigma'_z) = \tanh(\beta\mu_{\mathrm{B}}B).$$

This example shows explicitly that expectation values are representation-independent.

65. Canonical density matrix of a free particle in coordinate representation. Evaluate the canonical density matrix of a free particle in coordinate representation.

Solution. In coordinate representation, the Hamilton operator of a free particle with mass m is

$$H = -\frac{\hbar^2}{2m}\nabla^2.$$

Its eigenfunctions normalized to a volume $V = L^3$ are

$$\Psi_n(x) = \frac{1}{L^{3/2}}e^{i\mathbf{k}\cdot\mathbf{x}}, \quad E_n = \frac{\hbar^2 k^2}{2m}, \quad \mathbf{k} = \frac{2\pi}{n}\mathbf{n}, \quad n_i = 0, \pm 1, \pm 2, \ldots.$$

We therefore have

$$\langle x| e^{-\beta H}|x'\rangle = \sum_n \langle x|\Psi_n\rangle e^{-\beta E_n}\langle\Psi_n|x'\rangle$$

$$= \sum_n e^{-\beta E_n}\Psi_n(x)\Psi_n^*(x')$$

$$= \frac{1}{L^3}\sum_k \exp\left(-\frac{\beta\hbar^2}{2m}k^2 + i\mathbf{k}(\mathbf{x}-\mathbf{x}')\right)$$

$$\approx \frac{1}{(2\pi)^3}\int \exp\left(-\frac{\beta\hbar^2}{2m}k^2 + i\mathbf{k}(\mathbf{x}-\mathbf{x}')\right)d^3k$$

$$= \left(\frac{m}{2\pi\beta\hbar^2}\right)^{3/2}\exp\left(-\frac{m}{2\beta\hbar^2}(\mathbf{x}-\mathbf{x}')^2\right).$$

Here, we have replaced the sum by an integral and used quadratic completion. With

$$\text{tr}\left(e^{-\beta H}\right) = \int d^3x \left\langle x\right| e^{-\beta H} \left|x\right\rangle = V \left(\frac{m}{2\pi\beta\hbar^2}\right)^{3/2} ,$$

we find

$$\left\langle x\right| \rho \left|x'\right\rangle = \frac{\left\langle x\right| e^{-\beta H} \left|x'\right\rangle}{\text{tr}\left(e^{-\beta H}\right)} = \frac{1}{V}\exp\left(-\frac{m}{2\beta\hbar^2}(x - x')^2\right) .$$

As expected, the matrix elements are symmetric under exchange of x and x'. We can also understand that the diagonal elements $\left\langle x\right| \rho \left|x\right\rangle$ are independent of x since they represent the probability of finding the particle close to x. Note that the wave packet's extent is a purely quantum effect. In the limit of large temperatures, the extent becomes smaller and eventually turns into the δ-function coinciding with the classical solution. Finally, we use the density matrix ρ to calculate the energy:

$$\overline{H} = -\frac{\partial}{\partial\beta} \ln\text{tr}\left(e^{-\beta H}\right) = -\frac{\partial}{\partial\beta}\left[\ln V + \frac{3}{2}\ln\left(\frac{m}{2\pi\beta\hbar^2}\right)\right] = \frac{3}{2}kT .$$

4.7 Quantum Statistics

Following on from the previous section, where we discussed the classical Maxwell-Boltzmann statistics, we now turn our attention to quantum statistics. Here, the quantum nature of particles (bosonic or fermionic) is taken into account. As before, we restrict ourselves to systems of noninteracting particles, i.e., to ideal systems.

Using results from Section 3.9, we begin our discussion by reviewing some of the properties of systems composed of either fermionic or bosonic particles and introduce the *occupation number formalism*. With its help, many of the results of Fermi-Dirac, Bose-Einstein, and also Maxwell-Boltzmann statistics can be conveniently presented. We then look at the ideal Fermi gas and calculate its state equations for the classical limit of a very high temperature as well as for the purely quantum mechanical case of a very low temperature. Importantly, we shall find that due to Pauli's exclusion principle, there exist excited states of Fermi particles even at zero temperature $T = 0$, whose energy is below the *Fermi energy*. Following this, we discuss the ideal Bose gas and derive its equations of state in a similar way as we did for the Fermi gas. At very low temperatures below a critical value, we shall find an interesting phenomenon, the *Bose-Einstein condensation*.

4.7.1 General Formalism

Consider a quantum system containing N noninteracting particles, so that its Hamilton operator \boldsymbol{H} can be split into a sum of N one-particle operators \boldsymbol{H}_i:

$$H = \sum_{i=1}^{N} \boldsymbol{H}_i \ .$$

Having solved the eigenvalue problem of the one-particle operator,

$$\boldsymbol{H}_i \, | \, k_i \rangle = E_i \, | \, k_i \rangle \ ,$$

the total energy of the system follows as

$$E = \sum_i E_i \ .$$

The state vector of the complete system can be constructed via the tensor product of the normalized one-particle states $| \, k_i \rangle$. The detail of this construction depends on the nature of the particles at hand.

Fermi-Dirac statistics. If the particles are identical and have half-integer spin (fermions), we know from our previous discussion of quantum mechanical N-particle systems in Section 3.9 that the state of the system $| \, k_1, k_2, \ldots, k_n, \mathrm{A} \rangle$ must be antisymmetric with respect to the interchange of the degrees of freedom of any two particles. This antisymmetry can be achieved by the following construction:

$$
\begin{aligned}
| \, k_1, k_2, \ldots, k_N, \mathrm{A} \rangle &= \frac{1}{\sqrt{N!}} \sum_P \epsilon(P) \, | \, k_{P_1} \rangle \, | \, k_{P_2} \rangle \cdots | \, k_{P_N} \rangle \\
&= \frac{1}{\sqrt{N!}} \sum_P \epsilon(P) \, | \, k_{P_1}, k_{P_2}, \ldots, k_{P_N} \rangle \ ,
\end{aligned}
\tag{4.86}
$$

where

$$
P = \begin{pmatrix} 1 & \cdots & N \\ P_1 & \cdots & P_N \end{pmatrix} \ , \quad
\epsilon(P) = \begin{cases} +1 \text{ for } P \text{ even} \\ -1 \text{ for } P \text{ odd} \end{cases}
$$

denotes a permutation of $1, \ldots, N$. The factor $1/\sqrt{N!}$ in (4.86) is necessary since there are $N!$ such possible permutations. Note that due to Pauli's principle, all quantum numbers k_i are different. For fermionic states, we therefore have the following normalization:[16]

$$
\begin{aligned}
\langle k_1', &\ldots, k_N', \mathrm{A} | \, k_1, \ldots, k_N, \mathrm{A} \rangle \\
&= \frac{1}{N!} \sum_P \epsilon(P) \sum_{P'} \epsilon(P') \left\langle k_{P_1'}', \ldots, k_{P_N'}' \, \middle| \, k_{P_1}, \ldots, k_{P_N} \right\rangle
\end{aligned}
$$

[16] We assume that the k_i are discrete quantum numbers (bound states).

$$= \sum_P \epsilon(P) \langle k'_1, \ldots, k'_N | k_{P_1}, \ldots, k_{P_N} \rangle$$

$$= \sum_P \epsilon(P) \delta_{k'_1 k_{P_1}} \cdots \delta_{k'_N k_{P_N}} . \tag{4.87}$$

Here, we have used the fact that the double summation over all permutations is equal to $N!$ times the onefold sum over all permutations.

Bose-Einstein statistics. For identical particles with integer spin (bosons), the N-particle state is symmetric under exchange of the degrees of freedom of any two particles and is given by

$$| k_1, \ldots, k_N, \mathrm{S} \rangle = \frac{1}{\sqrt{N! n_1! \cdots n_N!}} \sum_P | k_{P_1}, \ldots, k_{P_N} \rangle .$$

Contrary to the fermionic case, there are no restrictions on the occupation numbers. If n_i particles have the same quantum numbers, the $n_i!$ permutations of the particles' quantum numbers do not yield a new physical state; this is taken into account by the normalization factor. The bosonic normalization corresponding to (4.87) is

$$\langle k'_1, \ldots, k'_N, \mathrm{S} | k_1, \ldots, k_N, \mathrm{S} \rangle$$

$$= \frac{1}{\sqrt{n'_1! \cdots n'_N! n_1! \cdots n_N!}} \sum_P \delta_{k'_1 k_{P_1}} \cdots \delta_{k'_N k_{P_N}} .$$

Maxwell-Boltzmann statistics. If the N particles can be treated as distinguishable, the total state of the system is

$$| k_1, k_2, \ldots, k_N \rangle = | k_1 \rangle | k_2 \rangle \cdots | k_N \rangle ,$$

with the normalization

$$\langle k'_1, \ldots, k'_N | k_1, \ldots, k_N \rangle = \delta_{k'_1 k_1} \cdots \delta_{k'_N k_N} ,$$

where the exchange of two different quantum numbers results in a new physical state. In the following, we will see that the Maxwell-Boltzmann statistics can often be used to describe approximately systems of identical particles, as we have already done in some of the preceding sections. In this case, the indistinguishability has to be taken into account by the ad-hoc introduction of Gibbs's correction factor $1/N!$, whereas in Fermi-Dirac and Bose-Einstein statistics, this factor follows naturally from the normalization of the state vectors.

Occupation number formalism. For what follows, it is best to represent the fermionic and bosonic system kets by occupation numbers of the one-particle states. If we order the possible quantum numbers k_i of the one-particle states in ascending order, the set of occupation numbers $\{n_0, n_1, \ldots\}$ with $\sum_{k=0}^{\infty} n_k = N$ in conjunction with the particle type qualification defines the N-particle state uniquely, and we can write

$$| n_0, n_1, \ldots, S, A \rangle = | k_1, k_2, \ldots, k_N, S, A \rangle .$$

The left-hand side should be read as: n_0 particles are in the lowest possible one-particle state $|0\rangle$, n_1 in the next-highest one-particle state $|1\rangle$, and so on. Using this identity, we also have

$$\left. \begin{aligned} H \, | n_0, n_1, \ldots, S, A \rangle &= E \, | n_0, n_1, \ldots, S, A \rangle \ , \quad E = \sum_{k=0}^{\infty} n_k E_k \\ N \, | n_0, n_1, \ldots, S, A \rangle &= N \, | n_0, n_1, \ldots, S, A \rangle \ , \quad N = \sum_{k=0}^{\infty} n_k . \end{aligned} \right\} \tag{4.88}$$

The last equation can be taken as the definition of the *particle operator* N. It determines all matrix elements of N in the $| n_0, n_1, \ldots, S, A \rangle$-basis. In the same vein, we can define the *occupation number operator* n_k by

$$n_k \, | n_0, n_1, \ldots, S, A \rangle = n_k \, | n_0, n_1, \ldots, S, A \rangle ,$$

with

$$n_k = \begin{cases} 0, 1 & \text{for fermions} \\ 0, 1, 2, \ldots & \text{for bosons .} \end{cases}$$

The normalization for the occupation number kets is

$$\langle n_0', n_1', \ldots, S, A | n_0, n_1, \ldots, S, A \rangle = \delta_{n_0' n_0} \delta_{n_1' n_1} \cdots .$$

According to this, two states are identical if all their occupation numbers coincide. Following (4.88), the matrix elements of the canonical density operator in the $| n_0, n_1, \ldots, S, A \rangle$-basis are obtained as

$$\begin{aligned} \langle n_0', n_1', \ldots, &S, A | \, \rho \, | n_0, n_1, \ldots, S, A \rangle \\ &= \frac{1}{Z} \left\langle n_0', n_1', \ldots, S, A | \, e^{-\beta H} \, | n_0, n_1, \ldots, S, A \right\rangle \\ &= \frac{1}{Z} \exp \left(-\beta \sum_k n_k E_k \right) \delta_{n_0' n_0} \delta_{n_1' n_1} \cdots , \end{aligned}$$

with

$$Z = \sum_{n_0, n_1, \ldots}^{\prime} \exp \left(-\beta \sum_k n_k E_k \right) .$$

The corresponding relations for the grand canonical density matrix are

$$\begin{aligned} \langle n_0', n_1', \ldots, &S, A | \, \rho \, | n_0, n_1, \ldots, S, A \rangle \\ &= \frac{1}{Y} \left\langle n_0', n_1', \ldots, S, A | \, e^{-\beta(H - \mu N)} \, | n_0, n_1, \ldots, S, A \right\rangle \\ &= \frac{1}{Y} \exp \left(-\beta \sum_k n_k (E_k - \mu) \right) \delta_{n_0' n_0} \delta_{n_1' n_1} \cdots , \end{aligned}$$

with

$$Y = \sum_{n_0, n_1, \dots} \exp\left(-\beta \sum_k n_k (E_k - \mu)\right) .$$

The symbol $'$ in the canonical partition function indicates that the sum is only taken over occupation numbers for which $\sum_k n_k = N$. In contrast, there is no such restriction for the grand canonical partition function. The diagonal elements of the Gibbs density matrix can be interpreted as the probability W to find the set of occupation numbers $\{n_0, n_1, \dots\}$ in the N-particle system:

$$W(n_0, n_1, \dots) = \langle n_0, n_1, \dots, S, A | \boldsymbol{\rho} | n_0, n_1, \dots, S, A \rangle .$$

Since there is no restriction in the grand canonical partition function with regard to the sum, it can be simplified further in the following way: for the bosonic case, we have

$$Y = \sum_{n_0, n_1, \dots = 0}^{\infty} \left[e^{-\beta(E_0 - \mu)}\right]^{n_0} \left[e^{-\beta(E_1 - \mu)}\right]^{n_1} \dots$$

$$= \prod_k \sum_{n_k = 0}^{\infty} \left[e^{-\beta(E_k - \mu)}\right]^{n_k} = \prod_k \frac{1}{1 - z e^{-\beta E_k}} , \quad z = e^{\beta \mu} .$$

For the fermionic case, the partition function becomes

$$Y = \sum_{n_0, n_1, \dots = 0}^{1} \left[e^{-\beta(E_0 - \mu)}\right]^{n_0} \left[e^{-\beta(E_1 - \mu)}\right]^{n_1} \dots$$

$$= \prod_k \sum_{n_k = 0}^{1} \left[e^{-\beta(E_k - \mu)}\right]^{n_k} = \prod_k \left(1 + z e^{-\beta E_k}\right) .$$

For completeness, we briefly discuss the corresponding expressions when using Maxwell-Boltzmann statistics in the occupation number formalism. For this case, the occupation numbers $\{n_0, n_1, \dots\}$ do not determine the system state $|k_1, k_2, \dots, k_N\rangle$ uniquely since it is not clear which particle is in which one-particle state. But, since all states in agreement with the set of occupation numbers $\{n_0, n_1, \dots\}$ have the same energy and thus the same probability, we only need to determine the number of these states. Now, there are $N!$ possible ways to number the N particles. However, if n_k of these are in a state $|k\rangle$, permutations of these particles do not lead to new physical states even on a classical level. Each set $\{n_0, n_1, \dots\}$ must therefore be given the weight $N!/(n_0! n_1! \cdots)$, and the canonical partition function can be written as

$$Z = \frac{1}{N!} \sum_{n_0, n_1, \dots}' \frac{N!}{n_0! n_1! \cdots} \exp\left(-\beta \sum_k n_k E_k\right) ,$$

where the Gibbs factor $1/N!$ for identical particles has been introduced by

hand.[17] For the grand canonical partition function (again identical particles), we find

$$
Y = \sum_{n_0,n_1,\ldots=0}^{\infty} \frac{1}{n_0! n_1! \cdots} \left[e^{-\beta(E_0-\mu)}\right]^{n_0} \left[e^{-\beta(E_1-\mu)}\right]^{n_1} \cdots
$$

$$
= \prod_k \sum_{n_k=0}^{\infty} \frac{1}{n_k!} \left[e^{-\beta(E_k-\mu)}\right]^{n_k}
$$

$$
= \prod_k \exp\left[z e^{-\beta E_k}\right] .
$$

Theorem 4.17: Grand canonical partition function for Bose-Einstein, Fermi-Dirac, and Maxwell-Boltzmann statistics

Given a system of N noninteracting identical particles, let E_k be the possible energies of the one-particle states. The grand canonical partition function is then given by [see (4.61)]

$$
\ln Y(T,V,\mu) = \frac{PV}{kT} = \frac{1}{\sigma} \sum_k \ln\left[1 + \sigma z e^{-\beta E_k}\right] ,
$$

with σ depending on the statistics as follows:

$$
\sigma = \begin{cases} -1 & \text{Bose-Einstein statistics} \\ +1 & \text{Fermi-Dirac statistics} \\ 0 & \text{Maxwell-Boltzmann statistics} . \end{cases}
$$

The case $\sigma = 0$ must be treated as the limiting case of $\sigma \to 0$. Average energy and particle number are obtained from this as

$$
E(T,V,\mu) = -\left(\frac{\partial \ln Y}{\partial \beta}\right)_{z,V} = \sum_k \frac{E_k}{\frac{1}{z} e^{\beta E_k} + \sigma} = \sum_k \overline{n_k} E_k
$$

$$
N(T,V,\mu) = \frac{1}{\beta}\left(\frac{\partial \ln Y}{\partial \mu}\right)_{T,V} = z\left(\frac{\partial \ln Y}{\partial z}\right)_{T,V} = \sum_k \overline{n_k} ,
$$

where the average occupation numbers are given by \triangleright

[17] Note that this equation reproduces the result (4.45):

$$
Z = \frac{1}{N!} \sum_{n_0,n_1,\ldots}' \frac{N!}{n_0! n_1! \cdots} \left[e^{-\beta E_0}\right]^{n_0} \left[e^{-\beta E_1}\right]^{n_1} \cdots
$$

$$
= \frac{1}{N!} \left[\sum_k e^{-\beta E_k}\right]^N = \frac{1}{N!} Z_1^N .
$$

$$\overline{n_k} = \frac{ze^{-\beta E_k}}{1 + \sigma z e^{-\beta E_k}} = \frac{1}{\frac{1}{z}e^{\beta E_k} + \sigma} .$$

For fermions, we have $0 \le \overline{n_k} \le 1$, whereas for bosons, the occupation numbers are not bounded from above. For bosons, this implies that the chemical potential is always smaller than the smallest one-particle energy E_0. The limit $T \to 0$ in the case of bosons has to be treated separately and leads to *Bose-Einstein condensation*.

In terms of the occupation numbers, the grand canonical partition function is

$$\ln Y = -\frac{1}{\sigma} \sum_k \ln(1 - \sigma \overline{n_k}) .$$

Using

$$E - \mu N = \sum_k \overline{n_k}(E_k - \mu) = kT \sum_k \overline{n_k} \ln \left(\frac{1 - \sigma \overline{n_k}}{\overline{n_k}} \right) ,$$

the entropy is obtained as

$$\begin{aligned}
S &= k \ln Y + \frac{1}{T}(E - \mu N) \\
&= -\frac{k}{\sigma} \sum_k \left[\ln(1 - \sigma \overline{n_k}) - \sigma \overline{n_k} \ln \left(\frac{1 - \sigma \overline{n_k}}{\overline{n_k}} \right) \right] \\
&= -k \sum_k \left[\overline{n_k} \ln \overline{n_k} + \frac{1}{\sigma}(1 - \sigma \overline{n_k}) \ln(1 - \sigma \overline{n_k}) \right] .
\end{aligned}$$

Consequently, unoccupied states do not contribute to entropy. This is also true for fermions with $\overline{n_k} = 1$. This type of behavior is also seen in the uncertainty of the occupation numbers,

$$(\Delta n_k)^2 = kT \frac{\partial \overline{n_k}}{\partial \mu} = \overline{n_k}(1 - \sigma \overline{n_k}) ,$$

which vanishes for $\overline{n_k} = 0$ and for fermions also for $\overline{n_k} = 1$. Note that for $\overline{n_k} \gg 1$ (bosons), the errors are proportional to $\overline{n_k}$, in contrast to the classical $\sqrt{\overline{n_k}}$-behavior.

State equations of the ideal quantum gas. We can simplify the partition function from Theorem 4.17 by replacing the sum over one-particle states by an integral. To this end, we proceed analogously to our discussion of the ideal classical gas in the microcanonical ensemble (Application 56) and write

$$\sum_k \longrightarrow \frac{V}{h^3} \sum_{m=-s}^{s} \int d^3p = \frac{4\pi g_s V}{h^3} \int p^2 dp = \int g(\epsilon)d\epsilon ,$$

with the energy density

$$g(\epsilon) = \frac{4\pi g_s V}{h^3} p^2(\epsilon) \left(\frac{\partial \epsilon}{\partial p}\right)^{-1} .$$

Here, the sum over possible spin degrees of freedom is taken into account by the degeneracy factor $g_s = (2s + 1)$. If we restrict ourselves to two special cases, nonrelativistic (NR) and ultrarelativistic (UR) particles, with

$$\epsilon_{NR} = \frac{p^2}{2m} , \quad \epsilon_{UR} = cp ,$$

the corresponding energy densities are given by

$$g(\epsilon) = \begin{cases} C_{NR}\sqrt{\epsilon} , & C_{NR} = \dfrac{2\pi g_s V (2m)^{3/2}}{h^3} \\[2mm] C_{UR}\epsilon^2 , & C_{UR} = \dfrac{4\pi g_s V}{c^3 h^3} . \end{cases}$$

Introducing the *distribution function*

$$f(\epsilon, T, \mu) = \frac{1}{\frac{1}{z}e^{\beta\epsilon} + \sigma} ,$$

we can obtain the grand canonical partition function and the average values of N and E from the following integrals:

$$\left. \begin{aligned} \ln Y(T, V, \mu) &= \frac{1}{\sigma} \int d\epsilon\, g(\epsilon) \ln\left(1 + \sigma z e^{-\beta\epsilon}\right) \\ N(T, V, \mu) &= \int dN = \int d\epsilon\, f(\epsilon, T, \mu) g(\epsilon) \\ E(T, V, \mu) &= \int \epsilon\, dN = \int d\epsilon\, \epsilon f(\epsilon, T, \mu) g(\epsilon) . \end{aligned} \right\} \tag{4.89}$$

In Subsections 4.7.2 and 4.7.3, we shall calculate these integrals explicitly. However, before we move on to this, we can already derive some useful relations for the above two special energy cases. For the nonrelativistic case, the grand canonical partition function is

$$\ln Y = \frac{C_{NR}}{\sigma} \int_0^\infty d\epsilon\, \epsilon^{1/2} \ln\left(1 + \sigma z e^{-\beta\epsilon}\right) . \tag{4.90}$$

This yields the expectation values

$$N = C_{NR} \int_0^\infty \frac{\epsilon^{1/2} d\epsilon}{\frac{1}{z}e^{\beta\epsilon} + \sigma} , \quad E = C_{NR} \int_0^\infty \frac{\epsilon^{3/2} d\epsilon}{\frac{1}{z}e^{\beta\epsilon} + \sigma} .$$

Due to $PV = kT \ln Y$, we find

$$PV = \frac{2}{3} \frac{C_{NR}}{\sigma} kT \left(\left[\epsilon^{3/2} \ln\left(1 + \sigma z e^{-\beta\epsilon}\right)\right]_0^\infty + \sigma\beta \int_0^\infty \frac{\epsilon^{3/2} d\epsilon}{\frac{1}{z}e^{\beta\epsilon} + \sigma} \right)$$

$$= \frac{2}{3} E ,$$

where we have used partial integration. This is the same relation as the one we found earlier for the ideal Boltzmann gas [see (4.30) and (4.31)]. The corresponding relations for the ultrarelativistic case are

$$\ln Y = \frac{C_{\mathrm{UR}}}{\sigma} \int_0^\infty \mathrm{d}\epsilon \epsilon^2 \ln\left(1 + \sigma z e^{-\beta\epsilon}\right)$$

$$N = C_{\mathrm{UR}} \int_0^\infty \frac{\epsilon^2 \mathrm{d}\epsilon}{\frac{1}{z}e^{\beta\epsilon} + \sigma} \;,\quad E = C_{\mathrm{UR}} \int_0^\infty \frac{\epsilon^3 \mathrm{d}\epsilon}{\frac{1}{z}e^{\beta\epsilon} + \sigma}\;,$$

and, using a similar line of calculation as above, we obtain

$$PV = \frac{1}{3}E\;.$$

Theorem 4.18: Energy-momentum relations of the ideal quantum gas

Independently of the type of statistics (Fermi-Dirac, Bose-Einstein, Maxwell-Boltzmann), an ideal quantum gas obeys

$$PV = \begin{cases} \dfrac{2}{3}E & \text{nonrelativistic} \\[2mm] \dfrac{1}{3}E & \text{ultrarelativistic}\;. \end{cases}$$

Equation of state for the classical limit. In the case of

$$z = e^{\beta\mu} \ll 1\;,$$

we can easily derive some further relations. For this, we set $x = \beta\epsilon$ and expand the logarithm in (4.90) up to second order around $z = 0$. This yields

$$\begin{aligned}
\ln Y &= \frac{C_{\mathrm{NR}}}{\sigma\beta^{3/2}} \int_0^\infty \mathrm{d}x\, x^{1/2} \left(\sigma z e^{-x} - \frac{\sigma^2 z^2}{2}e^{-2x}\right) \\[2mm]
&= \frac{C_{\mathrm{NR}}}{\sigma\beta^{3/2}} \left(\sigma z - \frac{\sigma^2 z^2}{2^{5/2}}\right) \int_0^\infty \mathrm{d}x\, x^{1/2} e^{-x} \\[2mm]
&= \frac{\sqrt{\pi}}{2}\frac{C_{\mathrm{NR}}}{\beta^{3/2}} z \left(1 - \frac{\sigma z}{2^{5/2}}\right) \\[2mm]
&= g_s V \left(\frac{2\pi m k T}{h^2}\right)^{3/2} z \left(1 - \frac{\sigma z}{2^{5/2}}\right)\;.
\end{aligned} \tag{4.91}$$

The first term in this equation is just the grand canonical partition function of the ideal classical gas and is in agreement with (4.48) from Application

58, as long as the degeneracy factor g_s is attributed for. The average particle number is found from (4.91) as

$$N = z\left(\frac{\partial \ln Y}{\partial z}\right)_{T,V} = g_s V\left(\frac{2\pi mkT}{h^2}\right)^{3/2} z\left[\left(1 - \frac{\sigma z}{2^{5/2}}\right) - \frac{\sigma z}{2^{5/2}}\right] , (4.92)$$

so that

$$\ln Y = N + g_s V\left(\frac{2\pi mkT}{h^2}\right)^{3/2}\frac{\sigma z^2}{2^{5/2}} .$$

Since (4.91) is of second order, we can insert in this equation the first-order term from (4.92):

$$z^2 = \frac{N^2}{g_s^2 V^2}\left(\frac{2\pi mkT}{h^2}\right)^{-3} .$$

This results in

Theorem 4.19: State equation of the ideal quantum gas in the classical limit

In the classical limit $z \ll 1$, the equation of state of the classical gas obeys an expansion of the type

$$E = \frac{3}{2}PV = \frac{3}{2}kT\ln Y = \frac{3}{2}NkT\left[1 + \frac{\sigma N}{g_s V 2^{5/2}}\left(\frac{h^2}{2\pi mkT}\right)^{3/2} + \cdots\right] .$$

The first-order correction is large for small temperatures. In the case of fermions, it implies an increased pressure (at constant density) and thus an effective mutual repulsion of the fermions. On the other hand, for bosons, the pressure is reduced, so that the bosons effectively attract each other.

To obtain any more useful results, it is necessary to solve the integrals (4.89) explicitly. This is the subject of the following two subsections on ideal Fermi and Bose gases.

4.7.2 Ideal Fermi Gas

Our starting point for this discussion is the Fermi-Dirac (FD) distribution function

$$f_{\mathrm{FD}}(\epsilon, T, \mu) = \frac{1}{\frac{1}{z}e^{\beta\epsilon} + 1}$$

as well as the nonrelativistic energy density

$$g(\epsilon) = C_{\mathrm{NR}}\sqrt{\epsilon} , \quad C_{\mathrm{NR}} = \frac{2\pi g_s V(2m)^{3/2}}{h^3} ,$$

which will be used to evaluate the integrals (4.89). In Fig. 4.12, the behavior of the distribution function f_{FD} is shown qualitatively for different

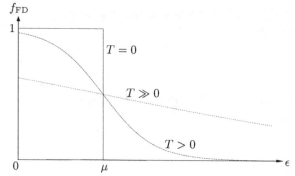

Fig. 4.12. Fermi-Dirac distribution functions for different temperatures.

temperatures. At absolute zero temperature, f_{FD} is a step function; all one-particle states below the *Fermi energy* $E_F = \mu$ are occupied, whereas all other states with $\epsilon > E_F$ are empty. As the temperature increases, more and more fermions are found in states of higher excitation, and the step-like character of the distribution function smears out more and more. Using partial integration and the substitution $x = \beta\epsilon$, the integrals in (4.89) can be transformed into the following expressions:

$$\left.\begin{aligned}
\ln Y(T,V,\mu) &= \frac{C_{NR}}{\beta^{3/2}}\frac{2}{3}\int_0^\infty \frac{x^{3/2}\mathrm{d}x}{\frac{1}{z}e^x + 1} = \frac{g_s V}{\lambda^3}f_{5/2}(z)\\
N(T,V,\mu) &= \frac{C_{NR}}{\beta^{3/2}}\int_0^\infty \frac{x^{1/2}\mathrm{d}x}{\frac{1}{z}e^x + 1} = \frac{g_s V}{\lambda^3}f_{3/2}(z)\\
E(T,V,\mu) &= \frac{C_{NR}}{\beta^{5/2}}\int_0^\infty \frac{x^{3/2}\mathrm{d}x}{\frac{1}{z}e^x + 1} = \frac{3}{2}\frac{g_s V}{\lambda^3\beta}f_{5/2}(z)\\
&= \frac{3}{2}NkT\frac{f_{5/2}(z)}{f_{3/2}(z)}\;,
\end{aligned}\right\} \quad (4.93)$$

with

$$\lambda = \left(\frac{h^2}{2\pi mkT}\right)^{1/2}$$

and the *Fermi-Dirac functions*

$$f_\nu(z) = \frac{1}{\Gamma(\nu)}\int_0^\infty \frac{x^{\nu-1}\mathrm{d}x}{\frac{1}{z}e^x + 1}\;,\quad zf'_\nu(z) = f_{\nu-1}(z)\;.$$

Generally, one wants to eliminate the fugacity z in (4.93) using the relation for N. Given the integral representation of the functions $f_\nu(z)$, this is not so easy. We shall therefore consider some special cases for which the elimination of z is possible.

Classical limit: $T \gg 0 \iff z \ll 1$. For small z, the functions f_ν can be expressed as a Taylor expansion:

$$f_\nu(z) = \sum_{n=1}^{\infty} (-1)^{n+1} \frac{z^n}{n^\nu} \, . \tag{4.94}$$

Substituting this expansion into the middle equation of (4.93) and setting

$$z = \sum_{l=1}^{\infty} a_l y^l \, , \quad y = \frac{N\lambda^3}{g_s V} \, ,$$

we see that

$$y = \sum_l a_l y^l - \frac{1}{2^{3/2}} \sum_{l,l'} a_l a_{l'} y^{l+l'} + \frac{1}{3^{3/2}} \sum_{l,l',l''} a_l a_{l'} a_{l''} y^{l+l'+l''} + \ldots$$

$$= a_1 y + \left(a_2 - \frac{a_1^2}{2^{3/2}} \right) y^2 + \left(a_3 - \frac{2 a_1 a_2}{2^{3/2}} + \frac{a_1^3}{3^{3/2}} \right) y^3 + \ldots \, .$$

This gives the first expansion coefficients of z:

$$a_1 = 1 \, , \quad a_2 = \frac{1}{2^{3/2}} \, , \quad a_3 = \frac{1}{4} - \frac{1}{3^{3/2}} \, .$$

Inserting this into the last equation of (4.93), we find a virial expansion of the form

$$E(T, V, N) = \frac{3}{2} NkT \sum_{l=1}^{\infty} b_l \left(\frac{N\lambda^3}{g_s V} \right)^{l-1} \, ,$$

with the first few coefficients

$$b_1 = 1 \, , \quad b_2 = \frac{1}{2^{5/2}} \, , \quad b_3 = \frac{1}{8} - \frac{2}{3^{5/2}} \, .$$

If we disregard all but the first term, we rediscover the caloric equation of state (4.30) of the ideal classical gas, whereas taking into account the first two terms leads back to Theorem 4.19.

Totally degenerate Fermi gas: $T = 0 \iff z \to \infty$. In the case of the *totally degenerate Fermi gas*, the expansion (4.94) does not converge. However, here we can use the fact that the distribution function is step-like:

$$f_{\mathrm{FD}}(\epsilon, 0, \mu) = \frac{1}{\frac{1}{z} e^{\epsilon/kT} + 1} = \begin{cases} 1 \text{ for } \epsilon \leq \mu_0 \\ 0 \text{ for } \epsilon > \mu_0 \end{cases} \, , \quad \mu_0 = \mu(T = 0) = E_{\mathrm{F}} \, .$$

With this, we calculate

$$N = C_{\mathrm{NR}} \int_0^{E_{\mathrm{F}}} d\epsilon \sqrt{\epsilon} = \frac{2}{3} C_{\mathrm{NR}} E_{\mathrm{F}}^{3/2} = \frac{4\pi g_s V (2m)^{3/2}}{3h^3} E_{\mathrm{F}}^{3/2} \, .$$

For the Fermi energy or the chemical potential, we find

$$E_{\mathrm{F}} = \mu_0 = \frac{h^2}{2m}\left(\frac{3N}{4\pi g_s V}\right)^{2/3} .$$

The ground-state energy follows as

$$E = C_{\mathrm{NR}}\int_0^{E_{\mathrm{F}}} \mathrm{d}\epsilon\,\epsilon^{3/2} = \frac{2}{5}C_{\mathrm{NR}}E_{\mathrm{F}}^{5/2} = \frac{4\pi g_s V(2m)^{3/2}}{5h^3}E_{\mathrm{F}}^{5/2} .$$

Theorem 4.20: Totally degenerate Fermi gas

For the totally degenerate Fermi gas, the average energy (*ground-state energy*) is given by

$$E(T = 0, V, N) = \frac{3}{5}NE_{\mathrm{F}} .$$

Due to Pauli's exclusion principle, not all particles are in the ground state. Rather, all states below the Fermi energy

$$E_{\mathrm{F}} = \frac{h^2}{2m}\left(\frac{3N}{4\pi g_s V}\right)^{2/3}$$

are occupied.

The case $T = 0$ is of practical interest since for many quantum statistical systems, typical excitation temperatures are far above the normal system temperatures, so that those systems are quite well described by their properties at zero temperature. Typical values for *Fermi temperatures* $T_{\mathrm{F}} = E_{\mathrm{F}}/k$ are 0.3 K in fluid ^3He, $5\cdot 10^4$ K for conductive electrons in simple metals, $3\cdot 10^9$ K in white dwarfs, and $3\cdot 10^{12}$ K in neutron stars.

Degenerate Fermi gas: $0 < T \ll T_{\mathrm{F}} \Longleftrightarrow z \gg 1$. As before, for a *degenerate Fermi gas*, the expansion (4.94) does not converge. However, we already know that the distribution function f_{FD} varies only slowly with the energy, with the exception of the small range around $\epsilon \approx \mu$. In other words: $\partial f_{\mathrm{FD}}/\partial \epsilon$ has a sharp maximum around $\epsilon = \mu$,

$$\left.\frac{\partial f_{\mathrm{FD}}}{\partial \epsilon}\right|_{\epsilon=\mu} = -\frac{1}{4kT} ,$$

and is very close to a δ-function. Consider, therefore, the function

$$F(\epsilon) = \int_0^{\epsilon} \mathrm{d}\epsilon'g(\epsilon') .$$

With this and using partial integration, we find that the average particle number evaluates to

$$N = \int\limits_0^\infty d\epsilon f_{\mathrm{FD}}(\epsilon) F'(\epsilon) = -\int\limits_0^\infty d\epsilon f'_{\mathrm{FD}}(\epsilon) F(\epsilon) \;.$$

Expansion of $F(\epsilon)$ around μ,

$$F(\epsilon) = F(\mu) + F'(\mu)(\epsilon - \mu) + \frac{1}{2} F''(\mu)(\epsilon - \mu)^2 + \dots \;,$$

gives

$$N = I_0 F(\mu) + I_1 F'(\mu) + I_2 F''(\mu) + \dots \;,$$

with

$$I_0 = -\int\limits_0^\infty d\epsilon f'_{\mathrm{FD}}(\epsilon) \;,\; I_1 = -\int\limits_0^\infty d\epsilon (\epsilon - \mu) f'_{\mathrm{FD}}(\epsilon)$$

$$I_2 = -\frac{1}{2} \int\limits_0^\infty d\epsilon (\epsilon - \mu)^2 f'_{\mathrm{FD}}(\epsilon) \;.$$

Since we are only considering low temperatures, we can move the lower integration limits of I_0, I_1, and I_2 to $-\infty$. Then, I_0 is one, and I_1 vanishes as $(\epsilon - \mu) f'_{\mathrm{FD}}(\epsilon)$ is an odd function in $x = \beta(\epsilon - \mu)$. We are left with

$$I_2 = \frac{1}{2\beta^2} \int\limits_{-\infty}^\infty dx \frac{x^2 e^x}{(e^x + 1)^2} = \frac{\pi^2}{6\beta^2} \;,$$

so that

$$N = F(\mu) + \frac{\pi^2}{6\beta^2} F''(\mu) + \dots = \int\limits_0^\mu d\epsilon g(\epsilon) + \frac{\pi^2}{6\beta^2} g'(\mu) + \dots \approx \int\limits_0^{E_{\mathrm{F}}} d\epsilon g(\epsilon)$$

$$\implies g(\mu)(E_{\mathrm{F}} - \mu) \approx \frac{\pi^2}{6\beta^2} g'(\mu) \;.$$

With $g(\epsilon) = C_{\mathrm{NR}} \sqrt{\epsilon}$, we finally see that

$$\mu(T) \approx \frac{E_{\mathrm{F}}}{2} + \sqrt{\frac{E_{\mathrm{F}}^2}{4} - \frac{\pi^2}{12\beta^2}} \approx E_{\mathrm{F}} \left(1 - \frac{\pi^2}{12\beta^2 E_{\mathrm{F}}^2} \right) \;.$$

The average energy is

$$E = \int\limits_0^\infty d\epsilon \, \epsilon f_{\mathrm{FD}}(\epsilon) g(\epsilon) \approx \int\limits_0^\mu d\epsilon \, \epsilon g(\epsilon) + \frac{\pi^2}{6\beta^2} \left[\frac{d}{d\epsilon} \epsilon g(\epsilon) \right]_{\epsilon = \mu}$$

$$\approx \int\limits_0^{E_{\mathrm{F}}} d\epsilon \, \epsilon g(\epsilon) + \int\limits_{E_{\mathrm{F}}}^\mu d\epsilon \, \epsilon g(\epsilon) + \frac{\pi^2}{4\beta^2} g(E_{\mathrm{F}})$$

$$\approx \frac{3}{5}NE_F + (\mu - E_F)E_F g(E_F) + \frac{\pi^2}{4\beta^2}g(E_F)$$

$$\approx \frac{3}{5}NE_F + \frac{\pi^2}{6\beta^2}g(E_F) = \frac{3}{5}NE_F\left[1 + \frac{5\pi^2}{12}\left(\frac{kT}{E_F}\right)^2\right].$$

Theorem 4.21: Degenerate Fermi gas

For $T \ll T_F = E_F/k$, we have the following correction to the ground-state energy:

$$E(T,V,N) = \frac{3}{5}NE_F\left[1 + \frac{5\pi^2}{12}\left(\frac{kT}{E_F}\right)^2 + \ldots\right].$$

The specific heat is

$$C_V(T,V,N) = \left(\frac{\partial E}{\partial T}\right)_V = \frac{Nk\pi^2}{2}\frac{T}{T_F},$$

which is much smaller than the classical specific heat $C_V = 3Nk/2$.

According to (4.61) and Theorem 4.18, the entropy of the degenerate gas is given by

$$S = \frac{1}{T}(E - F) = \frac{1}{T}\left(\frac{5}{3}E - \mu N\right) = \frac{Nk\pi^2}{2}\frac{T}{T_F}.$$

Contrary to the entropy of the classical gas [see (4.32) in Application 56], this is in agreement with the third law of thermodynamics: $T \to 0 \Longrightarrow S \to 0$.

Before finishing this subsection, we note that the above-presented expansion of

$$\int_0^\infty d\epsilon\, f_{FD}(\epsilon)g(\epsilon)$$

can be generalized for $z \gg 1$ and, following Sommerfeld, results in the following expansions of $f_\nu(z)$ in $\ln z$:

$$f_{5/2}(z) = \frac{8}{15\sqrt{\pi}}(\ln z)^{5/2}\left[1 + \frac{5\pi^2}{8}(\ln z)^{-2} + \ldots\right]$$

$$f_{3/2}(z) = \frac{4}{3\sqrt{\pi}}(\ln z)^{3/2}\left[1 + \frac{\pi^2}{8}(\ln z)^{-2} + \ldots\right]$$

$$f_{1/2}(z) = \frac{2}{\sqrt{\pi}}(\ln z)^{1/2}\left[1 - \frac{\pi^2}{24}(\ln z)^{-2} + \ldots\right].$$

Generally, only the first few terms need be considered, as the ratio of two subsequent terms is of the order of $(kT/\mu)^2$.

4.7.3 Ideal Bose Gas

We now discuss the ideal Bose gas and proceed analogously to the preceding subsection. Our starting points are the sums for $\ln Y$, E, and N given in Theorem 4.17 and the integrals (4.89), in conjunction with the Bose-Einstein (BE) distribution function

$$f_{\mathrm{BE}}(\epsilon, T, \mu) = \frac{1}{\frac{1}{z}e^{\beta\epsilon} - 1}$$

and the nonrelativistic energy density

$$g(\epsilon) = C_{\mathrm{NR}}\sqrt{\epsilon} \ , \quad C_{\mathrm{NR}} = \frac{2\pi g_s V (2m)^{3/2}}{h^3} \ .$$

Since $\overline{n_k} = \left(e^{\beta(E_k - \mu)} - 1\right)^{-1} \geq 0$, the Bose gas obeys

$$\mu \leq E_k \overset{E_0 = 0}{\Longrightarrow} \mu \leq 0 \ , \ 0 < z \leq 1$$

at all temperatures. The occupation number of the ground state is

$$\overline{n_0} = \frac{z}{1 - z}$$

and can grow arbitrarily large, in contrast to that of the Fermi gas. Clearly, this is not taken into account when naively replacing the sums of Theorem 4.17 by the integrals of (4.89) since the energy density $g(\epsilon)$ vanishes for $\epsilon = 0$ and this would result in a zero weighting of the ground state. Let us instead split off the ground state from the sums and approximate only the remaining expressions by integrals:

$$\ln Y = -\int d\epsilon\, g(\epsilon) \ln\left(1 - z e^{-\beta\epsilon}\right) - \ln(1 - z)$$

$$N = \int d\epsilon\, f(\epsilon, T, \mu) g(\epsilon) + N_0 \ , \quad N_0 = \overline{n_0} = \frac{z}{1 - z}$$

$$E = \int d\epsilon\, \epsilon f(\epsilon, T, \mu) g(\epsilon) \ .$$

Note that in the classical limit $z \ll 1$, N_0 can be disregarded, whereas for $z \approx 1$, this term contributes significantly. However, the term $-\ln(1 - z) = \ln(1 + N_0)$ is at most of the order of $\mathcal{O}(\ln N)$ and can therefore be dropped for all values $0 < z \leq 1$. As before, we now substitute $x = \beta E$ in these integrals and use partial integration. This gives the bosonic expressions corresponding to (4.93):

$$\left.\begin{aligned}
\ln Y(T, V, \mu) &= \frac{g_s V}{\lambda^3} g_{5/2}(z) \\[2mm]
N(T, V, \mu) &= \frac{g_s V}{\lambda^3} g_{3/2}(z) + N_0 \\[2mm]
E(T, V, \mu) &= \frac{3}{2}\frac{g_s V}{\lambda^3 \beta} g_{5/2}(z)
\end{aligned}\right\} \ , \ \lambda = \left(\frac{h^2}{2\pi m k T}\right)^{1/2} \ , \tag{4.95}$$

with the *Bose-Einstein functions*

$$g_\nu(z) = \frac{1}{\Gamma(\nu)} \int_0^\infty \frac{x^{\nu-1}dx}{\frac{1}{z}e^x - 1} \; , \; zg'_\nu(z) = g_{\nu-1}(z) \; . \tag{4.96}$$

Classical limit: $T \gg 0 \Longleftrightarrow z \ll 1$. For small z, we can use the expansion

$$g_\nu(z) = \sum_{n=1}^\infty \frac{z^n}{n^\nu} \; .$$

With its help, the middle equation of (4.95) can be expanded in z as in the fermionic case and, on insertion into the last equation of (4.95), yields the virial expansion

$$E(T,V,N) = \frac{3}{2}NkT \sum_{l=1}^\infty b_l \left(\frac{N\lambda^3}{g_s V}\right)^{l-1} \; , \tag{4.97}$$

with the coefficients

$$b_1 = 1 \; , \; b_2 = -\frac{1}{2^{5/2}} \; , \; b_3 = \frac{1}{8} - \frac{2}{3^{5/2}} \; .$$

The specific heat is

$$C_V(T,V,N) = \left(\frac{\partial E}{\partial T}\right)_{N,V} = \frac{3}{2}Nk \sum_{l=1}^\infty \frac{5-3l}{2} b_l \left(\frac{N\lambda^3}{g_s V}\right)^{l-1} \; ,$$

which, for $T \to \infty$, converges to its classical value of $C_V = 3Nk/2$ as expected. Note that the second term in the expansion is positive, so that for large but finite temperatures the specific heat is larger than its classical value. On the other hand, we already know that C_V must go to zero for $T \to 0$. We therefore conclude that there must exist a critical temperature T_c, for which C_V is maximal. As it turns out, the derivative of the specific heat with respect to the temperature is divergent at this point, which indicates a *phase transition of second order*.

Bose-Einstein condensation: T small $\Longleftrightarrow z \approx 1$. For small temperatures, the expansion (4.97) is no longer valid. In this case, we rewrite the middle equation of (4.95) as

$$N = N_\epsilon + N_0 \; , \; N_\epsilon = \frac{g_s V}{\lambda^3} g_{3/2}(z) \; ,$$

where N_0 and N_ϵ denote the number of particles in the ground state and in excited states, respectively. Since the function $g_{3/2}(z)$ grows monotonically in the interval $0 < z \le 1$, the number of excited states is restricted to the range

$$0 \le N_\epsilon \le N_\epsilon^{\max} = \frac{g_s V}{\lambda^3} g_{3/2}(1) \; , \; g_{3/2}(1) = 2.612 \; .$$

Therefore, for the two special cases $z < 1$ and $z = 1$, we can conclude:

Theorem 4.22: Bose-Einstein condensation

Depending on the fugacity z, the ideal Bose gas exhibits the following behavior:

- For $z < 1$, the term $N_0 = z/(1-z)$ does not contribute in the thermodynamic limit, and the fugacity is obtained from

$$N \approx N_\epsilon = \frac{g_s V}{\lambda^3} g_{3/2}(z) \ .$$

This means that all particles are in excited states.

- For $z = 1$, the number N_0 can be arbitrarily large and contributes significantly to the total number of particles. We then have

$$N > N_\epsilon^{\max} = \frac{g_s V}{\lambda^3} g_{3/2}(1) \ . \tag{4.98}$$

In this case, not all particles can be in excited states. Rather, $N_0 = N - N_\epsilon^{\max}$ particles *condensate* into the ground state.

Now we wish to study the phenomenon of Bose-Einstein condensation in more detail. According to (4.98), the condition for condensation to occur for a system with a constant number of particles and constant volume is given by

$$T < T_c = \frac{h^2}{2\pi mk} \left(\frac{N}{g_s V g_{3/2}(1)} \right)^{2/3} \ .$$

In general, for temperatures $T < T_c$, the system will consist of a mixture of both phases since we have

$$\frac{N_\epsilon}{N} = \begin{cases} 1 & \text{for } T \geq T_c \\ \left(\dfrac{T}{T_c} \right)^{3/2} & \text{for } T < T_c \end{cases}$$

$$\frac{N_0}{N} = \begin{cases} 0 & \text{for } T \geq T_c \\ 1 - \left(\dfrac{T}{T_c} \right)^{3/2} & \text{for } T < T_c \ . \end{cases}$$

These relations are shown in Fig. 4.13. If we use the first equation in (4.95), the pressure is calculated as

$$P(T,V,N) = \frac{kT}{V} \ln Y = \begin{cases} \dfrac{NkT}{V} \dfrac{g_{5/2}(z)}{g_{3/2}(z)} & \text{for } T > T_c \\[2ex] \dfrac{NkT_c}{V} \dfrac{g_{5/2}(1)}{g_{3/2}(1)} \approx 0.5134 \dfrac{NkT_c}{V} & \text{for } T = T_c \\[2ex] \dfrac{g_s kT}{\lambda^3} g_{5/2}(1) & \text{for } T < T_c \ , \end{cases}$$

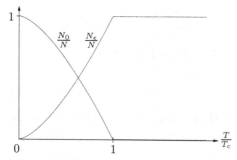

Fig. 4.13. Comparison of the numbers of bosonic ground states and excited states as a function of the temperature.

where we have used that for $T \geq T_c$

$$N_0 \approx 0 \implies \frac{N\lambda^3}{g_s V} = g_{3/2}(z) \,. \tag{4.99}$$

We see that at the critical point, the pressure is only half the size of that of the classical gas. Furthermore, below T_c, the pressure is independent of N and V. This is due to the fact that particles in the ground state do not contribute to the pressure at all. Next, we turn to the specific heat. Using

$$E = kT^2 \left(\frac{\partial \ln Y}{\partial T}\right)_{z,V} = \frac{3}{2}kT\frac{g_s V}{\lambda^3}g_{5/2}(z) \,,$$

we find

$$\frac{C_V}{Nk} = \frac{1}{Nk}\left(\frac{\partial E}{\partial T}\right)_{N,V} \,.$$

For $T < T_c$, $z = 1$, independently of T. We obtain

$$\frac{C_V}{Nk} = \frac{3}{2}\frac{g_s V}{N}g_{5/2}(1)\frac{\mathrm{d}}{\mathrm{d}T}\left(\frac{T}{\lambda^3}\right) = \frac{15}{4}\frac{g_s V}{N\lambda^3}g_{5/2}(1) \sim T^{3/2} \,.$$

On the other hand, for $T \geq T_c$, z is temperature-dependent. In this case, using (4.96) and (4.99), we find

$$\frac{C_V}{Nk} = \frac{3}{2}\frac{g_s V}{N}\frac{\mathrm{d}}{\mathrm{d}T}\left(\frac{T}{\lambda^3}g_{5/2}(z)\right) = \frac{15}{4}\frac{g_s V}{N\lambda^3}g_{5/2}(z) + \frac{3}{2}\frac{g_s V}{N}\frac{T}{\lambda^3}g_{5/2}'(z)\frac{\mathrm{d}z}{\mathrm{d}T}$$

$$= \frac{15}{4}\frac{g_{5/2}(z)}{g_{3/2}(z)} + \frac{3}{2}\frac{T}{z}\frac{\mathrm{d}z}{\mathrm{d}T} \,.$$

The derivative $\mathrm{d}z/\mathrm{d}T$ can be converted into

$$\frac{\mathrm{d}z}{\mathrm{d}T}\frac{1}{z}g_{1/2}(z) = \frac{\mathrm{d}z}{\mathrm{d}T}g_{3/2}'(z) = \frac{\mathrm{d}z}{\mathrm{d}T}\frac{\mathrm{d}g_{3/2}}{\mathrm{d}T}\frac{\mathrm{d}T}{\mathrm{d}z} = \frac{\mathrm{d}}{\mathrm{d}T}\left(\frac{N\lambda^3}{g_s V}\right) = -\frac{3}{2T}g_{3/2}(z)$$

so that overall

$$\frac{C_V}{Nk} = \frac{15}{4}\frac{g_{5/2}(z)}{g_{3/2}(z)} - \frac{9}{4}\frac{g_{3/2}(z)}{g_{1/2}(z)} \ .$$

From this, the classical limit $(z \to 0 \ , \ T \gg T_c)$ is found to be

$$C_V = Nk \left(\frac{15}{4} - \frac{9}{4}\right) = \frac{3}{2}Nk \ .$$

At the critical point $(z \to 1 \ , \ T = T_c)$, the function $g_{1/2}$ diverges, so that the specific heat is given by

$$C_V = \frac{15}{4}\frac{g_{5/2}(1)}{g_{3/2}(1)}Nk = 1.925Nk \ .$$

This is significantly larger than the value of the classical ideal Boltzmann gas $(C_V = 3Nk/2)$. Note that C_V, but not dC_V/dT, is continuous at the critical point.

Summary

- In contrast to Maxwell-Boltzmann statistics, **quantum statistics** takes into account the fermionic or bosonic character of particles.

- In quantum statistics, the **occupation number formalism** turns out to be very useful for systems of noninteracting particles. Here, the N-particle quantum state is specified by the occupation numbers of all possible one-particle states. This formalism yields relatively simple relations for the grand canonical partition function, the average energy, as well as the average particle number. For some specific cases, such as the classical limit (large T) and the purely quantum case (small T), we can eliminate the fugacity from these formulae.

- For an **ideal Fermi gas**, Pauli's exclusion principle implies that even at absolute zero temperature not all particles are in the ground state. Rather, all states with energy smaller than the **Fermi energy** are occupied.

- An **ideal Bose gas** exhibits **Bose-Einstein condensation** at small temperatures. This is due to the fact that excited states can only be occupied by a limited number of particles, so that all remaining particles are forced to condensate into the ground state.

Applications

66. Ideal photon gas. Calculate the free energy, the entropy, the energy, and the pressure of an ideal photon gas.

Solution. Photons are massless spin-1 particles and are therefore described by Bose-Einstein statistics. For these particles, the ultrarelativistic momentum-energy relation

$$\epsilon = c|\boldsymbol{p}| = \hbar\omega \ , \quad \boldsymbol{p} = \hbar\boldsymbol{k} \ , \quad \omega = c|\boldsymbol{k}|$$

holds, where ω denotes the frequency, and \boldsymbol{k} is the wave vector of the light particles. Since photons can be absorbed and emitted by atoms, the total number of photons is not constant, so that $\mu = 0$. The free energy is therefore

$$F = \mu N - PV = -PV \ .$$

On the other hand, using Theorem 4.17, we also have

$$PV = kT \ln Y = -kT \sum_{\boldsymbol{k}} \ln\left(1 - e^{\beta E_{\boldsymbol{k}}}\right) \ ,$$

where the sum is taken over all possible wave vectors \boldsymbol{k}. Using the ultrarelativistic energy density

$$g_{UR}(\epsilon) = C_{UR}\epsilon^2 \ , \quad C_{UR} = \frac{4\pi g_s V}{c^3 h^3} \ ,$$

we can write the sum as an integral:

$$F = kT \int_0^\infty d\epsilon g(\epsilon) \ln\left(1 - e^{-\beta\epsilon}\right) = kT \frac{8\pi V}{c^3 h^3} \int_0^\infty d\epsilon \epsilon^2 \ln\left(1 - e^{-\beta\epsilon}\right) \ .$$

Here, the degeneracy factor g_s is two (rather than three as for other spin-1 particles) due to the transversality of the electromagnetic field. Partially integrating and substituting $x = \beta\epsilon$, we find

$$F(T,V) = -\frac{(kT)^4 V}{3\pi^2 \hbar^3 c^3} \underbrace{\int_0^\infty \frac{dx\, x^3}{e^x - 1}}_{\pi^4/15} = -\frac{4\sigma}{3c} V T^4 \ , \quad \sigma = \frac{\pi^2 k^4}{60\hbar^3 c^2} \ ,$$

where σ is also called the *Boltzmann constant*. From the free energy, we obtain the other quantities as

$$P(T,V) = -\left(\frac{\partial F}{\partial V}\right)_T = P(T) = \frac{4\sigma}{3c} T^4 \qquad (\textit{Boltzmann law})$$

$$S(T,V) = -\left(\frac{\partial F}{\partial T}\right)_V = \frac{16\sigma}{3c} V T^3$$

$$E(T,V) = F + TS = \frac{4\sigma}{c} V T^4 = 3PV \ .$$

67. Ideal phonon gas. Calculate the specific heat of a solid in a model where the movement of the atoms around their equilibrium positions is described by quantized harmonic oscillations.

Solution. We need to describe N atoms of a crystalline solid. In the *harmonic approximation*, the movement of the atoms around their equilibria is given by a Taylor expansion of the potential energy up to and including the quadratic term. Since the linear term vanishes at equilibrium, the classical Hamilton function of the system consists of a constant ground-state energy, the kinetic energy $T = \frac{m}{2} \sum_{i=1}^{3N} \dot{x}_i^2$ and the potential energy $V = \sum_{i,j=1}^{3N} A_{ij} x_i x_j$. As we have seen in Application 7 in Section 1.2, this type of Hamilton function can always be transformed to normal form, so that the normal coordinates represent the normal vibrations of the lattice. These normal coordinates of the $3N$ decoupled linear harmonic oscillators can be formally quantized. The quantized lattice vibrations called *phonons* can be interpreted as an ideal ultrarelativistic Bose gas with the energy density

$$g_{\mathrm{UR}}(\epsilon) = C_{\mathrm{UR}} \epsilon^2 \; , \quad C_{\mathrm{UR}} = \frac{12 \pi V}{c^3 h^3} \; , \quad \epsilon = \hbar \omega \qquad (g_s = 3) \; .$$

Since phonons can be created in any number, their chemical potential is zero, and thus $z = 1$.

Einstein model. We initially calculate the specific heat in the Einstein approximation, in which the $3N$ oscillators are independent and have identical frequency ω_{E}. We have encountered this type of problem before in Subsection 4.6.3. According to (4.79), the energy of our system is given by

$$E = \frac{3 N \hbar \omega_{\mathrm{E}}}{2 \tanh \left(\frac{\beta \hbar \omega_{\mathrm{E}}}{2} \right)} \; .$$

The specific heat is therefore

$$C_V = \left(\frac{\partial E}{\partial T} \right)_V = \frac{3 N \hbar^2 \omega_{\mathrm{E}}^2}{k T^2} \frac{e^{\beta \hbar \omega_{\mathrm{E}}}}{\left(e^{\beta \hbar \omega_{\mathrm{E}}} - 1 \right)^2} = 3 N k \frac{x^2 e^x}{(e^x - 1)^2} \; ,$$

with

$$x = \frac{\hbar \omega_{\mathrm{E}}}{k T} \; .$$

For large temperatures, $x \ll 1$, we find

$$C_V = 3 N k \qquad (\textit{Dulong and Petit's law}).$$

For small temperatures, $x \gg 1$, it follows that

$$C_V = 3 N k x^2 e^{-x} \; .$$

Whereas the high-temperature result is verified experimentally, one observes a T^3-behavior for small temperatures, which is not explained in the simple Einstein model.

Debye model. In the Debye model, the single Einstein frequency ω_E is replaced by a continuous spectrum of oscillation modes. They can be calculated in a similar manner as the photon spectrum. The number of modes in the frequency interval $[\omega : \omega + d\omega]$ is given by

$$g_{UR}(\omega)d\omega = \frac{12\pi V}{c^3}\omega^2 d\omega \ .$$

The maximal frequency (*Debye frequency*) ω_D can be obtained from

$$\int_0^{\omega_D} g(\omega)d\omega = 3N$$

as

$$\omega_D^3 = \frac{3Nc^3}{4\pi V} \ .$$

This takes into account the fact that oscillations on the lattice are not possible with wavelengths smaller than the lattice constant. Such a constraint did not apply to the phonon gas. According to Theorem 4.17, the average energy is

$$E(T,V) = \sum_k \frac{\hbar\omega_k}{e^{\beta\hbar\omega_k} - 1} \ .$$

In the limit $V \to \infty$, this can be written as an integral,

$$E(T,V) = \int_0^{\omega_D} d\omega g_{UR}(\omega)\frac{\hbar\omega}{e^{\beta\hbar\omega} - 1} = 3NkTD(x_D) \ , \quad x_D = \beta\hbar\omega_D \ ,$$

where

$$D(x_D) = \frac{3}{x_D^3} \int_0^{x_D} dx \frac{x^3}{e^x - 1} = \begin{cases} 1 - \dfrac{3x_D}{8} + \dfrac{x_D^2}{20} + \dots & \text{for } x_D \ll 1 \\[2mm] \dfrac{\pi^4}{5x_D^3} + \mathcal{O}\left(e^{-x_D}\right) & \text{for } x_D \gg 1 \end{cases}$$

denotes the *Debye function.* Using

$$x_D = \frac{\hbar\omega_D}{kT} = \frac{T_D}{T} \ , \quad T_D = \frac{\hbar\omega_D}{k} \ ,$$

we find the energy

$$E = 3NkTD(x_D) = 3NkT \begin{cases} 1 - \dfrac{3}{8}\dfrac{T_D}{T} + \dfrac{1}{20}\left(\dfrac{T_D}{T}\right)^2 + \dots & \text{for } T \gg T_D \\[3mm] \dfrac{\pi^4}{5}\left(\dfrac{T}{T_D}\right)^3 + \mathcal{O}\left(e^{-T_D/T}\right) & \text{for } T \ll T_D \end{cases}$$

and the specific heat

$$\frac{C_V}{Nk} = \begin{cases} 3 - \dfrac{3}{20}\left(\dfrac{T_D}{T}\right)^2 + \cdots & \text{for } T \gg T_D \\[2ex] \dfrac{12\pi^4}{5}\left(\dfrac{T}{T_D}\right)^3 + \mathcal{O}(e^{-T_D/T}) & \text{for } T \ll T_D \ . \end{cases}$$

Thus, the Debye model is indeed capable of explaining the T^3-behavior of the specific heat at small temperatures.

Appendix A. Mathematical Appendix

This appendix recapitulates some basic mathematical relations from analysis and vector analysis frequently used in this book. Throughout, we apply minimal mathematical stringency in order to allow the reader to quickly find relevant formulae.

A.1 Vector Operations

Vector operators. In Cartesian representation with canonical orthonormal basis $\{e_x, e_y, e_z\}$, the three-dimensional vector operations gradient, divergence, and rotation (or curl) are defined as

Gradient:
$$\nabla \psi = e_x \frac{\partial \psi}{\partial x} + e_y \frac{\partial \psi}{\partial y} + e_z \frac{\partial \psi}{\partial z}$$

Divergence:
$$\nabla A = \frac{\partial A_x}{\partial x} + \frac{\partial A_y}{\partial y} + \frac{\partial A_z}{\partial z}$$

Curl:
$$\nabla \times A = e_x \left(\frac{\partial A_z}{\partial y} - \frac{\partial A_y}{\partial z} \right) + e_y \left(\frac{\partial A_x}{\partial z} - \frac{\partial A_z}{\partial x} \right)$$
$$+ e_z \left(\frac{\partial A_y}{\partial x} - \frac{\partial A_x}{\partial y} \right) ,$$

where $A = A_x(x, y, z)e_x + A_y(x, y, z)e_y + A_z(x, y, z)e_z$ denotes a vector field and $\psi = \psi(x, y, z)$ a scalar field. If the Cartesian coordinates depend on cylindrical coordinates through (Fig. A.1, right)

$$x = r \cos \varphi , \; y = r \sin \varphi , \; z = z ,$$

it holds that

$$\frac{\partial}{\partial x} = \cos \varphi \frac{\partial}{\partial r} - \frac{\sin \varphi}{r} \frac{\partial}{\partial \varphi}$$

$$\frac{\partial}{\partial y} = \sin \varphi \frac{\partial}{\partial r} + \frac{\cos \varphi}{r} \frac{\partial}{\partial \varphi}$$

$$\frac{\partial}{\partial z} = \frac{\partial}{\partial z} .$$

In the case of a dependency on spherical coordinates,

Fig. A.1. Relationship between Cartesian and spherical coordinates (*left*) and between Cartesian and cylindrical coordinates (*right*).

$$x = r \cos \varphi \sin \theta \ , \ y = r \sin \varphi \sin \theta \ , \ z = r \cos \theta \ ,$$

we have (Fig. A.1, left)

$$\frac{\partial}{\partial x} = \cos \varphi \sin \theta \frac{\partial}{\partial r} - \frac{\sin \varphi}{r \sin \theta} \frac{\partial}{\partial \varphi} + \frac{\cos \varphi \cos \theta}{r} \frac{\partial}{\partial \theta}$$

$$\frac{\partial}{\partial y} = \sin \varphi \sin \theta \frac{\partial}{\partial r} + \frac{\cos \varphi}{r \sin \theta} \frac{\partial}{\partial \varphi} + \frac{\sin \varphi \cos \theta}{r} \frac{\partial}{\partial \theta}$$

$$\frac{\partial}{\partial z} = \cos \theta \frac{\partial}{\partial r} - \frac{\sin \theta}{r} \frac{\partial}{\partial \theta} \ .$$

The corresponding equations for gradient, divergence, and curl in other basis systems, for example, the spherical basis $\{e_r, e_\theta, e_\varphi\}$ or the cylindrical basis $\{e_r, e_\varphi, e_z\}$, are not used in this book and hence not discussed here.

Frequently used formulae of vector calculus and analysis are

$$A(B \times C) = B(C \times A) = C(A \times B)$$

$$A \times (B \times C) = B(AC) - C(AB)$$

$$(A \times B)(C \times D) = (AC)(BD) - (AD)(BC)$$

$$\nabla \times \nabla \psi = 0$$

$$\nabla(\nabla \times A) = 0$$

$$\nabla \times (\nabla \times A) = \nabla(\nabla A) - \nabla^2 A$$

$$\nabla(\psi \phi) = \psi \nabla \phi + \phi \nabla \psi$$

$$\nabla(\psi A) = \psi \nabla A + A \nabla \psi$$

$$\nabla \times (\psi A) = \psi \nabla \times A - A \times \nabla \psi$$

$$\nabla(A \times B) = B(\nabla \times A) - A(\nabla \times B)$$

$$\nabla \times (A \times B) = (B\nabla)A - B(\nabla A) - (A\nabla)B + A(\nabla B)$$

$$\nabla(AB) = (B\nabla)A + B \times (\nabla \times A) + (A\nabla)B + A \times (\nabla \times B).$$

A.2 Integral Theorems

Gauss's law. Let $A(x)$ be a vector field and V a volume with a closed surface F, whose normal $dF = dFn$ is perpendicularly directed outward at each surface point. Then,

$$\int_V dV \nabla A = \oint_F dF A \ .$$

In particular, for $A = c\psi$ or $A = c \times B$, $c = \text{const}$, it follows that

$$\int_V dV \nabla \psi = \oint_F dF \psi \ , \quad \int_V dV \nabla \times B = \oint_F dF \times B \ .$$

Stokes's law. Let $A(x)$ be a vector field and C a closed curve with a directed circulation (curve element: dl), over which a regular surface F with oriented normal $dF = dFn$ is spanned. Then,

$$\int_F dF \nabla \times A = \oint_C dl A \ .$$

Replacing A by $c\psi$ or $c \times B$ leads to

$$\int_F dF \times \nabla \psi = \oint_C dl \psi \ , \quad \int_F (dF \times \nabla) \times B = \oint_C dl \times B \ .$$

Here, the orientation of F must be chosen such that the direction of circulation of C forms a right screw with the normal n of F (Fig. A.2).

Fig. A.2. A curve C with directed circulation, over which a surface F is spanned.

First Green identity. Setting $A = \phi \nabla \psi$, it follows from $\nabla(\phi \nabla \psi) = \phi \nabla^2 \psi + \nabla \phi \nabla \psi$ and Gauss's theorem

$$\int_V dV(\phi \nabla^2 \psi + \nabla \phi \nabla \psi) = \oint dF \phi \nabla \psi \ . \tag{A.1}$$

Second Green identity. Writing (A.1) again with ϕ and ψ interchanged, and then subtracting this equation from (A.1), yields

$$\int_V dV(\phi\nabla^2\psi - \psi\nabla^2\phi) = \oint_F dF(\phi\nabla\psi - \psi\nabla\phi) \ . \tag{A.2}$$

A.3 Partial Differential Quotients

Total differentials. In the following, we consider a function $f(x,y)$ of the variables x and y, which is at least twice differentiable, so that

$$\frac{\partial^2 f}{\partial x \partial y} = \frac{\partial^2 f}{\partial y \partial x} \tag{A.3}$$

(the extension to more than two variables is unproblematic). The differential df is called the total differential of f if

$$df = f(x + dx, y + dy) - f(x,y) = \left(\frac{\partial f}{\partial x}\right)_y dx + \left(\frac{\partial f}{\partial y}\right)_x dy \ .$$

Obviously, this expression is equivalent to the path independency of line integrals over df,

$$\int_{\substack{(x_1,y_1) \\ C_1}}^{(x_2,y_2)} df = \int_{\substack{(x_1,y_1) \\ C_2}}^{(x_2,y_2)} df \quad \Longleftrightarrow \quad \oint df = 0 \ ,$$

since after summing over the differences df, only the values at the endpoints remain. Due to (A.3), for any expression

$$df = A(x,y)dx + B(x,y)dy \ ,$$

it follows the forward direction of the statement

$$df \text{ is total differential} \Longleftrightarrow \left(\frac{\partial A}{\partial y}\right)_x = \left(\frac{\partial B}{\partial x}\right)_y \ .$$

The reverse direction can be shown with the help of Stokes's law:

$$\oint_C df = \oint_C dl V = \int_F dF \nabla \times V = \int_F dF \begin{pmatrix} 0 \\ 0 \\ \frac{\partial B}{\partial x} - \frac{\partial A}{\partial y} \end{pmatrix} \ ,$$

with

$$dl = dx e_x + dy e_y \ , \quad V = A e_x + B e_y \ .$$

Transformation of partial differential quotients. From

$$df = \left(\frac{\partial f}{\partial x}\right)_y dx + \left(\frac{\partial f}{\partial y}\right)_x dy \, ,$$

we can derive some useful relations between differential quotients. For example, holding the variable y fixed ($dy = 0$), it follows that

$$\left(\frac{\partial f}{\partial x}\right)_y = \left(\frac{\partial x}{\partial f}\right)_y^{-1} .$$

If f is held fixed ($df = 0$), we obtain

$$\left(\frac{\partial f}{\partial y}\right)_x = -\left(\frac{\partial f}{\partial x}\right)_y \left(\frac{\partial x}{\partial y}\right)_f . \tag{A.4}$$

Using this relation, fixed and free variables can be exchanged among one another. Turning to another variable, e.g., $y \to g(x, y)$, we find from

$$df = \left(\frac{\partial f}{\partial x}\right)_g dx + \left(\frac{\partial f}{\partial g}\right)_x \left[\left(\frac{\partial g}{\partial x}\right)_y dx + \left(\frac{\partial g}{\partial y}\right)_x dy\right] \tag{A.5}$$

that for $dy = 0$

$$\left(\frac{\partial f}{\partial x}\right)_y = \left(\frac{\partial f}{\partial x}\right)_g + \left(\frac{\partial f}{\partial g}\right)_x \left(\frac{\partial g}{\partial x}\right)_y .$$

This equation is used to choose a new fixed variable. For $dx = 0$, we obtain from (A.5) the partial version of the chain rule:

$$\left(\frac{\partial f}{\partial y}\right)_x = \left(\frac{\partial f}{\partial g}\right)_x \left(\frac{\partial g}{\partial y}\right)_x . \tag{A.6}$$

Furthermore, due to (A.4) and (A.6), we have

$$\left(\frac{\partial f}{\partial g}\right)_x \left(\frac{\partial g}{\partial f}\right)_y = \left(\frac{\partial f}{\partial x}\right)_g \left(\frac{\partial x}{\partial g}\right)_f \left(\frac{\partial g}{\partial y}\right)_f \left(\frac{\partial y}{\partial f}\right)_g$$

$$= \left(\frac{\partial x}{\partial y}\right)_f \left(\frac{\partial y}{\partial x}\right)_g .$$

With this, the pairs of variables (f, g) and (x, y) can be exchanged.

A.4 Complete Function Systems, Fourier Analysis

Let $\{g_n(x), n = 0, 1, 2, \ldots\}$ be a complete, real, or complex discrete function system, which is orthonormal with respect to the scalar product

$$\langle g_i, g_j \rangle = \int_a^{a+2L} dx\, g_i(x) g_j^*(x)$$

within the interval $[a : a + 2L]$, i.e., $\langle g_i, g_j \rangle = \delta_{ij}$. Assume further, a function f possesses the following properties:

- f and f' are piecewise continuous within $[a : a + 2L]$.
- f has a finite number of finite steps.
- f is periodic: $f(x) = f(x + 2L)$.

Then, f can be expanded as

$$f(x) = \sum_n a_n g_n(x) \; , \quad a_n = \langle f, g_n \rangle = \int_a^{a+2L} dx\, f(x) g_n^*(x) \; .$$

Fourier series. The complex functions

$$g_n(x) = \frac{1}{\sqrt{2L}} \exp\left(\frac{in\pi}{L} x\right) , \quad n = 0, \pm 1, \pm 2 \ldots$$

form a complete orthonormal system in the above sense. Using these functions, the series expansion (Fourier series) of a function f reads

$$f(x) = \frac{1}{\sqrt{2L}} \sum_{n=-\infty}^{\infty} a_n \exp\left(\frac{in\pi}{L} x\right) ,$$

with

$$a_n = \frac{1}{\sqrt{2L}} \int_a^{a+2L} dx\, f(x) \exp\left(-\frac{in\pi}{L} x\right) .$$

In particular, we have for the δ-function

$$\delta(x - x') = \sum_{n=-\infty}^{\infty} \frac{\exp(ik_n(x - x'))}{2L} \; , \quad k_n = \frac{n\pi}{L} \; . \tag{A.7}$$

Fourier integrals. Enlarging the periodicity interval to infinity, $L \to \infty$, (A.7) turns into

$$\delta(x - x') = \int_{-\infty}^{\infty} dk \frac{\exp(ik(x - x'))}{2\pi} = \int_{-\infty}^{\infty} dk \frac{\exp(ikx)}{\sqrt{2\pi}} \left(\frac{\exp(ikx')}{\sqrt{2\pi}}\right)^* ,$$

or, due to the symmetry in k and x,

$$\delta(k - k') = \int_{-\infty}^{\infty} dx \frac{\exp(ikx)}{\sqrt{2\pi}} \left(\frac{\exp(ik'x)}{\sqrt{2\pi}}\right)^* .$$

Obviously, the functions

$$g(k, x) = \frac{\exp(ikx)}{\sqrt{2\pi}}$$

form a complete continuous function system normalized to the δ-function. Therefore, the Fourier integral expansion for an arbitrary function f reads

$$f(x) = \int_{-\infty}^{\infty} dx'\delta(x-x')f(x') = \int_{-\infty}^{\infty} dk g(k,x) \int_{-\infty}^{\infty} dx' f(x')g(k,x')^*$$

$$= \frac{1}{\sqrt{2\pi}} \int_{-\infty}^{\infty} dk\, a(k)\exp(ikx) ,$$

with

$$a(k) = \frac{1}{\sqrt{2\pi}} \int_{-\infty}^{\infty} dx\, f(x)\exp(-ikx) .$$

The generalization to n dimensions is

$$f(\boldsymbol{x}) = \frac{1}{(2\pi)^{n/2}} \int d^n k\, a(\boldsymbol{k})\exp(i\boldsymbol{kx}) , \quad \boldsymbol{x} = \begin{pmatrix} x_1 \\ \vdots \\ x_n \end{pmatrix} , \quad \boldsymbol{k} = \begin{pmatrix} k_1 \\ \vdots \\ k_n \end{pmatrix}$$

$$a(\boldsymbol{k}) = \frac{1}{(2\pi)^{n/2}} \int d^n x\, f(\boldsymbol{x})\exp(-i\boldsymbol{kx}) .$$

A.5 Bessel Functions, Spherical Bessel Functions

Bessel functions. The Bessel differential equation is given by

$$\left[\frac{d^2}{dx^2} + \frac{1}{x}\frac{d}{dx} + \left(1 - \frac{m^2}{x^2}\right) \right] f(x) = 0 , \quad m \in \mathbb{R}.$$

Its solutions are the Bessel functions J_m and J_{-m}, with

$$J_m(x) = \left(\frac{x}{2}\right)^m \sum_{i=0}^{\infty} \frac{(-1)^i}{i!\Gamma(m+i+1)} \left(\frac{x}{2}\right)^{2i} .$$

If m is an integer, it holds that

$$J_m(x) = \left(\frac{x}{2}\right)^m \sum_{i=0}^{\infty} \frac{(-1)^i}{i!(m+i)!} \left(\frac{x}{2}\right)^{2i} , \quad J_{-m}(x) = (-1)^m J_m(x) .$$

Spherical Bessel functions. The spherical Bessel differential equation reads

$$\left[\frac{d^2}{dx^2} + \frac{2}{x}\frac{d}{dx} + 1 - \frac{l(l+1)}{x^2} \right] f(x) = 0 , \quad l = 0,1,2,\dots .$$

Its solutions are given by the spherical Bessel functions j_l, n_l (the latter are also called Neumann functions) and therefore also the Hankel functions $h_l^{(\pm)}$:

$$j_l(x) = \left(\frac{\pi}{2x}\right)^{1/2} J_{l+1/2}(x)$$

$$n_l(x) = (-1)^l \left(\frac{\pi}{2x}\right)^{1/2} J_{-l-1/2}(x)$$

$$h_l^{(\pm)}(x) = n_l(x) \pm i j_l(x) \;.$$

Their explicit forms are

$$\left.\begin{array}{rcl} j_l(x) & = & R_l(x)\dfrac{\sin x}{x} + S_l(x)\dfrac{\cos x}{x} \\[2ex] n_l(x) & = & R_l(x)\dfrac{\cos x}{x} - S_l(x)\dfrac{\sin x}{x} \\[2ex] h_l^{(\pm)}(x) & = & [R_l(x) \pm i S_l(x)]\dfrac{\mathrm{e}^{\pm i x}}{x} \;, \end{array}\right\} \qquad (\text{A.8})$$

with

$$R_l(x) + i S_l(x) = \sum_{s=0}^{l} \frac{i^{s-l}}{2^s s!} \frac{(l+s)!}{(l-s)!} x^{-s} \;,\quad R_l, S_l \in \mathbb{R} \;.$$

R_l and S_l are polynomials in $1/x$ of order l with real coefficients and parity $(-1)^l$ and $-(-1)^l$, respectively. For any linear combination $f_l = a j_l + b n_l$, a, b fixed, we have the recursion formulae

$$(2l+1) f_l(x) = x \left[f_{l+1}(x) + f_{l-1}(x) \right] \qquad (\text{A.9})$$

$$f_{l-1} = \left[\frac{\mathrm{d}}{\mathrm{d}x} + \frac{l+1}{x} \right] f_l = \frac{1}{x^{l+1}} \frac{\mathrm{d}}{\mathrm{d}x} \left(x^{l+1} f_l \right)$$

$$f_l = \left[-\frac{\mathrm{d}}{\mathrm{d}x} + \frac{l-1}{x} \right] f_{l-1} = -x^{l-1} \frac{\mathrm{d}}{\mathrm{d}x} \left(\frac{f_{l-1}}{x^{l-1}} \right) \;,$$

which implies that

$$f_l = \left[x^l \left(-\frac{1}{x} \frac{\mathrm{d}}{\mathrm{d}x} \right)^l \right] f_0 \;.$$

From (A.8), the first spherical functions are obtained as

$$j_0(x) = \frac{\sin x}{x} \;,\quad j_1(x) = \frac{\sin x}{x^2} - \frac{\cos x}{x}$$

$$n_0(x) = \frac{\cos x}{x} \;,\quad n_1(x) = \frac{\cos x}{x^2} + \frac{\sin x}{x}$$

$$h_0^{(\pm)}(x) = \frac{\mathrm{e}^{\pm i x}}{x} \;,\quad h_1^{(\pm)}(x) = \left(\frac{1}{x^2} \mp \frac{i}{x} \right) \frac{\mathrm{e}^{\pm i x}}{x} \;.$$

A.6 Legendre Functions, Legendre Polynomials, Spherical Harmonics

Legendre functions. The Legendre differential equation reads

$$\left[(1 - x^2)\frac{\mathrm{d}^2}{\mathrm{d}x^2} - 2x\frac{\mathrm{d}}{\mathrm{d}x} + l(l+1) - \frac{m^2}{1 - x^2} \right] f(x) = 0 \ ,$$

with $l = 0, 1, 2, \ldots$, $m = 0, \ldots, \pm l$. Its limited solutions within the interval $[-1:1]$ are the Legendre functions

$$P_{l,m}(x) = \frac{(1 - x^2)^{m/2}}{2^l l!} \frac{\mathrm{d}^{l+m}}{\mathrm{d}x^{l+m}} (x^2 - 1)^l \ . \tag{A.10}$$

$P_{l,m}$ is the product of $(1-x)^{m/2}$ with a polynomial of order $l - m$ and parity $(-1)^{l-m}$, which has $l - m$ zeros within the interval $[-1:1]$. We have the following recursion formulae ($P_{-1,\ldots} = 0$):

$$(2l + 1)xP_{l,m} = (l + 1 - m)P_{l+1,m} + (l + m)P_{l-1,m} \tag{A.11}$$

$$(1 - x^2)\frac{\mathrm{d}}{\mathrm{d}x}P_{l,m} = -lxP_{l,m} + (l + m)P_{l-1,m}$$

$$= (l + 1)xP_{l,m} - (l + 1 - m)P_{l+1,m}$$

as well as the orthonormality relations

$$\int\limits_{-1}^{1} \mathrm{d}x\, P_{l,m}(x)P_{l',m}(x) = \frac{2}{2l + 1}\frac{(l + m)!}{(l - m)!}\delta_{ll'} \ .$$

Legendre polynomials. In the case of $m = 0$, we obtain the Legendre polynomials from (A.10) as

$$P_l(x) = P_{l,0}(x) = \frac{1}{2^l l!}\frac{\mathrm{d}^l}{\mathrm{d}x^l}(x^2 - 1)^l \ .$$

P_l is a polynomial of order l with parity $(-1)^l$ and possesses l zeros within the interval $[-1:1]$. The Legendre polynomials can be obtained by expanding the functions $(1 - 2xy + y^2)^{-1/2}$ in powers of y:

$$\frac{1}{\sqrt{1 - 2xy + y^2}} = \sum_{l=0}^{\infty} y^l P_l(x) \ , \ |y| < 1 \ . \tag{A.12}$$

The first five Legendre polynomials read

$$P_0(x) = 1 \ , \ P_1(x) = x \ , \ P_2(x) = \frac{1}{2}(3x^2 - 1)$$

$$P_3(x) = \frac{1}{2}(5x^3 - 3x) \ , \ P_4(x) = \frac{1}{8}(35x^4 - 30x^2 + 3) \ .$$

Spherical harmonics. The spherical harmonics $Y_{l,m}$ are defined as the eigenfunctions of the quantum mechanical angular momentum operators \boldsymbol{L}^2 and \boldsymbol{L}_z:

$$\boldsymbol{L}^2 Y_{l,m} = \hbar^2 l(l+1) Y_{l,m} \,, \quad l = 0, 1, 2, \dots$$
$$\boldsymbol{L}_z Y_{l,m} = \hbar m Y_{l,m} \,, \quad m = 0, \dots, \pm l \,.$$

Their explicit forms are given by

$$Y_{l,m}(\theta, \varphi) = \frac{(-1)^l}{2^l l!} \sqrt{\frac{(2l+1)!}{4\pi}} \sqrt{\frac{(l+m)!}{(2l)!(l-m)!}}$$

$$\times e^{im\varphi} \sin^{-m} \theta \frac{d^{l-m}}{d(\cos\theta)^{l-m}} \sin^{2l} \theta \,.$$

They form a complete orthonormal function system on the unit circle. This means the following orthonormality and completeness relations hold:

$$\int Y_{l,m}^* Y_{l',m'} d\Omega = \int_0^{2\pi} d\varphi \int_0^\pi d\theta \sin\theta \, Y_{l,m}^*(\theta, \varphi) Y_{l',m'}(\theta, \varphi) = \delta_{ll'} \delta_{mm'}$$

$$\sum_{l=0}^\infty \sum_{m=-l}^l Y_{l,m}^*(\theta, \varphi) Y_{l,m}(\theta', \varphi') = \frac{\delta(\varphi - \varphi')\delta(\cos\theta - \cos\theta')}{\sin\theta} = \delta(\Omega - \Omega').$$

Further properties are

- Parity:

$$Y_{l,m}(\pi - \theta, \varphi + \pi) = (-1)^l Y_{l,m}(\theta, \varphi) \,.$$

- Complex conjugation:

$$Y_{l,m}^*(\theta, \varphi) = (-1)^m Y_{l,-m}(\theta, \varphi) \,. \tag{A.13}$$

- Relationship with Legendre functions:

$$Y_{l,m}(\theta, \varphi) = \sqrt{\frac{2l+1}{4\pi} \frac{(l-m)!}{(l+m)!}} P_{l,m}(\cos\theta) e^{im\varphi} \,, \quad m \geq 0 \,. \tag{A.14}$$

- Addition theorem: using

$$\boldsymbol{x} = r \begin{pmatrix} \cos\varphi \sin\theta \\ \sin\varphi \sin\theta \\ \cos\theta \end{pmatrix} \,, \quad \boldsymbol{x}' = r' \begin{pmatrix} \cos\varphi' \sin\theta' \\ \sin\varphi' \sin\theta' \\ \cos\theta' \end{pmatrix}$$

and

$$\boldsymbol{x}\boldsymbol{x}' = rr' \cos\alpha \,, \quad \cos\alpha = \sin\theta \sin\theta' \cos(\varphi - \varphi') + \cos\theta \cos\theta' \,,$$

it follows that

$$P_l(\cos\alpha) = \frac{4\pi}{2l+1} \sum_{m=-l}^{l} Y_{l,m}^*(\theta',\varphi') Y_{l,m}(\theta,\varphi) .$$

From this and respecting (A.12), we obtain

$$\frac{1}{|\boldsymbol{x}-\boldsymbol{x}'|} = \frac{1}{r\sqrt{1-2\frac{r'}{r}\cos\alpha + \left(\frac{r'}{r}\right)^2}} = \frac{1}{r} \sum_{l=0}^{\infty} \left(\frac{r'}{r}\right)^l P_l(\cos\alpha)$$

$$= \sum_{l=0}^{\infty} \sum_{m=-l}^{l} \frac{4\pi}{2l+1} \frac{r'^l}{r^{l+1}} Y_{l,m}^*(\theta',\varphi') Y_{l,m}(\theta,\varphi) . \tag{A.15}$$

The first spherical harmonics are

$$Y_{0,0}(\theta,\varphi) = \frac{1}{\sqrt{4\pi}} , \quad Y_{1,1}(\theta,\varphi) = -\sqrt{\frac{3}{8\pi}} e^{i\varphi} \sin\theta$$

$$Y_{1,0}(\theta,\varphi) = \sqrt{\frac{3}{4\pi}} \cos\theta , \quad Y_{2,2}(\theta,\varphi) = \sqrt{\frac{15}{32\pi}} e^{2i\varphi} \sin^2\theta \sin\theta \cos\theta$$

$$Y_{2,1}(\theta,\varphi) = -\sqrt{\frac{15}{8\pi}} e^{i\varphi} , \quad Y_{2,0}(\theta,\varphi) = \sqrt{\frac{5}{16\pi}} \left(3\cos^2\theta - 1\right) .$$

Appendix B. Literature List

B.1 General Textbooks

[1] R.P. Feynman, R.B. Leighton, M. Sands: *The Feynman Lectures on Physics*, Volumes 1–3, Addison-Wesley, 1971.

[2] L.D. Landau, E.M. Lifschitz: *Course in Theoretical Physics*, Volumes 1–10, Butterworth-Heineman, 1976–1987.

[3] A. Sommerfeld: *Lectures on Theoretical Physics*, Volumes 1–6, Academic Press, 1950–1964.

A very nice book is

[4] M.S. Longair: *Theoretical Concepts in Physics*, Cambridge University Press, 2003.

This book resulted from lectures given by Longair in Cambridge for students toward the end of their studies. It was Longair's goal to present the principles and methods of physics in a complementary manner to one usually chosen in the standard lectures. The author uses case studies to look at well-known results from a different angle.

For readers interested in an up-to-date discussion of "modern physics, philosophy and the meaning of quantum theory", we suggest

[5] J. Baggott: *Beyond Measure*, Oxford University Press, 2004.

B.2 Mechanics

[6] H. Goldstein, C. Poole, J. Safko: *Classical Mechanics*, Addison-Wesley, 2002.

Third edition of this classic. Updated and modernized.

[7] D. ter Haar: *Elements of Hamiltonian Mechanics*, Pergamon Press, 1971.

A rather short but concise book, which covers about the same as our first chapter, with the exception of relativity theory.

[8] F. Scheck: *Mechanics*, Springer, 2004.

A very nice and comprehensive book. In addition to the material covered here, this volume gives a readable introduction to the geometric aspects of mechanics as well as to the topics of stability and chaos. As Scheck says: "Mechanics is in no way a closed and archived topic."

[9] L. Hand and J. Finch: *Analytical Mechanics*, Cambridge University Press, 1998.

A good introductory alternative to Goldstein, which also covers chaos theory at a basic level.

[10] H. Iro: *A Modern Approach to Classical Mechanics*, World Scientific, 2003.

Another relatively new book that has taken recent developments in mechanics on board by including issues of nonintegrability from the start. Goes all the way to the KAM theorem.

[11] J.V. Jose and E.J. Saletan: *Classical Dynamics: A Contemporary Approach*, Cambridge University Press, 1998.

Yet another good introductory text that covers similar material to Goldstein. However, the authors use the more geometric language of manifolds and tangent bundles.

B.3 Electrodynamics

[12] J.D. Jackson: *Classical Electrodynamics*, Wiley and Sons, 1999.

One of the classics of this topic, which will have to be in any physicist's library.

[13] J.R. Oppenheimer: *Lectures on Electrodynamics*, Gordon and Breach Science Publishers, 1970.

These are lecture notes from lectures given by Oppenheimer between 1939 and 1947 in the United States. Its structure is similar to our second chapter, as Maxwell's equations are the starting point of the book. This allows Oppenheimer to move very swiftly to a discussion of more complex problems of electrodynamics (such as the self-energy of the electron). The second part of the book deals with relativity and discusses electrodynamics in this context.

[14] D. Griffiths: *Introduction to Electrodynamics*, Prentice Hall, 1998.

Very readable and didactic introduction for the undergraduate.

[15] J. Schwinger: *Classical Electrodynamics*, Perseus Books, 1998.

An advanced textbook – probably not the place to start with electrodynamics, but if you've got a good grounding, this makes for interesting reading.

[16] L.H. Ryder: *Quantum Field Theory*, Cambridge University Press, 1996.

For the reader wishing to make the transition from classical theories to modern quantum field theories, we suggest this book. Our discussion of the Noether theorem follows Ryder.

B.4 Quantum Mechanics

[17] S. Gasiorowicz: *Quantum Physics*, Wiley and Sons, 2003.

A very nice introduction to quantum mechanics. The mathematical complexity is kept to a minimum. Quantum mechanical principles are discussed mainly in the context of wave mechanics; at relevant sections, the algebraic structure of quantum theory is explained. The emphasis of this book lies in the many examples and applications.

[18] D. Griffiths: *Introduction to Quantum Mechanics*, Prentice Hall, 2004.

Very readable and didactic introduction for the undergraduate.

[19] D. Bohm: *Quantum Theory*, Dover Publications, 1989.

A classic. Very didactic and also contains an extensive discussion of measurement theory.

[20] E.S. Abers: *Quantum Mechanics*, Prentice Hall, 2003.

A recent textbook that is very comprehensive. Can be used to go all the way from an introduction to an advanced level.

[21] A. Messiah: *Quantum Mechanics*, Volumes 1–2, Dover Publications, 2000.

This is a two-in-one reprint of one of the most comprehensive classics of the quantum theory literature. The first part starts with a phenomenologically based motivation of wave mechanics. The author then moves the text onto a more abstract level revealing in a didactic manner more and more of the representation-independent algebraic structure of quantum physics. The second part contains many worked standard examples of quantum mechanics. There is also an extensive discussion of symmetry principles and approximation schemes. Finally, the author discusses relativistic quantum theory at a fairly high mathematical level.

[22] J.J. Sakurai: *Modern Quantum Mechanics*, Addison-Wesley, 1993.

This book contains a modern presentation of quantum mechanics. Sakurai explains clearly the symmetry principles, which also form the foundation of modern quantum field theories. A book for the slightly more advanced reader.

[23] P.A.M. Dirac: *The Principles of Quantum Mechanics*, Oxford University Press, 1982.

It's never a bad idea to go back to the originals.

[24] R. Shankar: *Principles of Quantum Mechanics*, Plenum Press, 1994.

A didactically very pleasing book. We borrowed from it the idea of having an introductory section on mathematical aspects of quantum mechanics using Dirac's notation. It also contains a nice introduction to the path integral method, which the author considers mandatory for an introductory text on quantum mechanics.

[25] A. Sudbery: *Quantum Mechanics and the Particles of Nature*, Cambridge University Press, 1989.

This book has the subtitle *An outline for mathematicians*. However, we nevertheless (or maybe: because of this) find this book very recommendable for physicists. The principles of quantum mechanics are explained using an axiomatic approach and whenever possible, Sudbery uses symmetry principles to simplify calculations. For example, the Runge-Lenz vector is used to calculate the spectrum of the hydrogen atom. Sudbery also has a discussion of quantum meta-physics.

B.5 Statistical Physics and Thermodynamics

[26] B.K. Agarwal, M. Eisner: *Statistical Mechanics*, Wiley and Sons, 1988.

In only 260 pages the authors manage to present clearly the concepts of Gibbs statistical mechanics. Thermodynamics is not a separate chapter, but is embedded throughout in the text. We liked the early introduction of quantum statistics.

[27] D. ter Haar: *Elements of Statistical Mechanics*, Butterworth-Heineman, 1995.

Very well-motivated introduction to statistical physics using the Gibbs ansatz. Also contains a detailed discussion of the Boltzmann \mathcal{H}-theorem.

[28] D. ter Haar, H. Wergeland: *Elements of Thermodynamics*, Addison-Wesley, 1966.

A very clear text on the principles of thermodynamics.

[29] K. Huang: *Statistical Mechanics*, Wiley and Sons, 1987.

A three-part textbook (thermodynamics, statistical physics, and special themes), where the last part goes beyond our content. One of the classics, which has been of great help to us.

[30] J. Kestin, J.R. Dorfman: *A Course in Statistical and Thermodynamics*, Academic Press, 1971.

A very nice introductory text. Contains thermodynamics and statistical physics for noninteracting systems. The relevant quantum mechanics is also presented.

[31] R. Kubo et al.: *Statistical Mechanics, an Advanced Course with Problems and Solutions*, North Holland Publishing Company, 1988.

Each chapter consists of a short presentation of the most important results, followed by an extensive collection of examples and solutions. All of the standard problems are solved here. Obviously good for students.

[32] R.K. Pathria: *Statistical Mechanics*, Butterworth-Heineman, 1996.

About the same content and structure as Huang's book and equally well readable. One of the books we liked best.

[33] F. Reif: *Fundamentals of Statistical and Thermal Physics*, Mc-Graw Hill, New York, 1965.

Quite extensive and didactic discussion of all the topics we treat. Very well-suited to get started on statistical physics.

[34] H.S. Robertson: *Statistical Thermophysics*, Prentice Hall, 1993.

The author chooses Shannon's information theory as the starting point for his discussion of Gibbs's ensemble theory. Criticisms of this approach are discussed. The book contains many examples, nicely presented, which are not necessarily part of the standard repertoire.

Index